Forest Ecosystem Services and Landscape Design

Guest Editor
Radu-Daniel Pintilii
Department of Human and
Economic Geography
University of Bucharest
Bucharest
Romania

Editorial Office
MDPI AG
Grosspeteranlage 5
4052 Basel, Switzerland

This is a reprint of the Special Issue, published open access by the journal *Forests* (ISSN 1999-4907), freely accessible at: https://www.mdpi.com/journal/forests/special_issues/B77128T09O.

For citation purposes, cite each article independently as indicated on the article page online and as indicated below:

Lastname, A.A.; Lastname, B.B. Article Title. *Journal Name* **Year**, *Volume Number*, Page Range.

ISBN 978-3-7258-6984-8 (Hbk)
ISBN 978-3-7258-6985-5 (PDF)
https://doi.org/10.3390/books978-3-7258-6985-5

Cover image courtesy of Radu-Daniel Pintilii

Contents

About the Editor

Radu-Daniel Pintilii

Radu-Daniel Pintilii (Associate Professor) is a geographer who earned a PhD in Geography from the University of Bucharest in 2008. He is an employee of the Faculty of Geography (Department of Human and Economic Geography) and a PhD advisor for two Doctoral Schools (the Faculty of Geography and the Interdisciplinary School of Doctoral Studies) at the same University. His main interests are natural resources, forestry, GISs, spatial planning, and statistics. He has attended more than 110 national and international scientific conferences and has published 27 ISI-indexed articles. He is also a scientific reviewer for prestigious journals, and his H-index, according to WoS, is 16. He developed essential collaborations with prestigious European higher education institutions through the CIVIS Network (Austria, Belgium, France, Germany, Great Britain, Greece, Italy, Spain, Sweden, and Switzerland).

forests

MDPI

Editorial

Recent Advances in Forest Ecosystem Services and Landscape Design

Radu-Daniel Pintilii

Department of Human and Economic Geography, Faculty of Geography, University of Bucharest, 1, Nicolae Bălcescu, 010041 Bucharest, Romania; radu.pintilii@geo.unibuc.ro; Tel.: +40-788990060

1. Introduction—Theoretical Background and Current Knowledge in the Field

In a world of constant change, ecosystem services are becoming increasingly important to support the rapid demographic growth of recent decades. From tangible ecosystem services, such as food, water, timber, fuel resources, to intangible ones with recreational value and spiritual enrichment, all have a vital role, especially since most of them are subject to increasing anthropogenic pressure. Ecosystem services provide an extensive series of direct and indirect benefits, designed to meet the diverse needs of humanity [1].

Specialists in the field identify four major categories of ecosystem services: provisioning, regulating, cultural, and supporting services [2].

Provisioning services refer to the various ecosystems, differing from one another, that cater to human needs. It comprises the essential elements for human life (fish, water, wood, minerals and rocks for construction, and a broad palette of medicinal plants such as berries and mushrooms) for human health [2].

Regulating services contribute to maintaining the environment in balanced parameters by providing climate regulation effects, filtering pollutants to improve air and water quality, preventing soil erosion, and enhancing soil fertility and crop production [3].

Cultural services create opportunities for recreational and well-being activities, such as hiking, birdwatching, fishing, and camping, which contribute to a high quality of mental and physical health, foster aesthetic appreciation and inspiration, promote overall well-being, and ultimately cultivate a connection with nature [4–6].

Supporting services, although having indirect benefits, play a crucial role by enabling interdependencies among ecology, economy, and society [7].

Landscape design is the art of combining multiple features by arranging and modifying the landscape to create new outdoor spaces that offer both functional and aesthetic pleasure while respecting general principles such as unity, balance, focalization, rhythm, line, proportion, and simplicity [8,9]. These actions involve a good combination of artistry and scientific planning to improve their functionality [10].

This Special Issue features 15 papers authored by 83 contributors from 11 countries, prominently including China, Australia, the USA, Sweden, Canada, Germany, Denmark, Norway, Bangladesh, and Romania, all within the fields of forestry and landscape design.

This Special Issue, "Forest Ecosystem Services and Landscape Design", aims to present recent advances in our knowledge of ecosystem services and landscape design in an integrated, multi-, and interdisciplinary approach. The fundamental core concepts are an essential component of understanding the following:

(1) Management of forest ecosystems in urban landscapes;
(2) Structured diversity and species composition;

 https://doi.org/10.3390/f16101527

(3) Effects of natural and anthropogenic disturbances on forests;
(4) Cultural ecosystem services in protected areas for human well-being.

2. Methodological Innovations and Strengths

Recent developments in forest ecosystem services and landscape design have led to continuous methodological improvement and adaptation. Research papers from this Special Issue have employed various methods, including advanced spatial statistics, remote sensing, and GIS, to investigate the value of ecosystem services (*ESV*) [4,11–13]. For big data collection, multiple sources were inquired about, including online questionnaires, surveys, and crowdsourced data collection. These were used as an alternative way to analyze and interpret the main results obtained (e.g., to quantify the tourists' perception of recreational value in large urban areas) [14–16]. Another example enhances traditional assessment methods such as the fuzzy Delphi method, the analytic hierarchy process (*AHP*), and the entropy method [7].

Allometric growth equations, considering multiple parameters for several dominant species, were developed using breast height diameter (*DBH*), tree height, and crown width to illustrate how tree structure influences composition [7]. Different diversity indices, such as Simpson and Shannon–Wiener, along with species evenness indices like Pielou and Alatalo, were employed to assess plant community diversity in both tropical and temperate forests [5,17].

Among the strengths of the methodology in this Special Issue is the application of forest landscape modeling to quantify seed harvesting effects at unprecedented spatial and temporal scales. The use of LANDIS PRO facilitated the simulation of the synergistic effects of population dynamics, seed dispersal, and harvesting across large temporal scales and spatial extents. Additionally, a creative method for simulating seed harvesting was developed in the model by harvesting young cohorts of Korean pine, described in an innovative disturbance simulation [8]. The disturbance index (*DI*) provides a more detailed assessment of perturbation levels by combining the percentage of cut stems, tree canopy openness, shrub/herb canopy openness, and the percentage of bare ground [3]. The pebble distribution method, used as a participatory rural appraisal tool, enables local participants to quantify the importance of the ecosystem services. The study shifts from monetary valuation toward a regional, perception-based, consensus-driven assessment, capturing complex, multidimensional valuation of services that often lack direct market equivalents [9]. Photovoice walks enable participants to visually express their preferences and perceptions by photographing elements they find, providing more detailed insights than forced-choice rankings alone, through digital storytelling for positive social change [10]. The density-dependent matrix growth model has been applied and calibrated not only for aboveground biomass (*AGB*), but also for simulating the recreation value, which is a primary output. A different approach to analyzing cultural ecosystem services is provided by structural equation modeling, which demonstrates the dimensions of visitors' recreation experiences [6,11].

After screening the 15 articles from this Special Issue, the research strengths on various topics were underlined. First, it began with the systematic approaches of local communities [7] and multidimensional aspects of ecosystem services [11], followed by quantitative and comparative research papers on landscape design [9] (Table 1).

Table 1. Strengths of the papers published in this Special Issue.

Strengths	Reference(s)
Assessing both above- and belowground services simultaneously provide a comprehensive perspective on the ecosystem.	[3]
Multimethod integration provides updated, region-specific ecosystem service value coefficients for a more accurate valuation.	[4]
A systematic approach was employed to investigate the relationships between structure and diversity through extensive field sampling.	[5]
Develop a technical method to monitor changes in mangrove cover.	[6]
Systematic approach—a holistic framework that considers protected areas (*PAs*), local communities, and tourism as an interconnected complex system.	[7]
Methodological rigor integrates empirical field data with advanced modeling techniques, supported by multiple replications.	[8]
Quantitative and comparative analysis using different statistical tests (Kruskal–Wallis).	[9]
Clear linkage of findings to practical forest management and communication strategies.	[10]
The multidimensional input-output approach is comprehensive, including capital, land, labor, climate, and five key ecosystem services.	[11]
Practical guidance for improving forest park management and sustainable tourism development practices based on specific experiences and gender considerations.	[12]
Addresses the challenge of low survival rates in the artificial restoration process by proposing a model-supported monitoring and evaluation framework for the future.	[13]
Innovative use of crowdsourced social media data and complex spatial and statistical analysis.	[14]
Innovative research design underlying ecological patterns with psychological profiling.	[15]
Provides a methodological guide for researchers and policymakers interested in the field of community forestry.	[16]
Silvicultural planning based on the public preferences in post-mining sites.	[17]

3. Conclusions

The collection of papers from this Special Issue provides valuable findings into the multidisciplinary pillars of forest management. The findings are gathered into four categories:

(1) Ecosystem services and social perception to quantify spatial patterns of recreational ecosystem services value (*ESV*) in urban environments;
(2) Forest structure and restoration are essential for tree diversity and ecological stability;
(3) Management of protected areas (*PAs*)—use different methods to increase the effectiveness of the forest ecosystem services through systematic monitoring and restoration for forest conservation;
(4) Human well-being and recreational activities—the multifunctional role of the urban green environment for both environmental education and a positive attitude toward landscape and recreational satisfaction.

Overall, these contributions advocate for a complex approach to ecosystem service value and forest management, emphasizing the crucial role of well-being for local communities.

Funding: This research received no external funding.

Conflicts of Interest: The author declares no conflicts of interest.

References

1. Asah, S.T.; Guerry, A.D.; Blahna, D.J.; Lawler, J.J. Perception, Acquisition and Use of Ecosystem Services: Human Behaviour, and Ecosystem Management and Policy Implications. *Ecosyst. Serv.* **2014**, *10*, 180–186. [CrossRef]
2. Savo, V.; D'Amato, L.; Bartoli, F.; Zappitelli, I.; Caneva, G. Evaluation of Main Regulating, Provisioning, and Supporting Ecosystem Services of Urban Street Trees: A Literature Review. *Ecosyst. Serv.* **2025**, *71*, 101690. [CrossRef]

3. Sultana, F.; Arfin-Khan, M.A.S.; Karim, M.R.; Mukul, S.A. Rainfall Modifies the Disturbance Effects on Regulating Ecosystem Services in Tropical Forests of Bangladesh. *Forests* **2023**, *14*, 272. [CrossRef]
4. Chen, S.; Liu, X.; Yang, L.; Zhu, Z. Variations in Ecosystem Service Value and Its Driving Factors in the Nanjing Metropolitan Area of China. *Forests* **2023**, *14*, 113. [CrossRef]
5. Li, Y.; Wang, X.; Zhang, C.; Li, J.; Jiang, H.; Sheng, L.; Li, Z. Stand Structure Management and Tree Diversity Conservation Based on Using Stand Factors: A Case Study in the Longwan National Nature Reserve. *Forests* **2023**, *14*, 750. [CrossRef]
6. Melichar, J.; Cienciala, E.; Albert, J.; Braun Kohlová, M.; Máca, V.; Pavelčík, P. Dynamics and Management of Restored Forests in Post-Mining Sites with Respect to Their Recreation Value: A Matrix Growth Model. *Forests* **2022**, *13*, 1519. [CrossRef]
7. Zhang, X.; Zhong, L.; Yu, H.; Wang, L.-E. Sustainability Assessment for the Protected Area Tourism System from the Perspective of Ecological-Economic-Social Coordinated Development. *Forests* **2023**, *14*, 890. [CrossRef]
8. Liu, K.; He, H.S.; Sun, H.; Wang, J. Evaluating the Legacy Effects of the Historical Predatory Seed Harvesting on the Species Composition and Structure of the Mixed Korean Pine and Broadleaf Forest from a Landscape Perspective. *Forests* **2023**, *14*, 402. [CrossRef]
9. Ahammad, R.; Stacey, N.; Sunderland, T.; Sangha, K.K. Land Use Preference for Ecosystem Services and Well-Being in Chittagong Hill Tracts of Bangladesh. *Forests* **2022**, *13*, 2086. [CrossRef]
10. Lupp, G.; Kantelberg, V.; Fäth, J.; Hirschbeck, T.; Käsbauer, C.; Ritter, A.; Schisslbauer, J.; Pauleit, S. Through the Eyes of Forest Visitors—Perception and Scenic Preferences of Munich's Urban Proximate Woodlands. *Forests* **2022**, *13*, 1584. [CrossRef]
11. Deng, Y.; Luo, J.; Wang, Y.; Jiao, C.; Yi, X.; Su, X.; Li, H.; Yao, S. Eco-Efficiency Evaluation of Sloping Land Conversion Program and Its Spatial and Temporal Evolution: Evidence from 314 Counties in the Loess Plateau of China. *Forests* **2023**, *14*, 681. [CrossRef]
12. Weng, L.; Zhu, Y.; Xu, X.; Yang, J.; Zheng, S.; Liu, H.; Wang, H.; Zhu, L. The Influence of Visitors' Recreation Experience and Environmental Attitude on Environmentally Responsible Behavior: A Case Study of an Urban Forest Park, China. *Forests* **2022**, *14*, 24. [CrossRef]
13. Jia, P.; Huang, W.; Zhang, Z.; Cheng, J.; Xiao, Y. The Carbon Sink of Mangrove Ecological Restoration between 1988–2020 in Qinglan Bay, Hainan Island, China. *Forests* **2022**, *13*, 1547. [CrossRef]
14. Xu, H.; Zhao, G.; Liu, Y.; Miao, M. Using Social Media Camping Data for Evaluating, Quantifying, and Understanding Recreational Ecosystem Services in Post-COVID-19 Megacities: A Case Study from Beijing. *Forests* **2023**, *14*, 1151. [CrossRef]
15. Kortmann, M.; Angelstam, P.; Mayer, M.; Leibl, F.; Reichert, J.; Thorn, C.; Thorn, S. Disturbance Severity and Human–Nature Relationships: A New Approach to Analyze People's Well-Being along a Bark Beetle Infestation Gradient. *Forests* **2022**, *13*, 1954. [CrossRef]
16. Hovis, M.; Frey, G.; McGinley, K.; Cubbage, F.; Han, X.; Lupek, M. Ownership, Governance, Uses, and Ecosystem Services of Community Forests in the Eastern United States. *Forests* **2022**, *13*, 1577. [CrossRef]
17. Tudoran, G.-M.; Cicșa, A.; Cicșa (Boroeanu), M.; Dobre, A.-C. Management of Recreational Forests in the Romanian Carpathians. *Forests* **2022**, *13*, 1369. [CrossRef]

forests

MDPI

Article

Using Social Media Camping Data for Evaluating, Quantifying, and Understanding Recreational Ecosystem Services in Post-COVID-19 Megacities: A Case Study from Beijing

Haiyun Xu [1], Guohan Zhao [2], Yan Liu [3] and Meng Miao [4,*]

1 College of Architecture and Urban Planning, Beijing University of Civil Engineering and Architecture, Beijing 100044, China; xuhaiyun@bucea.edu.cn
2 Research Center for Built Environment, Energy, Water and Climate, VIA University College, Banegårdsgades 2, 8700 Horsens, Denmark; gzha@via.dk
3 Department of Geoscience, Southwest University, Chongqing 400715, China; bluefire612@163.com
4 College of Finance, Renming University of China, Beijing 100872, China
* Correspondence: miaomeng@ruc.edu.cn

Abstract: Recreational ecosystem services (RESs) are the diverse recreational opportunities provided by nature to humans, which contribute to the improvement of public health and social well-being. The use of online social media is an efficient method for quantifying public perceptions of recreational ecosystem services (RESs) delivered by a given landscape. With the continuously changing demand for nature-focused outdoor recreational activities since COVID-19, camping has become the fastest-growing outdoor leisure activity in megacities and a key indicator for how people perceive and value the RESs provided by the landscape. Such unexpected changings triggered by COVID-19 have further led to an imbalance between demand and supply, which results in fierce conflicts in urban green space management. This study presents a spatial pattern analysis of how people perceive RESs in a megacity-scale case study of Beijing using geo-tagged camping notes posted on Little Red Book (LRB). We employed these camping notes in the context of a megacity to (i) map public camping behaviors patterns in urban green spaces, (ii) evaluate spatial clusters of high/low RESs, and (iii) investigate the relationship between RESs, local landscape features, and gender through correspondence analysis. Our results show that considerable spatial clustering of camping behaviors was observed in both suburban and urban green spaces. However, suburbs revealed a substantially higher RES value than central urban areas. In addition, water bodies were discovered to have remarkably low RES, while grassland and urban forests were found to have a close link with higher RES. In addition, significant gender preferences have been discovered, where female visitors prefer to camp in grassland, and male visitors favor bare ground and urbanized regions. Our findings would assist decision-makers in optimizing urban green space planning and management, adapting to fast-changing public camping demands in the context of the post-COVID-19 era. Findings also contribute to the literature by applying spatial analysis of social media data to understand public outdoor recreation activities and perceived value for megacities' green space management.

Keywords: recreational ecosystem service; camping; outdoor recreation; social media; urban green spaces planning; post-COVID-19; public well-belling

1. Introduction

Cultural Ecosystem Services (CESs) pertain to the notion of ecosystem services, which are considered the primary quantification tool for valuing the human benefits gained from nature [1,2]. CESs incorporate the complete range of non-material benefits related to human perceptions of natural landscapes, such as aesthetics and a sense of place. Among them, Recreational Ecosystem Services (RESs) can provide people with increased physical health (e.g., through exercise) and psychological and emotional well-being [3]. Hence, it is regarded as a specific recreational metric that quantifies the individual benefits gained from nature. RES spatial assessments have been explored from a variety of perspectives and sources during the past decades. For expert-based techniques employing theoretically-derived spatial indicators [3], a landscape quality index has been suggested to map the backdrop of outdoor leisure activities at various spatial scales [3]. In contrast to systems relying purely on the opinions of experts, a public-participation approach can generate a more precise spatial pattern of RESs provided by users. Consequently, geo-located data are often regarded as a more efficient source for assessing and analyzing recreational visiting rates [4,5], CES values, and motivations [6]. Here, some scholars [7] conducted a public participation GIS using both small-group workshops and larger-scale home sampling to obtain recreational information from the public for the design of Chugach National Forest. In addition, some scholars [8] evaluated the recreational values for cultural landscape corridor planning in China using PPGIS and local resident interviews and surveys.

Beyond traditional data sources, social media has attracted more attention than traditional data sources due to its crowdsourcing data and often updated data properties [9,10]. Indeed, numerous social media platforms provide geo-tagged information submission and sharing and then archive millions of user contributions into big data hubs [10], which enables a more efficient data collection technique than traditional human-to-human survey methods. In data-sparse environments, some authors [11] have discovered and characterized the areas of high recreational value in data-sparse environments by assessing low or infrequent visitor metrics represented in social media data. In addition, these crowdsourcing data have been updated in real-time, which means that new submission data incorporating real-world dynamics are constantly assimilated into available databases, thereby providing a more accurate and timely data-driven basis to reflect real-world dynamics than traditional data source approaches. Here, some scholars [11,12] have used social media data for RES evaluation to capture the temporal dynamics of recreation values, thereby shedding light on seasonal changes in RES with a spatial dimension.

With the demonstrated crowdsourcing and timely-updated features, we would further explore its potential on large-scale cases such as megacities, thereby providing a more comprehensive view and time-efficient solution to comprehending the ever-changing public perception and updated demands for a given urban landscape than traditional approaches. Since the COVID-19 pandemic, outdoor recreation activities have been severely restricted in the majority of megacities [13,14]. Meanwhile, such a global pandemic has had a profound effect on people's daily lives, extending its influence on the public perception and needs for urban green space. The substantially accumulated demands on natural experience have promoted camping transforming into the most popular outdoor recreational lifestyle. Nowadays, camping has also been shifted into a primary metric for RES assessments [15,16]. So far, little research has focused on the newly emergent "supply-and-demand" imbalance brought about by COVID-19. Consequently, it is important to collect data and evaluate how visitors perceive the RES value supplied by various urban green spaces in megacities in order to better meet public demands in the post-COVID-19 future. We believe that social media data can provide a practical answer for identifying and comprehending such newly developed supply and demand imbalances.

The primary objective of this study is to identify the public's perception of the recreation value of green spaces in the case of a megacity, Beijing. This is accomplished through the use of geo-tagged social media data of individuals' camping notes. Here, we define camping as outdoor recreational activities involving daytime or overnight stays in natural environments, which take occur in both urban and rural settings and provide people with a variety of recreational benefits. Camping in rural areas typically entails outdoor pursuits such as hiking, canoeing, climbing, fishing, swimming, kayaking, shooting, stone skipping, and bird watching. Urban camping activities include picnics, bicycling, Frisbee, dog walking, and football [17–19]. Thus, camping covers a variety of human recreations in natural environments and might be used as a major indicator for determining the RES given by green space. Using geo-tagged camping notes posted on Little Red Book (LRB), we performed a spatial pattern assessment of how people perceive RES in a megacity-scale case study of Beijing. We outline the research objective following three steps: (1) To map patterns of public camping behavior in verdant urban spaces. We can comprehend the general geo-differences of public camping behaviors and identify popular camping areas based on the spatial distribution pattern of public camping behaviors. Based on this outcome, planners and decision-makers will be able to identify the public's camping requirements and formulate appropriate policies to manage the overall supply and demand of green space and public outdoor recreation. (2) Evaluate the spatial concentrations of high/low RES. Through the spatial appraisal of public-perceived RES via social network data, we could investigate the high or low RES value they gained in each camping site and comprehend the geographic distribution of significant camping sites providing high or low RES value across the entire camping area. Planners could make targeted investments and differentiated enhancement policies in these sites with high or low values, which could have implications for the future management of green space. (3) Examine the relationship between RESs, local land cover characteristics, and gender. As previous CES studies have demonstrated that public perceptions of landscape value may be affected by local land cover characteristics [1,8], we investigate this relationship in RES value to determine which land cover characteristics in camping sites are associated with public preferences and perceptions of high or low RES value there. On the basis of this result, future planners will be able to adjust and manage various land cover characteristics in green space management policies in order to better satisfy the preferences of the public of different genders and enhance their camping experiences for the purpose of obtaining recreational value. Therefore, our study finally raises the following research questions:

(1) What is the spatial distribution of camping in Beijing?

(2) How does the public perceive the benefit of RES in these camping areas?

(3) What is the relationship between RES value and land cover characteristics and the gender of visitors?

Our findings would contribute to an updated understanding of the public's perception of RES provided by Beijing's green spaces in the camping trend influenced by the COVID-19 pandemic. The answers to these three primary queries may also contribute to the optimization of future green space planning and recreation policy management in response to changing public outdoor recreation demands.

2. Materials and Methods

2.1. Study Area and Camping Trend in Post-COVID-19

We selected Beijing, which has a total administrative area of 16,410.5 km^2, as the case study. It consists of two principal parts, comprising the central urban area and the suburbs (See Figure 1a). The central urban area (official Chinese name: Chengliuqu, means six distrscts in the central urban area) encompasses Beijing's ancient town, the urban environment with the highest population density [20]. The central urban area is comprised of six districts: Dongcheng, Xicheng, Haidian, Chaoyang, Fengtai, and Shijingshan (See Figure 1a in red). According to the population data from the Government of Beijing Municipality, we illustrated the distribution of population in each district in Figure 1b. Chaoyang district has the highest population (n = 3,449,000), followed by Haidian district (n = 3,130,000) (See Figure 1b). The historic old town is situated in the heart of the central urban area and is comprised of Dongcheng and Xicheng, which have respective populations of 110,000,000 and 700,000,000 (See Figure 1b). Beijing's suburbs consist of ten districts: Shunyi, Tongzhou, Daying, Fangshan, Mentougou, Changping, Pinggu, Miyun, Huairou, and Yanqing (Figure 1a, white text). The total population of the suburbs is 10,910,400 (Figure 1b). Figure 1b depicts that Changping has the greatest population (n = 2,270,000), followed by Daxing (n = 1,993,600) [20] (See Figure 1b). Generally, suburban areas are predominantly comprised of freshly developed metropolitan areas and abundant natural resources.

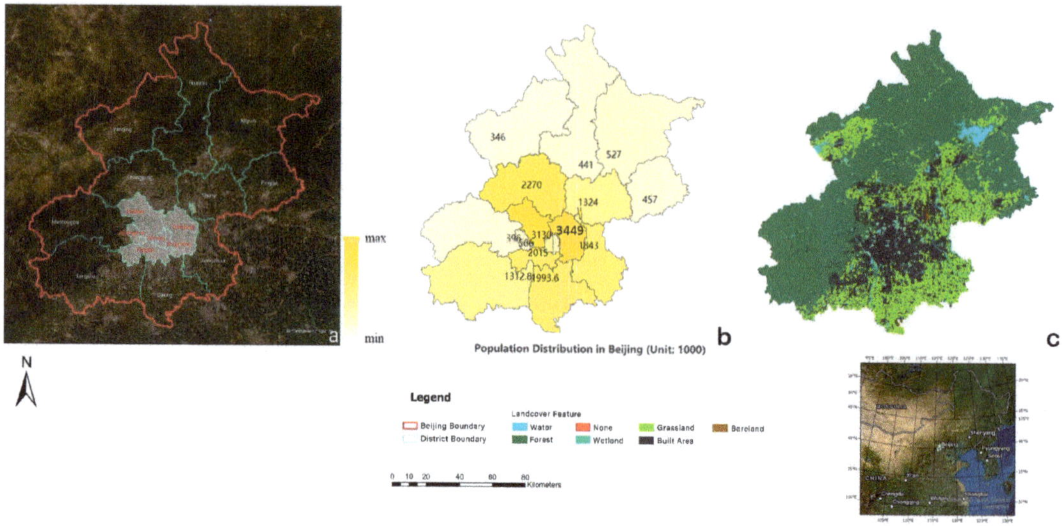

Figure 1. (**a**) the location and boundary of Beijing and its district divisions, where the grey regions with the central red fonts represent the central urban areas and the rest regions within the Beijing boundary represent the suburban areas. (**b**) The distribution of population in all districts. (**c**) the various green spaces in Beijing.

Beijing is the capital of China, a political, economic, and cultural hub, and the location of the royal historical garden green parks. Beijing has various green spaces in the central urban area and the suburban area (Figure 1c), such as urban parks, scenic areas, habitat conservation areas, forest parks, wetland parks, country parks, and diverse rural landscapes in the east and north mountain areas and plain area in the south, with land cover features such as forests, waterbodies, wetland, bare land, and grassland. These green places offered numerous CES, including recreational possibilities and RES values.

Since 2020, the pandemic has impacted public tourism activities and outdoor behaviors. In China, particularly after the COVID-19 outbreak, outward travel and outdoor recreational activities have been restricted, with safety being the top priority for outdoor activities [13,14]. Camping has become the most popular and fastest-growing outdoor lifestyle, especially for megacities in the post-COVID-19 age, particularly in Asia [15], as a result of a public need for a natural experience that has accumulated substantially throughout lengthy years of outdoor activity suppression. After that, camping flourished and gained significant public attention. As such, 2020 is called "the first year of Chinese camping" in social media by the tourism industry academy. In this trend, camping is becoming the most popular option for people to explore nature and gain leisure opportunities from the landscape in China in the post-COVID-19 era. According to Chinese social media data, search engine usage has surged by a factor of four. In 2021, the Chinese camping market is expected to rise by 75 percent and reach 29.90 billion yuan (4.69 billion US dollars) in value [17]. Beijing is one of the megacities with the fastest-growing camping tourism industry [17]. At the same time, despite the impact of the pandemic lockdown on construction projects in Beijing, the Beijing Municipal of Forestry and Park Bureau has promoted urban park enhancement projects since 2021. It included some micro-community green space initiatives, 26 new urban parks and 4 forest parks [21]. As one of the primary objectives of park construction projects in Beijing is to increase the green cover and park area per capita in this megacity, these park projects have a tendency to emphasize their ecological function. For micro-community green space improvement initiatives, decision-makers primarily focused on enhancing facilities for all ages, such as playgrounds for children and sports facilities for the elderly. However, camping recreation was rarely considered in the aforementioned new urban park and green space projects. In addition, few parks have been implemented for opening, constructing, and designing camping-specific areas, despite the fact that it is the fastest-growing outdoor activity in the post-COVID-19 era. Moreover, the lagging nature of existing park regulations has left camping in Beijing's urban park management in a grey area, as it is a relatively new form of public recreation and use of green space. Camping pursuits may be evicted, acquiesced, or permitted in urban parks and green spaces based on the discretion of park and site administrators. Consequently, it is time to address this emergent public demand for the use of green space and to optimize future green space planning and recreation policy management in response to the shifting outdoor recreation public needs of the post-COVID-19 era.

2.2. Data Collection and Preprocessing

We utilized LittleRedBook (LRB) social media data to acquire citizen-uploaded data from 2020 to 2022. LRB was founded in 2013 as a social media site for sharing photographs of personal lifestyle achievements. It has been referred to as "China's Instagram". In 2019, LRB had more than 300 million registered users and more than 85 million monthly active users [22]. People can contribute notes and photographs with names and descriptions, share them, and add information to them on the website. Once members have uploaded their notes to LRB, the notes will include geo-location data. In this work, we extracted the original note data using the Application Programming Interface (API) provided by LRB. The terms forest, park, water, wetland, river, lake, and grass were used in conjunction with the phrase "camping" to filter the database of the social network to collect only relevant notes. The selection of keywords was predicated on the premise that they pertain to natural resources with potential for RESs [18,19]. We collected all camping notes using a Python code that systematically mined the geographic coordinates included with the posts on LRB. Finally, we identified 3680 geo-tagged sites containing camping-related notes from 2020 to April 2022. Each of these camping notes is then shared together with the name of the point of interest (POI), its latitude and longitude, images such as photos and hiking maps, topics, detailed text contents of their camping experiences, and the number of "likes" it has received. The latitude and longitude data indicated the precise geo-referenced location where this letter was uploaded online in real-time. POI reveals the user's camping location in accordance with note-described occurrences. Thus, we defined the social media data we used in this study as geo-tagged camping notes with photos and text presenting public camping experiences. However, we discovered that not all recorded note locations definitely belonged to camping sites or green spaces, as some people may upload their notes from their homes. To address this issue, we manually reviewed the data that had been collected by looking at all POI names and removing notes whose real locations did not correspond with campgrounds or green spaces, such as notes posted in shopping malls and Central Business District (CBD) buildings. Moreover, considering the impact of the pictures themselves in camping notes on the public's perception and preferences, we deleted the notes containing images focusing on people (such as the user's own portrait during camping events) and kept only the images of the scenery and camping-related activities, to ensure that the public perceives the same type of camping notes as much as possible in an effort to reduce the interference caused by the people's favorite pictures. As a result, our final tally was reduced to 2971 notes following this data cleansing process. To see the geo-locations of the data over the entirety of Beijing, the data gathered from LittleRedbook (LRB) was converted into GIS point data based on each post's geo-tagged location (i.e., latitude and longitude) using the WGS 1982 projection.

2.3. Data Analysis

To understand the GIS-point data, the general data analysis was conducted in terms of (i) spatial pattern analysis, (ii) hot/cold point cluster analysis, and (iii) land-cover correspondence analysis.

2.3.1. Kernel Density Analysis

Kernel density analysis is an ArcGIS tool for spatial analysis that calculates the density of features in a neighborhood surrounding those features. It calculates a magnitude-per-unit area from point or polyline features by fitting a smoothly tapered surface to each point or polyline using a kernel function [18,23]. The kernel density analysis was used to determine the spatial cluster of GIS points in order to better understand the spatial camping pattern. Based on the point density of the scattered GIS points, the kernel density analysis generates a smoothly curved circular surface raster, where each cell value indicates the expected number of points [23]. This was accomplished in two steps. To begin, we gather the occurrences of camping events within a 1000 m search radius. Second, we use a decaying factor of 2 to extend their aggregate value into neighboring raster grid cells. As a result, such a technique makes an explicit depiction of the geographical clustering of our mapped GIS points possible. In this study, we calculated kernel density estimation with an output cell size of 10 m; thus each cell reflected the expected point (note) count in 10,000 m^2 cells.

2.3.2. Cluster/Outlier Analysis for Hot/Cold Point Classification

Cluster/outlier analysis (Anselin Local Moran's I) is a spatial analysis tool in ArcGIS that uses the Anselin Local Moran's I statistic to identify statistically significant hot spots, cold spots, and spatial outliers [24]. The Anselin Local Moran's I is a common statistical tool to identify local cluster and spatial auto-correlation in given datasets [24].In this case, the Cluster/outlier analysis based on Anselin local Moran's I allowed us to distinguish between high/low RES value provided by green spaces (i.e., hot/cold points) among transformed GIS points mapped by public geo-tagged camping notes. The values assigned to each geo-tagged camping note are determined by the public's perceived RES here. Each camping note that users submit has the potential to reflect the human-perceived RES value of established urban green spaces. Here, we chose the number of "likes" on each note as the metric to quantify the RES values corresponding to each geo-location. A "like" symbolizes the reception of these messages and their apparent interest in the uploaded information, according to the social media study [25–27]. It was feasible to measure user engagement and investigate individual nature tourism experiences by analyzing the "like" statistics for these posts. In this case, "like" data can convey people's perceptions and interests in their nature tourism experience. Thus, from a public standpoint, the number of "likes" can indicate the RES value of local landscapes.

Based hereon, the method produces a map that highlights local areas with high or low Anselin local Moran's I values ranging from -1 to 1. A high positive value indicates positive spatial autocorrelation, where the target value is identified as similar to its neighborhood, and thus their corresponding locations are considered spatial clusters, including high–high clusters or low–low clusters. Instead, a high negative local Moran's I value implies a potential spatial outlier differing from the value identified from its surroundings, including high–low outliers and low–high outliers. In this case, other than spatial outliers, we are particularly interested in spatial clusters, and thus points with positive values were chosen. In addition, to ensure a sufficient degree of significance, we implemented a p-value less than 0.05, where a 95% confidence interval was used to reject the null hypothesis that the observed pattern was created by a random process. Therefore, hot points in this context defined the regions with statistically significant high RES values surrounding a high RES value neighborhood. In contrast, cold points were statistically significant low RES values surrounding a low RES value neighborhood.

Furthermore, based on the locations of the clustering analysis's hot and cold points, we randomly followed and invited three users who posted camping notes in each hot and cold point location for online interviews. Typically, two or three out of three individuals responded. For those who responded to our online semi-structured interview surveys, we first confirmed that they had visited the location themselves. The interviews then centered on the query, "How is your camping experience there?" In addition to their camping notes, respondents were asked to describe their camping experiences at each location. We finally documented the text of these experiences and quoted some of them in subsequent sections.

2.3.3. Correspondence Analysis

We conducted correspondence analysis based on the identified hot/cold points to study further why hot/cold points were generated, taking into account the supplied land covers and the visitor's gender.

Correspondence analysis is a multivariate statistical tool developed by Jean-Paul Benzécri in 1973 to visualize the relationship between categories on a 2D map [28,29]. Differing from principal component analysis, it is applicable to categorical data as opposed to continuous data. Its purpose is to display any structure concealed by the multivariate data table setting in a biplot. Considered highly correlated are any two points, variables, or characteristics that are close together on the biplot. This analysis reveals the relative relationships between and within different groups of variables and is utilized in a variety of fields, such as medical research and marketing research [30]. In landscape management research, some authors have utilized correlation analysis to demonstrate the relationship between the public's perception of the CES and their diverse social backgrounds, such as age and education, in East Germany [1]. Some authors have investigated the relationships between public well-being, the socio-cultural characteristics of residents, and the perception of ecosystem service benefits in 13 European sites using correspondence analysis [31]. For landscape corridor planning, previous scholars conducted a correspondence analysis to examine how public perceptions of CES relate to local land-cover characteristics and public attitudes toward developing these lands [8]. On the basis of past practices, we intended to investigate the relationship between the public's perception of RES during their camping events, the land cover for their camping events, and their social characteristics. In this case, we postulate that different land-cover types may reflect on different key landscape elements in green space, which are linked to visitor decisions and experiences. Furthermore, gender differences in attitudes about camping activities may influence how the same urban green area is seen by men and women. So, based on the spatial relationship between hot/cold points and the land cover raster, as well as the spots in different districts, we created a cross-table that elaborates on the frequency of occurrence of hot/cold spots for various land cover types. The high-resolution land cover data was taken in 10-m resolution from Esri 2020 Land Cover (https://livingatlas.arcgis.com/landcover/, accessed on 1 June 2023). To ensure excellent data quality, the dataset was interpreted from Sentinel-2 Earth global observation photography, and the classifier was trained with billions of datasets from human-labeled image pixels. The land-cover dataset covers the entire city of Beijing and includes six major types of landscape features: water bodies, forests, built-up regions, grasslands, and bare land. Meanwhile, the original anonymous datasets were used to produce gender classifications for each note. Following the import of this data into Python's statistic correspondence analysis package, a correspondence analysis plot was generated, which shows the relative distance or closeness of the relationship between different points in the plot, reflecting lesser or greater degrees of correlation between the compared objects.

3. Results

3.1. The Spatial Pattern of Camping Behaviors

Figure 2a depicts the map of hotspots derived from kernel density analysis. In general, more camping behaviors (n = 2209) were seen in the suburbs than in the urban core. The Pinggu district drew the greatest number of campers (n = 494) of the regions surveyed (n = 452) (Figure 3b). According to the geo-referenced locations of 2971 camping notes gathered, six spatial camping clusters with the darkest color (in orange and red) were identified around Beijing. Table 1 lists the descriptive statistics for the areas with the highest population density. Similar to the spatial distribution, only one spatial grouping was found to be intersected within the central urban area, while the remaining seven spatial clusterings were located in suburban areas. The core area clustering is predominantly located to the northeast of the city's central limit. We compared the 2971 locations of camping notes obtained with the Beijing POI names to obtain the total number of camping notes within each clustering area. According to the names of the POIs, this hotspot was comprised of urban parks, such as Chaoyang Park, whose landscape characteristics included an artificial lake, forest, lawn, and farmland. In contrast, the most densely populated suburban neighborhood is the Jinhaihu area, which is renowned for its expansive lake, mountains with colorful forests, and rolling lawns.

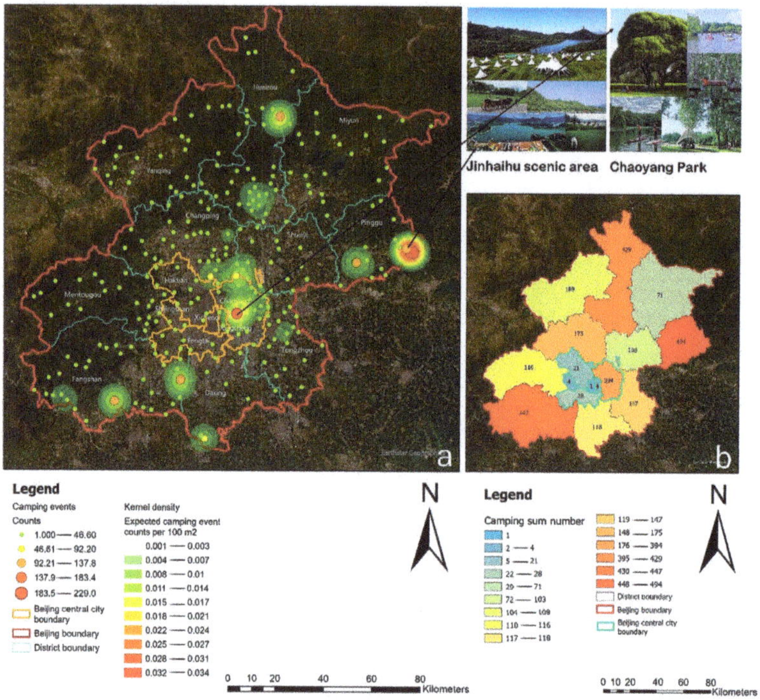

Figure 2. (**a**) kernel density heat maps of camping notes and (**b**) the spatial distribution of camping notes in Beijing.

3.2. Distribution of Hot and Cold Points of RES Perceived by the Public

Figure 3 shows the spatial distribution of hot/cold points classified based on the number of "likes" received from the public. The hot points are statistically significant clusters of sites with greater RES values (H-H value), whereas the cold points are statistically significant clusters of locations with lower RES values (L-L value). Using Anselin Local Moran's I in ArcGIS, this analysis identified 98 hot points and 427 cold points. The hot points were observed to be dispersed throughout Beijing (Figure 3a), indicating either suburban or central urban regions. In the suburban area, the hot points had an inward-to-outward spreading pattern: high RES values of hot spots were associated with scenic areas in beautiful and unique mountainous and water landscapes to the north of Beijing, including Jihaihu scenic area and Yudushan scenic areas (referred to Table 2). In the central urban area, hot points were also identified. According to POI (Table 2), these hots points were featured by forest parks and wetland parks, such as the Dayunhe Forest Park and Qinglonghu Wetland Park. Among these, Yudushan scenic region and Jinghaihu scenic area, with nearby camping sites, have the highest number of "likes" with 6661 and 5079, respectively (Table 2). Interestingly, both gorgeous regions accommodate different outdoor recreational activities. For instance, visitors are not only permitted to pitch tents and hammocks along the banks of large lakes for long-term stays; additionally, the recreational public facilities support picnicking, boating, carriage, horse riding, flying kites, walking dogs, and playing frisbee; and grassland concert events were also constructed accessible to the general public.

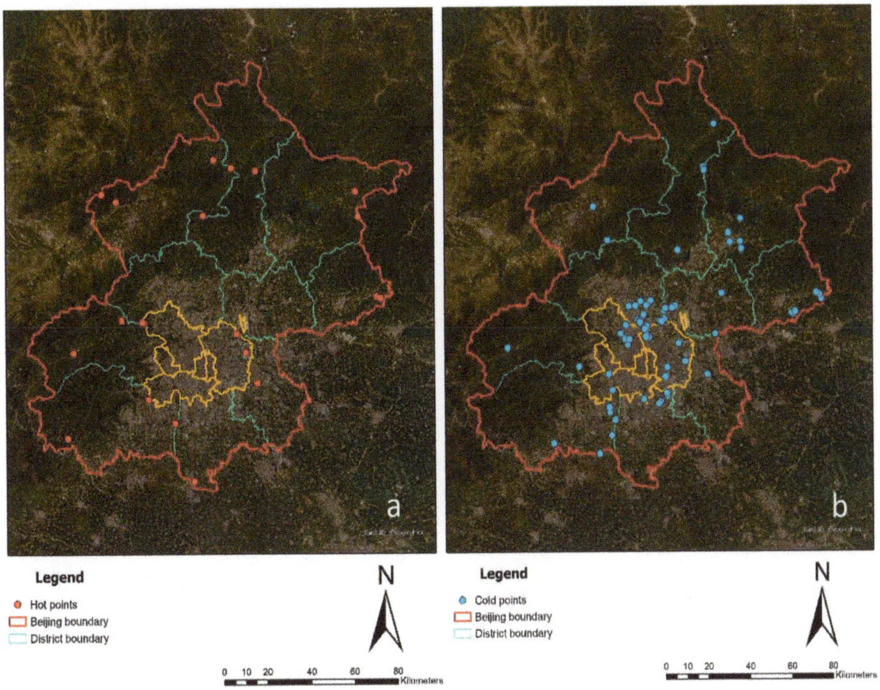

Figure 3. (**a**) Hot points per the RESs provided by camping notes. (**b**) Cold points per the RESs provided by camping notes.

Table 1. Descriptive statistics of the six highest-density areas.

No.	Locations	Districts	Landscapes	Total Camping Notes	Percentage	Kernel Density (Mean)
1	Jinghaihu scenic area; Weilan Valley camping park	Pinggu	Lake, mountain, forests, lawn	356	12.28%	0.0162
2	Chaoyang Park	Chaoyang (central urban area)	Urban park, lake, forest, lawn, European buildings	504	16.69%	0.0084
3	Baihewan scenic area	Huairou	Mountain, forests, riverside, bare land	183	6.15%	0.0093
4	Daxiangludao resort	Pinggu	Lawn, woods, lake,	107	3.60%	0.0070
5	Tiankai Reservoir, Tiankai Farm	Fanshan	Mountain, bare land, cultivated land, pond	158	5.32%	0.0081
6	Shuangying camping park	Fanshan	Grassland, bare land	137	4.61%	0.0083

Table 2. Descriptive statistics of the top three locations with the highest RES value as perceived by the public.

No.	Poi Name	Landscape Features	Recreation during Camping	n = Like
1	Yudushan scenic area and local Yudu camping park	Mountain, forests, lake, grassland, lawn, pool, waterfall	Tenting, picnicking, boating, hiking, dog walking, barbecues, sawanobori	6661
2	Jinhaihu scenic area and local Weilan Valley camping park	Lake, mountain, forests, lawn	Tenting, picnicking, boating, carriage, horse riding, flying kites, walking dogs, and playing Frisbee, and grassland concert events	5079
3	Darehuangye	Grassland, bare land, mountain	Tenting, picnicking, campfire, barbecues, motorbikes, concerts, and market events	2726

In contrast, the cold points were mainly found in the central urban areas, particularly in the northern Haidian and Chaoyang districts (Figure 3b). Intriguingly, we discovered that the cold points near the north Haidian, Chaoyang, and Changping districts (Table 3) indicated the presence of numerous local forest parks and country parks, including Olympic forest park, Wenyuhe park and Dongxiaokou country park. Numerous visitors have complained about local restrictions in these campgrounds, including prohibitions on dog walking, boating, horseback riding, and swimming, which weakens their recreational support for the public. Additionally, cold points were spotted in the suburban area. Those points mostly referred to scenic areas and various natural landscapes developed as camping parks by private companies or local village committees (Table 3). According to our survey of campers' submitted comments, this is owing to their differing degrees of facilities and design in relation to the area landscape, as well as their inadequate maintenance. For instance, the camping parks and villages around Jinhai town and Wensuhe tend to have inferior architecture, recreational facilities, and maintenance, such as a lack of water sources, restrooms, parking, and waste management.

Table 3. Descriptive statistics of the cold points of the top three high RES value locations as perceived by the public.

No.	Poi Name	Landscape Features	Recreation during Camping	n = Like
1	Huangye No.7 camping park, Jinhaihu town	Mountain, lake, grassland	Tenting, picnicking, hiking, dog walking	242
2	Xiaoqinghe	Woods, wetland, forests, riverside, bare land	Tenting, picnicking, BBQ	121
3	Wenyuhe	Woods, riverside, bare land, sand land	Tenting, picnicking, dog walking	117

3.3. Relationship between Hot/Cold Spots, Land Cover Features, and Visitor Genders

Correspondence analysis enabled us to evaluate the links between the six categories of land cover features, the sites where people perceive high-high RES values (hot points) and low-low RES values (cold points), and the gender of visitors. Figure 4 shows that component 0 explains 17.7% of the total variability along the x-axis, which proved to be the most important (major) contribution in this correspondence analysis. In contrast, component 1 explains 15.90% of the total variability along the *y*-axis, which comes in the second place. The outcome of Figure 4 indicates that the perceived hot points locations were significantly related to the grassland and forest. In contrast, perceived cold point locations tended to be associated with wetlands and waterbodies. In addition, female visitors tend to favor grassland during all of these activities, which results in high or low RES values as a result of their camping activity. Instead, males are significantly more interested in built-up areas and bare land.

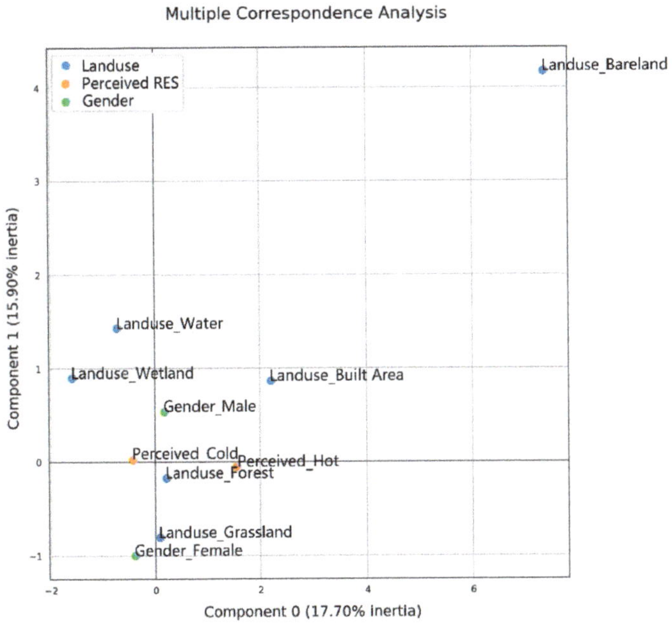

Figure 4. Correspondence analysis of hot and cold points relating to the six land cover types and all districts in Beijing.

4. Discussion

The study evaluated the public's perception of the RES given by urban green areas based on camping behaviors shown by social media data from the megacity Beijing. Here, we discuss our key findings: (i) the spatial pattern of camping activity, (ii) the public's perception of RES values, and (iii) the elements that influence RES values. In addition, we address the advantages and disadvantages of employing social media datasets for RES assessments in comparison to conventional approaches.

4.1. The Spatial Pattern of Camping Behaviors

Our spatial pattern of camping behaviors revealed that, excluding the urban area, a significantly greater number of camping behaviors were observed in the suburban area, which is comprised of picturesque locations and attractive natural landscapes. This is attributed by the author to the desire for privacy and tranquility, which inspires many people to go camping and explore nature, and as a result, empty and calm landscapes were favored globally [32]. Similarly, according to previous studies [33], a disproportionate number of social media posts are uploaded in landscapes with high scenic quality, such as national parks and river corridors. In addition, our study also addressed that the high density of camping behaviors is not only featured in suburban areas but also in the urban park in the central urban area. Similar to the findings of earlier studies, distance and transit variables could be primary factors that affect public activities such as visiting parks for natural recreation [34]. The proximity and accessibility of these urban parks in the central urban area also make them appealing camping locations for visitors.

4.2. The Spatial Pattern of Public Perceived RES Value

In accordance with the above finding, our hot/cold point analysis results revealed that the hot points concentrate around suburban areas. In addition, we discovered that other than the landscape elements provided, the local management policy for supporting camping and outdoor recreation can be another essential factor that determines high RES values. Similar findings were addressed by previous studies [32,34]. For example, the notes posted about the Jinhaihu Weilan Valley camping park (a private recreational park operated by a private company) demonstrated a high level of appreciation for available recreational activities such as camping, picnicking, boating, carriage and horseback rides, kite-flying, walking dogs, playing Frisbee, fishing, swimming, and kayaking in the lake, as well as concerts and an outdoor market that utilize the local landscape. In contrast, cold points were clustered in the central urban area featured with urban parks, forest parks, and country parks. Further, some cold points were discovered in an urban location with a high density of camping behavior. This could be attributed to recreation tensions between public recreational desires and the RES offered by urban parks. In contrast to previous conflicts between campers and local inhabitants [35], we believe that the conflicts between managers and campers in these places stem from the design and management philosophy of urban parks at the highest level. Apparently, the local urban park administration appears to have limited visitors' access to varied leisure possibilities, which is not the ideal option for attaining the maximum CES benefit for the public. For example, the camping notes located in the Olympic Park noted, "We came here for camping, and learned that tents, hammocks and picnic mats are not allowed in the area". "We had to camp secretly while the security staff were not patrolling". Furthermore, practically all urban park settings in Beijing have a pet prohibition, which dramatically reduces the perceived RES value. These leisure activities were labeled as "uncivilized" in the 2002 Beijing Park Regulations. Previous CES studies, on the other hand, found that dog walking and picnics are major recreation ecosystem service indicators that influence CES assessment patterns [36]. Picnic and camping areas are also seen as important features in the cultural landscape and green space management by several scholars [32,37,38]. In order to improve RES values for the majority of urban parks, it is therefore beneficial to reconsider urban park management regulations in order to suit the ever-changing aspirations of Beijing residents as the post-COVID-19 era continues.

4.3. Impact Factors towards Public-Perceived RES Value

According to the results of the correspondence analysis, the occurrence of hot points was highly associated with grasslands and woods. Several existing CES evaluation studies also demonstrate similar results. For instance, scholars [37,39,40] documented the significant contribution of grasslands and forestry lands towards RES value. In addition, previous research [1,8,40] identified water bodies as an important land cover that correlates to better recreation values and enhanced leisure possibilities. However, our results indicated that cold points were associated with wetlands and waterbodies, as camping notes indicated that visitors did not take advantage of the recreational opportunities given by surrounding waterbodies. It could be explained by the many limits on free enjoyment in public water bodies, in addition to the environmental protection of water quality in local water bodies. Since 2020, for instance, the Water Bureau has restricted not only swimming and fishing within water conservation zones but also gatherings, open-air barbecues, and picnicking around public waterbodies, such as rivers and lakes throughout the entire city [41]. Specialized prohibitions were placed on "wild swimming," which is criticized as "uncivilized behaviors" in the event of swimming in urban rivers and park lakes, as well as splashing water. This considerably restricts people's access to waterbodies; hence, the recreational value of public water places is significantly diminished. Conflicts between public leisure needs and public waterbody management have been the subject of a growing number of recent discussions and news reports [42–45]. In contrast to Beijing, many other cities, such as Copenhagen and Berlin, have permitted the establishment of dedicated outdoor water recreation areas, particularly for wild swimming, on a number of the city's natural waterbodies, such as urban rivers, canals, lakes, and harbors [46–48]. Numerous studies also highlight the multiple recreational benefits of urban water bodies [49,50], as well as the fact that water recreation activities such as swimming, fishing, and boating (including paddling and canoeing) can be used as an indicator for assessing the value and benefits derived from the public waterbodies' environment [51]. In order to increase the RES values in public waterbodies, we believe that the public sector should evaluate the reasonable requests of the public to swim and play in the water. In the future, a proper management system should be established to balance public expectations, public safety, water resource protection, and public safety.

Meanwhile, our findings suggested that gender significantly affects landscape preference, with female visitors preferring grassland and male visitors preferring bare ground and urban areas for camping enjoyment. Similar findings were reported in prior studies [52–55]. Particularly, the study regarding Phoenix revealed that women were much more averse to dry settings devoid of lush vegetation [55]. In addition, researchers have proposed gender variations in recreational use and views of nature [56–58]. "The lawn and grassland here is very good to be portrayed, with my newly purchased picnic cloth and picnic basket, very English style landscape oil painting scene feeling", stated a female visitor who posted her camping notes in a grassland area. In addition, the phrase "nice to be portrayed" appears frequently in the camping notes of female visitors. Similar to the study in Georgia [56], women emphasize photography substantially more than recreational activities such as hiking and bird viewing, which are more frequently addressed by men. Females are more likely to seek out campsites with luxuries, whereas males are more likely to focus on the activities the campsite has to offer and do duties peculiar to camping [34,59]. This could explain our observation that male visitors are more interested in barren land. "No campfire, no camping. Without vegetation, this gravel-filled location is suitable for campfires, and we also played mountain biking and motorbikes here", according to a male camper who enjoyed a unique camping experience on bare terrain.

In addition, we found that people camped in urban areas near pocket green spaces, their own residential districts, and universities. These sites provided convenient opportunities for people to experience nature in an urban context, although the camping behaviors of visitors may cause potential issues. Current various debates and news articles in China mention these conflicts regarding whether camping in open green spaces in residential districts and communities is detrimental to residents' public health and whether school administrators should prohibit students from camping on university grounds during school closures [60–62]. Previous research found that concerns about conflicts with others, personal safety, and harassment significantly reduced female participation in outdoor recreation, leading to the conclusion that women do not have as strong an interest in camping in built-up areas as men, despite the fact that it provided convenient green spaces for staying in nature, whereas males were not as concerned [63–65].

4.4. Advantages, Limitations, and Future Potentials for CES Evaluation Based on Mapping Geo-Tagged Camping Notes from Social Media Data

Our study demonstrated a crowdsourcing data approach to evaluating RES provided by urban green space throughout camping behavior. Such practices provide planners with a new resource for quantifying and comprehending the RES values given by existing green spaces. This enables decision-makers to maximize their policy-making actions, which better satisfies the public's rapidly changing expectations for outdoor leisure and enhances the well-being of individuals in densely populated megacities. In contrast to prior spatial assessments of RES, which were typically based on expert viewpoints or used land cover as a proxy [66,67]—our study spatially examined RES using public perspectives.

At this point, the public data collection activities allow citizens' preferences and concerns to be incorporated into the development and management of green space; this is known as public engagement. Such practices are regarded as quite valuable in China, where the mainstream planning process is viewed as a top-down technological engineering process based solely on the views and knowledge of decision-makers and experts, as opposed to a social process of cooperation with various stakeholders [68]. For many public participation studies conducted to date, the time-consuming and expensive nature of these surveys and interviews is viewed as a barrier that reduces planners' motives to involve public participation [69]. Currently, social media data is widely accessible and updated every minute as new data becomes available, which demonstrates considerable benefits. Consequently, a rising number of academics employ social media gathered from platforms such as Panoramio and Flickr [40] for tasks such as landscape value mapping, landscape aesthetics mapping, evaluations for establishing land use policy, and crowdsourcing data for CES assessment [40,70,71]. Nonetheless, the hunt for proxies that reliably identify RES for mapping the locations and attributes related to people's nature experiences and perceptions continues [3]. In this study, we used camping notes from the largest Chinese social media platform (LRB) for life sharing to explain the experiences of people who stay in the wilderness and acquire a variety of recreational activities from the local landscape. In addition, our study utilized social media data regarding camping notes and pertinent public perception for the evaluation of RES by giving a number of "likes". We consider not only the visitors' perceptions of RES value in their camping sites but also public perception and cognition of local RES affected by the geo-tagged photos and descriptions contents through the social media data, which could aid in the future exploration of the public's visiting preferences to various parks [32]. In contrast to conventional methods, we summarize the pros of utilizing social media camping data for public perception RES surveys as follows: (1) Social media data are readily accessible, enabling opportunistic sampling of visitors' expressed landscape values. (2) Social media data could reveal the public's perception of the landscape's value (e.g., cultural ecosystem services, recreational ecosystem services) over an extended period of time and on a broad scale than traditional interviews and surveys. (3) Social media data can capture large amounts of data in a brief amount of time to reflect public perceptions, making it a rapid and cost-effective method for identifying and mapping RES they gained in megacities.

However, the limits of social media data must also be acknowledged. For instance, geo-tagged photographs from social media sources may provide cellular signal strength uncertainties for electrical gadgets. According to Zielstra and Hochmair (2013), the accuracy of the geographical reference points for these devices can impact the quality of the data gathered [72]. The physical, perceptual, and evaluative aspects of landscapes are also difficult to measure from different user groups, and these are the limitations we must acknowledge. Although the Little Red Book (LRB) platform is currently the most popular social media platform in China, the majority of our collected data may only represent the perceptions of younger citizens regarding RESs and their natural experiences of green spaces, as the majority of LRB users (83.31%) are between the ages of 18 and 34 [73]. Citizens outside of this age range may not share as many remarks about RES locales on social media or other websites. The quantity of "likes" assigned to influencers (e.g., Key Opinion Leader (KOL) camping remarks) may further increase statistical uncertainty. These influencers may have a greater number of notes about recommended camping facilities than actual users, and their meticulously manipulated photographs and travel blogs rarely provide geo-tagged information about the camping place. Consequently, more stringent data cleaning is required to reduce the potential for bias in the outcomes resulting from correspondence. The statistical dependability and locational precision of crowdsourced camping notes must be examined in further research.

At the same time, even though we controlled the context of camping notes on the scenery and camping-related activities by deleting images of the user's personal features (such as the user's own portrait during camping events) in order to reduce the interference caused by the person rather than the context. There are still limitations to the public impression possessed by the social media image, as pictures published in camping notes may tend to avoid unsightly areas by emphasizing the best features, and posts may misrepresent something as positive when it is not. Therefore, additional field research validation of the information reflected in social media data demands further investigation, and field surveys and interviews with the general public will supplement the social media data in future studies. In addition, the drawbacks of social media data used in the RES evaluation included the loss of individual-level user-specific information. Due to users' right to privacy, the social demographic profile, including education, age, degree of literacy, and social class and income, cannot be distinguished in social media data. Even though the user profiles of some social networks can describe the general characteristics of the social background of users, additional interviews or questionnaires are required to investigate how social context influences the recreational value that individuals derive from the natural environment in future research.

5. Conclusions

This study uses social media data to assess the RES value of urban green spaces in the meta-city of Beijing. The present spatial patterns of camping behavior after COVID-19 were explored using geo-tagged camping posts, and hot/cold points linked with high/low RES values were identified and distinguished across the entire city. Based on this, the public's preference for urban green space was analyzed and comprehended in terms of land cover and gender inequalities. Our primary findings are as follows:

- According to the spatial pattern of camping habits, both suburban and urban central areas have given certain camping support services to the public since the implementation of COVID-19 restrictions. However, much more clusters were detected in the suburbs than in the central districts, indicating that existing urban green spaces do not provide enough camping facilities.
- Hot points associated with high-RES values were mainly identified in suburban areas. Despite the high density of camping noted in the central urban area, the green spaces in this area provide relatively low RES value for the public. We believe this is due to the restricted public leisure activities supported by these urban parks as a result of their current management policies. Therefore, additional optimization is required.
- Differing from previous studies, the lower RES value was correlated with waterbodies in Beijing. Here, we think this is due to local recreation restorations of public water bodies. Higher RES were closely related to grassland and forests. In addition, female visitors tend to camp in grassland while males tend to bare land and built-up areas, which could be explained by the gender differences in landscape preferences, recreational use of nature, and concerns about conflicts.

The last two years have witnessed a significant increase in camping activities, which has altered the most popular form of outdoor recreation in green spaces. This appears to be a fast-mind transition that alters how people perceive available green spaces. Here, RES value assessment based on social media data can provide a more up-to-date and in-depth understanding of public demands and perceptions, thereby providing better inspiration for decision-makers to improve current green space management in order to adapt to such a dynamic need from the public. In suburban areas, our study highlighted both the high density of public camping and the high perceived recreational value, which reflects the potential for rural tourism in suburban areas. Accordingly, we propose that management policies for suburban green space should be integrated with local rural development and tourism policies, with a focus on the new model of "camping+", such as camping+ outdoor adventure activities in wild lands, camping+ trip photographing, camping+ cultural recreational activities (e.g., open-air concerts on the lawn or by the waterside), and other products with different characteristics for public preferences, in order to inject new momentum into rural development. In the central region, our research revealed the public camping demand and the limited recreational value obtained by campers. It indicated conflicts between public new outdoor recreation and previous urban park management policies, such as the restriction on water sports recreation in blue spaces and the prohibition on picnicking in many urban parks. We propose that the management of urban green space reexamines the current one-size-fits-all park management regulations and that policies for open park use be formulated based on the landscape characteristics of different green spaces, such as forests, water bodies, grasslands, artificial ground, etc. For instance, open shared areas should be designated in areas such as park lawns and forested spaces, and supporting service facilities should be improved to better satisfy the needs of the public for outdoor activities such as camping and recreation close to nature, including the setting up of tents. In addition, the water surface's suitability for recreation could be evaluated, and water-opening zones and management policies should be developed to satisfy the public's demand for water-friendly outdoor sports and fitness activities. However, additional research is still needed to comprehend the long-term effects of these camping patterns on urban green areas since 2020. Beyond the RES value, these tendencies may have an impact on the other CES, such as aesthetics, natural education, and a sense of belonging, which may result in further difficulties. The following aspects are therefore proposed for future research:

(1) As green areas provide varied perceptions of CES, the conflict between campers and stakeholders could be investigated. (2) With different temporal scales, variations in RES value before and after camping would be interesting to investigate and compare in the future; (3) With a diverse range of group users (e.g., ages, education level, and agenda) in social media datasets, the different perceptions of different groups towards green space would be interesting to investigate and compare in the future.

Author Contributions: Conceptualization, H.X. and M.M.; methodology, H.X.; software, G.Z.; validation, Y.L.; formal analysis, G.Z. and M.M.; investigation, G.Z.; resources, H.X.; data curation, M.M.; writing—original draft preparation, H.X.; writing—review and editing, G.Z.; visualization, H.X.; supervision, H.X.; project administration, H.X.; funding acquisition, M.M. All authors have read and agreed to the published version of the manuscript.

Funding: This research was funded by [Research Enhancement Project for Young Scholars of BUCEA] grant number [x21044] and [National Natural Science Foundation] grant number [52008171].

Data Availability Statement: Not applicable.

Conflicts of Interest: The authors declare no conflict of interest.

References

1. Plieninger, T.; Dijks, S.; Oteros-Rozas, E.; Bieling, C. Assessing, mapping, and quantifying cultural ecosystem services at community level. *Land Use Policy* **2013**, *33*, 118–129. [CrossRef]
2. Hernández-Morcillo, M.; Plieninger, T.; Bieling, C. An empirical review of cultural ecosystem service indicators. *Ecol. Indic.* **2013**, *29*, 434–444. [CrossRef]
3. Hermes, J.; Van Berkel, D.; Burkhard, B.; Plieninger, T.; Fagerholm, N.; von Haaren, C.; Albert, C. Assessment and valuation of recreational ecosystem services of landscapes. *Ecosyst. Serv.* **2018**, *31*, 289–295. [CrossRef] [PubMed]
4. Tenkanen, H.; Di Minin, E.; Heikinheimo, V.; Hausmann, A.; Herbst, M.; Kajala, L.; Toivonen, T. Instagram, Flickr, or Twitter: Assessing the usability of social media data for visitor monitoring in protected areas. *Sci. Rep.* **2017**, *7*, 17615. [CrossRef]
5. Wood, S.A.; Guerry, A.D.; Silver, J.M.; Lacayo, M. Using social media to quantify nature-based tourism and recreation. *Sci. Rep.* **2013**, *3*, 2976. [CrossRef]
6. Oteros-Rozas, E.; Martín-López, B.; Fagerholm, N.; Bieling, C.; Plieninger, T. Using social media photos to explore the relation between cultural ecosystem services and landscape features across five European sites. *Ecol. Indic.* **2018**, *94*, 74–86. [CrossRef]
7. Brown, G.; Donovan, S.J.S.; Resources, N. Measuring change in place values for environmental and natural resource planning using public participation GIS (PPGIS): Results and challenges for longitudinal research. *Soc. Nat. Resour.* **2014**, *27*, 36–54. [CrossRef]
8. Xu, H.; Zhao, G.; Fagerholm, N.; Primdahl, J.; Plieninger, T. Participatory mapping of cultural ecosystem services for landscape corridor planning: A case study of the Silk Roads corridor in Zhangye, China. *J. Environ. Manag.* **2020**, *264*, 110458. [CrossRef]
9. Tang, J.; Liu, H. Trust in social media. *Synth. Lect. Inf. Secur. Priv. Trust.* **2015**, *10*, 1–129.
10. Liu, Y.; Sui, Z.; Kang, C.; Gao, Y. Uncovering patterns of inter-urban trip and spatial interaction from social media check-in data. *PLoS ONE* **2014**, *9*, e86026. [CrossRef]

11. Van Berkel, D.B.; Verburg, P. Spatial quantification and valuation of cultural ecosystem services in an agricultural landscape. *Ecol. Indic.* **2014**, *37*, 163–174. [CrossRef]
12. Schirpke, U.; Meisch, C.; Marsoner, T.; Tappeiner, U. Revealing spatial and temporal patterns of outdoor recreation in the European Alps and their surroundings. *Ecosyst. Serv.* **2018**, *31*, 336–350. [CrossRef]
13. Rice, W.L.; Mateer, T.J.; Reigner, N.; Newman, P.; Lawhon, B.; Taff, B.D. Changes in recreational behaviors of outdoor enthusiasts during the COVID-19 pandemic: Analysis across urban and rural communities. *J. Urban Ecol.* **2020**, *6*, juaa020. [CrossRef]
14. Venter, Z.S.; Barton, D.N.; Gundersen, V.; Figari, H.; Nowell, M.J.E. Urban nature in a time of crisis: Recreational use of green space increases during the COVID-19 outbreak in Oslo, Norway. *Environ. Res. Lett.* **2020**, *15*, 104075. [CrossRef]
15. Consumption Observation: Camping "Unlocks" a New Blue Ocean in the Domestic Tourism Market. *People's Daily Online*, 2 April 2021. Available online: https://baijiahao.baidu.com/s?id=1702610995321812573&wfr=spider&for=pc(accessed on 28 May 2023).
16. In the Past Two Years, the Entire Circle of Friends Has Been Camping. *Daily Economic News*, 15 June 2022. Available online: https://baijiahao.baidu.com/s?id=1728983702804391528&wfr=spider&for=pc(accessed on 28 May 2023).
17. Zhou, W. *The Rise of Camping Economy, the Development Status and Prospect Analysis of Camping Tent Industry*; China Industrial Research Institute: Beijing, China, 2022; p. 200.
18. Fischer, L.K.; Honold, J.; Botzat, A.; Brinkmeyer, D.; Cvejić, R.; Delshammar, T.; Elands, B.; Haase, D.; Kabisch, N.; Karle, S. Recreational ecosystem services in European cities: Sociocultural and geographical contexts matter for park use. *Ecosyst. Serv.* **2018**, *31*, 455–467. [CrossRef]
19. Wang, Z.; Jin, Y.; Liu, Y.; Li, D.; Zhang, B. Comparing social media data and survey data in assessing the attractiveness of Beijing Olympic Forest Park. *Sustainability* **2018**, *10*, 382. [CrossRef]
20. Govenrment of Beijing Municipality. Popluation of Beijing. Available online: https://www.beijing.gov.cn/renwen/bjgk/rk/rktj/202303/t20230322_2941861.html (accessed on 28 May 2023).
21. Govenrment of Beijing Municipality. 26 More Urban Recreational Parks to Be Added in 2022 the City's Forest Coverage Reaches 44.8%, Making It Easier for Citizens to Share the Benefits of Greenery. Available online: https://www.beijing.gov.cn/renwen/sy/whkb/202103/t20210319_2311413.html (accessed on 28 May 2023).
22. Pu, W. Little red book, let the user become the spreader. *Collecation* **2018**, *33*, 55–58. (In Chinese)
23. Okabe, A.; Satoh, T.; Sugihara, K. A kernel density estimation method for networks, its computational method and a GIS-based tool. *Int. J. Geogr. Inf. Sci.* **2009**, *23*, 7–32. [CrossRef]
24. Khan, D.; Rossen, L.M.; Hamilton, B.E.; He, Y.; Wei, R.; Dienes, E. Hot spots, cluster detection and spatial outlier analysis of teen birth rates in the US, 2003–2012. *Spat. Spatio-Temporal Epidemiol.* **2017**, *21*, 67–75. [CrossRef]
25. Chugh, R.; Patel, S.B.; Patel, N.; Ruhi, U. Likes, comments and shares on social media: Exploring user engagement with a state tourism Facebook page. *Int. J. Web Based Commun.* **2019**, *15*, 104–122.
26. Conti, E.; Heldt Cassel, S. Liminality in nature-based tourism experiences as mediated through social media. *Tour. Geogr.* **2020**, *22*, 413–432. [CrossRef]
27. Hermida, A.; Fletcher, F.; Korell, D.; Logan, D. Share, like, recommend: Decoding the social media news consumer. *J. Stud.* **2012**, *13*, 815–824. [CrossRef]
28. Greenacre, M. Tying up the loose ends in simple, multiple, joint correspondence analysis. In Proceedings of the Compstat 2006-Proceedings in Computational Statistics: 17th Symposium, Rome, Italy, 28 August–1 September 2006; pp. 163–185.
29. Dhalmahapatra, K.; Shingade, R.; Mahajan, H.; Verma, A.; Maiti, J. Decision support system for safety improvement: An approach using multiple correspondence analysis, t-SNE algorithm and K-means clustering. *Comput. Ind. Eng.* **2019**, *128*, 277–289. [CrossRef]
30. Greenacre, M.J.S. Correspondence analysis in medical research. *Stat. Methods Med. Res.* **1992**, *1*, 97–117. [CrossRef]
31. Plieninger, T.; Torralba, M.; Hartel, T.; Fagerholm, N. Perceived ecosystem services synergies, trade-offs, and bundles in European high nature value farming landscapes. *Landsc. Ecol.* **2019**, *34*, 1565–1581. [CrossRef]
32. van Zanten, B.T.; Zasada, I.; Koetse, M.J.; Ungaro, F.; Häfner, K.; Verburg, P.H. A comparative approach to assess the contribution of landscape features to aesthetic and recreational values in agricultural landscapes. *Ecosyst. Serv.* **2016**, *17*, 87–98. [CrossRef]
33. Martínez Pastur, G.; Peri, P.L.; Lencinas, M.V.; García-Llorente, M.; Martín-López, B. Spatial patterns of cultural ecosystem services provision in Southern Patagonia. *Landsc. Ecol.* **2016**, *31*, 383–399. [CrossRef]
34. Gursoy, D.; Chen, B.T. Factors influencing camping behavior: The case of Taiwan. *J. Hosp. Mark. Manag.* **2012**, *21*, 659–678. [CrossRef]
35. Caldicott, R.; Scherrer, P.; Jenkins, J. Freedom camping in Australia: Current status, key stakeholders and political debate. *Ann. Leis. Res.* **2014**, *17*, 417–442. [CrossRef]
36. Pueffel, C.; Haase, D.; Priess, J.A. Mapping ecosystem services on brownfields in Leipzig, Germany. *Ecosyst. Serv.* **2018**, *30*, 73–85. [CrossRef]
37. Pellaton, R.; Lellei-Kovács, E.; Báldi, A. Cultural ecosystem services in European grasslands: A systematic review of threats. *Ambio* **2022**, *12*, 2462–2477. [CrossRef] [PubMed]
38. Albert, C.; Bonn, A.; Burkhard, B.; Daube, S.; Dietrich, K.; Engels, B.; Frommer, J.; Götzl, M.; Grêt-Regamey, A.; Job-Hoben, B.; et al. Towards a national set of ecosystem service indicators: Insights from Germany. *Ecol. Indic.* **2016**, *61*, 38–48. [CrossRef]
39. Zisenis, M.; Richard, D.; Vačkář, D.; Lorencová, E.; Melichar, J.; Hönigová, I.; Oušková, V.; Hošek, M.; Chobot, K.; Götzl, M.; et al. Survey on grassland ecosystem services in the Czech Republic and literature review. *ETC/BD Rep. EEA* **2011**, *4*, 22–26.

40. Figueroa-Alfaro, R.W.; Tang, Z.J. Evaluating the aesthetic value of cultural ecosystem services by mapping geo-tagged photographs from social media data on Panoramio and Flickr. *J. Environ. Plan. Manag.* **2017**, *60*, 266–281. [CrossRef]
41. Beijing Centrally Announces No-Tour, No-Fishing Areas, Barbecue in the Riverside Fines up to 20,000 Yuan. *Guangming Daily*, 1 May 2020. Available online: https://m.gmw.cn/baijia/2020-05/01/1301195150.html(accessed on 28 May 2023).
42. Xia, Y. Shichahai and Yuyuantan Have People Swimming and Playing Every Day, How to Crack the Problem of Wild Swimming Management? *Beijing Daily*, 10 August 2021. Available online: https://baijiahao.baidu.com/s?id=1707685026212700302&wfr=spider&for=pc(accessed on 28 May 2023).
43. Swimming Is Forbidden, But They Take the Risk! Beijing's Zuo'anmen Moat is Frequently Filled with Wild Swimmers. *Renming Daily*, 13 June 2022. Available online: https://baijiahao.baidu.com/s?id=1735503615961625219&wfr=spider&for=pc(accessed on 28 May 2023).
44. Olympic Forest Park's 'Popular Creek'. A Cordon Can't Stop Kids from Playing in the Water! Southnet. 2022. Available online: https://travel.sohu.com/a/568486511_121123772 (accessed on 28 May 2023).
45. A Better Environment, Can You Paddle in the Beijing Moat? *Renming Daily News*, 26 July 2022. Available online: https://baijiahao.baidu.com/s?id=1739373589749025635&wfr=spider&for=pc(accessed on 28 May 2023).
46. Berlin: An Outdoor Swimming Tour of the Best Pools and Urban Lidos. Available online: https://www.theguardian.com/travel/2015/aug/07/berlin-outdoor-swimming-tour-best-pools-lidos (accessed on 28 May 2023).
47. Berlin's River Deemed "Excellent" for Swimming. Available online: https://www.thesourcemagazine.org/berlins-river-deemed-excellent-for-swimming/ (accessed on 28 May 2023).
48. Why Copenhagen Is One of the Greatest Cities for Summer Swimming. 2020. Available online: https://www.thrillist.com/travel/nation/best-swimming-spots-copenhagen (accessed on 28 May 2023).
49. Grzyb, T.; Kulczyk, S. How do ephemeral factors shape recreation along the urban river? A social media perspective. *Landsc. Urban Plan.* **2023**, *230*, 104638. [CrossRef]
50. Pearson, A.L.; Bottomley, R.; Chambers, T.; Thornton, L.; Stanley, J.; Smith, M.; Barr, M.; Signal, L. Measuring blue space visibility and 'blue recreation'in the everyday lives of children in a capital city. *Int. J. Environ. Res. Public Health* **2017**, *14*, 563. [CrossRef]
51. Vesterinen, J.; Pouta, E.; Huhtala, A.; Neuvonen, M. Impacts of changes in water quality on recreation behavior and benefits in Finland. *J. Environ. Manag.* **2010**, *91*, 984–994. [CrossRef]
52. Strumse, E. Demographic differences in the visual preferences for agrarian landscapes in western Norway. *J. Environ. Psychol.* **1996**, *16*, 17–31. [CrossRef]
53. Lyons, E. Demographic correlates of landscape preference. *Environ. Behav.* **1983**, *15*, 487–511. [CrossRef]
54. Jiang, B.; Chang, C.-Y.; Sullivan, W.C. A dose of nature: Tree cover, stress reduction, and gender differences. *Landsc. Urban Plan.* **2014**, *132*, 26–36. [CrossRef]
55. Yabiku, S.T.; Casagrande, D.G.; Farley-Metzger, E. Preferences for landscape choice in a Southwestern desert city. *Environ. Behav.* **2008**, *40*, 382–400. [CrossRef]
56. Burger, J.; Sanchez, J.; Gibbons, J.W.; Gochfeld, M. Gender differences in recreational use, environmental attitudes, and perceptions of future land use at the Savannah River site. *Environ. Behav.* **1998**, *30*, 472–486. [CrossRef]
57. Godtman Kling, K.; Margaryan, L.; Fuchs, M. (In) equality in the outdoors: Gender perspective on recreation and tourism media in the Swedish mountains. *Curr. Issues Tour.* **2020**, *23*, 233–247. [CrossRef]
58. Burger, J.; Gochfeld, M. Gender Differences in Resource Use and Evaluatuon of Attributes of Places of Resource Use by Native Americans and Caucasians from Western Idaho: Relevance to Risk Evaluations. *J. Toxicol. Environ. Health* **2010**, *73*, 1655–1664. [CrossRef]
59. Carlson, C.A. Gender Roles in the Camping Situation. Master's Thesis, Portland State University, Portland, OR, USA, 1977.
60. Is It Okay to Set Up a Tent on the Community Lawn? Available online: https://www.163.com/dy/article/H6JJM3370514R9KQ.html (accessed on 28 May 2023).
61. Holiday Camping Is Too Hot, Not Enough Lawn? Some People Take a Fancy to the Lawn of the Community, and the Owners and Netizens Are Arguing. Available online: https://baijiahao.baidu.com/s?id=1731891300241457401&wfr=spider&for=pc (accessed on 28 May 2023).
62. The University Held a "Star Camping" Festival to Unlock a New Way of Campus Life. *Xiaoxiang Daily*, 8 June 2022. Available online: https://baijiahao.baidu.com/s?id=1735062068840001058&wfr=spider&for=pc(accessed on 28 May 2023).
63. Ghimire, R.; Green, G.T.; Poudyal, N.C.; Cordell, H.K. An analysis of perceived constraints to outdoor recreation. *J. Park Recreat. Adm.* **2014**, *32*.
64. Johnson, C.Y.; Bowker, J.M.; Cordell, H.K. Outdoor recreation constraints: An examination of race, gender, and rural dwelling. *J. Rural. Soc. Sci.* **2001**, *17*, 6.
65. Sjögren, K.; Stjernberg, L. A gender perspective on factors that influence outdoor recreational physical activity among the elderly. *BMC Geriatr.* **2010**, *10*, 34. [CrossRef]
66. Cheng, X.; Van Damme, S.; Li, L.; Uyttenhove, P. Evaluation of cultural ecosystem services: A review of methods. *Ecosyst. Serv.* **2019**, *37*, 100925. [CrossRef]
67. Tratalos, J.A.; Haines-Young, R.; Potschin, M.; Fish, R.; Church, A. Cultural ecosystem services in the UK: Lessons on designing indicators to inform management and policy. *Ecol. Indic.* **2016**, *61*, 63–73. [CrossRef]

68. Xu, H.; Plieninger, T.; Zhao, G.; Primdahl, J. What Difference Does Public Participation Make? An Alternative Futures Assessment Based on the Development Preferences for Cultural Landscape Corridor Planning in the Silk Roads Area, China. *Sustainability* **2019**, *11*, 6525. [CrossRef]
69. Martin, V.Y.; Christidis, L.; Lloyd, D.J.; Pecl, G. Understanding drivers, barriers and information sources for public participation in marine citizen science. *J. Sci. Commun.* **2016**, *15*, A02. [CrossRef]
70. Langemeyer, J.; Calcagni, F.; Baró, F. Mapping the intangible: Using geolocated social media data to examine landscape aesthetics. *Land Use Policy* **2018**, *77*, 542–552. [CrossRef]
71. Retka, J.; Jepson, P.; Ladle, R.J.; Malhado, A.C.; Vieira, F.A.; Normande, I.C.; Souza, C.N.; Bragagnolo, C.; Correia, R.A. Assessing cultural ecosystem services of a large marine protected area through social media photographs. *Ocean Coast. Manag.* **2019**, *176*, 40–48. [CrossRef]
72. Zielstra, D.; Hochmair, H.H.; Neis, P. Assessing the Effect of Data Imports on the Completeness of OpenStreetMap–AU Nited S Tates Case Study. *Trans. GIS* **2013**, *17*, 315–334. [CrossRef]
73. LittleRedBook Active User Portrait Trend Report. 2021. Available online: https://finance.sina.com.cn/tech/2021-04-23/doc-ikmyaawc1278044.shtml (accessed on 28 May 2023).

forests

Article

Sustainability Assessment for the Protected Area Tourism System from the Perspective of Ecological-Economic-Social Coordinated Development

Xiangju Zhang [1], Linsheng Zhong [1,2,*], Hu Yu [1,2] and Ling-En Wang [1,2]

1 Institute of Geographic Sciences and Natural Resources Research, Chinese Academy of Sciences, Beijing 100101, China; 18288618224@163.com (X.Z.)
2 University of Chinese Academy of Sciences, Beijing 101408, China
* Correspondence: zhlsheng@263.net

Abstract: Tourism is a significant way for the public to enjoy the cultural ecosystem services provided by protected areas (PAs). However, with PAs being expected to make much wider ecological, social and economic contributions to sustainability and human well-being, PA managers face challenges in coordinating tourism with other goals, such as ecological conservation and local community development. To address this challenge, we developed a sustainability assessment framework that considers the PA, local community, and tourism as a complex system comprising social, economic, and ecological subsystems from the perspective of subsystem relationships. The coupling coordination degree model and the obstacle degree model were applied to assess sustainability of the tourism system in Qinghai Lake Nature Reserve of China. The assessment results indicate that the sustainability index fluctuated between 2010 and 2019, but generally exhibited an upward trend, undergoing three stages and reaching the stage in 2019 where ecological sustainability took the lead. At this stage, the coupling coordination degree between the economy and society subsystems was at its lowest, and the economic subsystem faced the highest obstacle degree. The study demonstrates that involving scholars and administrators in the index selection process and considering both index information and management concerns when determining index weight makes the coupling coordination degree model more suitable for PA tourism systems. The assessment method developed in this study effectively reflects the temporal evolution of PA tourism system sustainability and provides valuable implications for coordinated ecological-economic-social management by analyzing obstacle factors.

Keywords: protected area; tourism; complex system; coupling coordination degree; obstacle degree; Qinghai Lake Nature Reserve

1. Introduction

Protected areas (PAs) provide a most important and effective way to protect global biodiversity and ecological environments and contribute to human health and well-being [1–3]. Declining financial support for PAs in developing countries and even in some developed ones, such as Australia, the US and Canada, suggests that developing PAs by solely relying on government inputs is unsustainable [4]. Nature-based tourism is a popular type of cultural ecosystem service that can enhance the emotional connection between human beings and nature and contribute to the financial sustainability of PAs [5–8]. It is estimated that the annual tourist arrivals at the world's PAs reach 8 billion [9], and that the economic value of PAs derived from the improved mental health of visitors is US$6 trillion a year [10]. The wider benefits of park visits have not been quantified [11]. However, with PAs being expected to make much wider ecological, social and economic contributions to sustainability and human well-being, PA managers face challenges in coordinating tourism with other goals, such as nature conservation and local community development [4,12,13]. Therefore, both the International Union for Conservation of Nature (IUCN) and World

Tourism Organization (WTO) emphasize the importance of sustainability assessment and adaptive management of tourism in PAs, so as to bring into full play the role of tourism in poverty reduction, community development and biodiversity conservation [14,15].

PAs, as important nature-based tourism destinations, are complex adaptive systems that involve multiple stakeholders and are affected by social, economic and environmental factors [16–20]. Increasingly, the PA, the local community and the tourism within the PA are being recognized a complex system [21,22]. The significant impact of COVID-19 on PA tourism highlights the complex interdependencies among tourism, local communities and PAs, and such interdependencies should not be overlooked when seeking to improve PA sustainability [23,24]. A systematic way of thinking is therefore proposed to understand the interaction of key elements, the evolution of systems, and the assessment and management of PAs and local tourism [18,25]. Furthermore, Zhang et al. (2022) indicate that the interrelationships between subsystems provide an important and effective perspective for sustainability assessment of the PA tourism system [26].

Plummer and Fennell (2009) argued that sustainable tourism management in PAs should anticipate system dynamics and transformative changes [27]. However, traditional assessment methods tend to use sustainability indicators targeting current conditions and poor selection of the indicators often leads to the misidentification and misinterpretation of the changes over time. Research on systematic thinking suggested that future conditions may include more extreme and rapid changes than previously [21]. In addition, although previous studies have proposed many indicators on the sustainability of tourism in PAs, they have paid less attention to the coordination among subsystems [26]. Therefore, new methods acknowledging uncertainties, changes and frequent interactions of the subsystems are required.

The coupling coordination degree model (CCDD) is an effective method used to evaluate the consistency and positive interaction among systems, and can reflect the trend of complex systems transforming from disorder to order [28]. In recent years, it has been extensively used in studies on the relationship between tourism and other systems and among components of the tourism system, such as tourism and environment [29–31], and the social-ecological status of island tourism destinations [32]. These studies reveal the importance and applicability of a coupled coordination perspective in measuring complex tourism systems. But they mostly focus on city (prefectural), provincial and national scales, and smaller scale studies represented by PAs are limited. In addition, since the current indicators for the CCDD are generally selected by authors [29,30,33], with other experts or stakeholders seldom engaged, the availability of comparable data and information and the objectivity of the assessment results are inevitably undermined. As the WTO (2004) noted, a participatory process can be productive, especially when key stakeholders and potential data providers are involved [14].

The Tibetan Plateau is widely recognized for its abundant biodiversity and diverse ecosystems, which are intricately linked to the livelihoods of over one billion people [34]. To safeguard the rich flora and fauna in this region, numerous PAs have been established, many of which are renowned tourist destinations [35,36]. The Qinghai Lake Nature Reserve (QLNR) is a typical example of these PAs. Given the vulnerable ecology and underdeveloped economy of the Tibetan Plateau, tourism in these PAs is expected to play a greater role in poverty reduction, community prosperity and biodiversity conservation. Thus, coordinated ecological-economic-social development is not only essential for the sustainability of each PA but also of paramount importance for realizing the United Nations Sustainable Development Goals and, specifically, for promoting green development on the Tibetan Plateau [37].

Due to the gaps identified in both practical management and theoretical assessment methods in PA tourism, this study aims to enhance the sustainability of PA tourism by focusing on subsystem relationships using the CCDD. To achieve this goal, three sub-aims have been identified: (1) improving the applicability of CCDD to the PA tourism system and enhancing the objectivity of indicator selection, (2) evaluating subsystem relationships and their changes in the PA tourism system, and (3) identifying obstacles to the sustainable development of PA tourism.

2. Materials and Methods

2.1. The Assessment Framework

A growing body of research conceptualizes tourism as a complex adaptive system [21,38,39] or calls for systematic thinking in the conceptualization of the relationships among tourism, PAs, and the local communities [40,41]. Stone et al. (2021) believed that without clear identification of interacting variables, any study on PAs and tourism will reveal an incomplete and potentially confusing picture, as the complex interactions between system components will not be apparent [42]. Schianetz and Kavanagh (2008) also pointed out that systematic thinking is critical for assessing the sustainability of natural tourist destinations located in eco-environmentally fragile areas [25].

Sustainability indicators can provide managers with required information and are essential for improving tourism planning and management and promoting sustainable development [14,43]. Scholars have developed a series of indicators from one or more dimensions of ecology, economy and society to evaluate the sustainability of different scales and different types of tourism destinations [44–47]. However, more attention is paid to the sustainability of the ecological, social and economic dimensions themselves, and less to the relationship among the three [26]. In practical terms, the three dimensions are "pillars" of sustainable development with frequent interaction, and a balance must be struck between them [43]. For Bramwell and Lane (2011), the "balance" of economic, social and environmental sustainability is the cornerstone of sustainable tourism policies [48]. Systematic thinking makes it possible to analyze the relationship among the three dimensions.

We define the PA, local community and tourism within the PA as a complex adaptive system composed of the three subsystems of society, economy and ecology. The economic subsystem mainly includes tourism-related economic factors within and around the PA, such as tourism revenue and tourist arrivals. The social subsystem mainly encompasses social and cultural factors within the PA and the adjacent communities, such as community participation, cultural preservation, and environmental education. The ecological subsystem mainly consists of natural elements within and around the PA, such as environmental quality and biodiversity conservation. The three subsystems frequently interact with one another through the flow of capital, information, and tourists, among other factors, and this encourages the system to evolve. In order to assess the PA tourism from the perspective of ecological–economic–social coordinated development, this study calculates the coupling coordination degree among the subsystems based on sustainability evaluation (Figure 1). The evaluation covers two parts. The first is the sustainability of the subsystems, which includes the sustainability of the social, economic, and ecological subsystems. The social sustainability index, economic sustainability index, and ecological sustainability index are the three names given to the evaluation outcomes, accordingly. The second part of the evaluation concerns the coupling coordination degree among the subsystems, including the comprehensive coupling coordination degree of the three subsystems, and the coupling coordination degree between each pair of subsystems.

2.2. Study Area

QLNR is located in the Qinghai Province, northwest of China (Figure 2). Qinghai Lake, the most important tourist attraction of the PA, is the largest saline lake of China and a well-known tourist destination on the Tibetan Plateau. Furthermore, many people have lived by the lake for generations. Nature conservation, community prosperity, and sustainable tourism are three inseparable management objectives for QLNR [49]. The following are the reasons why we chose this reserve as a case study to assess the relationship between economic, social, and ecological subsystems of the PA tourism system.

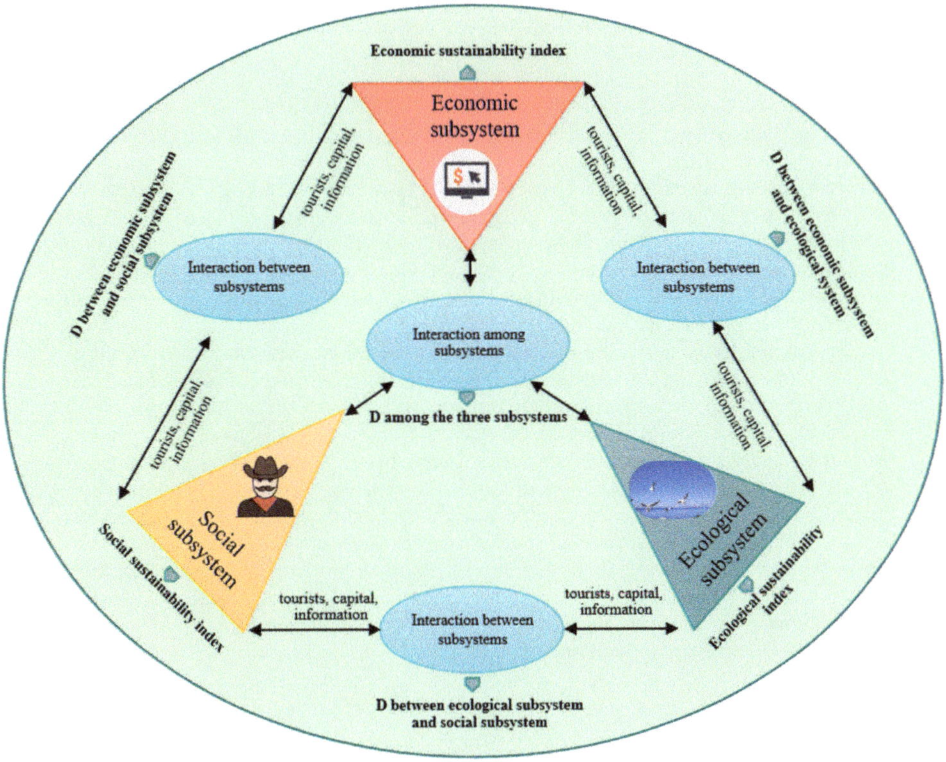

Figure 1. The assessment framework. (D represents the degree of coupling coordination).

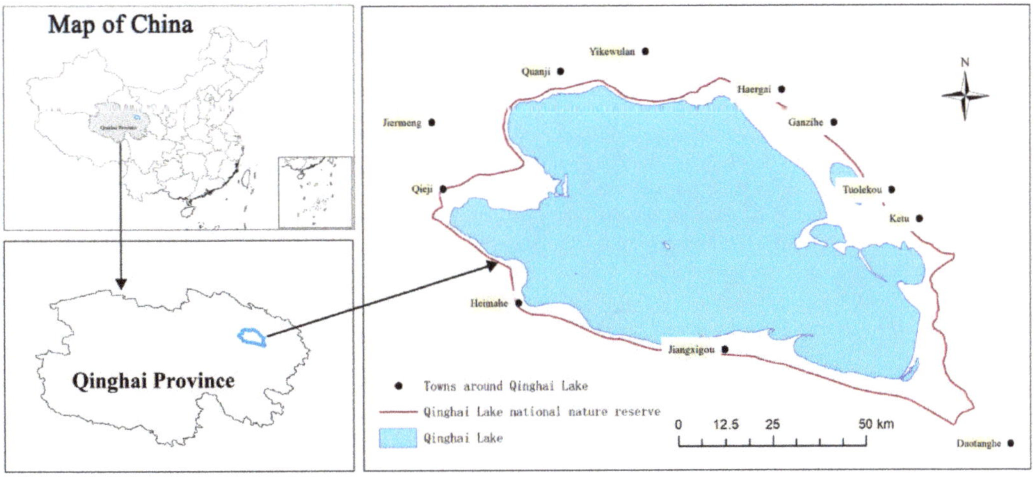

Figure 2. Location of the Qinghai Lake Nature Reserve.

First, Qinghai Lake is an important node of two international bird migration channels in East Asia and Central Asia and also the only habitat of the Przewalski's gazelle [50]. It was recognized as a "Wetland of International Importance" by the Ramsar Convention in

1992. However, the monitoring results of 25 largest lakes in the world from 2008 to 2010 by United Nations Environment Programme (UNEP) showed that the load of human activities in Qinghai Lake had reached 90% [51].

Second, Qinghai Lake has been a popular nature-based tourist destination since the 1980s. In 2019, it received 4.43 million tourists. According to estimates by Zhao (2018) [52], the per capita ecological deficit relating to tourists in Qinghai Lake showed an overall rising trend, and tourist overload was common from 2001 to 2015. In 2017, in response to the central government of China's environmental inspection, which aimed to supervise and enforce local-level environmental protection policies, several scenic spots were closed, and numerous tourist facilities, including tents and bed and breakfasts, were demolished within and around the QLNR for non-compliance with PA management regulations. As a result, the duration of visitor stays within or around the reserve decreased, and the number of overnight stays also declined.

Third, similar to most Chinese PAs, QLNR is home to many local people whose livelihoods and lives are closely linked to the reserve and its tourism development. There are 11 towns around the reserve with 5870 residents and 76.55 km^2 farmlands in the reserve. The establishment of the PA restricted local residents' activities such as grazing and planting and others that depend on natural resources. In order to supplement their income, many local residents have resorted to selling tourist souvenirs and taking on part-time jobs in nearby hotels and restaurants. Some individuals have even illegally opened access routes to the reserve and established small tourist attractions, offering paid services such as canola flower sightseeing, horse riding, and photography. However, managing these community residents poses significant challenges, and sometimes conflicts arise between the community and the authorities. The potential impacts of these changes in livelihoods on community resilience remain unclear.

2.3. Index System

Establishing the index system proceeded via the two key steps of selecting indicators and determining their weights. The process can be seen in the flow chart for index system establishment (Figure 3).

2.3.1. Selection of Indicators

This paper adopted the fuzzy Delphi method (FDM) to select indicators. The Delphi method is commonly used to select sustainability indicators, but its uncertainty, vagueness and subjectivity need be addressed [53]. The FDM applies fuzzy set theory to the Delphi method, which overcomes the shortcomings by reducing the number of questionnaire surveys, avoiding the distortion of individual expert opinions, and considering the fuzziness of the interview process [54]. The FDM with a dual-trigonometric fuzzy function especially uses a trigonometric fuzzy function and the grey zone testing method to integrate expert opinions, which is more objective than the calculation of geometric means [55]. Hence, this paper adopted the FDM with a dual-trigonometric fuzzy function to select sustainability indicators, and the process was as follows.

1. Step 1: Making a list of candidate indicators

Following the principles of practicality, comparability, objectivity and data availability, 28 candidate indicators were generated by referring to the current literature [14,54,56–61] and conducting semi-structured interviews with tourism stakeholders in QLNR (administrators, community residents and tourists).

2. Step 2: Establishment of the fuzzy Delphi expert group and questionnaire survey

The key to the Delphi method lies in the expertise of experts assigned and their familiarity with the subject matter, rather than the number of experts [53]. Saaty and Özdemir (2014) held that adding more experts who are less experienced may disturb the judgments of other experts and even lead to false conclusions [62]. Accordingly, 15 administrators and researchers familiar with tourism in PAs and having at least five years' professional experience in the related sectors were invited to fill in the expert questionnaire from June to July, 2020. Eliminating invalid questionnaires with obvious missing answers or no discrimination of scores (e.g., 10 for all the maximum values and 0 for all the minimum values), eight valid responses were considered. As revealed in Table 1, the experts were equally distributed among researchers on PA tourism (3), administrators of tourism in QLNR (2) and administrators for PA tourism at provincial or national level (3), and were thus representative.

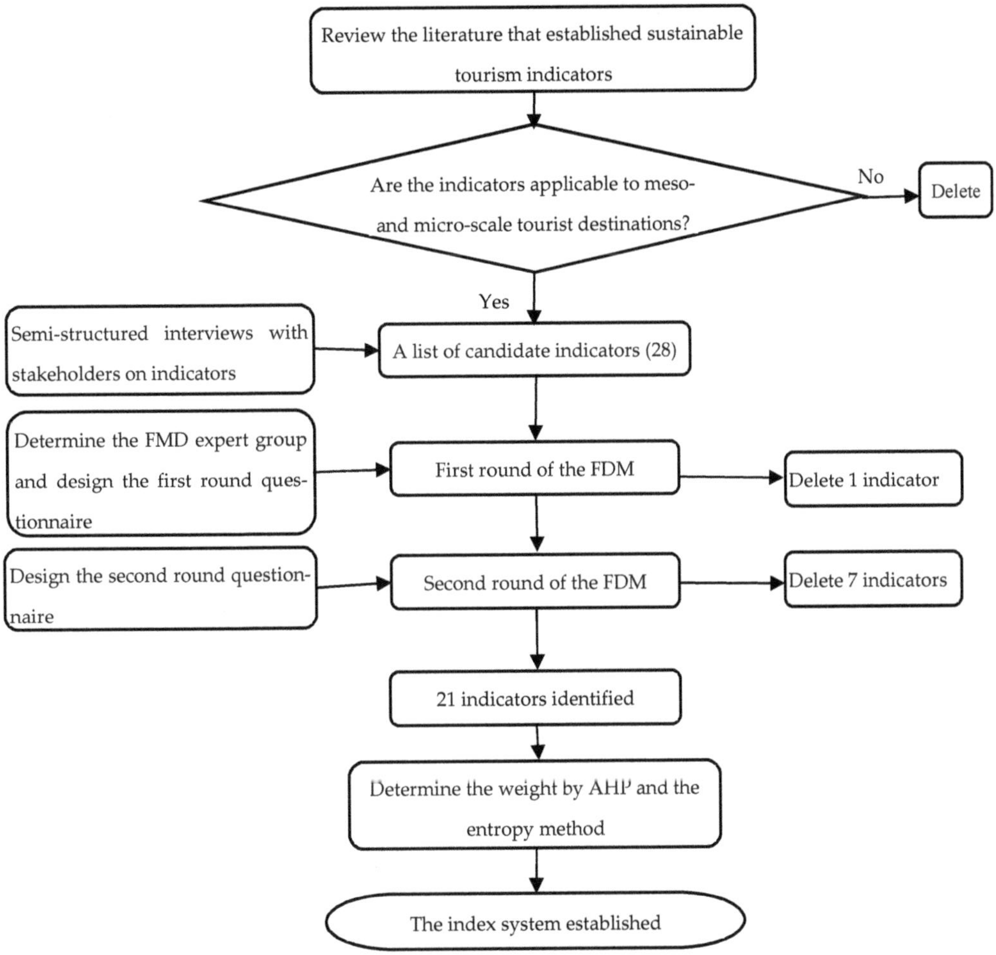

Figure 3. The flow chart for index system establishment. (FDM indicates the fuzzy Delphi method).

Table 1. Qualifications of the experts.

Organization/Institution	Number
Institute of Geographic Sciences and Natural Resources Research, Chinese Academy of Sciences	1
Northwest Plateau Institute of Biology, Chinese Academy of Sciences	2
Qinghai Lake Protection and Utilization Administration of Qinghai Province	2
Department of PAs Management, National Forestry and Grassland Administration in China	2
National Parks and PAs Management Division, Qinghai Forestry and Grassland Bureau	1

Note: The same group of experts was consulted in the analytic hierarchy process (AHP).

3. Step 3: Index selection

After two rounds of fuzzy Delphi questionnaire surveys, 21 indicators were generated in total (Table 2). The questionnaire and its data analysis process can be seen in Appendices A and B. Though no academic consensus on the number of sustainability indicators has been reached, the WTO (2004) pointed out after the summarization of global practice that 12–24 indicators are optimal, as an excessively large number of indicators may drive up the cost of data acquisition and be difficult to use, while use of only a few indicators tends to overlook economic, ecological or social issues. By this standard, the number of indicators in this paper is suitable [14].

Table 2. Results of two rounds of fuzzy Delphi questionnaires.

Subsystems	Indicators	O_L	O_U	C_L	C_U	M − Z	G	S
	The first round							
	Growth in local economy	6	10	1	7	3.36	6.30	5.05
	Growth in tourism revenue	8	10	1	7	5.56	6.56	6.41
	Growth in tourist numbers	6	10	1	8	2.23	6.56	5.23
	Rate of repeat visitors	2	10	1	7	**−1.99**	4.69	4.40
Economy	Distribution of tourism income	2	10	1	6	**−1.13**	4.32	4.26
	Spatial structure of tourism income	3	10	1	6	0.77	4.54	4.44
	Tourist seasonality	7	9	1	6	6.18	5.37	4.71
	Tourism revenue structure	6	10	1	7	3.24	6.32	5.33
	Tourist per capita consumption	5	10	1	8	1.45	5.99	5.04
	The per capita income of the local community	5	9	2	6	2.71	5.47	5.26
	Local prices	3	8	1	4	2.13	3.55	3.54
	Tourism transportation	3	10	1	8	**−1.44**	5.29	5.12
	Cultural places for tourists	4	10	1	8	**−0.20**	5.70	5.52
Society	Conflicts between community, PA, and tourists	2	10	1	7	**−2.15**	4.64	4.70
	Local cultural preservation	2	10	1	6	**−1.24**	4.35	4.22
	Community participation in tourism	3	8	1	5	0.65	4.12	4.17
	Environmental interpretation facility	5	9	1	7	1.92	5.76	5.14
	Environmental interpreters	5	10	1	6	2.76	5.33	4.61
	Capital input on environmental interpretation	5	10	1	8	0.91	6.12	5.65
	Key species protection	5	8	1	6	3.34	5.44	4.67
	Vegetation coverage area	5	10	1	9	0.71	6.55	5.47
	Water quality	9	10	2	9	4.11	9.00	7.27
	Tourism environment monitoring	5	10	3	9	**−1.27**	7.05	7.02
Ecology	Wildlife conservation	7	10	2	10	0.80	7.73	6.72
	The wetland area	3	10	2	9	**−2.68**	5.42	5.40
	Alien species	4	10	2	9	**−1.65**	5.67	5.67
	Vegetation protection	8	10	2	10	2.11	8.31	6.80
	Land use type change	6	10	1	7	4.03	6.34	5.06
	The second-round							
	Rate of repeat visitors	5	10	1	6	3.88	5.54	5.70
	Distribution of tourism income	4	10	1	5	2.87	4.57	4.64
	Tourism transportation	4	10	1	5	2.69	4.54	4.79
	Cultural places for tourists	6	10	1	6	4.59	5.56	5.64
	Conflicts between community, PA, and tourists	5	10	1	6	3.34	5.53	5.72
	Local cultural preservation	5	10	1	8	1.37	6.00	4.83
	Tourism environment monitoring	7	10	1	7	5.36	5.89	6.00
	The wetland area	4	10	1	8	0.52	5.67	4.88
	Alien species	8	10	3	8	4.04	7.45	7.70

Note: $M^i - Z^i < 0$ requires a second round of expert consultation (values in bold), and $G^i < S^i$ means the indicator should be deleted.

2.3.2. Calculation of Weights

Index data need to be standardized before weights are calculated. Formulas (1) and (2) were used to standardize the original data for a positive index

$$x'_{ij} = \frac{x_{ij} - \min\limits_{1 \le j \le n} x_{ij}}{\max\limits_{1 \le j \le n} x_{ij} - \min\limits_{1 \le j \le n} x_{ij}} \quad (1)$$

and for a negative index

$$x'_{ij} = \frac{\max\limits_{1 \le j \le n} x_{ij} - x_{ij}}{\max\limits_{1 \le j \le n} x_{ij} - \min\limits_{1 \le j \le n} x_{ij}} \quad (2)$$

where, x_{ij} and x'_{ij}, respectively, refer to the original value and the standardized value of indicator j in year i; $\max\limits_{1 \le j \le n} x_{ij}$ and $\min\limits_{1 \le j \le n} x_{ij}$ are the maximum and minimum value of indicator j among all years (2010–2019). An x'_{ij} whose standardized result is 0 is replaced by 0.0001 to avoid null value in the subsequent calculation with the entropy method (EM).

The analytic hierarchy process (AHP) is a common method to obtain the weight of sustainability indicators in the form of hierarchical data combined with experts' opinions [53,54]. It provides a way to systematize the complex issues of PA tourism with the advantage of being easy to operate and accommodating the views of different stakeholders [63]. This study used AHP to divide the indicator system into three hierarchical levels (Table 3), established the pairwise comparison matrix for each level, and invited the experts to compare each level of indicators pairwise on a scale of 1 to 9. Saaty and Özdemir (2014) found that in the use of AHP, engagement of no more than 7 or 8 experts is more likely to make for effective and consistent judgments [62]. The eight experts in Table 1 were therefore invited to participate, and seven of them eventually completed the expert questionnaire. Yaahp was used to process the AHP questionnaire data, using calculation and consistency checks to obtain the indicator weight w_j'.

The EM is commonly used to objectively calculate weights. Entropy is a measure of the uncertainty of indicator information. If the amount of information is higher, the uncertainty is lower and the entropy is smaller; if the amount of information is lower, the uncertainty is higher and the entropy is larger. Tang (2015) stated that the EM can avoid bias caused by subjective influence to a certain extent when determining the index weights by analyzing correlation degree and information among indexes [29]. The formulas are shown from (3) to (5).

$$y_{ij} = x_{ij}' / \sum_{i=1}^{m} x_{ij}' \quad (3)$$

$$d_j = 1 + \frac{1}{lnm} \sum_{i=1}^{m} y_{ij} ln y_{ij} \quad (4)$$

$$w_j'' = {d_j} \Big/ {\sum_{i=1}^{m} d_j} \quad (5)$$

In order to reduce the subjectivity of the AHP weight and make the assessment results more reliable, Formula (6) was used in combination of the EM and AHP weight to get the general weight w_j. The results are shown in Table 3.

$$w_j = \frac{W_j' + W_j''}{2} \quad (6)$$

Table 3. Indicators and the weights.

Subsystem	Weights by AHP	Weights by EM	General Weight	Dimension	Weights by AHP	Weights by EM	General Weight	Indicators	Sign	Weights by AHP	Weights by EM	General Weights	Verifiers
Economy (A)	0.3264	0.2186	0.2725	Economic structure (A_1)	0.1632	0.1254	0.1443	Tourism revenue structure (A_{11})	–	0.0360	0.0418	0.0389	The proportion of ticket revenue in total tourism revenue (%)
								Tourist per capita consumption (A_{12})	+	0.0640	0.0258	0.0449	Tourist per capita consumption (¥)
								Tourist seasonality (A_{13})	–	0.0138	0.0461	0.0300	Tourism seasonal intensity index
								Spatial structure of tourism income (A_{14})	–	0.0494	0.0117	0.0306	The income of Erlangjian scenic spot / the total income of QLNR (%)
				Economic growth (A_2)	0.1632	0.0932	0.1282	Growth in local economy (A_{21})	+	0.0898	0.0255	0.0577	Average annual GDP growth rate of the three counties involved in the reserve (%)
								Growth in tourism revenue (A_{22})	+	0.0342	0.0352	0.0347	Annual growth rate of tourism revenue (%)
								Growth in tourist numbers (A_{23})	+	0.0392	0.0325	0.0359	Annual growth rate of tourist numbers (%)
Society (B)	0.1396	0.4978	0.3187	Community development (B_1)	0.0718	0.0601	0.0660	Local economic development capacity (B_{11})	+	0.0340	0.0225	0.0283	Balance of household savings deposits (¥)
								Local prices (B_{12})	–	0.0108	0.0132	0.0120	Local consumer price index
								Community participation in tourism (B_{13})	+	0.0270	0.0244	0.0257	Proportion of working-age people participating in tourism
				Cultural protection (B_2)	0.0300	0.0580	0.0440	Local cultural preservation (B_{21})	+	0.0300	0.0580	0.0440	Number of state-level cultural relic protection units and intangible culture
				Nature education (B_3)	0.0378	0.3797	0.2088	Environmental interpretation facility (B_{31})	+	0.0126	0.0919	0.0523	Number of interpretive signs related to nature education
								Environmental interpreters (B_{32})	+	0.0126	0.1613	0.0870	Number of trained interpreters for nature education
								Capital input on on nature education (B_{33})	+	0.0126	0.1265	0.0696	Capital input on on nature education (¥)

Table 3. *Cont.*

Subsystem	Weights by AHP	Weights by EM	General Weight	Dimension	Weights by AHP	Weights by EM	General Weight	Indicators	Sign	Weights by AHP	Weights by EM	General Weights	Verifiers
Ecology (C)	0.5276	0.2836	0.4056	Ecological environment (C_1)	0.0878	0.1852	0.1365	The wetland area (C_{11})	+	0.0274	0.0621	0.0448	The wetland area (km^2)
								Vegetation coverage area (C_{12})	+	0.0274	0.0912	0.0593	Vegetation coverage area (km^2)
								The water quality (C_{13})	+	0.0256	0.0319	0.0288	Grade of the water quality
								Land use type change (C_{14})	−	0.0074	0.0000	0.0037	The proportion of urban land and agricultural land in the total area of QLNR (%)
				Biodiversity (C_2)	0.4398	0.0984	0.2691	Wildlife conservation (C_{21})	+	0.0786	0.0141	0.0464	The Shannon-Wiener diversity index of waterbirds in autumn
								Key species protection (C_{22})	+	0.3118	0.0606	0.1862	Population of Przewalski's gazelle
								Vegetation protection (C_{23})	+	0.0494	0.0237	0.0366	Mean total ground biomass

Note: "+" and "−" respectively represent positive indicators and negative indicators; Przewalski's gazelle serves as the flagship species in the reserve.

2.4. Data and Analysis Methods

2.4.1. Data

The indicator data on the ecology subsystem and the economy subsystem for the sustainability assessment were sourced from the Qinghai Lake Protection and Utilization Administration of Qinghai Province, mainly including the Monitoring Report on the National Nature Reserve of Qinghai Lake (2010–2019) and statistics on number of tourists and tourism income over the years. Data on social subsystem indicators and some of the local economic development indicators were obtained from China Statistical Yearbook (County-level) and China Statistical Yearbook (Township) from 2010–2019.

2.4.2. Coupling Coordination degree Model

Suppose $x_1', x_2', x_3' \cdots x_n'$ are the indicators of the economy subsystem and x is the corresponding standardized value of x', then the economic sustainability index is $f_1(x) = \sum_{i=1}^{n} w_i x_i$. w_i represents the weight of indicator i in the economy subsystem. Similarly, the social sustainability index and the ecological sustainability index are $f_2(x)$ and $f_3(x)$, respectively.

The coupling coordination degree among the subsystems was calculated using formulas (7) to (9).

$$C = \sqrt[n]{f_1(x) \times f_2(x) \cdots f_n(x) / (f_1(x) + f_2(x) \cdots f_n(x))^n} \quad (7)$$

$$T = \gamma_1 f_1(x) + \gamma_2 f_2(x) \cdots \gamma_n f_n(x) \quad (8)$$

$$D = \sqrt{C \times T} \quad (9)$$

where C represents the coupling degree, D represents the coupling coordination degree, γ is the weight coefficient of the corresponding subsystem, n is the number of subsystems. In the case of n = 3, T stands for the comprehensive sustainability index of the PA tourism system. By referring to the existing body of research [33,64,65], this paper defines the gradation criteria of the coupling degree and the coupling coordination degree, as shown in Table 4.

Table 4. Criteria for classification of the coupling degree and the coupling coordination degree.

Criteria for Classification of the Coupling Degree		Criteria for Classification of the Coupling Coordination Degree	
C	Coupling Level	D	Coupling Coordination Level
(0.0~0.3)	Seriously low coupling development	(0.0~0.1)	Extreme unbalanced development
		[0.1~0.2)	Seriously unbalanced development
		[0.2~0.3)	Moderately unbalanced development
[0.3~0.5)	Low coupling development	[0.3~0.4)	Slightly unbalanced development
		[0.4~0.5)	Imminently unbalanced development
[0.5~0.8)	High coupling development	[0.5~0.6)	Barely balanced development
		[0.6~0.7)	Favorably balanced development
		[0.7~0.8)	Intermediately balanced development
[0.8~1.0)	Superiorly high coupling development	[0.8~0.9)	Good balanced development
		[0.8~1.0)	Superiorly balanced development

2.4.3. Obstacle Degree Model

We used the obstacle degree model to identify obstacle factors of the tourism system in QLNR. The formulas are as follows [66]:

$$I_{ij} = 1 - x'_{ij} \quad (10)$$

$$O_j = F_j I_{ij} / \sum_{j=1}^{n} F_j I_{ij} \quad (11)$$

$$Q_j = \sum O_j \quad (12)$$

where x'_{ij} is the standardized value of indicator j in year i, I_{ij} represents the deviation degree of indicator j, F_j is the contribution degree of indicator j, which can be expressed by index weight, O_j represents the obstacle degree of indicator j, Q_j represents the obstacle degree of a subsystem.

3. Results and Discussion

3.1. Indicators and Weights

As shown in Table 3, the indicator system aligns with the sustainable tourism management principles for PAs put forward by IUCN, including indicators on nature conservation, communities' right to development and cultural authenticity, continuous and fair development of the tourism economy and provision of valuable recreational experience [15]. These principles also echo the functional orientation of China's PA system, which aims to protect nature, provide high-quality ecological products, and maintain harmonious coexistence between humans and nature for sustainable development [67]. Specifically, in the economy subsystem, A_1 and A_2 have the same weight, indicating that both economic growth and economic efficiency are critical for economic development. In the society subsystem, nature education is the most important, with the sum of the weights of the three indicators, namely environmental interpretation facility (B_{31}), environmental interpreters (B_{32}) and capital input on nature education (B_{33}), accounting for 65.50% of the whole subsystem. This reflects the importance of nature education for tourism in PAs in serving social functions. In the ecology subsystem, C_2 exerts the greatest influence, accounting for 66.35% of the entire subsystem. More specifically, protection of key species (C_{22}) was given the highest weight with AHP, occupying 59.10% of the ecology subsystem. Thus, the biodiversity conservation represented by key species is the most important factor for the ecology subsystem.

There is little difference in the weights of the three subsystems. The result that the ecology subsystem has the highest weight is consistent with the study by Yu (2006) in Tianmu Mountain Nature Reserve, which observed the principle of ecological conservation coming first [57]. What is different is that in the present study, the society subsystem carries more weight than the economy subsystem. Given the management objectives of promoting local development and ecological and cultural protection for PAs, we believe it rational to pay greater attention to social and cultural factors of Chinese PAs for two reasons. First, as many communities live in and around PAs in China, it is critical for sustainable tourism management in PAs to reduce conflicts between PAs and communities and win over the community support [68,69]. Second, unlike the western world's immersion in wilderness aesthetics, Chinese tourists uphold the traditional culture that the human is an integral part of nature and prefer landscapes with man and nature coexisting in harmony [70]. Cultural factors constitute one of the great appeals of tourism in PAs.

According to the analysis methods and their computational formulas in this study, it is evident that the weighting of indicators not only directly affects the sustainability index, but also influences the results of coupling coordination degree and obstacle degree calculations. Therefore, the method chosen for determining indicator weights is of great importance. As indicated in Table 3, it is observed that the weights of certain indicators differ significantly when obtained using the AHP compared to the EM. Some indicators are regarded as important by experts and thus heavily weighted, but offer limited information, such as A_{21}, C_{21} and C_{22}. For these indicators, weighting with the EM alone will not be able to reflect the importance of the indicators in practice. In contrast, some other indicators, such as B_{31}, B_{32} and B_{33}, which showed rapid changes in the study period, will be neglected if only weighted with the AHP. Therefore, it is appropriate and necessary to combine both methods in an indicator system reflecting the temporal changes.

3.2. Coupling Coordination Degree and the System Evolution

3.2.1. Sustainability Index

As shown in Figure 4, the sustainability index of QLNR tourism system and its subsystems fluctuated in 2010–2019, but generally trended upwards. The social sustainability index was at its lowest level in the three subsystems between 2010 and 2013, but has since maintained a steady upward trend overall since 2014. After 2017, it began to surpass the economic sustainability index. The ecological sustainability index exhibited fluctuations during the period between 2010 and 2016, but experienced a rapid increase after 2017, reaching a 10-year peak in 2019. The economic sustainability index continued to fluctuate over the decade and approached its lowest level in 2017. The gap in the sustainability index between the economic subsystem and the ecological subsystem widened further and further after 2017.

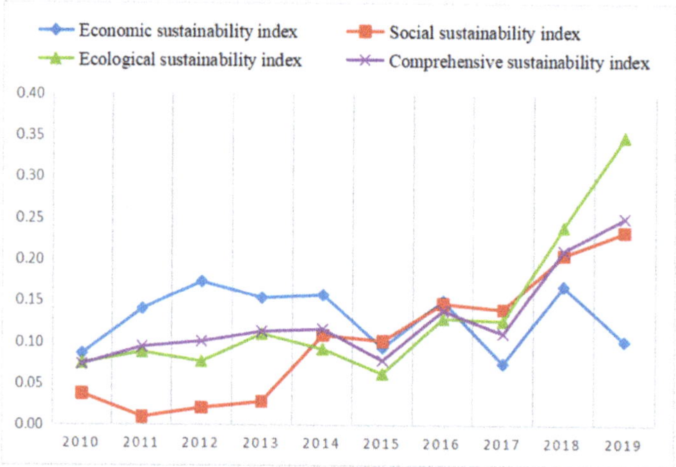

Figure 4. Sustainability index of QLNR tourism system and its subsystems from 2010 to 2019.

3.2.2. Coupling Degree

As revealed in Table 5, from 2010 to 2019, the comprehensive coupling degree among the three subsystems and the coupling degree between each pair of subsystems was averaged at 0.8 to 1.0, a "superiorly high" coupling level. It means the three subsystems were closely connected and frequently interacted with each other.

Table 5. The coupling degree of QLNR tourism system from 2010 to 2019.

Year	C between the Economic Subsystem and Social Subsystem	C between the Economic Subsystem and Ecological Subsystem	C between the Social Subsystem and Ecological Subsystem	Comprehensive C
2010	0.919	0.997	0.943	0.942
2011	0.486	0.973	0.592	0.616
2012	0.614	0.922	0.816	0.718
2013	0.722	0.986	0.804	0.802
2014	0.983	0.964	0.996	0.974
2015	0.999	0.979	0.970	0.978
2016	1.000	0.997	0.998	0.998
2017	0.952	0.966	0.999	0.965
2018	0.995	0.984	0.997	0.990
2019	0.918	0.835	0.980	0.886
The average	0.8588	0.9603	0.9095	0.8869

Note: C represents the coupling degree.

3.2.3. Coupling Coordination Degree

According to Figure 5, from 2010 to 2019, the comprehensive coupling coordination degree among the three subsystems and the coupling coordination degree between each pair of subsystems showed an overall upward trend, but the coordination level remained unbalanced until 2019. Only the coupling coordination degree between the society subsystem and ecology subsystem reached the "barely balanced" level in 2019, the highest score in a decade. Specifically, the coupling coordination degree between the ecological subsystem and the social subsystem remained at the lowest level before 2016. However, it rapidly increased thereafter and reached the best-coordinated level among the four groups. On the other hand, the coupling coordination degree between the economic and social subsystems significantly decreased after 2016, becoming the worst-coordinated level among them.

3.2.4. Stages of the System Evolution

Combination of the evaluation results of the subsystem sustainability index and the coupling coordination degree shows that the tourism system in QLNR evolved across three stages (Table 6). During the first stage (2010–2014), the economy subsystem was leading in development, whereas the society subsystem lagged behind. The relationships between the three subsystems were "moderately unbalanced" in general, with the coupling coordination degree between the society and ecology subsystems being the lowest. During the second stage (2015–2017), the society subsystem took the lead in development, while the ecology subsystem lagged behind. The coupling coordination degree among three subsystems was at the "slightly unbalanced" level, and the coupling coordination degree between the economy and the society subsystems was relatively higher. During the third stage (2018–2019), the ecological sustainability index rose rapidly, while the economic sustainability index declined. The coupling coordination degree between the society and the ecology subsystems was relatively higher, while that between the economy and the society subsystems was the poorest at this stage. Consequently, it is now urgent to improve the development level and efficiency of the economy subsystem and enhance the coupling coordination degree between the economy and the society and ecology subsystems.

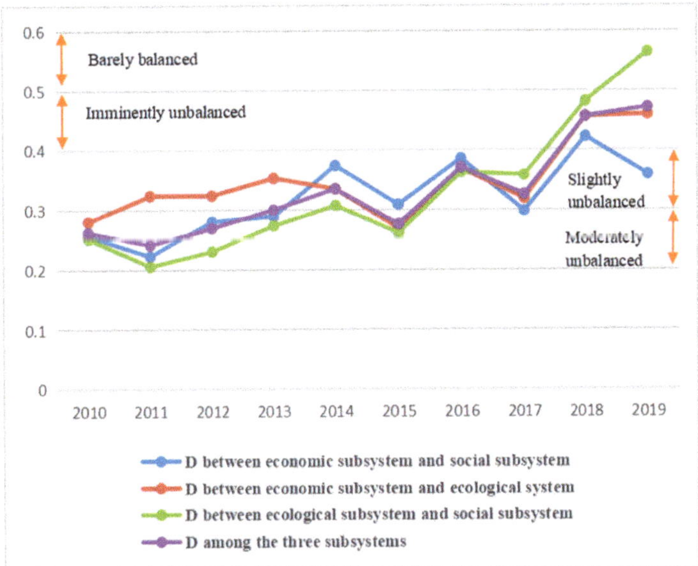

Figure 5. The coupling coordination degree of QLNR tourism system from 2010 to 2019. (D represents the degree of coupling coordination).

Table 6. Division of development stages of Qinghai Lake Nature Reserve tourism system.

Rankings	The Stage of Economic Dominance (2010–2014)	The Stage of Social Dominance (2014–2018)	The Stage of Ecological Dominance (2018–2019)
Rankings of subsystem sustainability index	$f_1(x) > f_3(x) > f_2(x)$	$f_2(x) > f_1(x) > f_3(x)$	$f_3(x) > f_2(x) > f_1(x)$
Rankings of coupling coordination degree between subsystems	$D_{13} > D_{12} > D_{23}$	$D_{12} > D_{13} > D_{23}$	$D_{23} > D_{13} > D_{12}$

Note: D_{12} refers to the coupling coordination degree of economic subsystem and social subsystem. D_{13} refers to the coupling coordination degree of economic subsystem and ecological subsystem. D_{23} refers to the coupling coordination degree of social subsystem and ecological subsystem.

3.3. Obstacle Factors for Sustainable Development and Management Implications

The obstacle model can help us identify the obstacle factors for the sustainable development of the system [71]. In order to promote coordinated development among subsystems, we conducted an analysis of the obstacle degree for each subsystem and identified the factors that caused them. Table 7 lists the obstacle degree values and the top three obstacle factors for each subsystem from 2010 to 2019. The social subsystem had the highest obstacle degree during 2010–2013, followed by the ecological subsystem during 2014–2018, and the economic subsystem in 2019. This is roughly consistent in time with the three stages that QLNR tourism system has gone through and explains the main obstacle factors to the system development in each stage. Specifically, over the decade, the most common obstacle factors in the social and ecological subsystems were the three natural education-related indicators (B_{33}, B_{32}, B_{31}), and the wetland area (C_{11}), vegetation coverage area (C_{12}) and key species protection (C_{22}), respectively. In contrast, obstacle factors in the economic subsystem were more dispersed, with the most common indicators being tourism revenue structure (A_{11}) and growth in tourist numbers (A_{23}). In 2019, the economic subsystem posed the greatest obstacle to the sustainable development of the QLNR tourism system. The top three obstacle indicators for achieving sustainable economic development were identified as the per capita tourist consumption level (A_{12}), local economic growth (A_{21}), and the spatial distribution of tourism income (A_{14}).

Table 7. The obstacle degree and the top three obstacle indicators in each subsystem from 2010 to 2019.

Year	Society		Ecology		Economy	
	Obstacle Degree	Dominant Obstacle Factors	Obstacle Degree	Dominant Obstacle Factors	Obstacle Degree	Dominant Obstacle Factors
2010	0.3767	B_{32}, B_{31}, B_{21}	0.3610	C_{22}, C_{11}, C_{23}	0.2623	A_{11}, A_{23}, A_{22}
2011	0.4739	B_{32}, B_{33}, B_{31}	0.3777	C_{22}, C_{11}, C_{12}	0.1484	A_{11}, A_{12}, A_{21}
2012	0.4612	B_{32}, B_{33}, B_{31}	0.4150	C_{22}, C_{11}, C_{23}	0.1238	A_{11}, A_{23}, A_{12}
2013	0.4336	B_{32}, B_{33}, B_{31}	0.4056	C_{22}, C_{11}, C_{21}	0.1608	A_{11}, A_{13}, A_{23}
2014	0.3237	B_{32}, B_{31}, B_{21}	0.4719	C_{22}, C_{12}, C_{11}	0.2044	A_{13}, A_{23}, A_{22}
2015	0.3174	B_{32}, B_{33}, B_{31}	0.4547	C_{22}, C_{12}, C_{23}	0.2279	A_{21}, A_{11}, A_{13}
2016	0.2943	B_{32}, B_{33}, B_{11}	0.5011	C_{22}, C_{12}, C_{11}	0.2046	A_{11}, A_{22}, A_{23}
2017	0.2858	B_{32}, B_{33}, B_{13}	0.4384	C_{22}, C_{12}, C_{11}	0.2758	A_{21}, A_{12}, A_{23}
2018	0.2254	B_{33}, B_{13}, B_{12}	0.5009	C_{22}, C_{12}, C_{21}	0.2737	A_{22}, A_{21}, A_{13}
2019	0.1568	B_{33}, B_{13}, B_{12}	0.3422	C_{12}, C_{23}, C_{13}	0.5009	A_{12}, A_{21}, A_{14}

As revealed by the assessment results, the tourism development in QLNR was in the leading stage of ecological sustainability. However, the coupling coordination degree between the economy and the society subsystems was the lowest, and the economic subsystem had the highest obstacle degree in 2019. Therefore, it is critical to improve the economy development efficiency and enhance the coupling coordination degree between economy and the other two subsystems for sustainability of the whole system.

Upon investigation, the decline in sustainability of the economic subsystem has been attributed to two significant events: the environmental inspection by central government in 2017 and the COVID-19 pandemic since 2019. The former led to a reduction in tourist attractions and reception facilities in and around the QLNR, resulting a change from a tourist destination to a transit point. The latter has caused a sharp decrease in tourists from outside Qinghai Province and a low motivation for tourism consumption within the province. With the aim of promoting the coordinated development of the economy, society, and ecology in Qinghai Lake Nature Reserve as a tourist destination, and based on the assessment of subsystem relationships and identification of obstacle factors, the following management insights can be derived.

Socially, community participation in tourism needs to be strengthened. On the one hand, local communities can engage in farming or herding on a flexible schedule when the PA tourism is suspended, which can alleviate the tourism operation pressure on hiring full-time staff under unexpected situations such as epidemics. On the other hand, local communities can gain knowledge, ability and income through participation that contributes to the goal of PA to promote community prosperity. Meanwhile, as livelihoods become less dependent on natural resource extraction and income increases, conflicts between communities and PA managers are expected to decrease. In addition, as one of the important functions of PAs, nature education, especially in terms of facilities (B_{31}), personnel (B_{32}) and input (B_{33}), requires more attention to enhance ecological awareness among the population and foster emotional connections between people and nature, in order to gain public support for PA efforts.

Ecologically, close attention should be paid to changes in some ecological indicators to reveal the influencing mechanisms of tourism. Restrictions on travel during the pandemic have created an opportunity for natural environmental restoration and less artificial interference with biodiversity [72]. Administrators and researchers can make use of the period to identify the ecological indicators that are most responsive to the weakened disturbance from tourism, such as animals, plants, and water (C_{12}, C_{13}, C_{13}). It is recommended to optimize tourism project planning by considering both the time of opening and spatial layout in light of the tourism-influenced mechanisms. This aims to identify the most favorable time and location for visits in order to mitigate the negative environmental impacts to key protection objects, such as waterbirds, Przewalski's gazelle, and plants.

Diversified environmentally-friendly tourism projects are suggested to improve the tourism economy efficiency on the premise of ecological conservation. For instance, long-distance birdwatching and sightseeing by bicycle as well as nature education and the Tibetan cultural experience can be designed to prolong the stay of tourists, increase per capita spending (A_{12}), and drive up tourism income from diversified sources. Furthermore, providing information about accommodations in nearby towns or partnering with local lodging services can also increase the overnight stay rate of tourists, thus contributing to the economic benefits of tourism. For the PA tourism development, it is crucial to rely on peripheral areas of the PA to provide accommodation, dining, and other services as much as possible, in order to minimize the impact of tourism on the PA's environment and biodiversity, while also promoting the development of local communities.

4. Conclusions

The sustainable tourism development in PAs is a complex process, in which economic, social and ecological factors interact with each other and in which resource administrators, tourists and local communities, among other stakeholders, participate. Systematic thinking has offered us a holistic perspective of analysis. From the case of the QLNR tourism system, it is evident that changes in external factors such as policies can significantly improve the sustainability of one subsystem while potentially reducing the sustainability of another subsystem. Hence, the assessment of relationships among subsystems should not be overlooked, as the sustainability of a PA tourism system depends not only on the sustainability level of its individual subsystems, but also on their balance.

In order to propose an integrated evaluation approach that reflects the temporal evolution of the relationship among subsystems from the perspective of ecological-economic-social coordinated development, we established a sustainability evaluation framework for the PA tourism system, which includes social, economic, and ecological subsystems, and identified a set of indicators in line with the development goals of sustainable tourism in the context of PAs by FDM. Subsequently, the CCDD and the obstacle degree model were used to reflect the temporal evolution of the sustainability of the reserve and identify the obstacle factors.

Our paper makes a significant contribution to the literature on three aspects. Firstly, the CCDD was introduced to assess the relationships among subsystems of a PA tourism system. While more studies have focused on large-scale tourist destinations such as cities (prefectures) and provinces or national-level destinations, our study specifically focuses on the PA tourism system. Secondly, we adopted the FDM to include scholars and administrators in the index selection process, which makes the CCDD more applicable to the PA tourism systems. This is a departure from the norm, as the indicators of CCDD are usually selected solely by authors, without engaging other stakeholders. Lastly, to improve the applicability and objectivity of the evaluation, we combined the analytic hierarchy process and the entropy method to determine the index weight, taking into account both index information and management concerns. The results show that this is necessary for diachronic evaluation and sustainability management of the PA tourism system.

5. Limitations and Future Research

Given the variety of PAs and their wide differences across countries and regions in natural ecological, social and cultural conditions, and tourist preference, the indicator system should be tailored to actual situations when applied in other PAs. In addition, a case study is not sufficient to draw general conclusions. Therefore, it is important to undertake more studies on tourism systems in different categories of PAs or PAs in different regions in the future to identify characteristics of subsystem relationships and obstacle factors.

For PA tourism systems, the coupling coordination degree assessment indicators should be adapted to the specific situations, such as conservation objectives and community conditions. Stakeholder participation is therefore crucial in selecting indicators. This paper involved administrators and related academics in the selection through the FDM. However, local residents in QLNR who generally speak the Tibetan language and have low literacy skills were not included due to difficulties in communicating and understanding of this method. Research in the future can include non-governmental organizations, tourists, local community residents and other stakeholders of tourism in PAs in selecting indicators and determining weights to better cater to local realities.

Author Contributions: Conceptualization, X.Z. and L.Z.; methodology, X.Z.; software, X.Z.; formal analysis, X.Z.; investigation, X.Z. and L.Z.; data curation, H.Y.; writing—original draft preparation, X.Z.; writing—review and editing, H.Y.; visualization, L.-E.W.; supervision, L.Z.; funding acquisition, L.Z. All authors have read and agreed to the published version of the manuscript.

Funding: This research was funded by the Strategic Priority Research Program of Chinese Academy of Sciences [XDA20020302], the Second Tibetan Plateau Scientific Expedition and Research of Ministry of Science and Technology of China [2019QZKK0401], and Sanjiangyuan National Park Joint Research Project of Chinese Academy of Sciences and Qinghai Province Government [LHZX-2020-05].

Data Availability Statement: The data presented in this study are available on request from the corresponding author.

Conflicts of Interest: The authors declare no conflict of interest.

Appendix A. The Questionnaire for Fuzzy Delphi Method

The questionnaire for experts on sustainability indicators of tourism in Qinghai Lake Nature Reserve.

Dear experts:

We are researchers from ***. Due to the research needs, we are conducting a questionnaire survey on the sustainable development of tourism in Qinghai Lake Nature Reserve (QLNR). Please feel free to fill in the questionnaire anonymously and for scientific research purposes only. Your true opinions are very important for us to get objective and meaningful research conclusions. Thank you for your support and cooperation!

Wish you good health, smooth work and a happy family!

❖ Instructions:

This questionnaire uses the assignment method (the score is divided from 0 to 10). The higher the number, the more you approve of using the indicator for evaluation. The smaller the number, the less suitable the index is for the sustainability evaluation of tourism in QLNR. Each indicator needs three numbers. Maximum value and Minimum Value indicate the applicability range of the indicator, and Confirmed value indicates a single value of the applicability degree of the indicator.

❖ The indicator assignment follows the following principles.

Relevance: This index is related to the development of tourism in protected areas.
Availability: The data required for indicators are relatively easy to obtain.
Comparability: The indicators are comparable in different years and protected areas.

Table A1. Index assignment table.

Number	Indicators	Maximum Value	Minimum Value	Confirmed Value
1	Growth in local economy			
2	Growth in tourism revenue			
3	Growth in tourist numbers			
4	Rate of repeat visitors			
5	Distribution of tourism income			
6	Spatial structure of tourism income			
7	Tourist seasonality			
8	Tourism revenue structure			
9	Tourist per capita consumption			
10	The per capita income of the local community			
11	Local prices			
12	Tourism transportation			
13	Cultural places for tourists			
14	Conflicts between community, PA, and tourists			
15	Local cultural preservation			
16	Community participation in tourism			
17	Environmental interpretation facility			
18	Environmental interpreters			
19	Capital input on environmental interpretation			

Table A1. *Cont.*

Number	Indicators	Maximum Value	Minimum Value	Confirmed Value
20	Key species protection			
21	Vegetation coverage area			
22	The water quality			
23	Tourism environment monitoring			
24	Wildlife conservation			
25	The wetland area			
26	Alien species			
27	Vegetation protection			
28	Land use type change			

Appendix B. Statistical Analysis of Questionnaire Results

In the expert consultation questionnaire, each expert is required to give a possible interval value $[C_i, O_i]$ and a definite value P_i between C_i and O_i for each indicator to be evaluated, where i is an indicator to be evaluated, the minimum value C_i is the "most conservative cognitive value" of i, and the maximum value O_i is the "most optimistic cognitive value" of i.

The steps of questionnaire analysis are as follows.

Step 1: Conducting statistical analysis for each index i

The extreme values other than "2 times standard deviation" were excluded, and then the minimum value (C^i_L, O^i_L), geometric mean value (C^i_M, O^i_M) and maximum value (C^i_U, O^i_U) were calculated. The conservative trigonometric fuzzy function $C^i = (C^i_L, C^i_M, C^i_U)$ and the optimistic trigonometric fuzzy function $O^i = (O^i_L, O^i_M, O^i_U)$ were established (Figure A1).

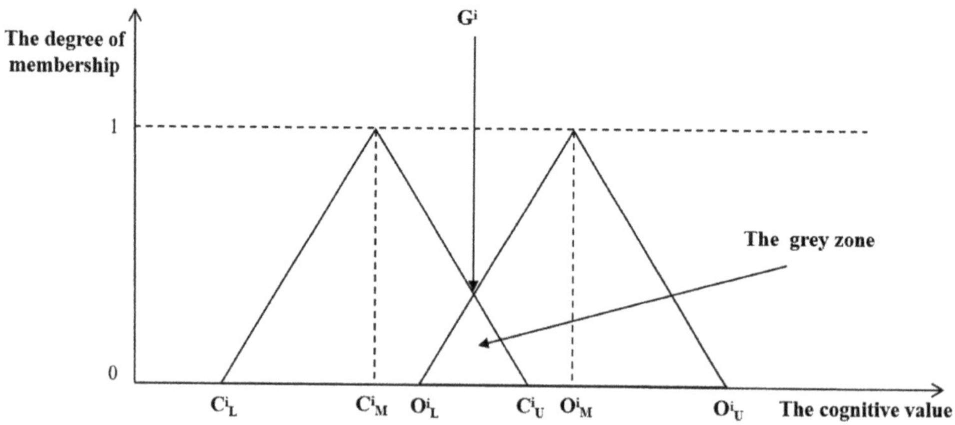

Figure A1. Dual-trigonometric fuzzy function map.

Step 2: Calculating the consistency degree of experts on indicators

The grey zone was used to judge whether the expert opinions reached convergence, and G^i (representing the consensus degree of experts) was determined according to different situations.

If $C^i_U \leq O^i_L$, $G^i = \frac{C^i_M + O^i_M}{2}$. It represents the non-overlapping interval of two trigonometric fuzzy functions, indicating that experts have reached a consensus on the index.

If ${C^i}_U > {O^i}_L$, it means that the two trigonometric fuzzy functions have overlapping intervals. When Z^i ($Z^i = {C^i}_U - {O^i}_L$) $< M^i$ ($M^i = {C^i}_M - {O^i}_M$), it indicates that there are different opinions among experts, but the difference is very small, and then $G^i = \frac{O^i_M C^i_U - C^i_M O^i_L}{Z^i + M^i}$. When $Z^i > M^i$, it tells that the opinions of experts differ greatly, and the above steps need to be repeated until the opinions of experts on all indicators converge.

Step 3: Calculating the threshold

There are three commonly used methods for determining the threshold value (S): (I) According to established experience, set the threshold value at 5–7; (II) Determine S of indicator i by calculating the geometric mean of C_i, O_i, and P_i, and then the geometric mean of the three geometric means; (III) Calculate the arithmetic mean value of P_i as the threshold value. This paper chooses the second method, which is relatively objective.

Step 4: Index selection

According to Table 2, after the first round of consultation, a total of nine indicators had $M_i - Z_i < 0$, indicating the expert opinions did not reach convergence. As a result of the second-round expert consultation, all the nine reached convergence, but seven with G value smaller than the threshold value S were deleted. After two rounds of fuzzy Delphi questionnaire surveys, 21 indicators were generated in total (Table 2).

References

1. Eagles, P.F.J.; McCool, S.F.; Haynes, C.D. *Sustainable Tourism in Protected Areas: Guidelines for Planning and Management*; IUCN: Gland, Switzerland, 2002.
2. Rahman, M.M.; Al Mahmud, A.; Shahidullah, M. Socioeconomics of biodiversity conservation in the protected areas: A case study in Bangladesh. *Int. J. Sustain. Dev. World Ecol.* **2017**, *24*, 65–72. [CrossRef]
3. Kabil, M.; Alayan, R.; Lakner, Z.; Dávid, L.D. Enhancing Regional Tourism Development in the Protected Areas Using the Total Economic Value Approach. *Forests* **2022**, *13*, 727. [CrossRef]
4. Ardoin, N.M.; Wheaton, M.; Bowers, A.W.; Hunt, C.A.; Durham, W.H. Nature-based tourism's impact on environmental knowledge, attitudes, and behavior: A review and analysis of the literature and potential future research. *J. Sustain. Tour.* **2015**, *23*, 838–858. [CrossRef]
5. Watson, J.E.M.; Dudley, N.; Segan, D.B.; Hockings, M. The performance and potential of protected areas. *Nature* **2014**, *515*, 67–73. [CrossRef] [PubMed]
6. Snyman, S.; Bricker, K.S. Living on the edge: Benefit-sharing from protected area tourism. *J. Sustain. Tour.* **2019**, *27*, 705–719. [CrossRef]
7. Zhang, H.; Huang, R.; Zhang, Y.; Buhalis, D. Cultural ecosystem services evaluation using geolocated social media data: A review. *Tour. Geogr.* **2020**, *24*, 646–668. [CrossRef]
8. Thapa, K.; King, D.; Banhalmi-Zakar, Z.; Diedrich, A. Nature-based tourism in protected areas: A systematic review of socio-economic benefits and costs to local people. *Int. J. Sustain. Dev. World Ecol.* **2022**, *29*, 625–640. [CrossRef]
9. Balmford, A.; Green, J.; Anderson, M.; Beresford, J.; Huang, C.; Naidoo, R.; Walpole, M.; Manica, A. Walk on the Wild Side: Estimating the Global Magnitude of Visits to Protected Areas. *PLoS Biol.* **2015**, *13*, e1002074. [CrossRef]
10. Buckley, R.; Brough, P.; Hague, L.; Chauvenet, A.; Fleming, C.; Roche, E.; Sofija, E.; Harris, N. Author Correction: Economic value of protected areas via visitor mental health. *Nat. Commun.* **2019**, *10*, 5005. [CrossRef]
11. Buckley, R.; Zhong, L.; Cooper, M.-A. Mental health and nature: More implementation research needed. *Nature* **2022**, *612*, 33. [CrossRef]
12. Soliku, O.; Schraml, U. Making sense of protected area conflicts and management approaches: A review of causes, contexts and conflict management strategies. *Biol. Conserv.* **2018**, *222*, 136–145. [CrossRef]
13. Karst, H.E.; Nepal, S.K. Conservation, development and stakeholder relations in Bhutanese protected area management. *Int. J. Sustain. Dev. World Ecol.* **2019**, *26*, 290–301. [CrossRef]
14. WTO (World Tourism Organization). *Indicators of Sustainable Development for Tourism Destinations: A Guidebook*; World Tourism Organization: Madrid, Spain, 2004.
15. Leung, Y.-F.; Spenceley, A.; Hvenegaard, G.; Buckley, R. *Tourism and Visitor Management in Protected Areas: Guidelines for Sustainability*; Best Practice Protected Area Guidelines Series No. 27; IUCN: Gland, Switzerland, 2018. [CrossRef]
16. Becken, S.; Job, H. Protected Areas in an era of global–local change. *J. Sustain. Tour.* **2014**, *22*, 507–527. [CrossRef]
17. Fisichelli, N.A.; Schuurman, G.W.; Monahan, W.B.; Ziesler, P.S. Protected Area Tourism in a Changing Climate: Will Visitation at US National Parks Warm Up or Overheat? *PLoS ONE* **2015**, *10*, e0128226. [CrossRef]
18. Cumming, G.S.; Allen, C.R. Protected areas as social-ecological systems: Perspectives from resilience and social-ecological systems theory. *Ecol. Appl.* **2017**, *27*, 1709–1717. [CrossRef]
19. Pásková, M.; Wall, G.; Zejda, D.; Zelenka, J. Tourism carrying capacity reconceptualization: Modelling and management of destinations. *J. Destin. Mark. Manag.* **2021**, *21*, 100638. [CrossRef]

20. Weng, L.; Wu, Y.; Han, G.; Liu, H.; Cui, F. Emotional State, Psychological Resilience, and Travel Intention to National Forest Park during COVID-19. *Forests* **2022**, *13*, 750. [CrossRef]
21. Strickland-Munro, J.K.; Allison, H.E.; Moore, S.A. Using resilience concepts to investigate the impacts of protected area tourism on communities. *Ann. Tour. Res.* **2010**, *37*, 499–519. [CrossRef]
22. Strickland-Munro, J.; Moore, S. Exploring the impacts of protected area tourism on local communities using a resilience approach. *Koedoe* **2014**, *56*, 1–10. [CrossRef]
23. McGinlay, J.; Gkoumas, V.; Holtvoeth, J.; Fuertes, R.F.A.; Bazhenova, E.; Benzoni, A.; Botsch, K.; Martel, C.C.; Sánchez, C.C.; Cervera, I.; et al. The Impact of COVID-19 on the Management of European Protected Areas and Policy Implications. *Forests* **2020**, *11*, 1214. [CrossRef]
24. Smith, M.K.S.; Smit, I.P.; Swemmer, L.K.; Mokhatla, M.M.; Freitag, S.; Roux, D.J.; Dziba, L. Sustainability of protected areas: Vulnerabilities and opportunities as revealed by COVID-19 in a national park management agency. *Biol. Conserv.* **2021**, *255*, 108985. [CrossRef]
25. Schianetz, K.; Kavanagh, L. Sustainability Indicators for Tourism Destinations: A Complex Adaptive Systems Approach Using Systemic Indicator Systems. *J. Sustain. Tour.* **2008**, *16*, 601–628. [CrossRef]
26. Zhang, X.; Zhong, L.; Yu, H. Sustainability assessment of tourism in protected areas: A relational perspective. *Glob. Ecol. Conserv.* **2022**, *35*, e02074. [CrossRef]
27. Plummer, R.; Fennell, D.A. Managing protected areas for sustainable tourism: Prospects for adaptive co-management. *J. Sustain. Tour.* **2009**, *17*, 149–168. [CrossRef]
28. Li, Y.; Li, Y.; Zhou, Y.; Shi, Y.; Zhu, X. Investigation of a coupling model of coordination between urbanization and the environment. *J. Environ. Manag.* **2012**, *98*, 127–133. [CrossRef]
29. Tang, Z. An integrated approach to evaluating the coupling coordination between tourism and the environment. *Tour. Manag.* **2015**, *46*, 11–19. [CrossRef]
30. Cong, X.L.; Huang, Y.; Liu, J.S. Spatial and temporal evolution of coupled coordination degree of ecotourism and tourism environment of Jilin Province. *Sci. Geogr. Sin.* **2019**, *39*, 496–505. [CrossRef]
31. Weng, G.M.; Tang, Y.B.; Pan, Y.; Mao, Y.Q. Spatiotemporal evolution and spatial difference of tourism-ecology-urbanization coupling coordination in Beijing-Tianjin-Hebei Urban Agglomeration. *Econ. Geogr.* **2021**, *41*, 196–204.
32. Kurniawan, F.; Adrianto, L.; Bengen, D.G.; Prasetyo, L.B. The social-ecological status of small islands: An evaluation of island tourism destination management in Indonesia. *Tour. Manag. Perspect.* **2019**, *31*, 136–144. [CrossRef]
33. Fei, J.; Lin, Y.; Jiang, Q.; Jiang, K.; Li, P.; Ye, G. Spatiotemporal coupling coordination measurement on islands' economy-environment-tourism system. *Ocean Coast. Manag.* **2021**, *212*, 105793. [CrossRef]
34. Pandit, M.K.; Manish, K.; Koh, L.P. Dancing on the Roof of the World: Ecological Transformation of the Himalayan Landscape. *Bioscience* **2014**, *64*, 980–992. [CrossRef]
35. Nepal, S.K. Tourism in protected areas. *Ann. Tour. Res.* **2000**, *27*, 661–681. [CrossRef]
36. Gao, J.; Li, J.; Fu, J.; Bao, Y.N.; Ma, S.T.; Li, W.Y.; Guo, X. Spatial patterns of ecoregions and protected areas in the Karakoram-Himalayan region. *Acta Ecol. Sin.* **2021**, *41*, 912–923. [CrossRef]
37. Zhong, L.S.; Zeng, Y.X.; Yu, H. The natural foundation and realization path of the recreation function of the Qinghai-Tibet Plateau National Park Group. *Acta Ecol. Sin.* **2021**, *41*, 861–873.
38. Farrell, B.H.; Twining-Ward, L. Reconceptualizing Tourism. *Ann. Tour. Res.* **2004**, *31*, 274–295. [CrossRef]
39. Lacitignola, D.; Petrosillo, I.; Cataldi, M.; Zurlini, G. Modelling socio-ecological tourism-based systems for sustainability. *Ecol. Model.* **2007**, *206*, 191–204. [CrossRef]
40. Nyaupane, G.P.; Poudel, S. Linkages among biodiversity, livelihood, and tourism. *Ann. Tour. Res.* **2011**, *38*, 1344–1366. [CrossRef]
41. Stone, M.T.; Nyaupane, G. Protected areas, tourism and community livelihoods linkages: A comprehensive analysis approach. *J. Sustain. Tour.* **2016**, *24*, 673–693. [CrossRef]
42. Stone, M.T.; Stone, L.S.; Nyaupane, G.P. Theorizing and contextualizing protected areas, tourism and community livelihoods linkages. *J. Sustain. Tour.* **2021**, *30*, 2495–2509. [CrossRef]
43. WTO; UNEP (United Nations Environment Programme and World Tourism Organisation). *Making Tourism More Sustainable: A Guide for Policy Makers*; United Nations Environment Programme and World Tourism Organisation: Paris, France; Madrid, Spain, 2005.
44. Torres-Delgado, A.; Saarinen, J. Using indicators to assess sustainable tourism development: A review. *Tour. Geogr.* **2014**, *16*, 31–47. [CrossRef]
45. Agyeiwaah, E.; McKercher, B.; Suntikul, W. Identifying core indicators of sustainable tourism: A path forward? *Tour. Manag. Perspect.* **2017**, *24*, 26–33. [CrossRef]
46. Qiu, H.; Fan, D.X.F.; Lyu, J.; Lin, P.M.C.; Jenkins, C.L. Analyzing the Economic Sustainability of Tourism Development: Evidence from Hong Kong. *J. Hosp. Tour. Res.* **2019**, *43*, 226–248. [CrossRef]
47. Leposa, N. Problematic blue growth: A thematic synthesis of social sustainability problems related to growth in the marine and coastal tourism. *Sustain. Sci.* **2020**, *15*, 1233–1244. [CrossRef]
48. Bramwell, B.; Lane, B. Critical research on the governance of tourism and sustainability. *J. Sustain. Tour.* **2011**, *19*, 411–421. [CrossRef]

49. Chen, D.; Zhong, L.; Fan, J.; Yu, H.; Yang, D.; Zeng, Y. Evaluation and structural analysis of the functions of the Tibetan Plateau National Park Cluster. *J. Geogr. Sci.* **2022**, *32*, 957–980. [CrossRef]
50. Li, X.Y.; Li, F.X.; Ma, Y.J.; Wu, X.M.; Sha, Z.J.; Wang, X.Q.; Wang, J.H. *Wetland Restoration and Biodiversity Conservation in Qinghai Lake Basin*; Science Press: Beijing, China, 2016.
51. UNEP (United Nations Environment Programme). *A Snapshot of the World's Water Quality: Towards a Global Assessment*; UNEP: Nairobi, Kenya, 2016.
52. Zhao, Y.Y. *Measurement and Dynamic Change of Tourism Ecological Footprint in Qinghai Lake Nature Reserve*; China Social Sciences Press: Beijing, China, 2018.
53. Ocampo, L.; Ebisa, J.A.; Ombe, J.; Escoto, M.G. Sustainable ecotourism indicators with fuzzy Delphi method—A Philippine perspective. *Ecol. Indic.* **2018**, *93*, 874–888. [CrossRef]
54. Lee, T.H.; Hsieh, H.-P. Indicators of sustainable tourism: A case study from a Taiwan's wetland. *Ecol. Indic.* **2016**, *67*, 779–787. [CrossRef]
55. Chen, M.G. *A Study on the Competitiveness of Chinese Insurance Brokerage Companies*; Zhongnan University of Economics and Law: Wuhan, China, 2017.
56. Ko, T.G. Development of a tourism sustainability assessment procedure: A conceptual approach. *Tour. Manag.* **2005**, *26*, 431–445. [CrossRef]
57. Yu, L. *Study on Evaluation Index System of Ecotourism Sustainability in Nature Reserves*; Beijing Forestry University: Beijing, China, 2006.
58. Blancas, F.J.; González, M.; Lozano, M.G.; Perez, F. The assessment of sustainable tourism: Application to Spanish coastal destinations. *Ecol. Indic.* **2010**, *10*, 484–492. [CrossRef]
59. Jitpakdee, R.; Thapa, G.B. Sustainability Analysis of Ecotourism on Yao Noi Island, Thailand. *Asia Pac. J. Tour. Res.* **2012**, *17*, 301–325. [CrossRef]
60. Pérez, V.; Guerrero, F.; González, M.; Pérez, F.; Caballero, R. Composite indicator for the assessment of sustainability: The case of Cuban nature-based tourism destinations. *Ecol. Indic.* **2013**, *29*, 316–324. [CrossRef]
61. Ashok, S.; Tewari, H.; Behera, M.; Majumdar, A. Development of ecotourism sustainability assessment framework employing Delphi, C&I and participatory methods: A case study of KBR, West Sikkim, India. *Tour. Manag. Perspect.* **2017**, *21*, 24–41. [CrossRef]
62. Saaty, T.L.; Özdemir, M.S. How many judges should there be in a group? *Ann. Data Sci.* **2014**, *1*, 359–368. [CrossRef]
63. Deng, J.; King, B.; Bauer, T. Evaluating natural attractions for tourism. *Ann. Tour. Res.* **2002**, *29*, 422–438. [CrossRef]
64. Zhou, C.; Feng, X.G.; Tang, R. Analysis and Forecast of Coupling Coordination Development among the Regional Economy-Ecological Environment-Tourism Industry—A Case Study of Provinces Along the Yangtze Economic Zone. *Econ. Geogr.* **2016**, *36*, 186–193. [CrossRef]
65. Zhao, J.; Liu, Y.; Zhu, Y.; Qin, S.; Wang, Y.; Miao, C. Spatiotemporal differentiation and influencing factors of the coupling and coordinated development of new urbanization and ecological environment in the Yellow River Basin. *Resour. Sci.* **2020**, *42*, 159–171. [CrossRef]
66. Yang, X.M.; Geng, J.; Xu, Y.; Zhuang, H.Y.; Chen, W.S. Spatio-temporal change and obstacle degree diagnosis of comprehensive land carrying capacity in Hainan Island based on TOPSIS model. *Acta Ecol. Sin.* **2022**, *42*, 9324–9334.
67. General Office of the Communist Party of China Central Committee; General Office of the State Council. A Guideline on Establishing a Protected Area System with National Parks as the Main Body. 2019. Available online: http://www.gov.cn/xinwen/2019-06/26/content_5403497.htm (accessed on 23 March 2022).
68. Wang, W.; Liu, J.; Innes, J.L. Conservation equity for local communities in the process of tourism development in protected areas: A study of Jiuzhaigou Biosphere Reserve, China. *World Dev.* **2019**, *124*, 104637. [CrossRef]
69. Zhang, Y.; Hu, Y.; Zhang, B.; Li, Y.; Zhang, X.; Xie, Y. Conflict between nature reserves and surrounding communities in China: An empirical study based on a social and ecological system framework. *Glob. Ecol. Conserv.* **2020**, *21*, e00804. [CrossRef]
70. Xu, H.; Cui, Q.; Sofield, T.; Li, F.M.S. Attaining harmony: Understanding the relationship between ecotourism and protected areas in China. *J. Sustain. Tour.* **2014**, *22*, 1131–1150. [CrossRef]
71. Cui, X.; Wang, Y.S.; Shi, L.N. Spatio-temporal characteristics and obstacle factors of human-land system coupling coordination in farming-pastoral ecotones of northern China. *J. Agric. Resour. Environ.* **2022**, *40*, 1–13.
72. Hentati-Sundberg, J.; Berglund, P.-A.; Hejdström, A.; Olsson, O. COVID-19 lockdown reveals tourists as seabird guardians. *Biol. Conserv.* **2021**, *254*, 108950. [CrossRef] [PubMed]

forests

MDPI

Article

Stand Structure Management and Tree Diversity Conservation Based on Using Stand Factors: A Case Study in the Longwan National Nature Reserve

Yue Li [1,2], Xue Wang [1,2], Chaofan Zhang [3], Jinpeng Li [4], Haibo Jiang [1,2], Lianxi Sheng [1,2,*] and Zhenxin Li [1,2,5,*]

1 State Environmental Protection Key Laboratory of Wetland Ecology and Vegetation Restoration, School of Environment, Northeast Normal University, Changchun 130117, China
2 Key Laboratory for Vegetation Ecology, Ministry of Education, Northeast Normal University, Changchun 130117, China
3 Jilin Provincial Academy of Forestry Sciences, Changchun 130033, China
4 Jilin Longwan National Nature Reserve Administration, Tonghua 135102, China
5 School of Environment and Ecology, Tibet University, Lhasa 850012, China
* Correspondence: shenglx@nenu.edu.cn (L.S.); lizx542@nenu.edu.cn (Z.L.)

Abstract: The management of stand structure and the protection of species diversity are crucial to forest ecosystem service functions. Changbai Mountain is one of the three major forest regions in Northeast China, and understanding the characteristics of stand structure and the allometric relationship between stand structure variables and the correlation between stand factors and species diversity is the basis for achieving effective forest management and ecological function improvement. In this study, the typical deciduous broad-leaved mixed forest and coniferous and broad-leaved mixed forest of the Longwan Nature Reserve in the Changbai Mountain were taken as the research objects, and the stand structure and diversity were investigated in detail. Allometric growth equations were established using the breast height diameter (DBH), tree height, and crown width of the main dominant species. Pearson correlation analysis was conducted on the stand structure and diversity of the forest community. The experimental results showed that young forests with small diameters and small individuals were the main part of the stands. Optimal allometric growth models of the main tree species (*Acer mandshuricum*, *Ulmus davidiana* var. *japonica*, *Juglans mandshurica*, *Acer mono*, and *Tilia amurensis*) were $0.807D^{0.646}$, $0608D^{0.381}H^{0.390}$, $0.502D^{0.533}H^{0.295}$, $0.795D^{0.540}H^{0.157}$, and $0.541D^{0.484}H^{0.301}$, respectively, with R^2 values ranging between 0.6 and 0.8, indicating a good fit for the models. Furthermore, tree density, crown width, and DBH were the main factors affecting tree diversity. The research results will provide theoretical support for the efficient management of forest stand structure and diversity conservation in the Longwan Reserve.

Keywords: deciduous broad-leaved mixed forest and coniferous and broad-leaved mixed forest; stand structure; species diversity; allometric growth equation; correlation analysis

1. Introduction

The plant community structure, and diversity characteristics have always been a hot issue in ecological research, which determines the niche and competition pattern of a species within the community and has an important impact on the sustainable supply of ecosystem functions [1–3]. Unlike ecosystems such as grasslands and farmlands, plant community structure and diversity play a more important role in evaluating the ecological functions of forests [4,5], and a reasonable and healthy community structure is widely recognized as the foundation for forests to perform a variety of ecological functions [6]. Arbors, as the primary component of forest vegetation, have drought and salinity tolerance, as well as various ecological service functions, such as species diversity maintenance, water conservation, and carbon fixation [7,8], and are the primary contributors for the forest ecosystem to perform ecological functions [9]. Therefore, analyzing the community structure characteristics and the effects of stand structure on tree growth and diversity is of great practical and scientific significance for forest conservation and the stability of an ecosystem structure and function [10,11].

The study of allometric growth is a crucial aspect of managing forest community structure [12], where stand structure can strongly influence forest growth [13]. Tree crown is a key parameter in characterizing individual tree growth [14], and studies have shown that DBH and tree height are important stand variables that affect tree growth and that all have significant allometric growth relationships with tree crowns [15–17]. Some studies have suggested that, compared to other models, the power function model is the best allometric growth model for stand variables and tree crowns [18,19]. Curtis and Reukema first developed a tree growth model using the tree height, DBH, and crown of *Pseudotsuga sinensis* [20]. Several researchers investigated the DBH and crown width of *Pinus sylvestris* L. located throughout the Czech Republic, and estimated the ideal spacing for tree growth through their allometric growth relationship [21]. In addition, research has shown that the relationship between DBH, tree height, and crown width of common tree species in Uganda and the northeastern United States conformed to the allometric growth model, which was used by local forest managers to estimate the growth of trees, thereby developing reasonable forest management strategies [22,23]. However, differences in forest type, growth environment, and species can greatly affect the relationship between tree growth and stand variables, even though there have been many allometric growth models for specific species in different regions based on DBH, tree height, and crown width [24]. Currently, there is limited research on forest allometric growth in the Changbai Mountains region of northeastern China.

In addition to the study of allometric growth of trees, the conservation of plant diversity is also an important part of forest ecosystem management [25]. Many ways to protect plant diversity are related to community structures [26], including height, crown width, canopy density and density [27–30]. For example, there was a significant negative correlation between the height of *Corylus cornuta* and the richness of understory species [31]. The differences in plant diversity in the herb layers of forests between evergreen and deciduous cover types were mainly related to crown width [32]. Similarly, tree density and canopy density were also important factors affecting species composition [33] when the planting density of trees was at a medium level, which was conducive to the increase in shrub diversity in the Mediterranean forest [34]. Additionally, one study emphasized that adjusting stand density would help enhance plant diversity [35]. However, there are few studies on the relationship between tree diversity and the stand structure of temperate forests in the Changbai Mountains, Northeast China.

Longwan Nature Reserve is a coniferous and broad-leaved mixed forest and deciduous broad-leaved mixed forest that formed after logging, and is an important object of natural forest protection and forest community restoration [36]. It has a unique plant community composition and biodiversity, dominated by northern temperate plants, with high vegetation coverage and good water and soil conservation [37]. Furthermore, the reserve contains a variety of rare wetlands with special causes, as well as numerous types of national rare and protected animals and plants that play an important role in maintaining biodiversity, water conservation, and carbon fixation, and have extremely important ecological significance and economic value [38]. In view of this area, where the relevant research on the community structure characteristics and the effects of stand structure on tree growth and diversity is still lacking, this study took Longwan Nature Reserve as the research object and addressed the following questions: (1) Which stand factors have significant effects on tree growth? Establish the optimal allometric growth models between stand variables and crown width of the main dominant species; and (2) What were the main stand factors affecting the tree layer species diversity?

2. Materials and Methods

2.1. Study Sites

Longwan National Nature Reserve is located in the Huinan County, Tonghua City, Jilin Province, China, and belongs to the middle section of the Longgang Mountains of Changbai Mountain, with an area of approximately 15,061 hm^2, geographical coordinates of 125°13′55″ E–126°32′02″ E, 42°16′20″ N–42°26′57″ N, and an altitude of 449–1233 m. The area belongs to the continental monsoon climate of the north temperate zone, with abundant rainfall in summer and minimum precipitation in winter, with an average annual precipitation of 704.2 mm, an average annual temperature of 4.1 °C, a frost-free period of 110–120 days, and dark brown zonal soil. In addition, according to the local resource archives, Longwan National Nature Reserve was formerly a coniferous and broad-leaved mixed forest that was excessively logged in the 1960s. At present, the community is a secondary forest formed by natural regeneration, and most trees are 40–60 years old. The forest types are mainly deciduous broad-leaved mixed forest and a small amount of coniferous and broad-leaved mixed forest. The composition of plant species is relatively rich, and the main trees are *Juglans mandshurica*, *Tilia amurensis*, *Ulmus japonica*, and *Acer mono* [39,40].

2.2. Survey of Plant Communities

In July–August 2019, 25 monitoring sample plots were set up in the reserve to investigate the species distribution and community structure characteristics of the arbor layer in Longwan Nature Reserve (Figure 1). The sample plots were of deciduous broad-leaved mixed forest and coniferous and broad-leaved mixed forest vegetation types. Moreover, trees with a DBH larger than 5 cm were investigated, and dead trees were not included in this study. Three sample squares of 20 m × 30 m were set up in each sample plot, and the species name, number, height, DBH, crown width, and other data of the species in the quadrat were recorded. The Chinese Virtual Herbarium (https://www.cvh.ac.cn/, accessed on 21 February 2021) was used to identify and review the species recorded in the survey.

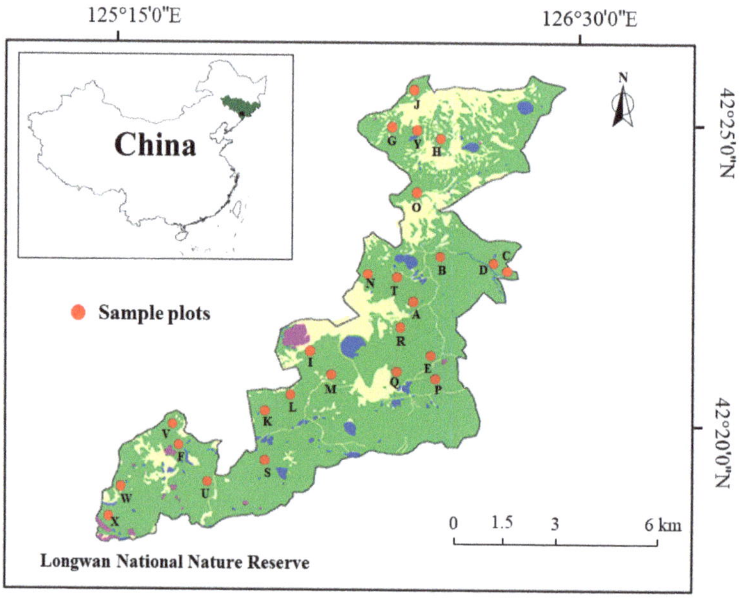

Figure 1. Distribution of survey sample plots in the Longwan Nature Reserve.

2.3. *Measurement Indicators and Measurement Methods*

2.3.1. Stand Structure

To describe the functional status of species in the community, the importance value (IV) was used as the dominance index of each species in the community and was calculated as follows [41]:

IV = (relative height + relative abundance + relative coverage)/3

The community structure of the tree layer was measured by density, DBH, tree height, crown width, basal area, and canopy density.

2.3.2. Construction of Allometric Growth Equation

Using stand variables such as DBH and tree height as independent variables and crown width as the dependent variable, scatter plots were created for the main dominant species (*Acer mandshuricum*, *Ulmus japonica*, *Juglans mandshurica*, *Acer mono* and *Tilia amurensis*) in the Longwan Nature Reserve. The trend line type and significance were determined to identify the main stand variables that affected crown width. A multiparameter allometric growth equation was established using the crown width as the dependent variable (y) and the stand structure factors as the independent variables (x). The model with the smallest residual sum of squares (RSS) and the largest determination coefficient (R^2) was selected as the optimal model. The expression of the multiparameter allometric growth equation is as follows:

$$y = a x_1^b x_2^c, \quad (1)$$

where y represents the crown width data of the dominant species and x represents the stand variables of the DBH and tree height. The model parameters are denoted by a, b, and c.

To estimate the parameters of the prediction model, multiple linear regression was used in this study. As the prediction model is nonlinear, it was transformed into a linear model. The equation after the transformation is $\ln(y) = \ln a + b \ln(x_1) + c \ln(x_2)$.

2.3.3. Plant Community Species Diversity

The species diversity index of trees adopted the richness index, diversity index and evenness index, and the calculation formulas were as follows [42]:

Richness index:

$$R = S, \quad (2)$$

where S represents the number of species in the plant community.

The species diversity indexes were expressed by the Simpson index (D) and the Shannon–Wiener index (H).

Simpson index:

$$D = 1 - \sum_{i=1}^{S} P_i^2, \quad (3)$$

where S represents the number of species in the plant community, and P_i is the proportion of the ith species in the total number of species.

Shannon–Wiener index:

$$H = -\sum_{i=1}^{S} P_i \ln(P_i), \quad (4)$$

where S represents the number of species in the plant community, and P_i is the proportion of the ith species in the total number of species.

Species evenness indexes were expressed by the Pielou index (J_{SW}) and the Alatalo index (E_a).

Pielou index:

$$J_{sw} = \frac{-\sum_{i=1}^{S} P_i \ln(P_i)}{\ln S}, \quad (5)$$

where S represents the number of species in the plant community, and P_i is the proportion of the ith species in the total number of species.

Alatalo index:

$$E_a = \frac{\frac{1}{\sum_{i=1}^{S} P_i^2} - 1}{\exp(-\sum_{i=1}^{S} P_i ln P_i) - 1}, \quad (6)$$

where S represents the number of species in the plant community, and P_i is the proportion of the ith species in the total number of species.

2.4. Data Analysis

Data sorting and mapping were carried out using Excel 2019 (Microsoft Corporation, Redmond, WA, USA). Multiple linear regression was used to calculate the parameters of the allometric growth equation, and Pearson correlation was used to analyze the relationship between stand factors and plant diversity indexes (SPSS 22.0, IBM Corporation, Chicago, IL, USA).

3. Results

3.1. Species Composition and Stand Structure

All 25 sample plots in the Longwan Nature Reserve belonged to the mixed deciduous broad-leaved forest and coniferous and broad-leaved mixed forest community types, with 14 families, 22 genera, and 36 species of trees (Table S1). *Juglans mandshurica* was an important species in several sample plots, including A, B, C, D, M, R, T, and Y, with proportions ranging from 23.47 to 46.73%. The dominant species in sample plots F, K, and L was *Tilia amurensis*, which accounted for 29.10%, 20.53%, and 24.85%, respectively. The dominant species in sample plots H, O, and Q was *Ulmus japonica*, with proportions of 18.21%, 25.82%, and 42.28%, respectively. Sample plots G and J had *Acer mono* as the dominant species, with proportions of 23.65% and 25.11%, respectively. Sample plots I and P had *Larix gmelinii* (Rupr.) Kuzen. as the dominant species, with proportions of 27.24% and 33.91%, respectively. Sample plots N and V had *Acer mandshuricum* as the dominant species, with proportions of 35.77% and 28.65%, respectively. Sample plots S and W had *Fraxinus mandschurica* Rupr. as the dominant species, with proportions of 30.25% and 30.21%, respectively. The dominant species in sample plots E, U, and X were *Betula platyphylla* Suk., *Tilia mandschurica* Rupr.et Maxim., and *Quercus mongolica* Fisch.exLedeb., respectively. The proportions of the dominant species in these sample plots were 22.60%, 23.61%, and 22.10%, respectively (Table S1).

The distribution range of tree layer density in the surveyed sample plots was 367–1400 trees/ha (Figure 2a). About 60% of the sample plots had a density between 600 and 1100 trees/ha. Most sample plots had a basal area concentrated between 15 and 35 m^2/ha. However, sample plots I and W had relatively low basal areas of only 9.22 m^2/ha and 11.53 m^2/ha, respectively (Figure 2b). More than 70% of the sample plots had a canopy density concentrated between 0.4 and 0.8 (Figure 2c). The DBH of the surveyed sample plots ranged from 5 to 25 cm (Figure 2d). Most sample plots had trees with a height between 5 and 17 m (Figure 2e), and the crown width was concentrated between 3 and 7 m (Figure 2f).

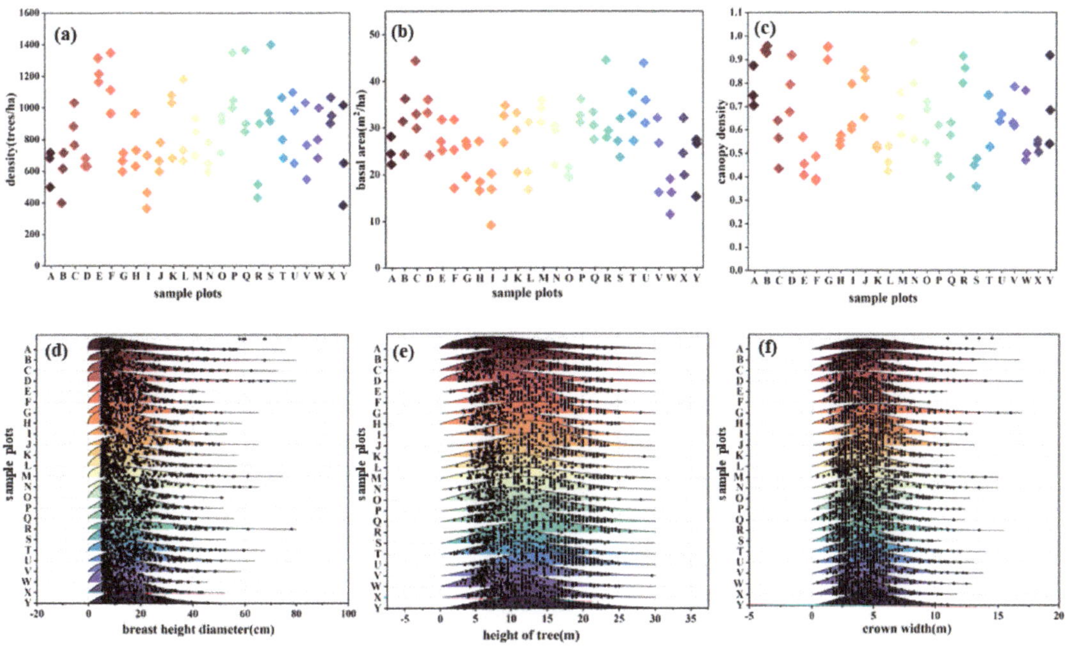

Figure 2. Community structure characteristics of the arbor layer. (**a–c**) Data distribution of density, basal area and canopy density of three quadrats in sample plots; (**d–f**) data distribution and Weibull curve of DBH, tree height, and crown width. In (**d–f**), the vertical axes represent the plot numbers A–Y of the survey sample.

3.2. Distribution Characteristics of Species Diversity

The distribution range of species richness was 4–18, and the number of species in most sample plots was 7–13 (Figure 3a). The Simpson index was basically above 0.65 (Figure 3b); the distribution characteristics of the Shannon–Wiener index and the richness index in various plots were similar (Figure 3c), and the Shannon–Wiener index in most sample plots was mainly concentrated in the range of 1.5–2.5. The Pielou index was basically above 0.7 (Figure 3d), and the Alatalo index was concentrated in the range of 0.6–0.9 (Figure 3e).

3.3. Allometric Growth Models of DBH, Tree Height, and Crown Width

The DBH, tree height, and crown width of the main dominant species (*Acer mandshuricum, Ulmus japonica, Juglans mandshurica, Acer mono* and *Tilia amurensis*) in the Longwan Nature Reserve all had significant correlations ($p < 0.01$) (Figures 4 and 5). To better understand these correlations, we established two allometric growth models: $CW = aD^b$ and $CW = aD^bH^c$. By comparing the R^2 and RSS values of the two models, the optimal models for *Acer mandshuricum, Ulmus japonica, Juglans mandshurica, Acer mono,* and *Tilia amurensis* were $0.807D^{0.646}$, $0.608D^{0.381}H^{0.390}$, $0.502D^{0.533}H^{0.295}$, $0.795D^{0.540}H^{0.157}$, and $0.541D^{0.484}H^{0.301}$, respectively (Table 1), with R^2 values between 0.6 and 0.8, indicating that the constructed model had good fitting effects.

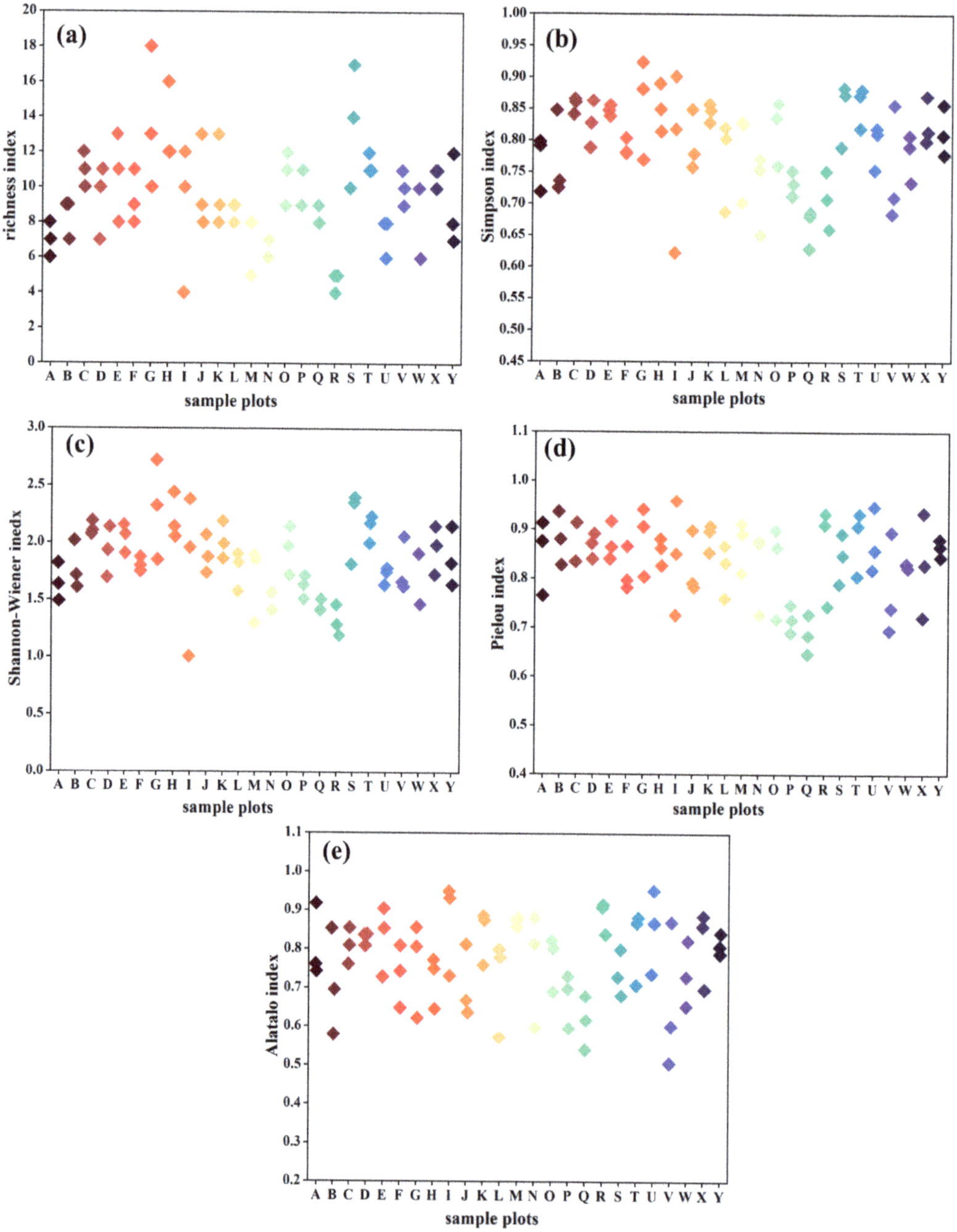

Figure 3. Distribution characteristics of species diversity in three quadrats of the sample plots. (**a**) Species richness index of three quadrats in sample plots; (**b**) Simpson index of three quadrats in sample plots; (**c**) Shannon–Wiener index of three quadrats in sample plots; (**d**) Pielou index of three quadrats in sample plots; (**e**) Alatalo index of three quadrats in sample plots.

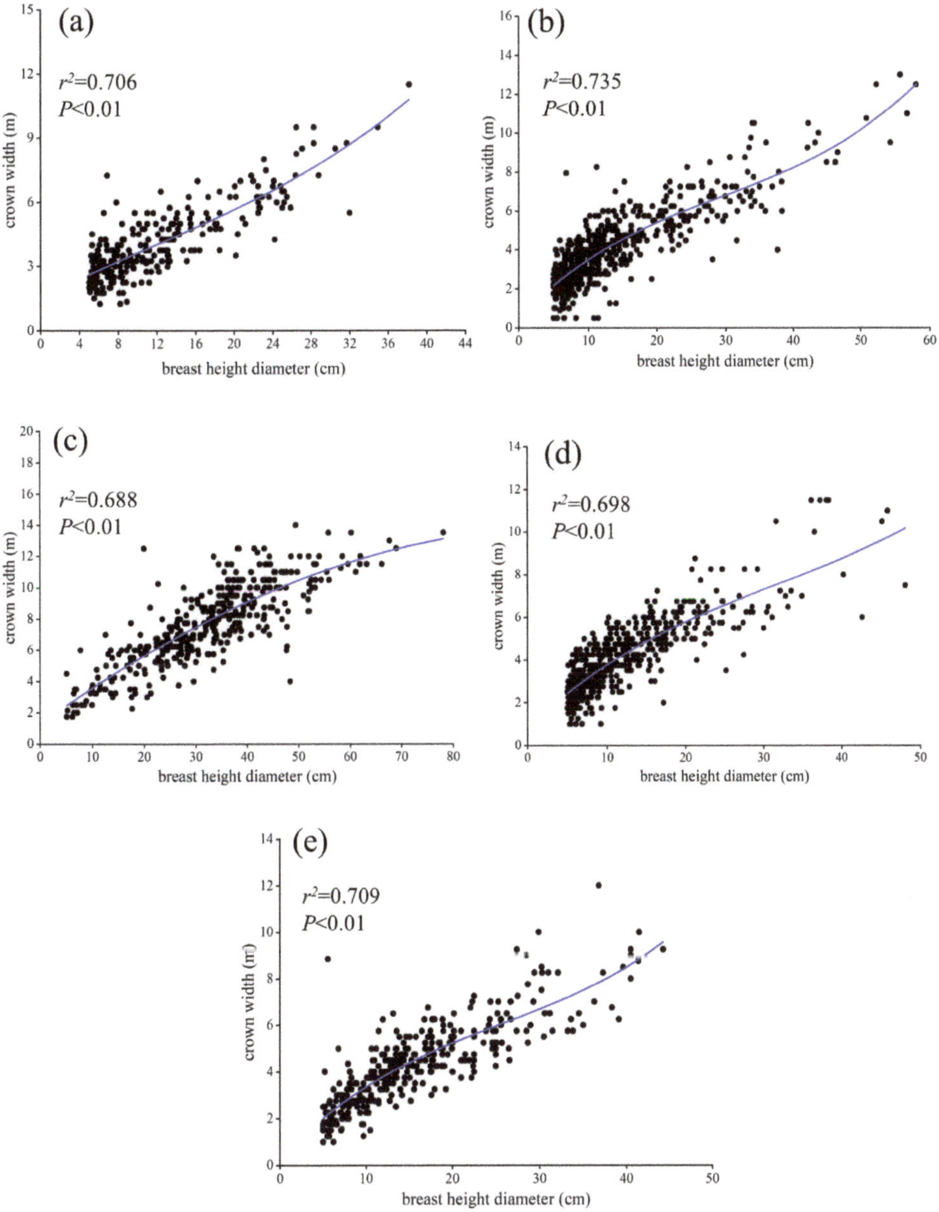

Figure 4. Scatterplot of DBH and crown width. The lines in (**a**–**e**) represent the optimal curve fitting between DBH and crown width of the main dominant species. (**a**) *Acer mandshuricum*; (**b**) *Ulmus japonica*; (**c**) *Juglans mandshurica*; (**d**) *Acer mono*; (**e**) *Tilia amurensis*.

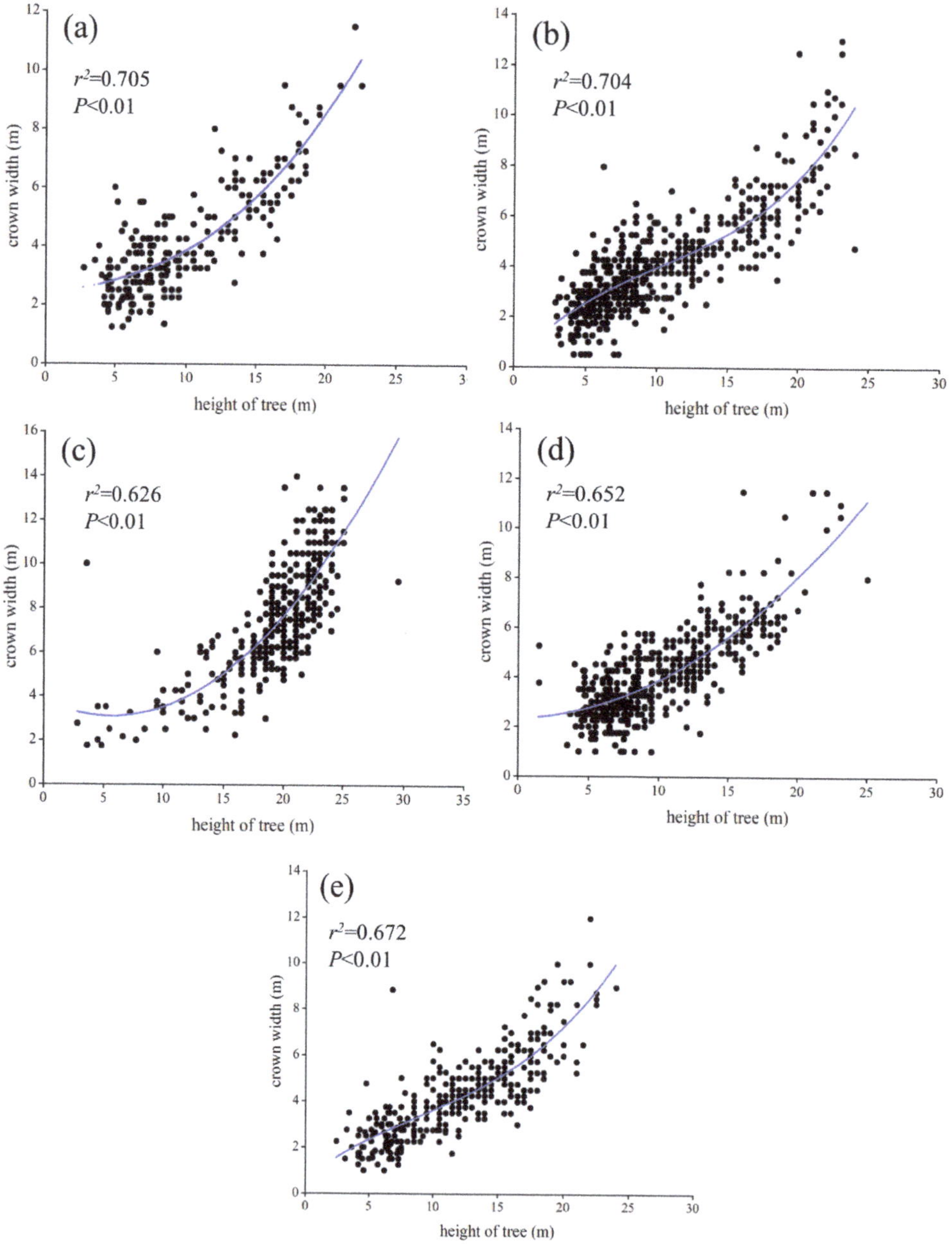

Figure 5. Scatterplot of tree height and crown width. The lines in figures (**a**–**e**) represent the optimal curve fitting between the height of the tree and crown width of the main dominant species. (**a**) *Acer mandshuricum*; (**b**) *Ulmus japonica*; (**c**) *Juglans mandshurica*; (**d**) *Acer mono*; (**e**) *Tilia amurensis*.

Table 1. Parameter estimation of prediction models and significance testing of optimal models.

Species	Model	Parameter Estimation			Evaluation Index		Significance Test	
		a	*b*	*c*	R^2	*RSS*	*F*-Value	*p*-Value
Acer mandshuricum	$CW = aD^b$	0.807	0.646		0.624	15.750	382.590	0.000
	$CW = aD^bH^c$	0.505	0.716	0.273	0.551	17.773		
Ulmus japonica	$CW = aD^b$	0.685	0.682		0.579	57.574		
	$CW = aD^bH^c$	0.608	0.381	0.390	0.607	53.785	422.307	0.000
Juglans mandshurica	$CW = aD^b$	0.728	0.678		0.716	16.549		
	$CW = aD^bH^c$	0.502	0.533	0.295	0.738	14.264	534.089	0.000
Acer mono	$CW = aD^b$	0.827	0.644		0.618	34.490		
	$CW = aD^bH^c$	0.795	0.540	0.157	0.636	30.290	477.96	0.000
Tilia amurensis	$CW = aD^b$	0.637	0.698		0.708	19.521		
	$CW = aD^bH^c$	0.541	0.484	0.301	0.782	17.926	460.426	0.000

Note: *D*: breast height diameter; *H*: height of tree; *CW*: crown width.

3.4. Correlation Analysis of Community Structure and Species Diversity

Tree density had a significant positive correlation with species richness ($p < 0.01$) and had a significant negative effect on the Pielou index and Alatalo index ($p < 0.05$); DBH had a significant negative impact on species richness ($p < 0.05$) and had a significant positive correlation with the Pielou index and Alatalo index ($p < 0.05$); crown width had a significant negative correlation with species richness ($p < 0.05$) and a significant positive correlation with the Pielou index and Alatalo index ($p < 0.01$) (Table 2).

Table 2. Correlation analysis of community structure and species diversity.

	R_a	D_a	H_a	J_{sw}	E_a	Dy	DBH	TH	CW	BC	CD
R_a	1										
D_a	0.658 **	1									
H_a	0.853 **	0.919 **	1								
J_{sw}	0.108	0.630 **	0.466 **	1							
E_a	−0.204	0.430 **	0.137	0.672 **	1						
Dy	0.321 **	0.060	0.128	−0.271 *	−0.248 *	1					
DBH	−0.265 *	−0.077	−0.161	0.243 *	0.232 *	−0.346 **	1				
TH	−0.215	0.136	−0.194	0.061	0.192	−0.096	0.700 **	1			
CW	−0.229 *	−0.028	−0.043	0.298 **	0.253 *	−0.677 **	0.561 **	0.392 **	1		
BC	−0.033	−0.006	−0.035	0.092	0.044	0.322 **	0.546 **	0.369 **	0.139	1	
CD	−0.081	0.067	0.028	0.196	0.186	−0.481 **	0.762 **	0.480 **	0.721 **	0.191	1

Note: * indicates a significant correlation at the 0.05 level, ** indicates a significant correlation at the 0.01 level. R_a: richness index; D_a: Simpson index; H_a: Shannon–Wiener index; J_{SW}: Pielou index. E_a: Alatalo index; D_y: density; DBH: breast height diameter; TH: height of tree; CW: crown width; BC: basal area; CD: canopy density.

4. Discussion

4.1. Tree-Layer Community Structure and Diversity

The community structure and diversity of plants are important biological characteristics at the level of ecological organization, reflecting the composition and spatial configuration of a species [43]. The survey found that the Longwan Nature Reserve was relatively rich in plant species, with a total of 36 species of trees belonging to 14 families and 22 genera in the sample plots. Among these species, *Juglans mandshurica*, *Ulmus japonica* and *Tilia amurensis* were the main dominant species, and the number of individuals such as *Alnus cremastogyne*, *Acer komarovii*, and *Salix matsudana* was small, but they enriched the composition of species and increased the species diversity of arbors (Table S1). The density of the 25 sample plots was 367–1400 trees/hm^2, the distribution range of the DBH was 5–25 cm, and the tree height was mainly concentrated at 5–17 m (Figure 2), which was different from the typical broad-leaved mature forest in the Changbai Mountain area (the density of trees was 400–600 trees/hm^2, the average DBH was 24–28 cm, and the height was 15–20 m) [44]. The analysis of plant diversity showed that the richness index of the tree layer was 4–18, the Simpson index was concentrated in the range of 0.65–0.9, the Shannon–Wiener index was 1.32–2.30, and the Pielou and Alatalo indexes were concentrated at 0.7–0.95 and 0.6–0.9, respectively (Figure 3), which were basically consistent with the species diversity of the broad-leaved mixed forest in the Changbai Mountain [44]. In addition, the differences in species diversity among the plots were mainly related to the proportion of dominant species, the differences in understory environments and nutrient limitation, resulting in some differences in species richness, diversity, and evenness among the surveyed sample plots [45–48]. In summary, young forests with small diameters and small individuals were the main part of the stands in the Longwan Nature Reserve, and the degree of fluctuation in DBH, tree height, and canopy density was large, indicating that these sample plots were in a relatively intense succession stage and had not yet recovered to the zonal vegetation—coniferous and broadleaved mixed forest in the secondary succession process after logging.

4.2. Allometric Growth Models of DBH, Tree Height, and Crown Width

Research on individual tree growth is crucial for managing forest community structure and stand management [49]. Several studies have demonstrated that the DBH and tree height are the most significant factors in the allometric growth relationship between tree crown and stand variables. Both factors have a significant allometric growth relationship with crown width [50]. The allometric growth model has become a standard for evaluating forest growth scientifically and guiding forest management decisions [51]. For instance, Jenkins, Sawadogo, and others investigated the stand structure of common tree species in the eastern United States and west Africa. They estimated the growth and space required for tree growth in the region by creating allometric growth equations for DBH, tree height, and tree crown [52,53]. Similarly, a study investigated stand variables such as DBH and tree height of ash trees growing in Oakville, Canada, and established allometric growth equations for DBH, tree height, and crown width. This model may be used to estimate the growth of ash trees in Oakville with similar climate and management environments [54]. Our study also found that the DBH and tree height significantly affect crown width. The crown gradually increased with increasing DBH and tree height (Figures 4 and 5). Among them, the optimal models for *Acer mandshuricum, Ulmus japonica, Juglans mandshurica, Acer mono* and *Tilia amurensis* were $0.807D^{0.646}$, $0.608D^{0.381}H^{0.390}$, $0.502D^{0.533}H^{0.295}$, $0.795D^{0.540}H^{0.157}$, and $0.541D^{0.484}H^{0.301}$, respectively, and the R^2 values were between 0.6 and 0.8 (Table 1), indicating that the models have a good fitting effect. In summary, allometric growth models can provide guidance for nurturing and thinning trees and forest management in the Longwan Nature Reserve.

4.3. Impact of Plant Community Structure on Diversity

There is a coupling relationship between forest community structure and species diversity. For instance, tree density is an important factor affecting species composition and diversity [33]. There was a negative correlation between the stand density and species richness of *Pinus tabulaeformis* Carr. and *Pinus massoniana* Lamb. forests, and plant diversity was the highest at medium density [55,56]. In addition, a study showed that as the Tumbesian dry forest stand density decreased, species richness decreased significantly [57]. Our results suggested that stand density had a positive effect on species richness and had a significant negative correlation with species evenness (Table 2). With the increase in tree density, although species diversity did not change significantly, the species composition of the tree layer changed, and the evenness decreased. Moreover, one study suggested that the shading of Sitka spruce and western hemlock was detrimental to the survival and growth of vascular plants [58], which was closely related to the crown width of arbors. Similarly, the average DBH was also a major factor affecting species composition. One study of mountain forests in western Hungary found that the DBH of trees significantly influenced species composition and diversity in the understory [59]. Our experimental results showed that the DBH and crown width had negative effects on species richness, while both had a significant positive correlation with evenness (Table 2): with the increase in the DBH and crown width, the evenness of the tree layer increased. It is worth noting that in this study, the stand factors (density, DBH, crown width) were only significantly related to species richness and evenness, but had no significant correlation with the Simpson and Shannon–Wiener diversity indexes, which may be due to its plentiful water and heat resources, soil fertility, and the fact that the species with similar niches had a harmonious interspecific relationship. Consequently, there was no significant impact on the species diversity indexes of the tree layer. Some studies have indicated that the individual size of arbors significantly affects species diversity in forest ecosystems. For instance, it has been shown that the average height of plants decreases, which is conducive to the improvement of plant diversity [30], and a significant correlation between tree height and plant diversity has been found in subtropical montane forests [60]. In contrast, there was no significant correlation between species diversity and tree height in our study, which indicated that stand size was not the limiting factor for tree species diversity in the Longwan Nature Reserve, probably because most of the species were young stands and the difference in tree heights was not significant. The comprehensive analysis of the stand structural characteristics revealed that tree density, DBH and crown width were the main factors influencing temperate forest diversity in the Changbai Mountain region of Northeast China. This study can identify ways to effectively improve species diversity by exploring the coupling relationship between community structure and plant diversity, which is beneficial to the conservation of plant diversity and the stability of forest ecological functions.

5. Conclusions

The tree layer community of the forest ecosystem in the Longwan Nature Reserve belongs to the secondary forest formed by natural regeneration, and the forest types are mainly deciduous broad-leaved mixed forest and a small amount of coniferous broad-leaved mixed forest. The trees with small diameters and small individuals were the main part of the stand, and more importantly, DBH, tree height and canopy density fluctuated greatly, which was in an intense succession stage. The results indicated that the optimal allometric growth models of the main tree species (*Acer mandshuricum*, *Ulmus japonica*, *Juglans mandshurica*, *Acer mono* and *Tilia amurensis*) were $0.807D^{0.646}$, $0608D^{0.381}H^{0.390}$, $0.502D^{0.533}H^{0.295}$, $0.795D^{0.540}H^{0.157}$, and $0.541D^{0.484}H^{0.301}$, respectively. In addition, density, DBH, and crown width were the main factors affecting the richness index and evenness index of the tree layer. Considering that forest ecosystems have many important ecological functions, this study could provide a scientific basis for the optimal allocation of forest resources, the protection of plant diversity, and the stability of ecological functions.

Supplementary Materials: The following supporting information can be downloaded at: https://www.mdpi.com/article/10.3390/f14040750/s1, Table S1: Proportion (%) of importance value of species at different sample plots.

Author Contributions: Conceptualization, Y.L.; data curation, X.W.; formal analysis, C.Z.; investigation, J.L.; software, H.J.; writing—original draft, Z.L.; writing—review and editing, L.S. All authors have read and agreed to the published version of the manuscript.

Funding: This work was supported in part by the National Natural Science Foundation of China (Grant No. 32271624) and the National Key R&D Program of China (Grant No. 2022YFF1300900).

Data Availability Statement: The data presented in this study are available on request from the corresponding author.

Conflicts of Interest: The authors declare no conflict of interest.

References

1. Symstad, A.J.; Chapin, F.S.; Wall, D.H.; Gross, K.L.; Huenneke, L.F.; Mittelbach, G.G.; Peters, D.P.; Tilman, D. Long-term and large-scale perspectives on the relationship between biodiversity and ecosystem functioning. *Bioscience* **2003**, *53*, 89–98. [CrossRef]
2. Xu, M.; Ma, L.; Jia, Y.; Liu, M. Integrating the effects of latitude and altitude on the spatial differentiation of plant community diversity in a mountainous ecosystem in China. *PLoS ONE* **2017**, *12*, e0174231. [CrossRef]
3. Lasch, P.; Lindner, M.; Erhard, M.; Suckow, F.; Wenzel, A. Regional impact assessment on forest structure and functions under climate change—The Brandenburg case study. *For. Ecol. Manag.* **2002**, *162*, 73–86. [CrossRef]
4. Pastorella, F.; Paletto, A. Stand structure indices as tools to support forest management: An application in Trentino forests (Italy). *J. Trop. For. Sci.* **2013**, *59*, 159–168. [CrossRef]
5. Ali, A. Forest stand structure and functioning: Current knowledge and future challenges. *Ecol. Indic.* **2019**, *98*, 665–677. [CrossRef]
6. Zhang, T.; Chen, Y.; Wang, W.; Chen, Y.; Liu, X. Characteristics of Plant Community and Its Relationship with Groundwater Depth of the Desert Riparian Zone in the Lower Reaches of the Ugan River, Northwest China. *Water* **2022**, *14*, 1663. [CrossRef]
7. Li, J.; Ren, Z.; Zhou, Z. Ecosystem services and their values: A case study in the Qinba mountains of China. *Ecol. Res.* **2006**, *21*, 597–604. [CrossRef]
8. Guo, Z.; Xiao, X.; Gan, Y.; Zheng, Y. Ecosystem functions, services and their values–a case study in Xingshan County of China. *Ecol. Econ.* **2001**, *38*, 141–154. [CrossRef]
9. Mori, A.S.; Lertzman, K.P.; Gustafsson, L. Biodiversity and ecosystem services in forest ecosystems: A research agenda for applied forest ecology. *J. Appl. Ecol.* **2017**, *54*, 12–27. [CrossRef]
10. Staudhammer, C.L.; LeMay, V.M. Introduction and evaluation of possible indices of stand structural diversity. *Can. J. For. Res* **2001**, *31*, 1105–1115. [CrossRef]
11. Ouyang, S.; Xiang, W.; Gou, M.; Chen, L.; Lei, P.; Xiao, W.; Forrester, D.I. Stability in subtropical forests: The role of tree species diversity, stand structure, environmental and socio-economic conditions. *Glob. Ecol. Biogeogr.* **2021**, *30*, 500–513. [CrossRef]
12. Cifuentes Jara, M.; Henry, M.; Réjou-Méchain, M.; Wayson, C.; Zapata-Cuartas, M.; Piotto, D.; Westfall, J. Guidelines for documenting and reporting tree allometric equations. *Ann. For. Sci.* **2015**, *7*, 763–768. [CrossRef]
13. Forrester, D.I. Linking forest growth with stand structure: Tree size inequality, tree growth or resource partitioning and the asymmetry of competition. *For. Ecol. Manag.* **2019**, *447*, 139–157. [CrossRef]
14. Pretzsch, H.; Biber, P.; Uhl, E.; Dahlhausen, J.; Rötzer, T.; Caldentey, J. Crown size and growing space requirement of common tree species in urban centres, parks, and forests—ScienceDirect. *Urban For. Urban Green.* **2015**, *14*, 466–479. [CrossRef]
15. Subedi, T.; Bhandari, S.K.; Pandey, N. Form factor and volume equations for individual trees of Shorea robusta in Western low land of Nepal. *Austrian J. For. Sci.* **2021**, *138*, 143–166.
16. Víquez, E.; Pérez, D. Effect of pruning on tree growth, yield, and wood properties of Tectona grandis plantations in Costa Rica. *Silva Fenn.* **2005**, *39*, 381. [CrossRef]
17. Laubhann, D.; Sterba, H.; Reinds, G.J.; De Vries, W. The impact of atmospheric deposition and climate on forest growth in European monitoring plots: An individual tree growth model. *For. Ecol. Manag.* **2009**, *258*, 1751–1761. [CrossRef]
18. Ruark, G.A.; Martin, G.L.; Bockheim, J.G. Comparison of constant and variable allometric ratios for estimating Populus tremuloides biomass. *For. Sci.* **1987**, *33*, 294–300. [CrossRef]
19. Kusmana, C.; Hidayat, T.; Tiryana, T.; Rusdiana, O. Allometric models for above-and below-ground biomass of *Sonneratia* spp. *Glob. Ecol. Conserv.* **2018**, *15*, e00417. [CrossRef]
20. Curtis, R.O.; Reukema, D.L. Crown Development and Site Estimates in a Douglas-Fir Plantation Spacing Test. *For. Sci.* **1970**, *16*, 287–301. [CrossRef]
21. Sharma, R.P.; Bílek, L.; Vacek, Z.; Vacek, S. Modelling crown width–diameter relationship for Scots pine in the central Europe. *Trees* **2017**, *31*, 1875–1889. [CrossRef]
22. Tumwebaze, R. Allometric biomass equations for tree species used in agroforestry systems in Uganda. *Agrofor. Syst.* **2013**, *87*, 781–795. [CrossRef]
23. Troxel, B.; Piana, M.; Ashton, M.S.; Murphy-Dunning, C. Relationships between bole and crown size for young urban trees in the northeastern USA. *Urban For. Urban Green.* **2013**, *12*, 144–153. [CrossRef]
24. Semenzato, P.; Cattaneo, D.; Dainese, M. Growth prediction for five tree species in an Italian urban forest. *Urban For. Urban Green.* **2011**, *10*, 169–176. [CrossRef]
25. Rametsteiner, E.; Simula, M. Forest certification—An instrument to promote sustainable forest management? *J. Environ. Manag.* **2003**, *67*, 87–98. [CrossRef] [PubMed]
26. Berger, A.L.; Puettmann, K.J. Overstory composition and stand structure influence herbaceous plant diversity in the mixed aspen forest of northern Minnesota. *Am. Midl. Nat.* **2000**, *143*, 111–125. [CrossRef]
27. Martins, A.C.; Willig, M.R.; Presley, S.J.; Marinho-Filho, J. Effects of forest height and vertical complexity on abundance and biodiversity of bats in Amazonia. *For. Ecol. Manag.* **2017**, *391*, 427–435. [CrossRef]
28. Ishii, H.T.; Tanabe, S.I.; Hiura, T. Exploring the relationships among canopy structure, stand productivity, and biodiversity of temperate forest ecosystems. *For. Sci.* **2004**, *50*, 342–355. [CrossRef]
29. Kern, C.C.; Montgomery, R.A.; Reich, P.B.; Strong, T.F. Harvest-created canopy gaps increase species and functional trait diversity of the forest ground-layer community. *For. Sci.* **2014**, *60*, 335–344. [CrossRef]
30. Brokaw, N.; Busing, R.T. Niche versus chance and tree diversity in forest gaps. *Trends Ecol. Evol.* **2000**, *15*, 183–188. [CrossRef]

31. Spehn, E.M.; Joshi, J.; Schmid, B.; Diemer, M.; Körner, C. Above-ground resource use increases with plant species richness in experimental grassland ecosystems. *Funct. Ecol.* **2000**, *14*, 326–337. [CrossRef]
32. Auclair, A.N.; Goff, F.G. Diversity relations of upland forests in the western Great Lakes area. *Am. Nat.* **1971**, *105*, 499–528. [CrossRef]
33. Pitkänen, S. Correlation between stand structure and ground vegetation: An analytical approach. *Plant Ecol.* **1997**, *131*, 109–126. [CrossRef]
34. Oliver, P.A.T. Patterns of shrub diversity and tree regeneration across topographic and stand-structural gradients in a Mediterranean forest. *For. Syst.* **2015**, *24*, 11. [CrossRef]
35. Barbier, S.; Gosselin, F.; Balandier, P. Influence of tree species on understory vegetation diversity and mechanisms involved—A critical review for temperate and boreal forests. *For. Ecol. Manag.* **2008**, *254*, 1–15. [CrossRef]
36. Ma, L.; Sun, G.Q.; Qu, Y.; Li, J.Q. Applying systematic conservation planning to constitute a protection strategy for broad-leaved Korean pine forests in Changbai Mountains, China. *J. Mt. Sci.* **2016**, *13*, 493–507. [CrossRef]
37. Song, H.-J.; Ye, J.; Shi, S.; Zhang, Z.-C.; Kuang, X.; Xing, D.-L.; Yuan, Z.-Q.; Lin, F.; Wang, X.-G.; Hao, Z.-Q. Woody plant species composition and community structure in residual fragments of broad-leaved Korean pine mixed forests in Changbai Mountains area. *J. Appl. Ecol.* **2014**, *25*, 1239–1249.
38. Xue, D.; Tisdell, C. Valuing ecological functions of biodiversity in Changbaishan Mountain Biosphere Reserve in northeast China. *Biodivers. Conserv.* **2001**, *10*, 467–481. [CrossRef]
39. Li, Y.; Wang, X.; He, C.; Jiang, H.; Sheng, L. Multi-environment factors dominate plant community structure and diversity in an ombrotrophic bog: The water level is the main regulating mechanism. *Front. Environ. Sci.* **2022**, *10*, 1032068. [CrossRef]
40. Wang, X.; Bai, X.; Ma, L.; He, C.; Jiang, H.; Sheng, L.; Luo, W. Snow depths' impact on soil microbial activities and carbon dioxide fluxes from a temperate wetland in Northeast China. *Sci. Rep.* **2020**, *10*, 8709. [CrossRef]
41. Curtis, J.T.; McIntosh, R.P. An upland forest continuum in the prairie-forest border region of Wisconsin. *Ecology* **1951**, *32*, 476–496. [CrossRef]
42. Ma, K.; Huang, J.; Yu, S.; Chen, L. Studies on the Diversity of Plant Communities in Dongling Mountain Area of Beijing: II Abundance, Evenness and Species Diversity. *J. Ecol.* **1995**, *15*, 268–277. (In Chinese)
43. Zhang, L.; Du, H.; Yang, Z.; Song, T.; Zeng, F.; Peng, W.; Huang, G. Topography and Soil Properties Determine Biomass and Productivity Indirectly via Community Structural and Species Diversity in Karst Forest, Southwest China. *Sustainability* **2022**, *14*, 7644. [CrossRef]
44. Wu, X.; Zhu, B.; Zhao, S.; Piao, S.; Fang, J. Comparison of community structure and species diversity of mixed forests of deciduous broad-leaved tree and Korean pine in Northeast China. *Biol. Divers.* **2004**, *12*, 174–181. (In Chinese)
45. Zhang, Y.; Chen, H.Y.; Taylor, A. Multiple drivers of plant diversity in forest ecosystems. *Glob. Ecol. Biogeogr.* **2014**, *23*, 885–893. [CrossRef]
46. Hejda, M.; Pyšek, P.; Jarošík, V. Impact of invasive plants on the species richness, diversity and composition of invaded communities. *J. Ecol.* **2009**, *97*, 393–403. [CrossRef]
47. Pekin, B.K.; Boer, M.M.; Wittkuhn, R.S.; Macfarlane, C.; Grierson, P.F. Plant diversity is linked to nutrient limitation of dominant species in a world biodiversity hotspot. *J. Veg. Sci.* **2012**, *23*, 745–754. [CrossRef]
48. Wright, J.S. Plant diversity in tropical forests: A review of mechanisms of species coexistence. *Oecologia* **2002**, *130*, 1–14. [CrossRef]
49. Özçelik, R.; Diamantopoulou, M.J.; Brooks, J.R.; Wiant, H.V.L., Jr. Estimating tree bole volume using artificial neural network models for four species in Turkey. *J. Environ. Manag.* **2010**, *91*, 742–753. [CrossRef]
50. Santos, A. Tree allometry and crown shape of four tree species in Atlantic rain forest, south-east Brazil. *J. Trop. Ecol.* **2002**, *18*, 245–260. [CrossRef]
51. Yang, J.; Zhang, M.; Zhang, J.; Lu, H.; Hauer, R.J. Allometric Growth of Common Urban Tree Species in Qingdao City of Eastern China. *Forests* **2023**, *14*, 472. [CrossRef]
52. Jenkins, J.C.; Chojnacky, D.C.; Heath, L.S.; Birdsey, R.A. National-scale biomass estimators for United States tree species. *For. Sci.* **2003**, *49*, 12–35. [CrossRef]
53. Sawadogo, L.; Savadogo, P.; Tiveau, D.; Dayamba, S.D.; Zida, D.; Nouvellet, Y.; Oden, P.C.; Guinko, S. Allometric prediction of above-ground biomass of eleven woody tree species in the Sudanian savanna-woodland of West Africa. *J. For. Res.* **2020**, *21*, 475–481. [CrossRef]
54. Peper, P.J.; Alzate, C.P.; McNeil, J.W.; Hashemi, J. Allometric equations for urban ash trees (*Fraxinus* spp.) in Oakville, Southern Ontario, Canada. *Urban For. Urban Green.* **2014**, *13*, 175–183. [CrossRef]
55. Sun, S.; Gao, X.; Bao, W.; Wang, Z. Effects of Plantation Density of Pinus tabulaeformis on the Growth and Community Structure of Pinus tabulaeformis in the Upper Reaches of Minjiang River. *J. Appl. Environ. Biol.* **2005**, *11*, 8–13. (In Chinese)
56. Li, F.; Pan, P.; Ni, J.; Lai, G.; Ouyang, X.; Xu, H.; Guo, L.; Wu, Z.; Yi, Z. Effects of spatial structure of aerial seeding *Pinus massoniana* forest on understory vegetation diversity. *J. Northeast. For. Univ.* **2016**, *44*, 6. (In Chinese)
57. Espinosa, C.I.; Cabrera, O.; Luzuriaga, A.L.; Escudero, A. What Factors Affect Diversity and Species Composition of Endangered Tumbesian Dry Forests in Southern Ecuador? *Biotropica* **2011**, *43*, 15–22. [CrossRef]
58. Schoonmaker, P.; McKee, A. Species composition and diversity during secondary succession of coniferous forests in the western Cascade Mountains of Oregon. *For. Sci.* **1988**, *34*, 960–979. [CrossRef]

59. Király, I.; Ódor, P. The effect of stand structure and tree species composition on epiphytic bryophytes in mixed deciduous–coniferous forests of Western Hungary. *Biol. Conserv.* **2010**, *143*, 2063–2069. [CrossRef]
60. Zhang, C.; Li, X.; Chen, L.; Xie, G.; Liu, C.; Pei, S. Effects of topographical and edaphic factors on tree community structure and diversity of subtropical mountain forests in the Lower Lancang River Basin. *Forests* **2016**, *7*, 222. [CrossRef]

forests

MDPI

Article

Eco-Efficiency Evaluation of Sloping Land Conversion Program and Its Spatial and Temporal Evolution: Evidence from 314 Counties in the Loess Plateau of China

Yuanjie Deng [1,2], Ji Luo [1,2], Ying Wang [1,2], Cuicui Jiao [1,2], Xiaobo Yi [1,2], Xiaosong Su [1,2], Hua Li [3] and Shunbo Yao [3,*]

1 School of Economics, Sichuan University of Science and Engineering, Zigong 643000, China; ecodyj@suse.edu.cn (Y.D.); steelseek@suse.edu.cn (J.L.); hywangying@suse.edu.cn (Y.W.); jiaocuicui@suse.edu.cn (C.J.); yixiaobo@suse.edu.cn (X.Y.); suxiaosong@suse.edu.cn (X.S.)

2 Research Center of Agricultural Economy, Sichuan University of Science and Engineering, Zigong 643000, China

3 College of Economics and Management, Northwest A&F University, Xianyang 712100, China; lihua7485@nwafu.edu.cn

* Correspondence: yaoshunbo@nwafu.edu.cn

Abstract: China's Sloping Land Conversion Program (SLCP) is the largest ecological restoration program (ERP) in the world. Since its full implementation in China in 2002, it has achieved remarkable eco-effects worldwide. However, few researchers have paid attention to the efficiency behind the ecological achievements. Understanding the eco-efficiency of the Sloping Land Conversion Program (EEoSLCP) and its spatial and temporal evolution is necessary for the design and implementation of other ERPs. Therefore, we took the counties reflecting the basic implementation units of the SLCP as the research samples, and evaluated and analyzed the EEoSLCP on the Loess Plateau (LP) and its spatial and temporal evolution based on remote sensing data and county statistics. Our results reveal that: (1) The SLCP in LP has achieved good eco-effects, but the eco-efficiency is generally low. (2) The EEoSLCP of the LP is increasing year by year in time and the spatial distribution pattern is "high in the southeast and low in the northwest" with a gradual decrease in efficiency from southeast to northwest. (3) The EEoSLCP in each county of the LP has a positive spatial autocorrelation and this correlation increases with the passage of time. (4) The EEoSLCP in each county of the LP shows relatively stable geographical spatial agglomeration characteristics of "H-H" and "L-L" in local spatial autocorrelation, and there are spatial neighboring companion effects and spatial neighboring spillover effects in the EEoSLCP in each county of the LP. (5) Natural conditions and redundancy of input and output are important reasons that affect the level of EEoSLCP. Our study will not only provide a general approach and methodological framework for evaluating the eco-efficiency of ERPs and their spatial and temporal evolution, but also provide better guidance and inspiration for the implementation of large-scale ERPs in the background of "The UN Decade on Ecosystem Restoration" and the "carbon peaking and carbon neutrality" strategy.

Keywords: ecological restoration programs (ERPs); Sloping Land Conversion Program (SLCP); eco-effects; eco-efficiency; spatial–temporal evolution; Loess Plateau (LP)

1. Introduction

Over the past 40 years of reform and opening-up, China has experienced far-reaching social changes. Although China's overall economic volume has continued to rise at a rapid pace, it has also revealed a slew of issues, including environmental degradation, resource depletion, and unequal wealth distribution, all of which have hampered the country's economic and social progress [1,2]. Among them, the contradiction between socio-economic development and ecological environment deterioration is particularly prominent.

Nationwide large-scale deforestation and land clearing, and unreasonable use of land resources have led to intensified soil erosion, increased land desertification, degraded soil quality, reduced biodiversity, and rapid ecological environment deterioration [3]. To this end, the Chinese government has launched several ecological restoration programs (ERPs) to respond actively, including implementing the Natural Forest Protection Program, the Sloping Land Conversion Program (SLCP), the Beijing–Tianjin source of wind and sand control program, etc. [4]. Among them, SLCP has been successfully piloted in 3 provinces in China since 1999 and fully implemented in China in 2002, involving a total of 2435 counties in 25 provinces in China [5,6]. By 2020, the Chinese government has invested a total of CNY 517.4 billion (the CNY is the base unit of a number of former and present-day currencies in Chinese; 1 USD equals CNY 6.70 as of January 13, 2023) into the SLCP and has completed a total of 515 million mu (mu is the Chinese unit of land measurement; 1 mu is equal to $1/15$ hm^2) of the SLCP, with 41 million farmers participating in the program, making the SLCP the world's largest ERP [4,7,8].

After implementing the SLCP, researchers have carried out a great deal of research on its ecological performance. The study shows that the SLCP has significantly increased the vegetation coverage [9–11], curbed soil erosion [12,13], increased terrestrial carbon sequestration [14–16], and enhanced the water conservation service [17,18]. There is no doubt that the ecological performance of the SLCP is outstanding in effect, and it has brought substantial ecological benefits. However, few scholars have evaluated the eco-efficiency of the Sloping Land Conversion Program (EEoSLCP), leading us to be unclear about the ecological performance of the SLCP in terms of efficiency. More importantly, considering the low development cost of forest carbon sequestration and great potential for emission reduction, China has made a commitment to the world to a "carbon peaking and carbon neutrality" target, and has clearly indicated that it will continue to implement SLCP and will plan to implement a large number of other ERPs to increase forest area and enhance forest carbon sequestration to help achieve the carbon neutrality target by 2060. In this context, compared with the huge demand for ERP, the capital, land, and other factors that the Chinese government can invest are pretty limited. On the one hand, China, as the largest developing country in the world, still has a strong demand for financial resources in all aspects of socioeconomic development. On the other hand, China is the world's most populous country with a severe shortage of cultivated land per capita and has implemented the most stringent cultivated land protection system to ensure food security. Therefore, the evaluation of the EEoSLCP can provide a reference for improving the new round of SLCP and guide China and other countries in the world to implement other ERPs, thus improving the overall efficiency of the allocation of resources they invest in ERP.

In the evaluation of the ecological performance of the SLCP, most researchers have focused their studies on evaluating the eco-effects of the SLCP, while neglecting to evaluate the EEoSLCP. However, there are still a few scholars who have evaluated the EEoSLCP from two aspects. Firstly, the EEoSLCP was evaluated from the perspective of "cost–benefit". Wang et al. [19] evaluated the "cost–benefit" of the SLCP on ecosystem carbon sequestration based on farmer survey data, and found that the neglect of land productivity and environmental heterogeneity in the program implementation was an important reason for the low EEoSLCP. Chen et al. [20] not only confirmed Wang's conclusion, but also pointed out that ignoring the spatial heterogeneity of opportunity cost and environmental benefit in the SLCP is also an essential reason for the low EEoSLCP. Ning et al. [21] evaluated the EEoSLCP of Yan'an City using the "cost–benefit" method based on remote sensing and GIS technology, and showed that the EEoSLCP of Yan'an City was low from 2000 to 2015. Xian et al. [6] evaluated EEoSLCP from a provincial perspective in China using an improved cost–benefit approach, and found that planting trees in unsuitable locations or not selecting suitable tree species in some provinces was the root cause of the low EEoSLCP. Secondly, some scholars have evaluated EEoSLCP by constructing econometric models. Lu et al. [22] analyzed the influence of SLCP financial inputs on forest carbon sinks using the least squares dummy variable method in Yunnan Province and showed that, although SLCP helps increase forest carbon sinks, the EEoSLCP is low because the impact of natural resource endowments on reforestation of the program is not considered in the implementation of SLCP in each locality. Zhang et al. [23] used a geo-weighted regression (GWR) model to explore the impact of investment in SLCP on the Enhanced Vegetation Index (EVI) in Zhidan County and Wuqi County, China. The results showed that there were significant spatial differences in EEoSLCP among villages, and the spatial mismatch between investment and local resource endowment is an important reason for low EEoSLCP. Liu et al. [24] used the fixed-effect model to investigate the impact of financial input on the Normalized Difference Vegetation Index (NDVI) in Shaanxi Province, and further analyzed the conversion rate between financial input and ecological output using the panel threshold model. The results show that the EEoSLCP in Shaanxi Province is characterized by "high in the middle and low at both ends", and the conversion rates between financial inputs and ecological outputs varied widely under different precipitation conditions.

Although the studies of previous scholars can provide us with a reference for evaluating EEoSLCP, there are three main shortcomings. First of all, in EEoSLCP evaluation methods, although the "cost–benefit" method is simple and easy to implement and has a wide range of application, it is difficult to provide decision makers with an optimal set of "cost–benefit" ratios for improving efficiency from the perspective of optimizing the allocation of input and output factors [25,26]. Although the researchers used econometric models to compensate for the shortcomings of the "cost–benefit" method to a certain extent, the results can only reflect the direction and intensity of the relationship between the inputs and outputs of the SLCP. Furthermore, the efficiency measured by econometric models generally belongs to the parametric method, which relies on the choice of production function and is difficult to solve the efficiency calculation of multiple inputs and multiple outputs. Secondly, in the selection of indicators for efficiency measurement, researchers mostly take the financial expenditure of the SLCP as an input, while ignoring other elements such as land and labor. In the selection of output indicators, vegetation cover or vegetation NDVI is often taken as the output indicator, ignoring that the core objectives of SLCP are to reduce soil erosion and increase ecosystem service functions. It can be concluded that the researchers' unreasonable selection of indicators for calculating the EEoSLCP will make it difficult to make a scientific and accurate assessment of the EEoSLCP. Finally, in terms of an efficiency analysis perspective, most of the existing literature is based on the description of phenomena and mechanism explanation from management or economics perspectives, and it lacks the identification of the spatial pattern evolution and the portrayal of the temporal dynamic evolution of the EEoSLCP from a two-dimensional perspective in time and space.

To bridge these gaps, this study makes the following contributions. First, the data envelopment analysis (DEA) model was used to evaluate the EEoSLCP from the perspective of multiple inputs and multiple outputs, which not only made up for the methodological shortcomings of previous studies, but also pointed out the reasons for the non-DEA-effective or weak-DEA-effective counties in the EEoSLCP and the direction and extent of improvement using the projection method. Second, based on multi-type data such as land use, vegetation NDVI, and soil type, various eco-effects of SLCP were obtained with the support of GIS and calculated by models such as pixel dichotomy and the revised universal soil loss equation (RUSLE). Furthermore, they are used as ecological output indicators of the SLCP in order to be able to evaluate the EEoSLCP scientifically and accurately. Third, kernel density estimation (KDE) and exploratory spatial data analysis (ESDA) were used to portray and identify the temporal dynamic evolution and spatial correlation of the EEoSLCP, respectively, which comprehensively reveals the spatial and temporal evolution characteristics of the EEoSLCP.

2. Materials and Methods

2.1. Study Area

LP is located in the middle and upper reaches of the Yellow River Basin in China (107°28′~111°15′ E 35°21′~39°34′ N), spanning seven provinces and regions in Shanxi,

Ningxia, Shaanxi, Gansu, Inner Mongolia, Qinghai, and Henan, with a total area of about 6.49×10^5 km^2, accounting for 6.7% of China's land area (Figure 1a). LP terrain slopes from northwest to southeast, with elevation differences exceeding 3000 m. The climate is temperate continental monsoon climate with cold winters and warm humid summers. The average annual temperature ranges from 3.7 to 14.0 °C, and the annual precipitation is 144 to 812 mm, mainly between June and September [27]. The LP is one of the most severe soil erosion and fragile ecosystems in the world due to frequent heavy rainfall, steep topography, low vegetation cover and loose soils [28]. At the same time, the over-exploitation and unreasonable utilization of the resources on the LP have aggravated the soil erosion and led to the serious degradation of land and ecosystem [29,30]. At the end of 1990s, the LP took the lead in implementing the SLCP as a pilot. Through afforestation and other measures, the vegetation coverage was effectively increased, the serious problem of soil erosion was alleviated, and the ecosystems have been greatly restored and improved [31]. By 2015, a total of 154 million mu of cultivated land had been converted to forests on the LP (Figure 1c), with a cumulative financial investment of CNY 61 billion (Figure 1b) and a cumulative participation of 0.9 billion households (Figure 1d).

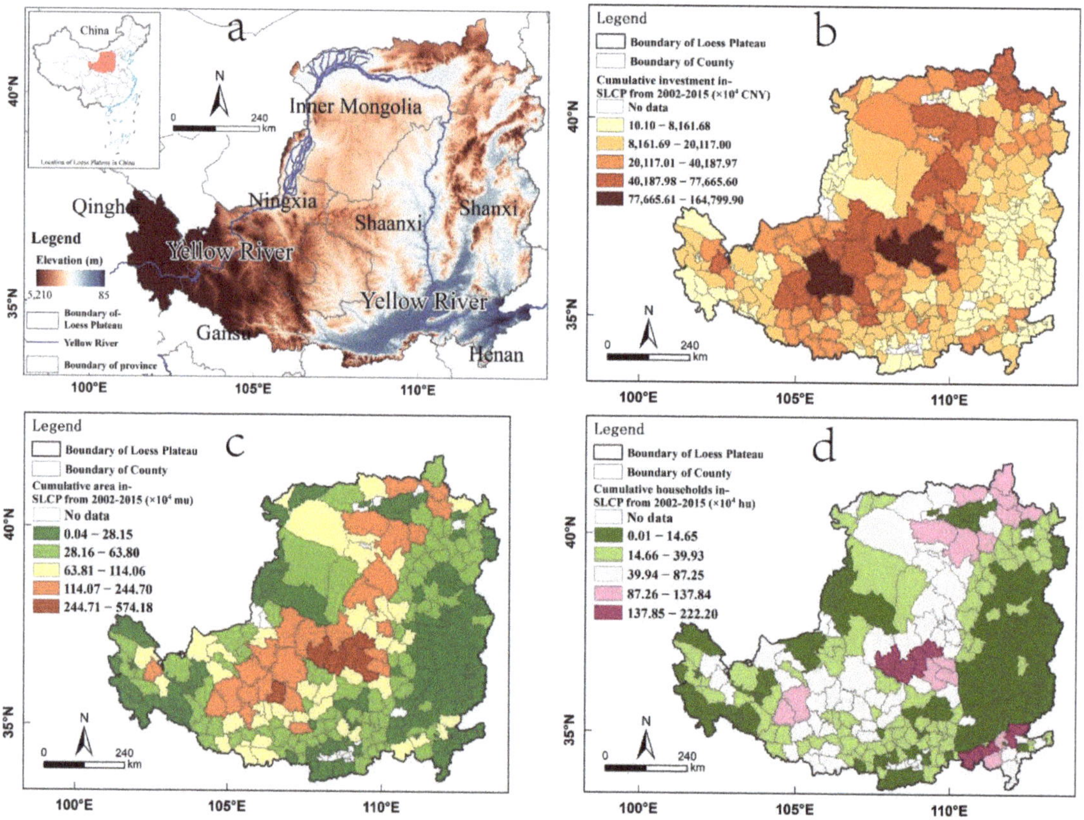

Figure 1. (**a**) Location and topography of the LP; (**b**) cumulative investment in SLCP; (**c**) cumulative area in SLCP, and (**d**) cumulative households in SLCP, 2002 to 2015.

2.2. Methodology

Figure 2 shows the research framework of this study for the evaluation of EEoSLCP and its spatial and temporal variation. Firstly, based on the definition of the EEoSLCP, the input-output indexes for evaluating the EEoSLCP were determined. Secondly, based on multi-type data such as land use, vegetation NDVI and soil type, various eco-effects of SLCP were obtained with the support of GIS and calculated by models such as pixel dichotomy and RUSLE. Finally, based on the input-output data of SLCP, the DEA-BCC model is applied to measure EEoSLCP, and based on this, the KDE and ESDA models are applied to analyze the spatial and temporal changes of EEoSLCP comprehensively, respectively.

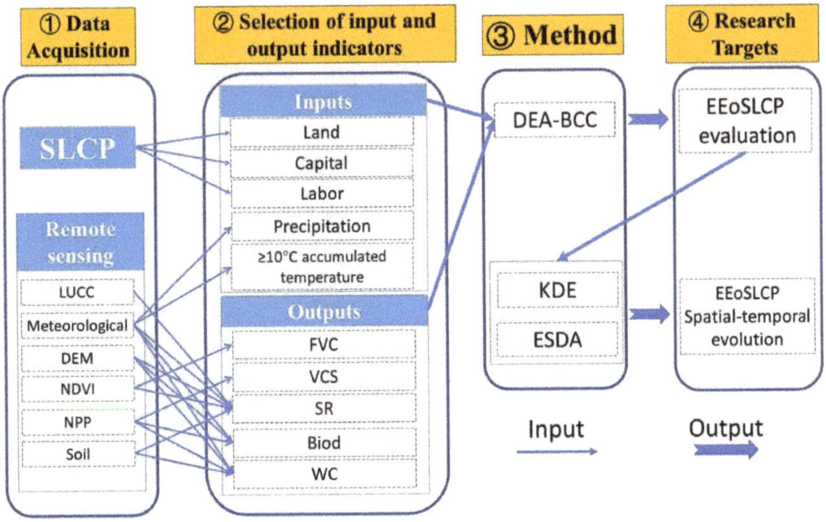

Figure 2. EEoSLCP evaluation and its spatial and temporal variation research framework.

2.2.1. Construction of EEoSLCP Evaluation Index System

Based on the definition of eco-efficiency and the characteristics of the SLCP in practice, we define EEoSLCP as: restoring degraded ecosystems with less resource input in SLCP implementation, producing more ecosystem services to meet human needs, and making the ecosystem environment better. Based on this, we constructed the EEoSLCP evaluation index system in terms of both inputs and outputs (Table 1). For the inputs, we chose three indicators closely related to SLCP: capital, land, and labor. At the same time, considering SLCP as a reforestation program, its ecological benefits are not only influenced by the input of the program itself but also by the natural conditions [32,33]. Numerous studies have shown that temperature and precipitation are the main natural factors that determine the effectiveness of SLCP on the LP [34,35]. For this reason, we also selected precipitation and temperature as input indicators. In terms of output, we selected five indicators that best reflect the ecological changes on the LP, namely: vegetation fraction cover (VFC), vegetation carbon sequestration (VCS), soil retention (SR), biodiversity (Biod) and water conservation (WC). The specific process and reasons for the selection of EEoSLCP evaluation indicators are explained in detail in our published paper [33]. The references involved in the selection of evaluation indicators are: [10,13–16,18,32,36–41].

2.2.2. Eco-Effects Calculation of SLCP

We calculated and analyzed the spatial and temporal changes of eco-effects such as VFC and VCS on the LP from 2002 to 2018 after the implementation of SLCP by collecting various remote sensing data obtained from land use, vegetation NDVI, DEM, and meteorology, using GIS technology. Please refer to the supplementary materials [33] of our published paper [33] for the specific formulae and results of the five eco-effects of VFC, VCS, SR, WC, and Biod in the Loess Plateau from 2002 to 2018. The references for the calculation of the five eco-effects of the Loess Plateau are: [42–59].

Table 1. Input and output indicators of EEoSLCP.

Indicator	Variable	Variable Description	Unite
Input	Land	Cumulative area of SLCP implementation	mu *
	Capital	SLCP cumulative financial investment	CNY *
	Labor	Cumulative number of SLCP participating households	hu *
	≥10 °C accumulated temperature	Average annual ≥10 °C accumulated temperature	°C
	Precipitation	Average annual precipitation	mm
Output	VFC	Cumulative increase in average VFC compared to 2002	%
	SR	Cumulative increase in average SR compared to 2002	$t{\cdot}hm^{-2}{\cdot}yr^{-1}$
	VCS	Cumulative increase in average VCS compared to 2002	$gC{\cdot}m^{-2}{\cdot}yr^{-1}$
	WC	Cumulative increase in average WC compared to 2002	dimensionless, value range 0–1
	Biod	Cumulative increase in average Biod compared to 2002	dimensionless, value range 0–1

*: mu is the Chinese unit of land measurement; 1 mu is equal to 1/15 hm^2. The CNY is the base unit of a number of former and present-day currencies in Chinese; 1 USD equals CNY 6.70 as of 13 January 2023 (https://www.federalreserve.gov/releases/h10/current/ (accessed on 13 January 2023)). According to the 7th national census bulletin released by the China Statistics Bureau, 1 hu equals approximately 2.62 people. (http://www.stats.gov.cn/tjsj/tjgb/rkpcgb/qgrkpcgb/202106/t20210628_1818821.html (accessed on 13 January 2023)).

2.2.3. Ecological Efficiency Calculation of SLCP

(1) DEA model

The core idea of the DEA model is to keep the input or output values of the decision units constant, project the decision units onto the effective frontier surface with the help of the linear programming principle, and then compare the deviation of the decision units from the effective frontier surface to evaluate their relative effectiveness. Using this method, we can not only derive the efficiency between input and output variables, but also the direction from which we need to improve each input variable in order to enhance the overall efficiency. DEA models are divided into CCR models with constant returns to scale and BCC models with variable returns to scale [60–62]. Considering that the ecological benefits obtained from the implementation of SLCP are characterized by variable payoffs to scale, we will complete the measurement of EEoSLCP for the LP using the DEA-BCC model. The specific model is as follows.

$$\min\theta_z - \varepsilon\left(\sum_{i=1}^{m} s_i^- + \sum_{r=1}^{s} s_r^+\right) \tag{1}$$

$$s.t.\sum_{j=1}^{n} \lambda_j x_{ij} + s_i^- = \theta_z i x_{rz}, i = 1,2,\cdots,m \tag{2}$$

$$\sum_{j=1}^{n} \lambda_j y_{rj} - s_i^+ = y_{rz}, r = 1,2,\cdots,s \tag{3}$$

$$\sum_{j=1}^{n} \lambda_j = 1 \tag{4}$$

$$\theta_0, \lambda_j, s_i^-, s_i^+ \geq 0 \tag{5}$$

where θ_z is the EEoSLCP in the LP under VRS, where n = 314 is the number of counties in the LP, j represents the jth county, which can also be called the jth DMU. m and s, respectively, represent the number of input–output indicators of SLCP, x_{ij} represents the input of the i factor in the jth county, y_{rj} represents the r output in the jth county, and λ_j

is the weight coefficient of the input index of a certain factor in the jth county. When the sum of S_i^- and S_r^+ is 0, all input and output are in a relaxed state, $\theta_z = 1$, and the EEoSLCP is in complete efficiency. When the sum of S_i^- and S_r^+ is not 0, $\theta_z < 1$, the EEoSLCP is in incomplete efficiency, and the complete efficiency can be achieved again by adjusting the level of input or output.

(2) Kernel Density Estimation (KDE)

KDE is an important method to study spatial disequilibrium distribution, which is surface interpolation through discrete sampling points [63]. It is a nonparametric estimation method in which the position, shape, and extensibility of random variables are described by continuous smooth density curves instead of histograms. The kernel function mainly includes linear kernel function, polynomial kernel function, Gaussian kernel function, etc. In this paper, we choose the more commonly used Gaussian kernel function. Assuming that the density function of the random variable x is $f(x)$, the kernel density is estimated as [64]:

$$f(x) = \frac{1}{Nh}\sum_{i=1}^{N} K\left(\frac{X_i - \overline{x}}{h}\right) \tag{6}$$

$$K(x) = \frac{1}{\sqrt{2\Pi}} \exp\left(-\frac{x^2}{2}\right) \tag{7}$$

where $K(\cdot)$ is the kernel function; N is the number of observations; X_i is the independent observation with the same distribution; x is the mean; h is the bandwidth, which determines the smoothness and accuracy of the estimated density curve. The larger the bandwidth, the smaller the variance of the kernel estimation and the smoother the curve.

(3) Exploratory Spatial Data Analysis (ESDA)

ESDA is a collection of a series of spatial data analysis methods and techniques, which takes spatial association measure as the core, and distinguishes spatial agglomeration and spatial differentiation by describing and visualizing the spatial distribution pattern of things or phenomena, so as to reveal the spatial dependence among objects. ESDA consists of global spatial autocorrelation (GSA) and local spatial autocorrelation (LSA) [65].

GSA is mainly used to judge whether a phenomenon has clustering characteristics in space, which is generally measured by Moran's I index, and its calculation formula is [66]:

$$I = \frac{\sum_{i-1}^{n}\sum_{j-1}^{n} W_{ij}(Y_i - \overline{Y})(Y_j - \overline{Y})}{S^2 \sum_{i=1}^{n}\sum_{j=1}^{n} W_{ij}} \tag{8}$$

Among them, $S^2 = \frac{1}{n}\sum_{i=1}^{n}(Y_i - \overline{Y})^2, \overline{Y} = \frac{1}{n}\sum_{i=1}^{n} Y_i$, Y_i is the EEoSLCP for county i, n is the number of samples, and the sample number in this study refers to the 314 counties in the LP; Y is the average of EEoSLCP in all counties; W_{ij} is the spatial weight matrix, which represents the influence factors between counties i and j and constitutes a complete set of spatial relationships. Moran's I value range is $[-1,1]$. When I is greater than 0, it indicates that there is a positive spatial correlation between regions and, the closer it is to 1, the more similar observed values (high or low values) tend to gather in space. When I is less than 0, it shows that there is a spatial negative correlation between regions and, the closer it is to -1, the more scattered the similar observations tend to be in space. When I is equal to 0, it shows that there is no spatial correlation between regions and the observed values tend to be randomly distributed.

GSA mainly explains the average correlation degree of a phenomenon in space, but cannot clearly reflect which areas in the study area have agglomeration. LSA is an index to measure the spatial correlation of the research object from the local perspective, which measures the similarity between the regional observations in this area and the observations in neighboring areas, and can be used to identify the agglomeration and discrete characteristics of the local spatial pattern [65]. It is usually measured by local Moran's I index and its calculation formula is [66]:

$$I_i = Z_i \sum_{j=1}^{n} W_{ij} Z_j \quad (9)$$

where $Z_i = Y_i - \overline{Y}$, $Z_j = Y_j - \overline{Y}$, Y_i, Y_j denote the observed values of the ith and jth region, which is the EEoSLCP in this study, n is the number of samples, the sample number in this study refers to the 314 counties in the LP, and W_{ij} is the spatial weight matrix. The local Moran's I index measures the degree of correlation between the ith region and other surrounding areas. If $I_i > 0$, it means that the local area neighboring units are spatially clustered with similar values; if $I_i < 0$, it means that the local area neighboring units are not spatially clustered with similar values; if $I_i = 0$ means that the local area neighboring units are spatially randomly distributed. Specifically, LSA analysis can be divided into the following four types: high–high cluster (H-H), high–low outlier (H-L), low–high outlier (L-H), and low–low cluster (L-L).

2.2.4. Data Sources

The data of this study include remote sensing data, land use data, meteorological data, soil data, and other multi-source data sets. All the raster data below are processed with a resolution of 500 m and the coordinate system is Krasovsky_1940_Albers. The data sources of this study are as follows:

The land use map data of 2000, 2005, 2010, and 2018 were downloaded from the Data Center for Resources and Environmental Sciences, Chinese Academy of Sciences (RESDC) (https://www.resdc.cn/ (accessed on 5 May 2021)), including cultivated land, forests, grassland, water area, construction land, and unused land. The spatial resolution of the data was 30 m. The land use data of each period were verified and modified by field survey, and the combined precision of the data in 2000 was 92.9% [67], in 2005 was 95% [68], in 2010 was 94.3% [69], and in 2018 was 95.53% [70].

In the meteorological data, precipitation and temperature from 2002 to 2018 was obtained from the Chinese National Meteorological Science Data Service Center (http://data.cma.cn/ (accessed on 6 September 2020)). Moreover, ≥10 °C accumulated temperature from 2002 to 2015 was obtained from the National Ecosystem Science Data Center (http://www.nesdc.org.cn (accessed on 5 May 2021)). Furthermore, via ArcGIS software using the Kriging interpolation method, we obtained the meteorological raster dataset.

The digital elevation model (DEM) data used in the study is ASTER GDEM V2, with a spatial resolution of 30 m, downloaded from Japan Space Systems.

The NDVI from 2002 to 2018 was derived from the MOD13A1 product synthesized by the maximum value composite (MVC) method 16d and downloaded from the NASA MODIS website with a spatial resolution of 500 m (https://ladsweb.modaps.eosdis.nasa.gov/search (accessed on 5 May 2021)).

The net primary productivity (NPP) from 2002 to 2018 was derived from the MOD17A3HGF database of the Terra Net Primary Production and downloaded from the NASA MODIS website with a spatial resolution of 500 m (https://ladsweb.modaps.eosdis.nasa.gov/search (accessed on 5 May 2021)).

Soil attribute data were obtained from the China soil map based harmonized world soil database (HWSD) (v1.1) (2009) shared by the National Tibetan Plateau/Third Pole Environment Data Center (https://data.tpdc.ac.cn/ (accessed on 5 May 2021)), which was provided by the Second National Land Survey data from the Nanjing Institute of Soil Research, Chinese Academy of Sciences.

The data of SLCP financial investment and project implementation area of 314 counties in the LP from 2002 to 2015 were obtained from the Central South Survey and Planning Institute under the National Forestry and Grassland Administration of China (http://www.forestry.gov.cn/sites/zny/zny/ (accessed on 5 May 2021)). The number of farm households participating in the SLCP is based on provincial data published in the China Forestry Statistical Yearbook (2003–2016) and obtained by proportional conversion of financial investment data in each county.

Administrative boundaries of the cities are obtained from the national 1:1,000,000 database of China Geographic Information Monitoring Platform (https://www.webmap.cn (accessed on 5 May 2021)).

3. Results

3.1. Eco-Effects of SLCP

From 2002 to 2018, after implementing SLCP on the LP, the ecological afforestation effect was remarkable, the ecological environment was significantly improved, and the growth of each eco-effect was remarkable. Firstly, from the temporal changes of various eco-effects (Figure 3a–e), the average VFC increased from 50.12% to 64.15%, with a growth rate of 27.99%; the average VCS increased from 883.54 $gC{\cdot}m^{-2}{\cdot}yr^{-1}$ to 1296.35 $gC{\cdot}m^{-2}{\cdot}yr^{-1}$, with a growth rate of 46.72%; the average SR increased from 221.11 $t{\cdot}hm^{-2}{\cdot}yr^{-1}$ to 228.41 $t{\cdot}hm^{-2}{\cdot}yr^{-1}$, with a growth rate of 3.3%; the average WC increased from 0.1458 to 0.1643, with a growth rate of 12.73%; and the average Biod increased from 0.1549 to 0.1815, with a growth rate of 17.12%. Secondly, from the spatial change trend of the eco-effects of the LP from 2002 to 2018 (Figure 3f–j), the eco-effects of the southwest, central, and northeast of the LP have been significantly improved.

3.2. Temporal Changes and Evolution of EEoSLCP

The EEoSLCP of the LP and its 314 counties from 2002 to 2015 was measured using the DEA-BCC model selected by MaxDEA software. Figure 4 shows the average score of EEoSLCP for LP from 2002 to 2015. Firstly, it was found that the average EEoSLCP score for the LP from 2002 to 2015 was low, with an overall average score of only 0.413. Secondly, the EEoSLCP presented an overall growth trend from 0.305 in 2002 to 0.531 in 2015, indicating that EEoSLCP has been effectively improved over time. Finally, based on the trend of EEoSLCP changes on the LP during 2002–2015, the temporal changes of EEoSLCP can be divided into three stages in conjunction with the implementation of SLCP during the study period: the first stage was from 2002 to 2006, in which the EEoSLCP increased year by year; the second stage was from 2006 to 2010, in which the EEoSLCP decreased first (2006–2007) and then increased (2007–2010); the third stage was from 2010 to 2015, in which the EEoSLCP also decreased first (2010–2012) and then increased (2012–2015).

Figure 5 is obtained after kernel density calculation by Matlab R2018a (Matlab software from MathWorks, Inc., Natick, MA, USA).

(1) In terms of the distribution position, the overall EEoSLCP on the LP shows a "rightward" distribution from left to right, with peaks from high to low, and the curve tends to flatten out with increasing years, with a small shift to the right, indicating that the number of counties with high eco-efficiency is increasing and the number of counties with relatively low efficiency is decreasing, implying that the eco-efficiency of each county is steadily improving over time.

(2) From the distribution pattern, the overall distribution curve of EEoSLCP on the LP shows the evolution pattern of "two peaks standing side by side" composed of "one main peak plus one side peak". It shows that the EEoSLCP in each county of the LP is always in the pattern of polarization during the study period and the eco-efficiency of some counties is concentrated at a higher level, while that of other counties is concentrated at a lower level. From 2002 to 2015, the evolutionary dynamics of the main peak height showed an overall decrease over time and an increase in the width of the main peak, indicating that the absolute gap in efficiency tends to widen in counties clustered at a lower level of EEoSLCP. The evolution of the height of the lateral peaks shows an overall increase over time and a gradual decrease in width, indicating that the absolute difference in efficiency tends to decrease in counties clustered at a higher level of EEoSLCP.

(3) In terms of polarization characteristics, the EEoSLCP in the LP from 2002 to 2015 shows the phenomenon of "bimodality" and polarization, with relatively large distances between the side peaks and the main peaks, and obvious differences in eco-efficiency between counties on the LP, gradually forming a "bimodal" evolution pattern of "low–low agglomeration and high–high agglomeration", which is similar to "club convergence".

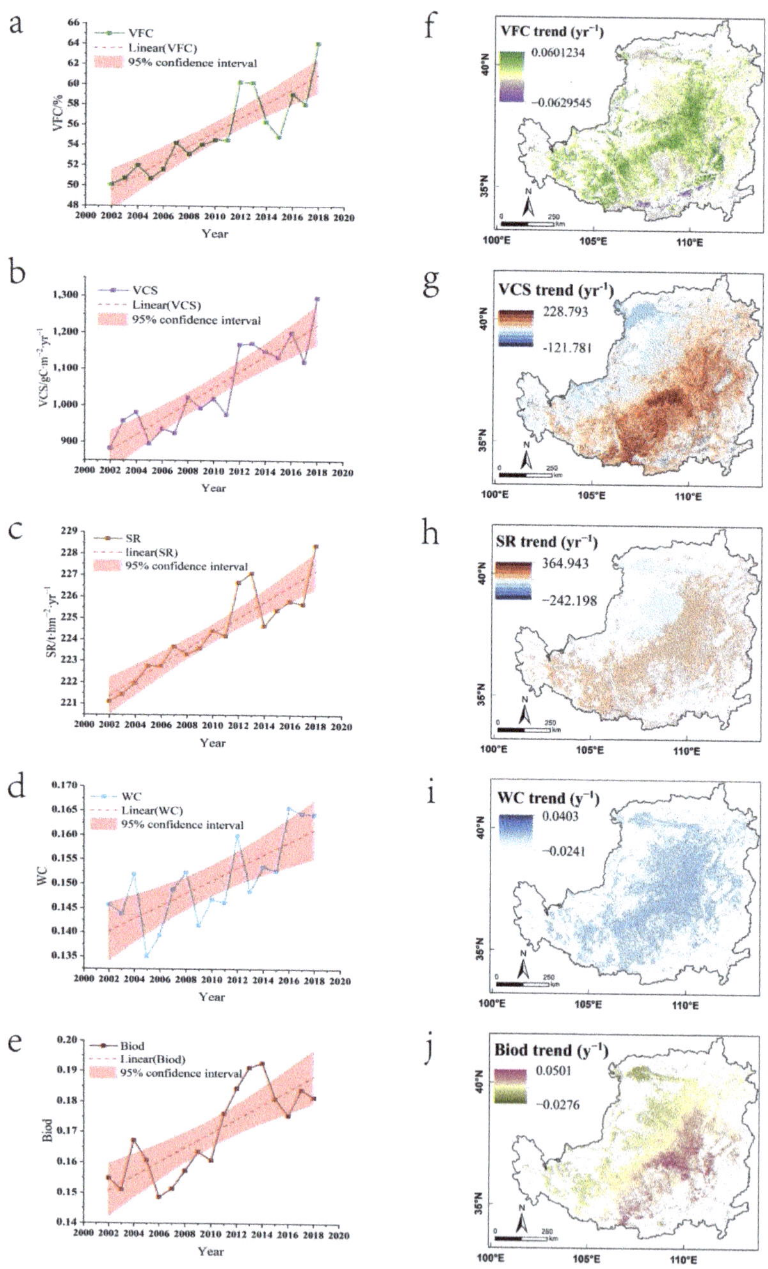

Figure 3. Variation of (**a**) VFC, (**b**) VCS, (**c**) SR, (**d**) WC, and (**e**) Biod from 2002 to 2018; (**f**) VFC, (**g**) VCS, (**h**) SR, (**i**) WC, and (**j**) Biod have changed significantly ($p < 0.05$) from 2002 to 2018.

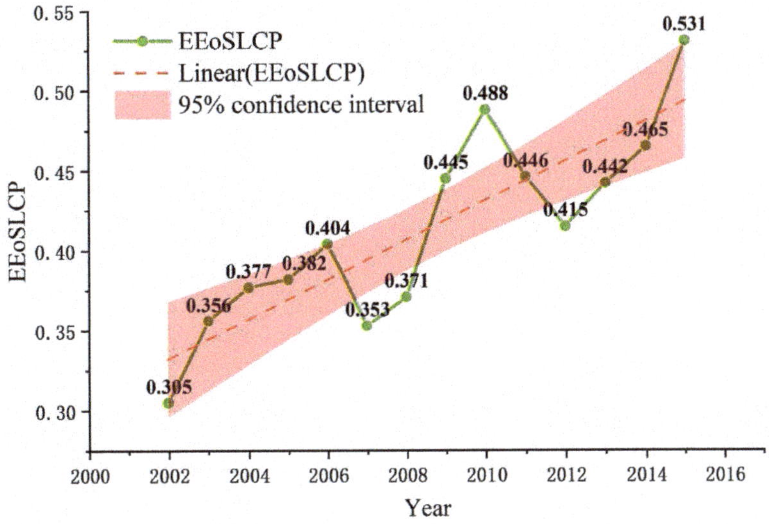

Figure 4. The average value of EEoSLCP, 2002 to 2015.

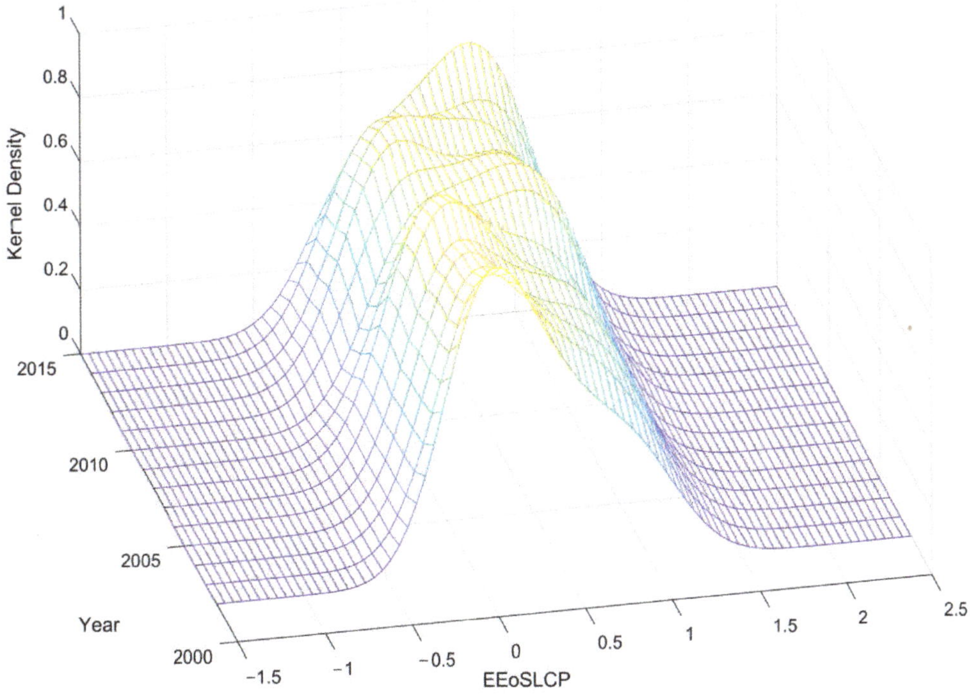

Figure 5. Three-dimensional kernel density map of EEoSLCP from 2002 to 2015.

3.3. Spatial Distribution of EEoSLCP

Based on the EEoSLCP of each county in the LP from 2002 to 2015, the EEoSLCP of each county was classified into four states of low efficiency (0, 0.25], medium–low efficiency (0.25, 0.5], medium–high efficiency (0.5, 0.75], and high efficiency (0.75, 1] with reference to previous studies [33]. From Figure 6, it can be seen that the EEoSLCP of most counties in the LP is low-efficiency during the study period, mainly distributed in the central, northern, and western regions in a contiguous manner. There are fewer medium–low efficiency counties in the study period, which are mainly distributed in the middle of high-efficiency and medium-high efficiency counties in the southern and eastern regions, distributed in a northeast–southwest strip. Fewer counties have medium–low efficiency status, and they are mainly concentrated in the middle of high efficiency and medium–high efficiency counties in the southern and eastern regions, distributed in a northeast–southwest strip. The spatial distribution areas of high efficiency and medium–high efficiency counties are roughly the same, mainly distributed in the eastern, southern, and western regions. Overall, although the EEoSLCP has obvious spatial heterogeneity among counties in the LP, it presents a distribution trend that gradually decreases from southeast to northwest and from south to north.

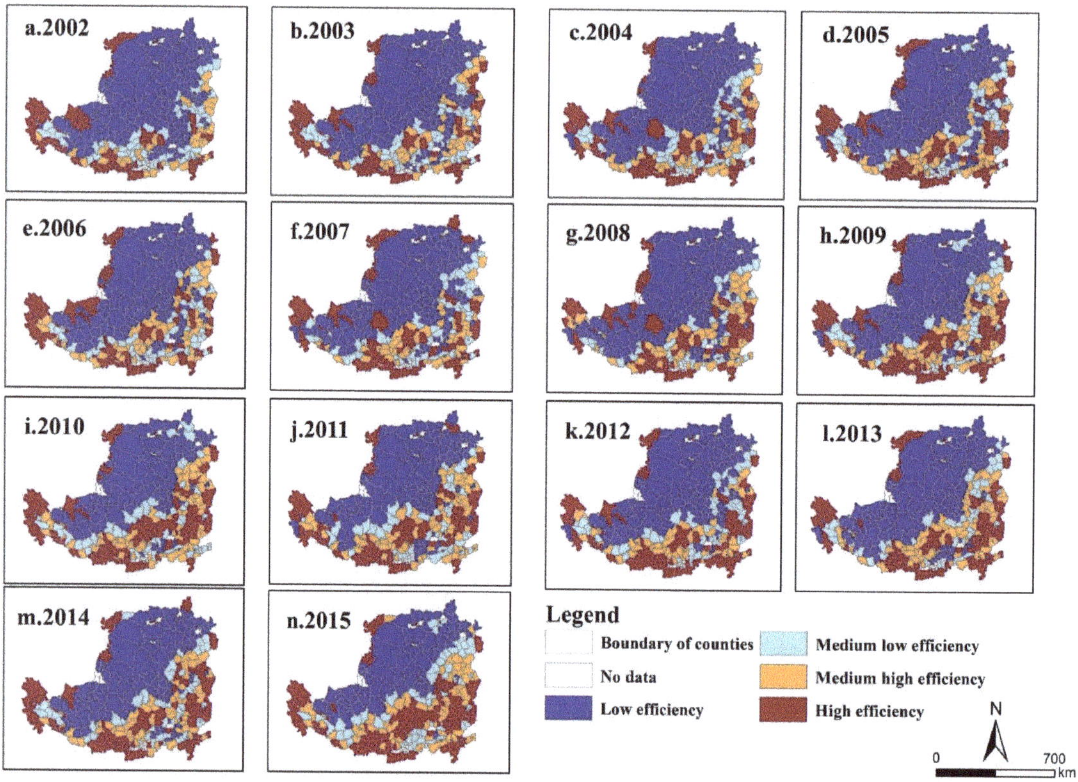

Figure 6. Spatial distribution of EEoSLCP by counties in the LP from 2002 to 2015.

From the number of counties divided into efficiency status intervals (Appendix A, Figure A1), the number of counties with low EEoSLCP showed a fluctuating downward trend, from 161 in 2002 to 76 in 2015, mainly concentrated in the southeastern part of the LP, and the range of counties with low efficiency shrank from southeast to northwest year by year. The number of counties with high efficiency and medium efficiency showed a fluctuating upward trend, from 60 and 35 in 2002 to 114 and 69 in 2015, respectively, mainly located in the central and eastern regions of the LP. The number of medium–low efficiency counties shows a stable and unchanging trend, basically stable at about 50.

3.4. Spatial Correlation of EEoSLCP

3.4.1. GSA Analysis

Based on the ESDA model, ArcGIS software was used to calculate the global Moran's I index of EEoSLCP in the LP from 2002 to 2015. From Figure 7, it can be concluded that the global Moran's I index was positive, ranging from 0.41 to 0.57, and all of them passed the significance test at the 1% level (Appendix A, Table A1), indicating that the EEoSLCP in each county of the LP showed positive spatial autocorrelation. Its spatial connection characteristics are as follows: the counties with higher eco-efficiency tend to be adjacent to the counties with higher eco-efficiency, while the counties with lower eco-efficiency tend to be adjacent to the counties with lower eco-efficiency; that is, the adjacent counties mostly show the spatial agglomeration characteristics of "high–high" or "low–low", but seldom show the spatial agglomeration characteristics of "high–low" or "low–high". The spatially positive autocorrelations of the EEoSLCP in each county of the LP show a relatively stable trend. The positive spatial correlation of the EEoSLCP decreased from 2002 to 2006, increased from 2006 to 2007, decreased again from 2007 to 2008, first increased, then decreased and then increased from 2008 to 2012, then decreased from 2012 to 2013, and finally increased from 2013 to 2015. Its spatial pattern is more obviously clustered in all years except for two years in 2006 and 2008 when it showed weak clustering distribution. The global Moran's I value was the lowest in 2006 at 0.41 and the highest in 2015 at 0.57. This indicates that the spatial agglomeration state of the EEoSLCP of the Loess is relatively stable but also has some fluctuations. Specifically, the relative positions of the spatial pattern of the EEoSLCP in neighboring counties are basically constant, but their relative differences are variable and show certain strong and weak variations.

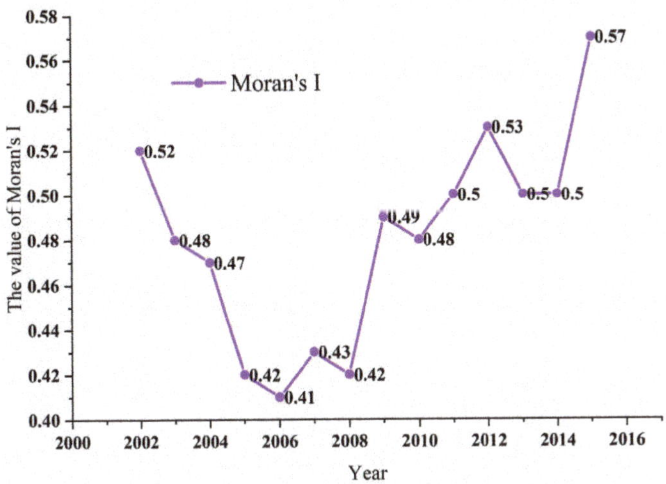

Figure 7. Global Moran's I of the EEoSLCP, 2000 to 2015.

3.4.2. LSA Analysis

The global Moran's I index can prove that there is a positive spatial correlation in the EEoSLCP on the LP, but it cannot reflect its local spatial characteristics. Therefore, the local Moran's I index is needed to identify the local spatial characteristics of the EEoSLCP in each county of the LP. According to Figure 8 and Appendix A, Figure A2, the specific local spatial pattern and distribution quantity can be classified into the following four categories:

(1) H-H clustering, which indicates that, if a county has high EEoSLCP, then its neighboring counties also have high efficiency, mainly concentrated in the western, southern, and eastern regions, which is a high level of the spatial equilibrium associated agglomeration state of "high in the center and high around". The number of H-H in 2002 to 2015 showed an overall fluctuating upward trend in time and, in space, it showed that the range of the H-H clusters in the west remained basically unchanged, while the H-H clusters in the south expanded and spread to the north and northeast, and the H-H clusters in the southeast corner region expanded and spread to the south and to the west.

(2) L-H outlier, which indicates that, although the EEoSLCP of a county is low, the efficiency of its neighboring counties is high, sporadically distributed in the western and eastern parts of the LP, showing a spatially unbalanced correlated agglomeration of "low in the center and high in the surroundings". This trend is spatially distributed in the western and eastern regions.

(3) L-L clustering, which indicates that not only is the EEoSLCP of a county low but also the efficiency of its neighboring counties is low, mainly concentrated in the central, northern, and northeastern areas of the LP, with an overall "southwest–northeast" distribution, showing a low level of spatial equilibrium associated with the clustering state of "low in the center and low around". The number of L-L in 2002–2015 showed a fluctuating decline in time and a gradual decrease in space from the southeast to the northwest, but the L-L cluster in the southwest showed a continued spreading and expanding trend to the southwest.

(4) H-L outlier, which indicates that, although a county has high EEoSLCP, its neighboring counties have low efficiency, mainly scattered in the central and eastern parts of the LP, showing a spatially unbalanced correlated clustering state of "high in the center and low around". The number of such counties is small, and the temporal change increases then decreases, and this trend is spatially distributed in the central region.

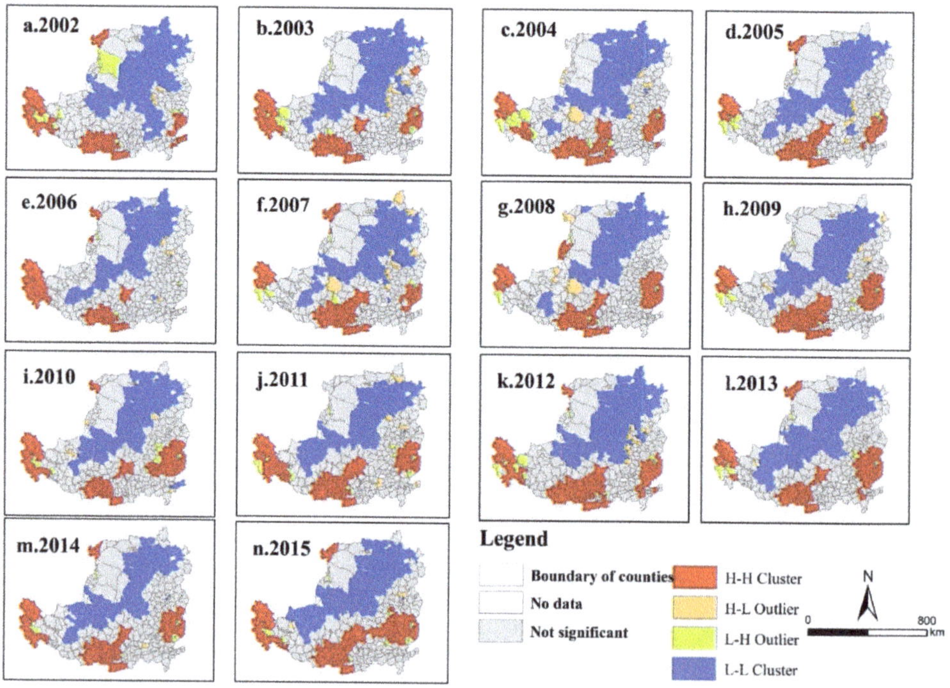

Figure 8. The LISA cluster distribution of EEoSLCP, 2002 to 2015.

Overall, the number of local spatial characteristics of the EEoSLCP in the LP during the study period was L-L > H-H > L-H > H-L. Further analysis of the local spatial pattern of EEoSLCP in each county shows that there is a strong spatial neighboring companion effect of EEoSLCP among the counties in the LP. Specifically, in the LP, when the EEoSLCP around a county is at a high (low) level, its own efficiency tends to reach a high (low) level.

4. Discussion

4.1. Analysis of Temporal Changes in EEoSLCP

Our research shows that the ecological environment of the LP has been greatly improved after the implementation of SLCP. Good ecological benefits have been achieved in VFC, VCS, SR, and other aspects, which is consistent with the research results of previous scholars [9–18]. However, from the perspective of efficiency, we found that the average EEoSLCP score for the LP from 2002 to 2015 was low, with an overall average score of only 0.413. The reason for this is that SLCP is a top-down, government-led ERP [7], and the implementation of the program involves a wide range of tasks and high construction requirements, although the construction of the project has achieved good eco-effects. However, the lack of scientific proof, empirical guidance, and a scientific supervision system for project implementation, coupled with the generally low compensation standard for households, makes it difficult to create a sustainable incentive for the households to participate in the program, which eventually leads to low EEoSLCP [71,72]. However, from the temporal changes of EEoSLCP, it can be obtained that, from 2002 to 2015, the overall EEoSLCP on LP has been growing and the growth rate has gradually accelerated. The reason is that 2002–2006 is still the initial stage of the SLCP implementation, although a lot of investment, land, and households have been input; however, because of the irrational planting structure of economic forests (the planting proportion of economic forests is far greater than ecological forests) [73], the low quality of afforestation [74,75] and the low subsidy standard [71,74], etc. As a result, eco-efficiency is always on the low side at this stage and the growth rate is slow. The period 2007–2010 was the optimization and adjustment phase of SLCP and the problems that existed before were greatly solved by learning the lessons from the implementation of SLCP in the previous phase. More importantly, the State Council decided to increase cash subsidies for participating SLCP households and to prolong the subsidy period for those who are enjoying the subsidy expiration [76,77], and also decided to give households a certain degree of autonomy in SLCP participation [78], which greatly motivated households to participate in the SLCP, improved the afforestation quality of SLCP implementation, and finally achieved greater ecological benefits [79]. Therefore, EEoSLCP was able to ramp up rapidly during this stage. The period 2011–2015 was the consolidation and improvement stage of the SLCP, and the EEoSLCP decreased and then increased in this stage. The reason is that, since 2011, the main task of SLCP has been adjusted to consolidate the ecological achievements made in the previous period but, in practice, due to the lack of subsidy distribution, lack of supervision, and backward management techniques, some of the trees planted in the previous period have died and some districts and counties have even experienced the phenomenon of deforestation and replanting [74,80,81]. However, the central government still provides financial subsidies according to the completed area of SLCP in each county, resulting in a decrease in the input–output ratio and a decline in eco-efficiency from 2011 to 2012. After 2013, local governments began to subsidize households participating in SLCP with their own funds so as to further improve the results of SLCP and strengthened the management of the results of SLCP implementation. Meanwhile, the implementation of a new round of SLCP after 2014 not only led to the expansion of afforestation area but also further improved and consolidated the achievements of the previous SLCP, which eventually led to a renewed increase in EEoSLCP between 2013 and 2015.

4.2. Analysis of the Spatial Pattern of EEoSLCP

Low–low agglomeration area in the northern part of LP. A large part of this area is located in sandy land and desert areas, which belong to arid and semi-arid regions. The growth and survival of vegetation are strictly limited by the supply of water resources [82,83]. Although the SLCP has input a certain amount of area (Figure 1c), investment (Figure 1b), and households (Figure 1d) in the region, the precipitation (Appendix A, Figure A3a) and ≥10 °C accumulated temperature (Appendix A, Figure A3b) conducive to the growth of vegetation are very limited, coupled with the unreasonable selection of tree species, leading to a low afforestation survival rate, and it is difficult to achieve the expected eco-effects of afforestation in sandy and desert areas [84]. In addition, the presence of sparse trees can concentrate airflow between trees, thereby increasing wind speed and erosion forces, and increasing soil drying and erosion when trees are unable to block strong winds, instead accelerating the rate of land degradation within sandy desert areas and further exacerbating the degree of desertification within the region [85,86]. As mentioned above, due to the limitation of precipitation and accumulated temperature, it is difficult to effectively convert the input of SLCP into eco-effects in the northern part of LP, which leads to the low EEoSLCP in this region.

Low–low agglomerations in the central and western parts of the LP. Most of this area is located in the hill and gully region of the LP in Shaanxi and Gansu, where soil and water loss is the most serious, so the counties in this area have always been the key areas for the implementation of SLCP. From 2002 to 2015 the region was the most invested in the LP Retirement Forestry Project. The counties belonging to this region were the areas that received the most input from the SLCP from 2002 to 2015 (Figure 1b–d). In terms of effect, the VFC, VCS, SR, WC, and Biod in the area have been significantly improved (Figure 3f–j), indicating that the implementation of SLCP in the area has achieved significant eco-effects. However, in terms of efficiency, the EEoSLCP in the above areas has been low. In order to find the reasons, we conducted statistics on the average input redundancy and average output deficiency of the SLCP in 314 counties of the LP. The reason for this is that the above-mentioned areas have higher redundancy in terms of area (Appendix A, Figure A3c), investment (Appendix A, Figure A3d), and households (Appendix A, Figure A3e) inputs for the SLCP, and at the same time higher deficiency in terms of eco-effects outputs such as VFC, VCS, and SR (Appendix A, Figure A4a–e). Therefore, although the aforementioned areas have achieved good eco-effects with a large amount of investment, land, and labor resources invested, excessive input redundancy and deficiencyoutput have led to low EEoSLCP in the area.

Medium–high and high-efficiency agglomerations in the eastern and southern parts of the LP. The region is in a semi-humid area and has precipitation (Appendix A, Figure A3a) and ≥10 °C temperature accumulation (Appendix A, Figure A3b) conditions that are more suitable for vegetation growth than other areas of the LP. Secondly, the above-mentioned areas have low redundancy in the investment of area (Appendix A, Figure A3c), investment (Appendix A, Figure A3d), and households (Appendix A, Figure A3e) for SLCP and, at the same time, the deficiency in eco-effects such as VFC, VCS, and SR is also low (Appendix A, Figure A4a–e). As a result, the conversion rate between input and eco-effects output of SLCP in the above areas is higher, so the EEoSLCP is also higher.

4.3. Analysis of the Spatial Correlation of EEoSLCP

ESDA of the EEoSLCP showed that there was a significant positive spatial correlation between the EEoSLCP among counties in the LP. This not only proves once again that there is an obvious spatial clustering characteristic of the EEoSLCP in each county (Figure 6) but also further indicates that there is a strong spatial dependence of the EEoSLCP in some counties. However, according to the change of global Moran's I index (Figure 7), the global Moran's I index showed a downward trend from 2002 to 2006, indicating that the spatial dependence of EEoSLCP in each county of the LP showed a weakening trend. This may be because, in the initial stage of the implementation of SLCP, various places are still in the exploratory stage and, with the implementation of SLCP moving forward, the gap between the implementation technology and management level of the program gradually appears [87,88], which leads to the increasing relative difference in the spatial EEoSLCP implementation in each county, and the spatial dependence is gradually weakened. The global Moran's I index showed an increasing trend from 2006 to 2015, indicating the increasing spatial dependence of the EEoSLCP in each county of the LP. This may be due to the fact that all localities have learned from the previous program implementation experience at this stage and, by actively learning from the surrounding counties with high eco-efficiency [78,89], the level of program implementation technology and management has been improved, and the eco-efficiency has been improved. Thus, the relative spatial difference of eco-efficiency in each county continues to narrow and the spatial dependence is gradually enhanced.

Further, from the local autocorrelation results of the EEoSLCP in each county on the LP, it can be seen that the neighboring counties mostly show "H-H" or "L-L" clustering, and the number of counties with "H-H" clustering is expanding, while the number of counties with "L-L" clustering is decreasing (Figure 8). The range of the H-H cluster area has been expanding along the H-H cluster area at the early stage of SLCP implementation (Figure 8), probably because the southern and southeastern counties of the LP have attracted neighboring counties to learn afforestation and management experience through the demonstration effect [78,89], which has led to the increasing conversion rate of the inputs to the eco-effects output of SLCP, thus promoting the improvement of the EEoSLCP. The scope of the L-L cluster area gradually shrinks along the L-L cluster area at the beginning of SLCP implementation, which may be related to the warning effect brought by the low efficiency counties. Officials in some counties have been punished for poor implementation of SLCP and corruption [73]. The resulting warning effect will motivate officials in L-L catchment counties to take the initiative to learn from the experience of neighboring counties with high eco-efficiency, improve engineering and management levels, and increase the conversion rate of SLCP inputs eco-efficiency, which in turn has led to improved eco-efficiency.

4.4. Limitation and Future Directions

Nonetheless, this study has several limitations. First, in terms of the temporal selection of the study, due to the limitation of data acquisition, we only assessed the eco-efficiency of the first round of SLCP implemented in 314 counties in the LP from 2002 to 2015. However, the Chinese government has started to implement a new round of SLCP after 2015, so what is the eco-efficiency of the new round of SLCP implemented from 2015 to the present? How does the eco-efficiency of the two different periods compare? All of the above questions need to be addressed in our future research. Second, in the selection of eco-efficiency output indicators of SLCP, we have selected the core indicators that are commonly used and easily quantified by academia. However, a comprehensive measurement of the EEoSLCP may need to include more eco-efficiency output indicators of SLCP, such as sandstorm prevention, flood mitigation, air purification, etc., which will be the direction of further research on the measurement of EEoSLCP. Third, in the analysis of the drivers of the ecological efficiency of SLCP, this study uses a combination of qualitative and quantitative methods to analyze to some extent the possible reasons that may affect the spatial and temporal evolution of the EEoSLCP, but the SLCP is a complex socioeconomic ecological system project, which requires us to explore the inner mechanism and driving mechanism of the spatial and temporal evolution of the EEoSLCP through the establishment of econometric models in future studies.

5. Conclusions and Policy Implications

5.1. Main Conclusions

"The UN Decade on Ecosystem Restoration 2021–2030" strategy and China's "carbon peaking and carbon neutrality" strategy marked the beginning of a large-scale ERP. At the same time, the full-scale ecological restoration project also implies a large investment of capital, labor, and other resources. However, the scarcity of resources determines that we have limited resources to invest in ERPs and failure to convert the limited resources into ecological restoration effectiveness in the implementation of ERPs will hinder the realization of the above two strategic goals. To this end, we take the world's largest ERP, the SLCP, as an example and evaluate its eco-efficiency, with a view to summarizing the lessons learned and then providing reference for improving the resource utilization efficiency in the implementation of ERPs under resource constraints, and ultimately promoting the realization of the above two strategic goals. This study takes LP, the core region of SLCP implementation, as the study area, and uses the DEA-BCC model to measure the EEoSLCP of LP counties based on the SLCP input–output data of 314 counties in LP and, on this basis, the KDE and ESDA methods are used to portray and identify the time-series dynamic evolution pattern and spatial pattern evolution characteristics of EEoSLCP from a two-dimensional perspective in time and space, respectively. The major findings were as follows:

(1) The ecosystem of LP has been greatly restored after the implementation of SLCP and the ecological environment in the territory has been greatly improved. Using 2002 as the base period for SLCP implementation, VFC, VCS, SR, WC, and Biod in the region were all enhanced to varying degrees during 2002–2018.
(2) The EEoSLCP of LP is low and there is still more room for improvement. In terms of the temporal variation of EEoSLCP, although EEoSLCP has been maintained at a low level, it has generally shown a fluctuating upward trend. In terms of the time-series dynamic evolution of EEoSLCP, there were marked variations in the EEoSLCP between LP counties, with obvious polarization, and a "bimodal" evolution pattern of "low–low clustering and high–high clustering", which is similar to "club convergence", is gradually developed with time.
(3) The EEoSLCP in the LP is spatially distributed with regular differences. In terms of spatial distribution, high-efficiency counties are mainly concentrated in the eastern, southern, and western regions, while low-efficiency counties are mainly concentrated in the central, northern, and western regions, and the overall spatial distribution of eco-efficiency is gradually decreasing from southeast to northwest, and from south to north. In terms of spatial change, the number of high-efficiency and medium-high efficiency counties is increasing, while the number of low-efficiency counties is decreasing.
(4) The EEoSLCP in each county of the LP has a strong spatial dependence and a relatively stable trend in terms of time. In terms of local autocorrelation, the EEoSLCP in each county of the LP shows two correlation trends: "H-H" cluster and "L-L" cluster. H-H clusters are mainly concentrated in the western, southern, and northeastern regions, while L-L clusters are mainly concentrated in the central, northern, and northeastern regions. In terms of spatial and temporal changes, the H-H cluster area is expanding and the L-L cluster area is shrinking.

5.2. Policy Implication

The results of this study can provide three policy implications for the implementation of ERPs around the world. The details are as follows:

(1) Do the top-level strategy for the implementation of ERPs. Through scientific demonstration, reasonable planning, and farmers' will, determine the ERPs' implementation area and, on this basis, determine the ERPs' implementation tree species according to the natural conditions of local vegetation growth. For example, the EEoSLCP in the northwestern part of the LP is low due to improper selection of SLCP tree species and, in order to improve eco-efficiency, drought-tolerant shrubs (grasses) should be planted and the original vegetation restored in this area.
(2) Establish a dynamic evaluation mechanism for the eco-effects and ecological efficiency of ERPs, increase the matching of effect and efficiency, and improve the efficiency of resource utilization. For example, in order to improve the EEoSLCP, the SLCP input

should be reduced for the central and western counties of the LP, and increased for the eastern and southern counties of the LP.

(3) Set up a typical example of a successful ecological restoration area to play a demonstration and guiding role. In the implementation of ERPs, it is necessary to make good use of the spatial dependence between neighboring regions, strengthen cooperation and communication between regions, and build an ecological restoration community. For example, in the LP region, H-H cluster type counties should continue to make use of the agglomeration advantages of strong alliances to move towards the goal of higher ecological efficiency of the SLCP, promote the high-quality development of projects, and play a leading role in demonstration; L-H outlier type counties should proactively learn from neighboring high-efficiency counties about SLCP management techniques and experiences. In the implementation of the SLCP, the L-L cluster type counties in the northwest need to base their strategy on local resource endowment and natural conditions, plan tree planting areas, follow the principle of tree species suitability, plant trees in places suitable for tree growth, plant shrubs in places suitable for shrub growth, and reasonably choose trees and shrubs to be planted together to create a compound ecosystem of multiple symbiosis. Meanwhile, the L-L cluster type counties in the center need to reduce land inputs to avoid wasting resources; H-L cluster type counties, while continuing to maintain their own ecological efficiency of the SLCP, should also make use of their accumulated experience of SLCP implementation to guide neighboring low efficiency counties, jointly improve the ecological efficiency of SLCP, and be a good role model.

Author Contributions: Y.D.: Conceptualization, Methodology, Data curation, Formal analysis, Writing—original draft. J.L.: Writing—review and editing. Y.W.: Writing—review and editing. C.J.: Writing—review and editing, Data processing. X.Y.: Writing—review and editing. X.S.: Writing—review and editing. H.L.: Conceptualization, Writing—review and editing, Supervision, Funding acquisition. S.Y.: Conceptualization, Writing—review and editing, Investigation, Supervision, Funding acquisition. All authors have read and agreed to the published version of the manuscript.

Funding: This study benefited from joint financial support by the programs of the National Natural Science Foundation of China (No. 71473195, 71773091) and the Special Research Project for Forestry Public Welfare Industry of the State Forestry and Grassland Administration of China (No. 201504424).

Data Availability Statement: Data are available from the first author (Yuanjie Deng) upon a reasonable request.

Acknowledgments: The first author of the paper, Yuanjie Deng, is very grateful to his family for their continuous support and encouragement.

Conflicts of Interest: The authors declare that they have no known competing financial interests or personal relationships that could have appeared to influence the work reported in this paper.

Appendix A

Table A1. Global Moran's I value and its test results of EEoSLCP in the LP from 2002 to 2015.

Year	Moran's I	Z	*p*
2002	0.518 ***	14.532	0.000
2003	0.481 ***	13.490	0.000
2004	0.470 ***	13.192	0.000
2005	0.424 ***	11.911	0.000
2006	0.414 ***	11.637	0.000
2007	0.432 ***	12.123	0.000
2008	0.420 ***	11.807	0.000
2009	0.494 ***	13.857	0.000
2010	0.484 ***	13.588	0.000
2011	0.504 ***	14.137	0.000
2012	0.534 ***	14.964	0.000

Table A1. *Cont.*

Year	Moran's I	Z	*p*
2013	0.496 ***	13.924	0.000
2014	0.499 ***	13.985	0.000
2015	0.565 ***	15.852	0.000

Note: *** denote statistical significance at 1%, 5%, and 10%, respectively.

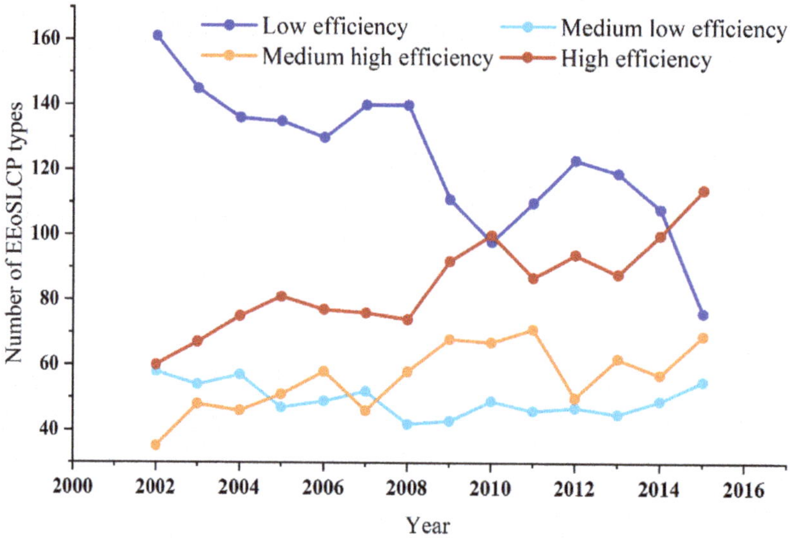

Figure A1. Quantitative changes of four EEoSLCP types in counties, 2002 to 2015.

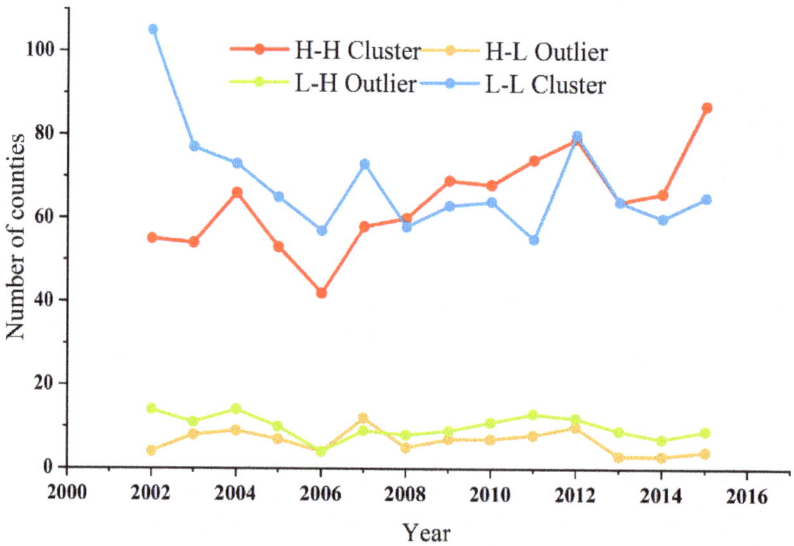

Figure A2. Quantitative changes of four LSA states of EEoSLCP in counties, 2002 to 2015.

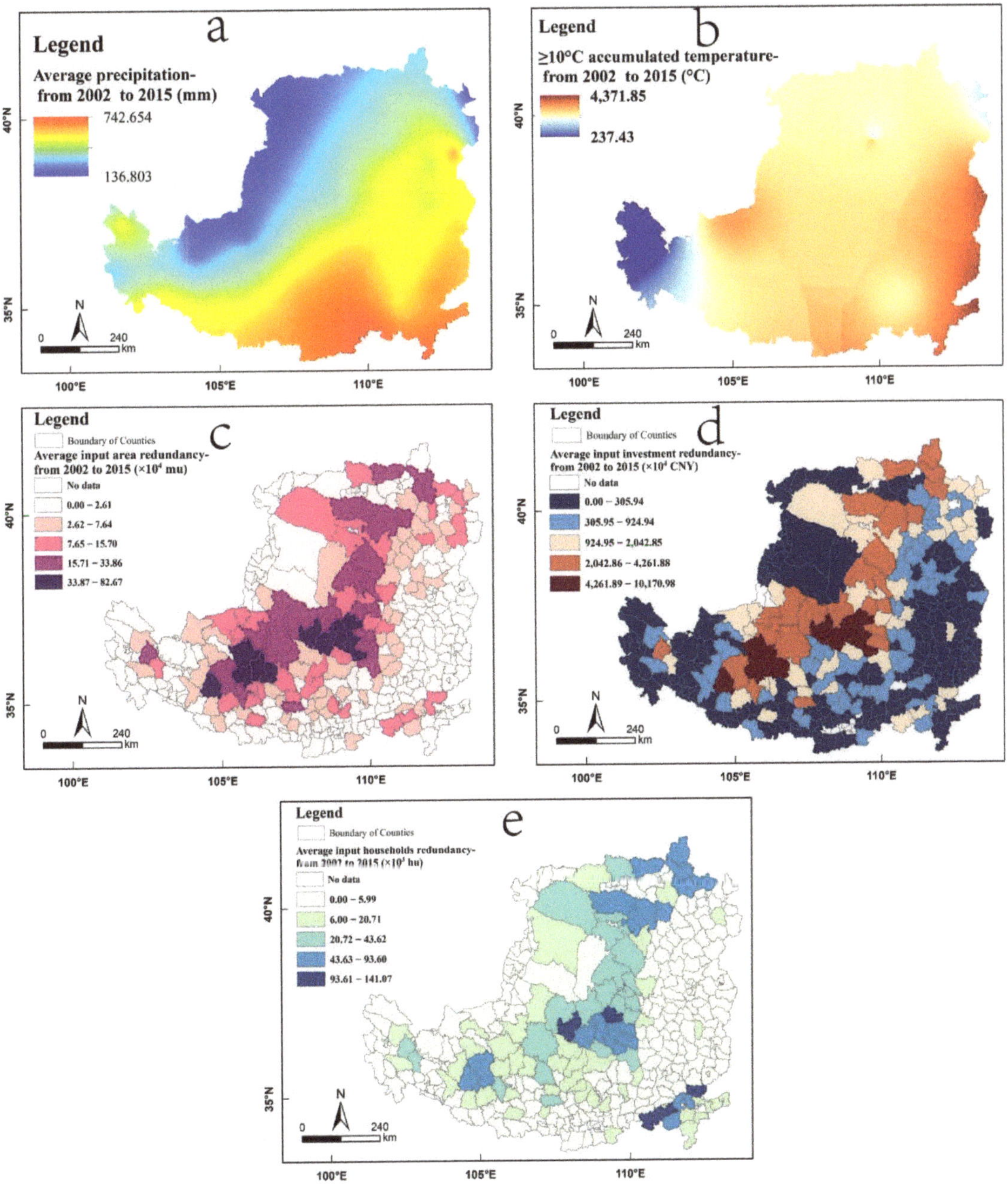

Figure A3. (**a**) Average precipitation; (**b**) average ≥10 °C accumulated temperature; (**c**) average input area redundancy; (**d**) average input investment redundancy; (**e**) average input households participating redundancy, 2000 to 2015.

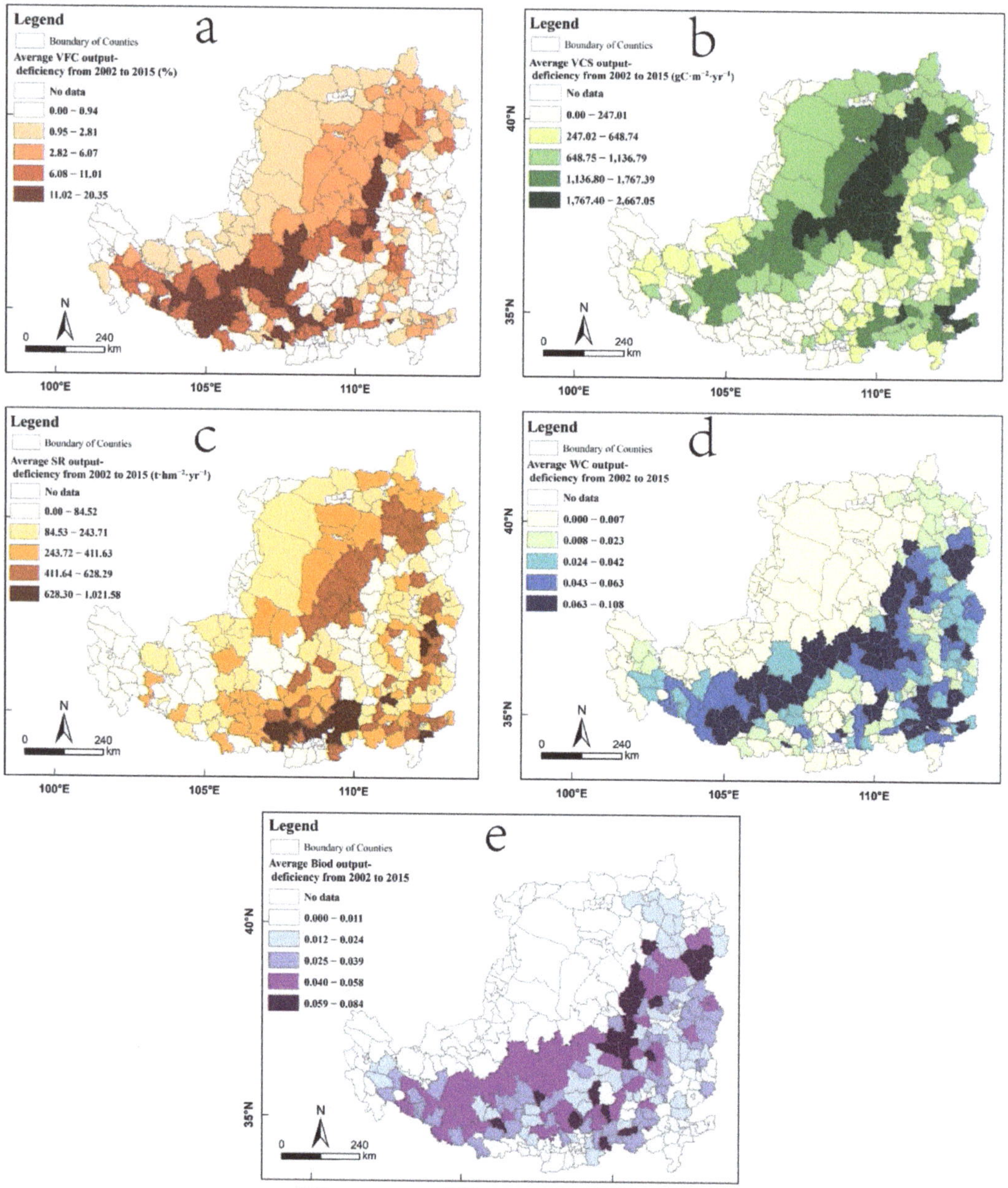

Figure A4. (**a**) Average output VFC deficiency; (**b**) average output VCS deficiency; (**c**) average output SR deficiency; (**d**) average output WC deficiency; (**e**) average output biod deficiency, 2002 to 2015.

References

1. Bryan, B.A.; Gao, L.; Ye, Y.; Sun, X.; Connor, J.D.; Crossman, N.D.; Stafford-Smith, M.; Wu, J.; He, C.; Yu, D.; et al. China's Response to a National Land-System Sustainability Emergency. *Nature* **2018**, *559*, 193–204. [CrossRef] [PubMed]
2. Sun, Y.; Ding, W.; Yang, Z.; Yang, G.; Du, J. Measuring China's Regional Inclusive Green Growth. *Sci. Total Environ.* **2020**, *713*, 136367. [CrossRef]

3. Liu, J.; Li, S.; Ouyang, Z.; Tam, C.; Chen, X. Ecological and Socioeconomic Effects of China's Policies for Ecosystem Services. *Proc. Natl. Acad. Sci. USA* **2008**, *105*, 9477–9482. [CrossRef] [PubMed]
4. Yin, R.; Yin, G.; Li, L. Assessing China's Ecological Restoration Programs: What's Been Done and What Remains to Be Done? *Environ. Manag.* **2010**, *45*, 442–453. [CrossRef] [PubMed]
5. Wang, X.; Bennett, J. Policy Analysis of the Conversion of Cropland to Forest and Grassland Program in China. *Environ. Econ. Policy Stud.* **2008**, *9*, 119–143. [CrossRef]
6. Xian, J.; Xia, C.; Cao, S. Cost–Benefit Analysis for China's Grain for Green Program. *Ecol. Eng.* **2020**, *151*, 105850. [CrossRef]
7. Bennett, M.T. China's Sloping Land Conversion Program: Institutional Innovation or Business as Usual? *Ecol. Econ.* **2008**, *65*, 699–711. [CrossRef]
8. Lu, G.; Yin, R. Evaluating the Evaluated Socioeconomic Impacts of China's Sloping Land Conversion Program. *Ecol. Econ.* **2020**, *177*, 106785. [CrossRef]
9. Chen, C.; Park, T.; Wang, X.; Piao, S.; Xu, B.; Chaturvedi, R.K.; Fuchs, R.; Brovkin, V.; Ciais, P.; Fensholt, R.; et al. China and India Lead in Greening of the World through Land-Use Management. *Nat. Sustain.* **2019**, *2*, 122–129. [CrossRef]
10. Li, J.; Peng, S.; Li, Z. Detecting and Attributing Vegetation Changes on China's Loess Plateau. *Agric. For. Meteorol.* **2017**, *247*, 260–270. [CrossRef]
11. Li, P.; Wang, J.; Liu, M.; Xue, Z.; Bagherzadeh, A.; Liu, M. Spatio-Temporal Variation Characteristics of NDVI and Its Response to Climate on the Loess Plateau from 1985 to 2015. *Catena* **2021**, *203*, 105331. [CrossRef]
12. Chen, Y.; Wang, K.; Lin, Y.; Shi, W.; Song, Y.; He, X. Balancing Green and Grain Trade. *Nat. Geosci.* **2015**, *8*, 739–741. [CrossRef]
13. Jin, F.; Yang, W.; Fu, J.; Li, Z. Effects of Vegetation and Climate on the Changes of Soil Erosion in the Loess Plateau of China. *Sci. Total Environ.* **2021**, *773*, 145514. [CrossRef] [PubMed]
14. Deng, L.; Liu, S.; Kim, D.G.; Peng, C.; Sweeney, S.; Shangguan, Z. Past and Future Carbon Sequestration Benefits of China's Grain for Green Program. *Glob. Environ. Chang.* **2017**, *47*, 13–20. [CrossRef]
15. Feng, X.; Fu, B.; Lu, N.; Zeng, Y.; Wu, B. How Ecological Restoration Alters Ecosystem Services: An Analysis of Carbon Sequestration in China's Loess Plateau. *Sci. Rep.* **2013**, *3*, 2846. [CrossRef]
16. Wang, K.; Hu, D.; Deng, J.; Shangguan, Z.; Deng, L. Biomass Carbon Storages and Carbon Sequestration Potentials of the Grain for Green Program-Covered Forests in China. *Ecol. Evol.* **2018**, *8*, 7451–7461. [CrossRef] [PubMed]
17. Wang, B.; Gao, P.; Niu, X.; Sun, J. Policy-Driven China's Grain to Green Program: Implications for Ecosystem Services. *Ecosyst. Serv.* **2017**, *27*, 38–47. [CrossRef]
18. Wang, Y.; Zhao, J.; Fu, J.; Wei, W. Effects of the Grain for Green Program on the Water Ecosystem Services in an Arid Area of China-Using the Shiyang River Basin as an Example. *Ecol. Indic.* **2019**, *104*, 659–668. [CrossRef]
19. Wang, C.; Ouyang, H.; Maclaren, V.; Yin, Y.; Shao, B.; Boland, A.; Tian, Y. Evaluation of the Economic and Environmental Impact of Converting Cropland to Forest: A Case Study in Dunhua County, China. *J. Environ. Manag.* **2007**, *85*, 746–756. [CrossRef]
20. Chen, X.; Lupi, F.; Viña, A.; He, G.; Liu, J. Using Cost-Effective Targeting to Enhance the Efficiency of Conservation Investments in Payments for Ecosystem Services. *Conserv. Biol.* **2010**, *24*, 1469–1478. [CrossRef]
21. Ning, J.; Zhang, D.; Yu, Q. Quantifying the Efficiency of Soil Conservation and Optimized Strategies: A Case-Study in a Hotspot of Afforestation in the Loess Plateau. *Land Degrad. Dev.* **2021**, *32*, 1114–1126. [CrossRef]
22. Lu, Y.; Yao, S.; Ding, Z.; Deng, Y.; Hou, M. Did Government Expenditure on the Grain for Green Project Help the Forest Carbon Sequestration Increase in Yunnan, China? *Land* **2020**, *9*, 54. [CrossRef]
23. Zhang, Y.; Zhang, D. Efficiency Measurement and Influencing Factors of Ecological Compensation: A Case Study from Wuqi and Zhidan on the Loess Plateau. *Nat. Resour. Res.* **2021**, *30*, 1905–1921. [CrossRef]
24. Liu, S.; Yao, S. The Effect of Precipitation on the Cost-Effectiveness of Sloping Land Conversion Program: A Case Study of Shaanxi Province, China. *Ecol. Indic.* **2021**, *132*, 108251. [CrossRef]
25. Neto, J.Q.F.; Walther, G.; Bloemhof, J.; Van Nunen, J.; Spengler, T. A Methodology for Assessing Eco-Efficiency in Logistics Networks. *Eur. J. Oper. Res.* **2009**, *193*, 670–682. [CrossRef]
26. Wang, Y.; Zhang, T.; Yao, S.; Deng, Y. Spatio-Temporal Evolution and Factors Influencing the Control Efficiency for Soil and Water Loss in the Wei River Catchment, China. *Sustainability* **2019**, *11*, 216. [CrossRef]
27. Li, G.; Sun, S.; Han, J.; Yan, J.; Liu, W.; Wei, Y.; Lu, N.; Sun, Y. Impacts of Chinese Grain for Green Program and Climate Change on Vegetation in the Loess Plateau during 1982–2015. *Sci. Total Environ.* **2019**, *660*, 177–187. [CrossRef]
28. Li, Z.; Zheng, F.; Liu, W.; Flanagan, D.C. Spatial Distribution and Temporal Trends of Extreme Temperature and Precipitation Events on the Loess Plateau of China during 1961–2007. *Quat. Int.* **2010**, *226*, 92–100. [CrossRef]
29. Wang, L.; Shao, M.; Wang, Q.; Gale, W.J. Historical Changes in the Environment of the Chinese Loess Plateau. *Environ. Sci. Policy* **2006**, *9*, 675–684. [CrossRef]
30. Wang, J.; Liu, Z.; Gao, J.; Emanuele, L.; Ren, Y.; Shao, M.; Wei, X. The Grain for Green Project Eliminated the Effect of Soil Erosion on Organic Carbon on China's Loess Plateau between 1980 and 2008. *Agric. Ecosyst. Environ.* **2021**, *322*, 107636. [CrossRef]
31. Wu, X.; Wang, S.; Fu, B.; Feng, X.; Chen, Y. Socio-Ecological Changes on the Loess Plateau of China after Grain to Green Program. *Sci. Total Environ.* **2019**, *678*, 565–573. [CrossRef] [PubMed]
32. Amacher, G.S.; Ollikainen, M.; Koskela, E. *Economics of Forest Resources*; MIT Press: Cambridge, UK, 2009.

33. Deng, Y.; Cai, W.; Hou, M.; Zhang, X.; Xu, S.; Yao, N.; Guo, Y.; Li, H.; Yao, S. How Eco-Efficiency Is the Forestry Ecological Restoration Program? The Case of the Sloping Land Conversion Program in the Loess Plateau, China. *Land* **2022**, *11*, 712. [CrossRef]
34. Deng, Y.; Jia, L.; Guo, Y.; Li, H.; Yao, S.; Chu, L.; Lu, W.; Hou, M.; Mo, B.; Wang, Y.; et al. Evaluation of the Ecological Effects of Ecological Restoration Programs: A Case Study of the Sloping Land Conversion Program on the Loess Plateau, China. *Int. J. Environ. Res. Public Health* **2022**, *19*, 7841. [CrossRef] [PubMed]
35. Sun, W.; Song, X.; Mu, X.; Gao, P.; Wang, F.; Zhao, G. Spatiotemporal Vegetation Cover Variations Associated with Climate Change and Ecological Restoration in the Loess Plateau. *Agric. For. Meteorol.* **2015**, *209–210*, 87–99. [CrossRef]
36. Wang, S.; Zhao, D.; Chen, H. Government Corruption, Resource Misallocation, and Ecological Efficiency. *Energy Econ.* **2020**, *85*, 104573. [CrossRef]
37. He, L.; Guo, J.; Jiang, Q.; Zhang, Z.; Yu, S. How Did the Chinese Loess Plateau Turn Green from 2001 to 2020? An Explanation Using Satellite Data. *Catena* **2022**, *214*, 106246. [CrossRef]
38. Zhang, D.; Pearse, P.H. *Forest Economics*; UBC Press: Vancouve, BC, Canada, 2011.
39. Li, J.; Wang, X.; Shao, M.; Zhao, Y.; Li, X.; Wu, S.; Li, J.; Zhang, R. Simulation of Biomass and Soil Desiccation of Robinia Pseudoacacia Forestlands on Semi-Arid and Semi-Humid Regions of China's Loess Plateau. *Chin. J. Plant Ecol.* **2010**, *34*, 330–339.
40. Qian, C.; Shao, L.; Hou, X.; Zhang, B.; Chen, W.; Xia, X. Detection and Attribution of Vegetation Greening Trend across Distinct Local Landscapes under China's Grain to Green Program: A Case Study in Shaanxi Province. *Catena* **2019**, *183*, 104182. [CrossRef]
41. Hua, F.; Wang, X.; Zheng, X.; Fisher, B.; Wang, L.; Zhu, J.; Tang, Y.; Yu, D.W.; Wilcove, D.S. Opportunities for Biodiversity Gains under the World's Largest Reforestation Programme. *Nat. Commun.* **2016**, *7*, 12717. [CrossRef]
42. Wang, F.; Pan, X.; Gerlein-Safdi, C.; Cao, X.; Wang, S.; Gu, L.; Wang, D.; Lu, Q. Vegetation Restoration in Northern China: A Contrasted Picture. *Land Degrad. Dev.* **2020**, *31*, 669–676. [CrossRef]
43. Chen, J.; Fan, W.; Li, D.; Liu, X.; Song, M. Driving Factors of Global Carbon Footprint Pressure: Based on Vegetation Carbon Sequestration. *Appl. Energy* **2020**, *267*, 114914. [CrossRef]
44. Peng, S.; Wen, D.; He, N.; Yu, G.; Ma, A.; Wang, Q. Carbon Storage in China's Forest Ecosystems: Estimation by Different Integrative Methods. *Ecol. Evol.* **2016**, *6*, 3129–3145. [CrossRef] [PubMed]
45. Xu, L.; Yu, G.; He, N.; Wang, Q.; Gao, Y.; Wen, D.; Li, S.; Niu, S.; Ge, J. Carbon Storage in China's Terrestrial Ecosystems: A Synthesis. *Sci. Rep.* **2018**, *8*, 2806. [CrossRef] [PubMed]
46. Xu, L.; He, N.; Yu, G. A dataset of carbon density in Chinese terrestrial ecosystems (2010s). *Science Data Bank*, 29 July 2018. [CrossRef]
47. Renard, K.G. *Predicting Soil Erosion by Water: A Guide to Conservation Planning with the Revised Universal Soil Loss Equation (RUSLE)*; United States Government Printing: Washington, DC, USA, 1997.
48. Wischmeier, W.H.; Smith, D.D. *Predicting Rainfall Erosion Losses: A Guide to Conservation Planning*; Department of Agriculture, Science and Education Administration: Washington, DC, USA, 1978.
49. Williams, J.R.; Renard, K.G.; Dyke, P.T. EPIC: A New Method for Assessing Erosion's Effect on Soil Productivity. *J. Soil Water Conserv.* **1983**, *38*, 381–383.
50. Zhang, H.; Yang, Q.; Li, R.; Liu, Q.; Moore, D.; He, P.; Ritsema, C.J.; Geissen, V. Extension of a GIS Procedure for Calculating the RUSLE Equation LS Factor. *Comput. Geosci.* **2013**, *52*, 177–188. [CrossRef]
51. Cai, C.; Ding, S.; Shi, Z.; Huang, L.; Zhang, G. Study of applying USLE and geographical information system IDRISI to predict soil erosion in small watershed. *J. Soil Water Conserv.* **2000**, *14*, 19–24.
52. Yan, R.; Zhang, X.; Yan, S.; Chen, H. Estimating Soil Erosion Response to Land Use/Cover Change in a Catchment of the Loess Plateau, China. *Int. Soil Water Conserv. Res.* **2018**, *6*, 13–22. [CrossRef]
53. Wu, X.; Wang, S.; Fu, B.; Liu, Y.; Zhu, Y. Land Use Optimization Based on Ecosystem Service Assessment: A Case Study in the Yanhe Watershed. *Land Use Policy* **2018**, *72*, 303–312. [CrossRef]
54. Kong, L.; Zheng, H.; Rao, E.; Xiao, Y.; Ouyang, Z.; Li, C. Evaluating Indirect and Direct Effects of Eco-Restoration Policy on Soil Conservation Service in Yangtze River Basin. *Sci. Total Environ.* **2018**, *631–632*, 887–894. [CrossRef]
55. Liu, W.; Liu, J.; Kuang, W. Spatiotemporal patterns of soil protection effect of the Grain for Green Project in northern Shaanxi. *Acta Geogr. Sin.* **2019**, *74*, 1835–1852.
56. Carreño, L.; Frank, F.C.; Viglizzo, E.F. Tradeoffs between Economic and Ecosystem Services in Argentina during 50 Years of Land-Use Change. *Agric. Ecosyst. Environ.* **2012**, *154*, 68–77. [CrossRef]
57. Liu, Y.; Lü, Y.; Jiang, W.; Zhao, M. Mapping Critical Natural Capital at a Regional Scale: Spatiotemporal Variations and the Effectiveness of Priority Conservation. *Environ. Res. Lett.* **2020**, *15*, 124025. [CrossRef]
58. Duan, Y.; Di, B.; Ustin, S.L.; Xu, C.; Xie, Q.; Wu, S.; Li, J.; Zhang, R. Changes in Ecosystem Services in a Montane Landscape Impacted by Major Earthquakes: A Case Study in Wenchuan Earthquake-Affected Area, China. *Ecol. Indic.* **2021**, *126*, 107683. [CrossRef]
59. Barral, M.P.; Oscar, M.N. Land-Use Planning Based on Ecosystem Service Assessment: A Case Study in the Southeast Pampas of Argentina. *Agric. Ecosyst. Environ.* **2012**, *154*, 34–43. [CrossRef]
60. Banker, R.D.; Charnes, A.; Cooper, W.W. Some Models for Estimating Technical and Scale Inefficiencies in Data Envelopment Analysis. *Manag. Sci.* **1984**, *30*, 1078–1092. [CrossRef]

61. Charnes, A.; Cooper, W.W.; Rhodes, E. Measuring the Efficiency of Decision Making Units. *Eur. J. Oper. Res.* **1978**, *2*, 429–444. [CrossRef]
62. Farrell, M.J. The Measurement of Productive Efficiency. *J. R. Stat. Soc. Ser. A* **1957**, *120*, 253–281. [CrossRef]
63. Parzen, E. On Estimation of a Probability Density Function and Mode. *Ann. Math. Stat.* **1962**, *33*, 1065–1076. [CrossRef]
64. Davis, R.A.; Lii, K.-S.; Politis, D.N. Remarks on Some Nonparametric Estimates of a Density Function. In *Selected Works of Murray Rosenblatt*; Davis, R.A., Lii, K.-S., Politis, D.N., Eds.; Selected Works in Probability and Statistics; Springer: New York, NY, USA, 2011; pp. 95–100. ISBN 978-1-4419-8339-8.
65. Anselin, L. Interactive Techniques and Exploratory Spatial Data Analysis. In *Regional Research Institute Working Paper, 200*; West Virginia University: Morgantown, WV, USA, 1996.
66. Anselin, L. Local Indicators of Spatial Association—LISA. *Geogr. Anal.* **1995**, *27*, 93–115. [CrossRef]
67. Liu, J.; Liu, M.; Zhuang, D.; Zhang, Z.; Deng, X. Study on Spatial Pattern of Land-Use Change in China during 1995–2000. *Sci. China Ser. D-Earth Sci.* **2003**, *46*, 373–384.
68. Liu, J.; Zhang, Z.; Xu, X.; Kuang, W.; Zhou, W.; Zhang, S.; Li, R.; Yan, C.; Yu, D.; Wu, S.; et al. Spatial Patterns and Driving Forces of Land Use Change in China during the Early 21st Century. *J. Geogr. Sci.* **2010**, *20*, 483–494. [CrossRef]
69. Liu, J.; Kuang, W.; Zhang, Z.; Xu, X.; Qin, Y.; Ning, J.; Zhou, W.; Zhang, S.; Li, R.; Yan, C.; et al. Spatiotemporal Characteristics, Patterns, and Causes of Land-Use Changes in China since the Late 1980s. *J. Geogr. Sci.* **2014**, *24*, 195–210. [CrossRef]
70. Kuang, W.; Zhang, S.; Du, G.; Yan, C.; Wu, S.; Li, R.; Lu, D.; Pan, T.; Ning, J.; Guo, C.; et al. Remotely sensed mapping and analysis of spatio-temporal patterns of land use change across China in 2015–2020. *Acta Geogr. Sin.* **2022**, *77*, 1056–1071. (In Chinese)
71. Xu, J.; Yin, R.; Li, Z.; Liu, C. China's Ecological Rehabilitation: Unprecedented Efforts, Dramatic Impacts, and Requisite Policies. *Ecol. Econ.* **2006**, *57*, 595–607. [CrossRef]
72. Yin, R.; Yin, G. China's Primary Programs of Terrestrial Ecosystem Restoration: Initiation, Implementation, and Challenges. *Environ. Manag.* **2010**, *45*, 429–441. [CrossRef] [PubMed]
73. Yu, X. Central–Local Conflicts in China's Environmental Policy Implementation: The Case of the Sloping Land Conversion Program. *Nat. Hazards* **2016**, *84*, 77–96. [CrossRef]
74. Qu, M.; Liu, G.; Lin, Y.; Driedger, E.; Peter, Z.; Xu, X.; Cao, Y. Experts' Perceptions of the Sloping Land Conversion Program in the Loess Plateau, China. *Land Use Policy* **2017**, *69*, 204–210. [CrossRef]
75. Trac, C.J.; Harrell, S.; Hinckley, T.M.; Henck, A.C. Reforestation Programs in Southwest China: Reported Success, Observed Failure, and the Reasons Why. *J. Mt. Sci.* **2007**, *4*, 275–292. [CrossRef]
76. Delang, C.O.; Wang, W. Chinese Forest Policy Reforms after 1998: The Case of the Natural Forest Protection Program and the Slope Land Conversion Program. *Int. For. Rev.* **2013**, *15*, 290–304. [CrossRef]
77. He, J.; Sikor, T. Notions of Justice in Payments for Ecosystem Services: Insights from China's Sloping Land Conversion Program in Yunnan Province. *Land Use Policy* **2015**, *43*, 207–216. [CrossRef]
78. He, J. Governing Forest Restoration: Local Case Studies of Sloping Land Conversion Program in Southwest China. *For. Policy Econ.* **2014**, *46*, 30–38. [CrossRef]
79. He, J. Situated Payments for Ecosystem Services: Local Agencies in the Implementation of the Sloping Land Conversion Programme in Southwest China. *Dev. Chang.* **2020**, *51*, 73–93. [CrossRef]
80. Chen, C.; König, H.J.; Matzdorf, B.; Zhen, L. The Institutional Challenges of Payment for Ecosystem Service Program in China: A Review of the Effectiveness and Implementation of Sloping Land Conversion Program. *Sustainability* **2015**, *7*, 5564–5591. [CrossRef]
81. Liu, J.; Chen, H.; Hou, X.; Zhang, D.; Zhang, H. Time to Adopt a Context-Specific and Market-Based Compensation Scheme for a New Round of the Grain for Green Program. *Land Use Policy* **2021**, *108*, 105675. [CrossRef]
82. Cao, S.; Chen, L.; Shankman, D.; Wang, C.; Wang, X.; Zhang, H. Excessive Reliance on Afforestation in China's Arid and Semi-Arid Regions: Lessons in Ecological Restoration. *Earth-Sci. Rev.* **2011**, *104*, 240–245. [CrossRef]
83. Donohue, R.J.; Roderick, M.L.; McVicar, T.R. On the Importance of Including Vegetation Dynamics in Budyko's Hydrological Model. *Hydrol. Earth Syst. Sci.* **2007**, *11*, 983–995. [CrossRef]
84. Jiang, G.; Han, X.; Wu, J. Restoration and Management of the Inner Mongolia Grassland Require a Sustainable Strategy. *AMBIO A J. Hum. Environ.* **2006**, *35*, 269–270. [CrossRef]
85. Cao, S. Why Large-Scale Afforestation Efforts in China Have Failed To Solve the Desertification Problem. *Environ. Sci. Technol.* **2008**, *42*, 1826–1831. [CrossRef]
86. Cao, S.; Tian, T.; Chen, L.; Dong, X.; Yu, X.; Wang, G. Damage Caused to the Environment by Reforestation Policies in Arid and Semi-Arid Areas of China. *AMBIO* **2010**, *39*, 279–283. [CrossRef]
87. Bullock, A.; King, B. Evaluating China's Slope Land Conversion Program as Sustainable Management in Tianquan and Wuqi Counties. *J. Environ. Manag.* **2011**, *92*, 1916–1922. [CrossRef]

88. Tao, R.; Xu, Z.; Xu, J. Grain for green project, grain policy and sustainable development. *Soc. Sci. China* **2004**, *150*, 25–38. (In Chinese)
89. Xu, J.; Tao, R.; Xu, Z.G. Sloping land conversion program: Cost-effectiveness, structural effect and economic sustainability. *China Econ. Q.* **2004**, *4*, 139–162. (In Chinese)

forests

MDPI

Article

Evaluating the Legacy Effects of the Historical Predatory Seed Harvesting on the Species Composition and Structure of the Mixed Korean Pine and Broadleaf Forest from a Landscape Perspective

Kai Liu [1,2], Hong S. He [3,*], Hang Sun [1] and Jingfeng Wang [4]

1 Key Laboratory of Geographical Processes and Ecological Security in Changbai Mountains, Ministry of Education, School of Geographical Sciences, Northeast Normal University, Changchun 130024, China
2 Key Laboratory for Vegetation Ecology, Ministry of Education, Northeast Normal University, Changchun 130024, China
3 School of Natural Resources, University of Missouri, 203 ABNR Bldg, Columbia, MO 65211, USA
4 Heilongjiang Yichun Forest Industry Refco Group Ltd., Yichun 153000, China
* Correspondence: heh@missouri.edu

Abstract: Adequate seed provenance is an important guarantee for the restoration of the mixed Korean pine *(Pinus koraiensis)* and broadleaf forest (MKPBF). However, the commercial harvest exclusion in natural forests has led to a sharp decline in economic income from timber. Given the economic value of Korean pine seeds, predatory seed harvesting (PSH) has become increasingly serious in the past 20 years and has significantly reduced the seed and seedling bank, which might seriously threaten the restoration and sustainable management of the MKPBF. How the historical PSH has affected the future of the MKPBF is unclear at the landscape scale. In this study, we quantified the effects of seed harvesting at the landscape scale by a forest landscape model LANDIS PRO, and then assessed the legacy effects of the historical PSH on the composition and structure of the MKPBF in the Xiaoxing'an Mountains, Northeast China. Our results showed that the historical PSH decreased the Korean pine basal area of all age cohorts, with an average decrease of 0.06 to 0.19 m^2 ha^{-1} but insignificantly altered the age structure diversity of Korean pine throughout the simulation. Our results indicated that the historical PSH remarkably decreased the dominance of Korean pine by 11.1%, but significantly increased the dominance of spruce (*Picea koraiensis* and *Picea jezoensis*) and fir *(Abies nephrolepis*) by 3.8% and 4.5%, respectively, and had an insignificant effect on the other tree species over the simulation. We found that the historical PSH evidently changed the succession trajectories of the disturbed stands, which would result in the transition from the succession pattern dominated by Korean pine to that dominated by spruce and fir. The historical PSH decreased the importance value of Korean pine by 12.2% on average but increased it by 5.1% and 6.0% for spruce and fir, respectively, and resulted in an average 33.2% increase in the dissimilarity index compared with the initial state during the whole simulation period. Future forest management should strictly limit the intensity and rotation of seed harvesting to protect the seed provenance of Korean pine and consider how to ensure the recovery and sustainable management of the MKPBF through direct seeding or seedling planting.

Keywords: seed harvesting; forest landscape model; mixed Korean pine and broadleaf forest

1. Introduction

The mixed Korean pine (*Pinus koraiensis*) and broadleaf forest (MKPBF) provides important ecosystem services, such as timber and seed products [1,2]. However, the primary MKPBF has degraded into a secondary forest due to excessive harvesting in the past half of the century; meanwhile, the abundance of the dominant species Korean pine has dropped sharply. The Chinese government implemented the Natural Forest Conservation Program (NFCP) in 1998 and the Commercial Harvest Exclusion Policy (CHEP) in 2014 to increase forest stock and restore the MKPBF [3]. Although the above measures restored forest timber stock, they resulted in a significant reduction in income for local residents [4,5]. In order to compensate for the decreasing income from timber harvest, Korean pine seed was intensively harvested because of its high economic value. A previous study showed that the predatory seed harvesting (PSH) declined seed abundance and further hindered Korean pine regeneration, which increased the challenges of forest restoration [6].

The key issue of restoring the MKPBF was to promote Korean pine regeneration, which was mediated by seed sources [7,8]. Seed harvesting, with a long history, is an important anthropogenic disturbance that reduces Korean pine seed abundance in the MKPBF [9–11]. Especially after the NFCP, the effect of PSH was serious in the general ecological welfare forests and the commercial forests due to the decrease in timber income [4,6,11,12]. At present, the PSH has lasted for about 20 years in these regions. Although the PSH had had a nonobvious influence on the Korean pine forest at the landscape scale, it has caused a lagging effect on its regeneration for about 20 years. Korean pine was characterized by poor natural regeneration [13], which may be exacerbated by the PSH. A recent study found that provenance limitations, which were caused by the disappearance of Korean pine due to the historical disturbance and caused by the seed harvesting, reduced the effectiveness of seed dispersal, and thereby decreased the chance of regeneration of young Korean pine in the secondary forest [14]. It remained unknown whether the adverse effect of the historical PSH on Korean pine young age cohort would be transferred to older age cohorts and would further reduce the age structure diversity of Korean pine.

Disturbance would alter the interspecific relationship between Korean pine and its associated tree species in the MKPBF. Korean pine was often associated with fir (*Abies nephrolepis*), spruce (*Picea koraiensis and Picea jezoensis*), Amur linden (*Tilia amurensis*), Manchurian ash (*Fraxinus mandschurica*), Ribbed birch (*Betula costata*) and Mongolian oak (*Quercus mongolica*), and formed a typical Korean pine mixed forest [15]. Korean pine and its associated tree species were both symbiotic and competitive. The PSH would increase the challenges for Korean pine regeneration in the disturbed forest landscape, but it would indirectly promote the regeneration of its associated tree species, which might change the species composition of the MKPBF in the long term and might even alter its succession trajectory. A previous study based on gap model found that once there was no Korean pine provenance, the successional trajectory of forest would change to the mixed broadleaf forest rather than the MKPBF [16]. Currently, it is unclear whether such a cumulative anthropogenic disturbance would significantly change the species composition and succession trajectory of the MKPBF.

It was challenging to assess the legacy effects of seed harvesting on the MKPBF at the landscape scale. Seed harvesting, a landscape process, could synergize with seed dispersal at the landscape scale and competition at the stand scale to affect Korean pine regeneration and further impact the future of the MKPBF. To date, most of the studies focused on investigation and monitoring at the plot or stand scale [6,10–12]. Such research reflects the adverse effect of seed harvesting on some trees or stands at the individual and stand scales, but it is hard to clarify its effect on the MKPBF over a larger temporal and spatial scale. However, the forest landscape model could simulate the synergistic effects among forest population dynamics, seed dispersal and timber harvesting on the MKPBF at the large temporal and spatial scales [4,17–19]. The previous investigation found that the PSH for 20 years decreased the seedling bank by at least 80% [12], and we assumed that the relationship between seed harvesting and seedling bank would also hold in the future. We attempted to simulate harvesting Korean pine seedlings by the forest landscape model to consider the effect of seed harvesting. Thus, the forest landscape model provided a feasible tool to evaluate the effect of seed harvesting at the landscape scale.

In this study, we assessed how the PSH in the past 20 years would influence the species composition and structure of the mixed Korean pine and broadleaf forest in the

Xiaoxing'an Mountains at the landscape scale through a forest landscape model, LANDIS PRO. Specifically, we addressed the following question: (1) whether the negative effect of the historical PSH on Korean pine young age cohort would be delivered to older age cohorts, and then would decrease the age structure diversity of Korean pine; (2) the dominance of which tree species would be promoted by the historical PSH; and (3) whether the historical PSH would change the succession trajectory dominated by the MKPBF into the broadleaf forest or the other mixed coniferous and broadleaf forest. We attempted to evaluate the long-term effect of seed harvesting on the MKPBF at the landscape scale by LANDIS PRO and provided a new view for quantifying its effect on forest.

2. Materials and Methods

2.1. Study Area

The study area is located in the Xiaoxing'an Mountains, with a longitude and latitude range of 127°50′ E to 130°10′ E and 47°05′ N to 49°10′ N, covering an area of about 1.5 million hectares and an altitude between 139 m and 1429 m (Figure 1). This area has a temperate continental monsoon climate, and annual mean temperature, and total precipitation decreases from 1.0 °C and 700 mm in the south to −1.0 °C and 550 mm in the north, respectively. The study area includes eight forestry bureaus and one natural reserve. Overharvesting in the past half century has resulted in the degradation of the primary mixed Korean pine and broadleaf forest (MKPBF) to a secondary forest, and the current stand is dominated by a middle-age forest. The major tree species are listed in Table 1. Since the Natural Forest Conservation Program (NFCP) in 1998, forests in this region have been divided into the special ecological welfare forests (SEWFs), the general ecological welfare forests (GEWFs), and the commercial forests (CFs) for classified management [5]. We will parameterize the management area map required for LANDIS according to these three forest types in the Section 2.2. After 2014, commercial harvest was excluded in the area [3]. At present, the seed harvesting of Korean pine is an important human disturbance in the GEWFs and the CFs.

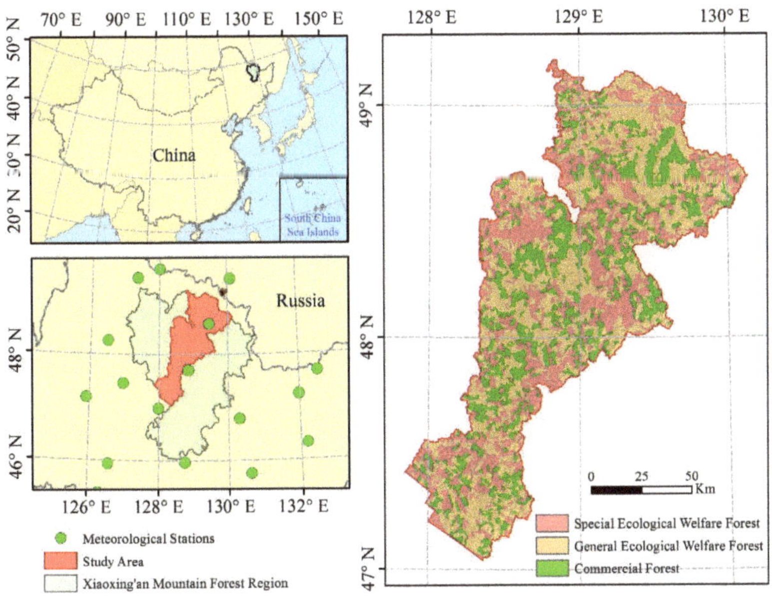

Figure 1. Geographical location and three categories of management zones of the study area.

Table 1. The main species biological traits adapted from liu et al., 2020 [20].

Species	MT	LG	ST	MD	MDBH	MSDI	NPGS
Korean pine, *Pinus koraiensis*	40	300	4	150	110	550	20
Korean spruce, *Picea koraiensis and Picea jezoensis*	30	300	4	150	90	600	20
Khingan fir, *Abies nephrolepis*	30	300	4	150	85	650	20
Larch, *Larix gmelinii*	20	300	2	300	95	650	30
Manchurian ash, *Fraxinus mandschurica*	30	250	3	300	100	600	25
Manchurian walnut, *Juglans mandshurica*	20	250	2	200	90	650	25
Amur corktree, *Phellodendron amurense*	20	250	3	300	95	650	25
Mongolian oak, *Quercus mongolica*	20	300	2	200	95	600	20
Black elm, *Ulmus davidiana*	20	250	3	800	90	600	25
Mono maple, *Acer mono*	20	200	3	200	60	700	25
Ribbed birch, *Betula costata*	20	250	3	800	90	650	25
Dahur Birch, *Betula dahurica*	15	150	2	800	50	750	25
Amur linden, *Tilia amurensis*	30	300	3	200	85	650	20
White birch, *Betula platyphylla*	15	150	1	2000	50	800	30
Poplar, *Populus davidiana*	15	150	1	2000	60	800	30

MT, age of tree species sexual maturity (year); LG, mean maximum age of tree species (year); ST, shade tolerance class (including 1–5, 1 and 5 mean the least and the most tolerance, respectively); MD, maximum dispersal distance (m); MDBH, maximum diameter at breast height (cm); MSDI, maximum stand density index (number of standard trees per hectare, which is 25.5cm tree); NPGS, number of potential germination seeds per mature tree (number/one raster cell).

2.2. Forest Landscape Model and Parameterization

LANDIS PRO tracked changes in the absence or presence of tree species, and recorded the tree number and basal area by age cohorts on raster cells [18,21]. LANDIS can simulate the effects of population dynamics, seed dispersal, and harvesting on forest composition and structure over large spatial and temporal extents with flexible resolutions. The population dynamics driven by species ecological traits include growth, fecundity, colonization, and mortality (Table 1). Seed dispersal, a spatial process, establishes spatial interaction among different cells [22]. Harvesting is an anthropogenic disturbance, and it could release growing space by thinning from above or below. The input parameters for the LANDIS model contained that the spatial data mainly included a species composition map, management area map, harvest unit map and land-type map, and that the non-spatial data mainly comprised the species biological traits, species establishment probability (SEP), maximum growing space occupied (MGSO), and growth curve. Individual growth was simulated by a growth curve, which was determined by the relationship between age and DBH (diameter at breast height). Fecundity was regulated by seed yield and dispersal. Once the species' SEP exceeded a random number ranging from 0 to 1, and, meanwhile, the growing space occupied (GSO) was less than MGSO on one cell, then the individual colonized there. Competition was triggered once the GSO reached the MGSO, and the mortality induced by competition was simulated based on Yoda's thinning theory, which was characterized as the larger diameter class and the lower mortality rate [23]. Additionally, LANDIS incorporated the mortality caused by disturbances or by excessive maximum longevity [24]. Harvesting was conducted by the management area map and harvest unit map, which provided the boundaries for the harvest event and the minimum harvest unit, respectively.

We considered the effect of seed harvesting on the MKPBF by harvesting a Korean pine young cohort with the temporal and spatial resolutions of 5 years and 100 m. We used the forest inventory data and a stand composition map to generate an initial species composition map, which detailed the species distribution and abundance by age cohort on each pixel. We derived the management area map according to forest function type described in the stand composition map, and the map was divided into three classes. We only simulated seed harvesting in the general ecological welfare forest and commercial forest and excluded it in the special ecological welfare forest. We used the boundaries of the stand composition map to generate a harvesting unit map. We divided this area into eight land types based on elevation, aspect and the ≥10 °C cumulative temperature line, and we assumed SEP and MGSO were the same for a certain land type but different among other land types. We evaluated the SEP and MGSO in different land types on the basis of the outputs from the forest ecosystem process model LINKAGES [20]. We parameterized species' ecological traits based on the previous ones in this region [20,25]. The species growth curve was derived from the previous study and the forest inventory data that described the relationship between the number and width of tree rings [20,25]. We used 94 forest inventory plots in 2010 and extracted the corresponding basal area from the LANDIS output to verify the simulation. We used a scatter plot of inventory basal area with LANDIS output at the 10th year of simulation because the initial year of parameterization was 2000 and conducted a linear fit (Figure 2). There was a significant linear correlation between the inventory basal area and LANDIS output ($p < 0.01$, R2 = 0.57), while the ANOVA result showed no significant difference between the inventory basal area and LANDIS output ($p = 0.17 > 0.05$)

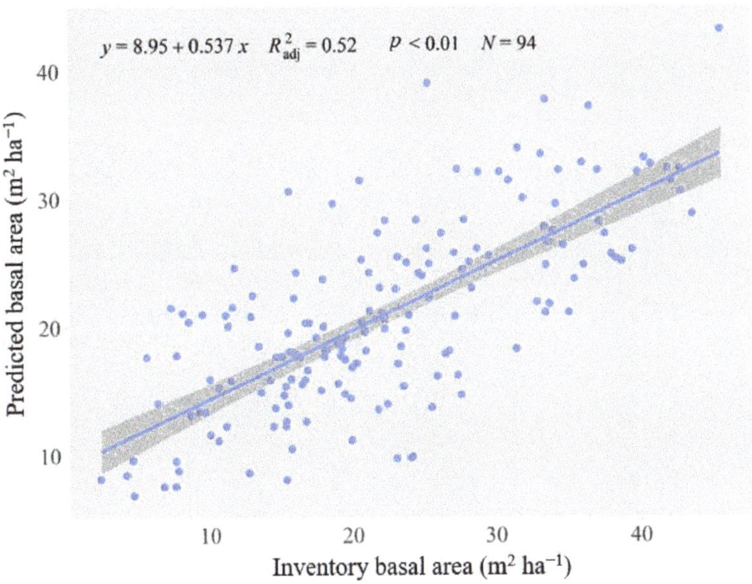

Figure 2. Comparison of the total basal area between forest inventory data and LANDIS outputs in 2010. The grey area indicates the 95% confidence interval of the regression equation.

2.3. Experimental Design

We designed the baseline scenario (BH) and the historical predatory seed harvesting (PSH) scenario (PH) to address how the historical PSH affected the species composition and structure of the mixed Korean pine and broadleaf forest (Table 2). In the BH scenario, we harvested 5% of the distribution area of the Korean pine young cohort every 5 years as a background disturbance to consider the mortality due to animal feeding. A previous inventory found that the PSH lasting for 20 years resulted in the loss of the Korean pine seedling bank by at least 80% [12]. Besides the above background disturbance, we also harvested 80% of the distribution area of the Korean pine young cohort in the first 20 years in the PH scenario to incorporate the effects of the historical PSH. Each scenario was repeatedly simulated five times to account for the uncertainty of the simulation, and the 10 simulations were conducted by LANDIS PRO with the same initial status to predict the variations in the species composition and structure of the MKPBF for 150 years.

Table 2. Forest management scenarios.

Scenarios	Removal Method	Percent Area for Background Harvesting	Percent Area for Historical Predatory Seed Harvesting
Baseline harvest (BH)	Thinning from below	5%	—
Predatory harvest (PH)	Thinning from below	5%	80%

2.4. Data Analysis

We obtained all cells disturbed by the PSH through subtracting the basal area in the BH scenario from that in the PH scenario at each time step for the first 20 years of the simulation. We then summarized the changes in species composition and structure on these cells to reflect the legacy effects of the historical PSH. We divided the Korean pine forest into 5 age cohorts [3], namely young forest (YF, 0–40 years), middle-aged forest (MIF, 40–60 years), near mature forest (NMF, 60–80 years), mature forest (MF, 80–120 years) and aged forest (OGF > 120 years). We counted the basal area of each age cohort output by the two scenarios, respectively, and then used them to calculate the Shannon diversity index in the short-(0–50 years), medium-(55–100 years), and long terms (105–150 years) [26], in order to answer how the historical PSH impacted the age structure of the Korean pine. We calculated the absolute and relative changes in the basal area of tree species under the two scenarios severally to clarify which tree species' dominance the historical PSH improved. Relative changes (increase or decrease) were defined as (PH scenario -BH scenario)/BH scenario*100. We computed the absolute and relative changes in the importance value of tree species in both of the two scenarios, and then we calculated the Bray–Curtis dissimilarity index [27] of the tree species composition in the two scenarios compared with the initial species composition, respectively, to address whether the historical PSH altered the succession trajectory. The importance value (IV), one of the outputs from LANDIS PRO, is described as Equation (1). We used ANOVA to determine whether there were significant differences between the BH scenario and the PH scenario or not. Before executing ANOVA, we tested the normality and variance homogeneity of the data. All data processing and statistical analyses were carried out in R (Version 3.2.2, Vienna, Austria) [28]:

$$IV = \frac{SD}{2 \times TD} + \frac{SBA}{2 \times TBA} \quad (1)$$

where *SD*, *SBA*, *TD* and *TBA* represent species *i* density, species *i* basal area, total density and total basal area, respectively.

3. Results

3.1. Change in the Age Structure of Korean Pine

The historical predatory seed harvesting (PSH) averagely decreased the Korean pine basal area irrespective of age cohorts throughout the simulation. Compared with the baseline scenario (BH), the historical PSH scenario (PH) averagely decreased the basal area of young forest (YF), middle-age forest (MIF), near-mature forest (NMF), mature forest (MF) and old-growth forest (OGF) by 0.06, 0.18, 0.13, 0.19 and 0.15 m^2 ha^{-1}, respectively (Figure 3a,b). The historical PSH insignificantly altered the age structure diversity of Korean pine ($p > 0.05$). In comparison with the BH scenario, the PH scenario increased the age structure Shannon–Wiener index of Korean pine by 0.007 on average, with a relative increase of 0.5% in the short term (Figure 4a), while it averagely reduced the index by 0.009 and 0.012, with relative decreases of 0.6% and 1.1% in the medium and long terms (Figure 4b,c).

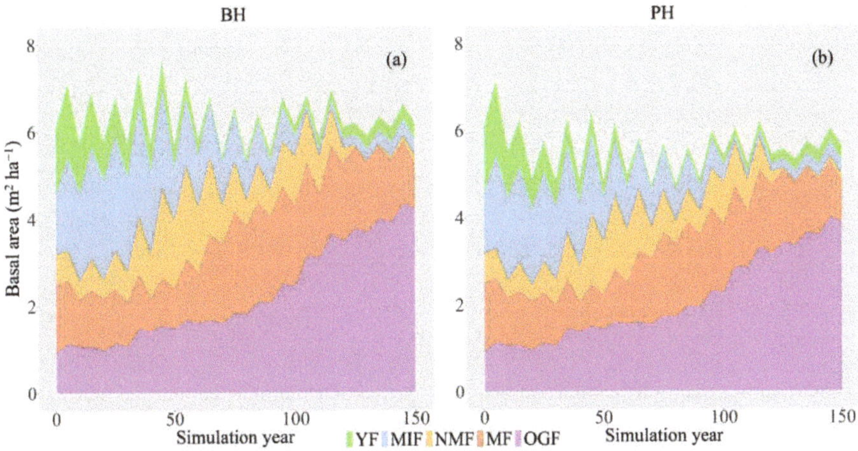

Figure 3. Each age cohort basal area of Korean pine across the simulation period under the BH and PH scenarios. BH, the baseline scenario; PH, the historical PSH scenario. YF, young forest; MIF, middle-age forest; NMF, near-mature forest; MF, mature forest; OGF, old-growth forest.

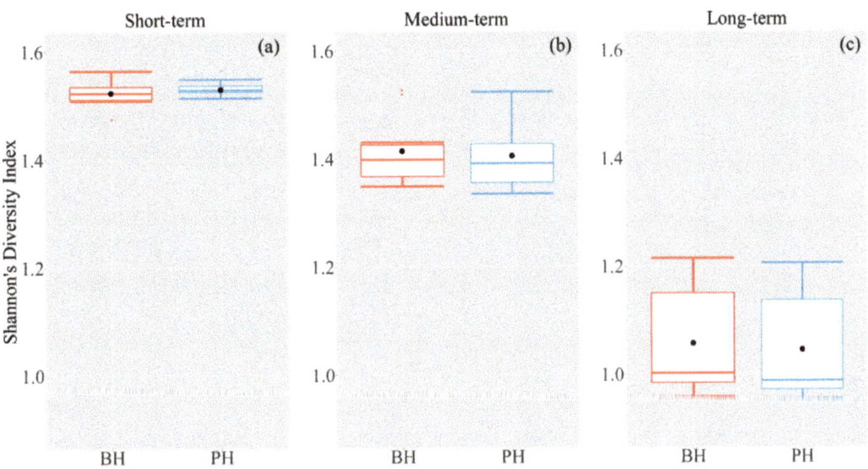

Figure 4. The age structure Shannon–Wiener index of Korean pine in the short-, medium-, and long term under the BH and PH scenarios. The black dots represented the average. The variance referred to the variance among five replicates of Shannon–Wiener index on all cells disturbed by seed harvesting in different terms under the two scenarios, respectively. BH, the baseline scenario; PH, the historical PSH scenario.

3.2. Change in Species Composition

The historical PSH significantly decreased the Korean pine basal area ($p < 0.01$) but remarkably increased spruce and fir ($p < 0.05$) throughout the simulation. In comparison with the BH scenario, the PH scenario averagely decreased the Korean pine basal area by 0.63 m^2 ha^{-1} (Figure 5a), with a relative decrease of 11.1% (Figure 6a), but it increased the basal area of spruce and fir by 0.19 and 0.23 m^2 ha^{-1} (Figure 5b,c), with a relative increase of 3.8% and 4.5% respectively (Figure 6b,c), while it little improved the basal area of larix, ash, walnut, corktree, oak, elm, maple, ribbed birch, dahur birch, linden, white birch, and poplar, with a range from 0.0003 to 0.0536 m^2 ha^{-1} (Figure 5d–o) and a relative change from 0.6% to 2.7% (Figure 6d–o).

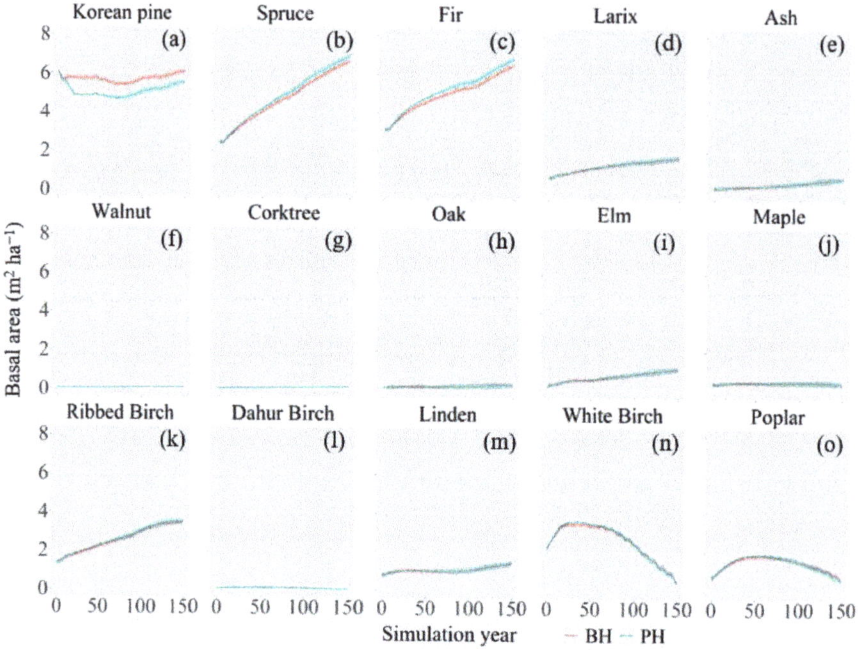

Figure 5. Species basal area under the BH and PH scenarios during the simulation. BH, the baseline scenario; PH, the historical PSH scenario. Error bands indicate one standard deviation of the five replicates.

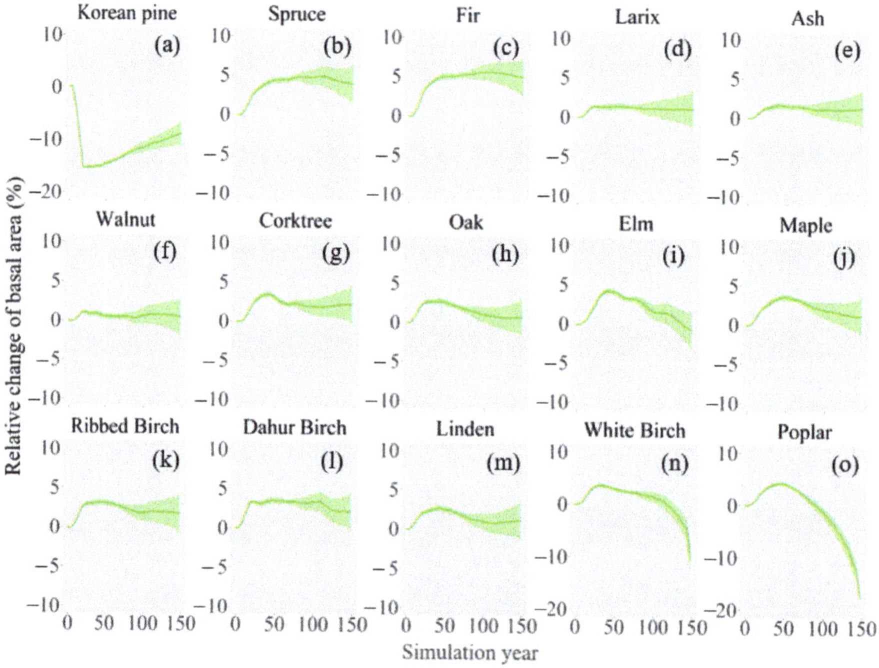

Figure 6. Relative changes in the basal area of each species under the PH scenario compared with the BH scenario over the simulation. Error bands indicate one standard deviation of the five replicates.

3.3. *Change in Succession Trajectory*

The historical PSH amplified the difference in species composition and significantly altered forest succession trajectory ($p < 0.01$). Compared with the initial year, the Bray–Curtis dissimilarity index of species composition gradually increased under both of the BH and PH scenarios over the simulation, and the index reached 0.21 and 0.24 at the end of simulation, respectively (Figure 7). In comparison with the BH scenario, the PH scenario averagely increased the Bray–Curtis dissimilarity index by 0.03 with a relative increase of 33.2% during the simulation period.

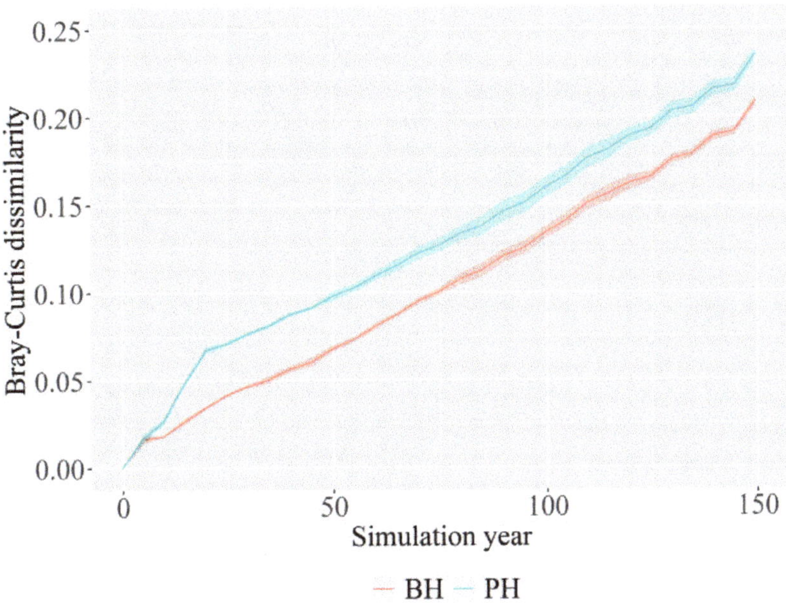

Figure 7. The Bray–Curtis dissimilarity index of species composition under the BH and PH scenarios over the simulation compared with the initial year. BH, the baseline scenario; PH, the historical PSH scenario. Error bands indicate one standard deviation of the five replicates.

The historical PSH significantly decreased the importance of Korean pine, while it remarkably increased spruce and fir in the disturbed forest. Compared with the BH scenario, the PH scenario decreased the importance value of Korean pine by 0.05 on average with a relative decrease of 12.2% (Figures 8a and 9a, $p < 0.01$) but increased both spruce and fir by 0.02 with relative increases of 5.1% and 6.0% (FIgures 8b,c and 9b,c, $p < 0.01$), respectively. The PH scenario barely increased the importance value of larix, ash, walnut, corktree, oak, elm, maple ($p < 0.05$), ribbed birch ($p < 0.05$), dahur birch, linden, and white birch in comparison with the BH scenario, while little decreasing poplar, with a range from 0.001 to 0.003 and a relative change from 0.5% to 2.8% (Figures 8d–o and 9d–o).

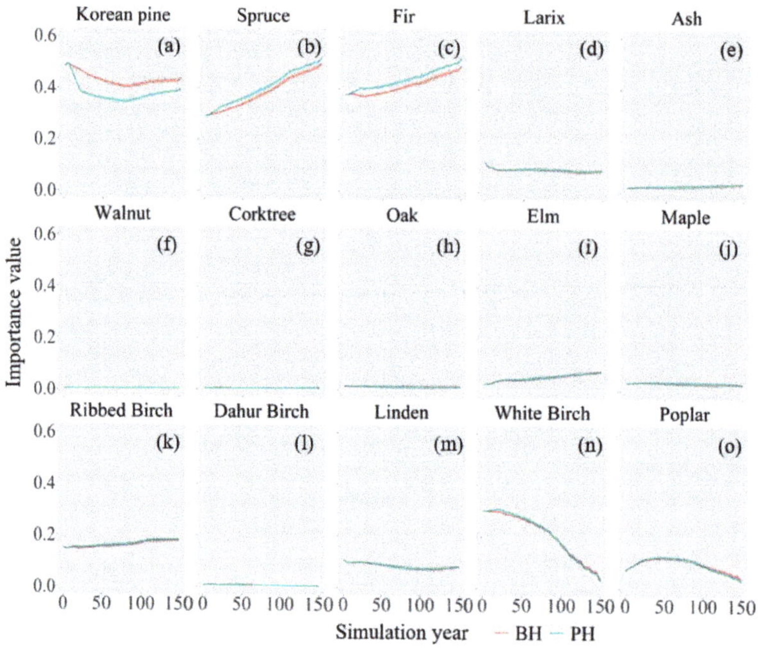

Figure 8. The importance value of each species during the simulation period under the BH and PH scenarios. BH, the baseline scenario; PH, the historical PSH scenario. Error bands indicate one standard deviation of the five replicates.

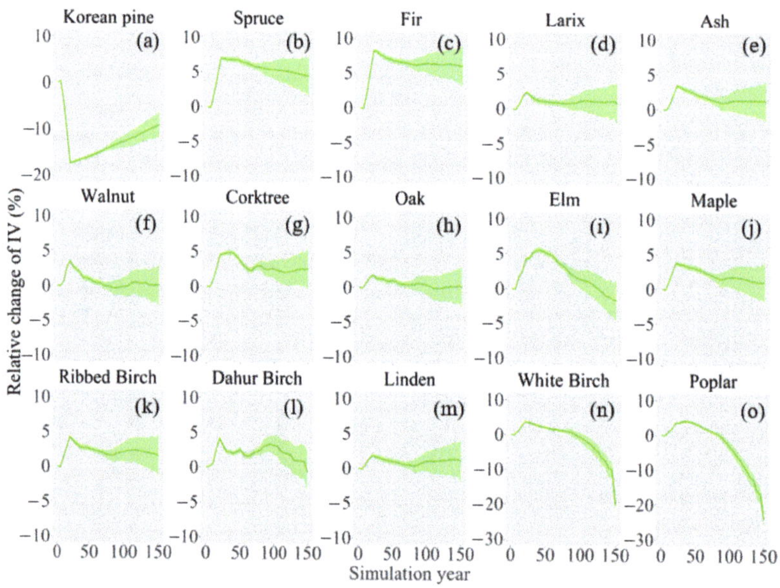

Figure 9. Relative changes in the importance value of each species under the PH scenario compared with the BH scenario over the simulation. Error bands indicate one standard deviation of the five replicates.

4. Discussion

The previous studies showed that seed harvesting decreased the quantities of seed bank and seedling bank at the site or stand scale [10–12], while it was unclear how such a disturbance affected the MKPBF at the landscape scale. We applied a forest landscape model to quantify the legacy effect of seed harvesting on the mixed Korean pine and broadleaf forest (MKPBF) at the landscape scale, which broadened understanding of seed harvesting at the large temporal and spatial scales. Seed harvesting, a spatially landscape process, can occur to a large spatial extent, such as several to hundreds of hectares, which not only affects the disturbed landscape, but also has impacts on the landscape adjacent to the disturbed landscape by seed dispersal. Seed harvesting directly reduces provenance and also indirectly synergizes with the stand-scale competition to limit the effectiveness of seed dispersal in these direct or indirect ways, which further adversely affects the regeneration and recovery of target tree species. A recent study found that the provenance restriction of Korean pine decreased the effectiveness of seed dispersal and impeded the Korean pine recruitment and the recovery of secondary forests to the MKPBF [14], which was consistent with our result.

Our result showed that the historical predatory seed harvesting (PSH) unremarkably altered the age structure of Korean pine throughout the simulation. Compared with the baseline scenario, although the historical PSH insignificantly decreased the age structure diversity of Korean pine during the whole simulation period, it resulted in a decreasing trend of diversity in the medium- and long terms (Figure 3). However, if the PSH continued for a long time, it would evidently reduce the age structure diversity of Korean pine in the long run. A previous study found that such high seed harvesting significantly decreased the distribution of the Korean pine young age cohort [4], which was consistent with our result. In addition, we found that the adverse effect of the historical PSH on young Korean pine had a significant cascade effect and lessened the abundance of older age cohorts (Figure 3).

The legacy effects of the historical PSH are long lasting, and the differences in tree species composition and succession trajectory in future forests have mainly been caused by the PSH in the past 20 years. Our result showed that the historical PSH remarkably decreased the dominance of Korean pine while significantly increasing spruce and fir (Figures 5 and 6). This is due to the fact that the PSH reduces the provenance of Korean pine and decreases its chances of colonizing and establishing, which further releases the growth space that would otherwise be occupied by Korean pine. Spruce and fir seedlings occupy the released growth space since they are the climax species in the mixed Korean pine and broadleaf forest region and they share similar ecological niches to Korean pine [29], which thus contributes to their dominance. Our result found that the succession trajectory of the disturbed forest landscape was significantly different, and the trajectory was altered from that dominated by the MKPBF to that dominated by the spruce, fir and broadleaf forest (Figures 7–9). The previous study based on the gap model found that the abundance of Korean pine significantly decreased with the reduction of saplings [30]. We infer that once the PSH lasts for a long time, it will be bound to change the succession trajectory of the MKPBF to a greater extent. However, a previous study based on the gap model found that the MKPBF would eventually evolve into a mixed broadleaved forest in the absence of Korean pine provenance [16], which is inconsistent with our result. This is because the initial stand condition in their sampling sites excludes species such as spruce and fir that occupied similar ecological niches to Korean pine, and because the gap model can only simulate forest succession at the stand scale but cannot spatially simulate seed dispersal. However, the forest landscape model is used to consider more tree species and seed dispersal process at the landscape level in this study, which can more truly reflect the forest succession dynamics.

The PSH has become an important obstacle to the regeneration and restoration of the MKPBF. The previous studies found that seed predation and seedling damage by terrestrial small mammals were often primary limitations that adversely influenced the regeneration and community dynamic of the coniferous forest, particularly in the case of an insufficient seed supply due to excessive human seed harvesting [10,31,32]. Korean pine relies on the rodents with a propensity to bury food to finish seed dispersal, and they are both predators and dispersers. When seeds are in short supply, the rodents prefer to eat them in situ rather than spread and bury them [32]. Due to the historically excessive timber harvesting of Korean pine and the continued PSH, the stock of Korean pine seed in the secondary forest is difficult to meet the consumption levels by predators, and therefore, few Korean pine seeds escape predation and successfully germinate, which seriously hinders the regeneration of Korean pine [14]. Thus, the intensity and rotation of seed harvesting should be severely restricted in the future to ensure that there are enough seeds to sustain the regeneration of Korean pine and the recovery of the MKPBF.

Future forest management should also focus on how to promote the restoration of the MKPBF through artificial silvicultural treatments, such as direct seeding or seedling planting so as to ensure the sustainable management of the MKPBF. A recent study found that only 9% of Korean pine seedlings survived after direct seeding due to large amount of animal feeding, while seedlings planting in forest gaps had a higher survival proportion, which was a more effective method to promote the regeneration of Korean pine and the recovery of the MKPBF [32]. However, planting seedlings is costly, and the costs connected with seedling plantation (such as transportation, planting, weeding, watering, and long-period management) might significantly restrict the potential application of seedling plantation over the large spatial extent. Additionally, the planted seedlings might initially have insufficient resistance to cope with extreme natural or anthropogenic disturbances. In comparison with planting seedling, direct seeding is a simple and cheap treatment [33,34]; although the survival ratio of seedlings is not high, it can also promote tree regeneration to a certain extent [35]. Therefore, direct seeding might be more appropriate for restoring the MKPBF within a large spatial extent.

Our study also had some limitations. Ignoring that the population dynamics of animals that feed on Korean pine seeds might increase uncertainty about the impact of seed harvesting on the regeneration of Korean pine, generally, Korean pine mainly depends on the seeds buried by animals to complete regeneration [32]. However, seed harvesting would reduce the food source of these animals, which would decrease their population and might subsequently reduce the magnitude of buried seeds, and therefore adversely affect the regeneration of Korean pine. We assumed that the boundaries of the current management area and stand remained consistent throughout the simulation, but forest management policies might adjust in the future, and changes in policies might alter these boundaries. Although these neglected factors might increase the uncertainty of the model prediction, our results still reflect the legacy effects of the historical PSH on the composition and structure of the MKPBF in the future and provide a reference for the restoration of the MKPBF at the landscape scale.

5. Conclusions

We demonstrated how to quantify the effects of seed harvesting at the landscape scale by a forest landscape model, LANDIS PRO, in this study. We evaluated the legacy effects of the historical predatory seed harvesting (PSH) on species composition and structure of the mixed Korean pine and broadleaf forest (MKPBF) for 150 years in the future. We found that the historical PSH had a cascading effect on the age structure of Korean pine, and the adverse effect was transmitted from the young cohort to the older cohort. Although the historical PSH had negative effects on all age cohorts of Korean pine, it altered insignificantly the diversity of the age cohort of Korean pine. The historical PSH had a long-lasting effect and would decrease the dominance of Korean pine but increase spruce and fir. Additionally, we found that the historical PSH would change the succession trajectories of the disturbed stand and would transform the succession pattern from Korean pine as the predominant species to spruce and fir. Future forest management should reduce the intensity and rotation of seed harvesting and consider using direct seeding or seedling planting to promote the recovery and sustainable development of the MKPBF.

Author Contributions: Conceptualization, K.L. and H.S.H.; formal analysis, K.L. and H.S.; funding acquisition, K.L. and H.S.H.; resources, K.L. and J.W.; supervision, H.S.H.; writing—original draft, K.L.; writing—review and editing, K.L., H.S.H., H.S. and J.W. All authors have read and agreed to the published version of the manuscript.

Funding: This research was funded by the National Natural Science Foundation of China, grant number 42101107, the Natural Science Foundation of Jilin Province, China, grant number YDZJ202201ZYTS487, the Joint Fund of National Natural Science Foundation of China, grant number U19A2023, the National Key Research and Development Program of China, grant number 2017YFA0604403-3, the Fundamental Research Funds for the Central Universities.

Data Availability Statement: The data presented in this study are available on request from the corresponding author.

Acknowledgments: We thank Jiangtao Xiao for the assistance about parameterization of LANDIS PRO and thank Guohua Song for the introduction of the mixed Korean pine and broadleaf forest.

Conflicts of Interest: The authors declare no conflict of interest.

References

1. Yu, D.; Zhou, W.; Zhou, L.; Dai, L. Exploring the history of the management theory and technology of broad—Leaved Korean pine (*Pinus koraiensis* Sieb. et Zucc.) forest in Changbai Mountain Region, Northeast China. *Chin. J. Appl. Ecol.* **2019**, *30*, 1426–1434.
2. Zheng, X.; Yu, D.; Wang, A.; Dai, G. Long-term Monitoring and Field Control Experiments Reveal the Structure, Function and Evolution of Typical Temperate Forests. *Bull. Chin. Acad. Sci.* **2019**, *34*, 948–961.
3. Liu, K.; Liang, Y.; He, H.S.; Wang, W.J.; Huang, C.; Zong, S.; Wang, L.; Xiao, J.; Du, H. Long-term impacts of China's new commercial harvest exclusion policy on ecosystem services and biodiversity in the temperate forests of Northeast China. *Sustainability* **2018**, *10*, 1071. [CrossRef]
4. Zhao, F.; He, H.; Dai, L.; Yang, J. Effects of human disturbances on Korean pine coverage and age structure at a landscape scale in Northeast China. *Ecol. Eng.* **2014**, *71*, 375–379. [CrossRef]
5. Dai, L.; Li, S.; Zhou, W.; Qi, L.; Zhou, L.; Wei, Y.; Li, J.; Shao, G.; Yu, D. Opportunities and challenges for the protection and ecological functions promotion of natural forests in China. *For. Ecol. Manag.* **2018**, *410*, 187–192. [CrossRef]
6. Ji, L.; Liu, Z.; Hao, Z.; Wang, Q.; Wang, M. Effect of Cones Picking on Broad-leaved *Pinus koraiensis* Forest in Changbai Mountian. *Chin. J. Ecol.* **2002**, *21*, 39–42.
7. Yu, D.; Zhou, W.; Bao, Y.; Qi, L.; Zhou, L.; Dai, L. Forest management of Korean pine and broadleaf mixed forest in Northeast China since the implementation of Natural Forest Protection Project. *Acta Ecol. Sin.* **2015**, *35*, 10–17.
8. Zhou, G.; Liu, Q.; Xu, Z.; Du, W.; Yu, J.; Meng, S.; Zhou, H.; Qin, L.; Shah, S. How can the shade intolerant Korean pine survive under dense deciduous canopy? *For. Ecol. Manag.* **2020**, *457*, 117735. [CrossRef]
9. Liu, Z.; Ji, L.; Hao, Z.; Zhu, J.; Kang, H. Effect of cone-picking on natural regeneration of Korean pine in Changbai Mountain Nature Reserve. *Chin. J. Appl. Ecol.* **2004**, *15*, 958–962.
10. Piao, Z.; Tang, L.; Swihart, R.K.; Wang, S. Human–wildlife competition for Korean pine seeds: Vertebrate responses and implications for mixed forests on Changbai Mountain, China. *Ann. For. Sci.* **2011**, *68*, 911–919. [CrossRef]
11. Li, Y.-B.; Mou, P.; Wang, T.-M.; Ge, J. Evaluation of regeneration potential of *Pinus koraiensis* in mixed pine-hardwood forests in the Xiao Xing'an Mountains, China. *J. For. Res.* **2012**, *23*, 543–551. [CrossRef]
12. Jin, G.; Yang, G.; Ma, J.; Li, L.; Xu, Z.; Zhao, X.; Hong, M. Effect of Anthropogenic Cone-picking on Seed Bank and Seedling Bank of Korean Pine in the Major Forest Types in Lesser Hing'an Mountains. *J. Nat. Resour.* **2010**, *25*, 1845–1854.
13. Li, J.; Li, J. Regeneration and restoration of broad-leaved Korean pine forests in Lesser Xing'an Mountains of Northeast China. *Acta Ecol. Sin.* **2003**, *23*, 1268–1277.
14. Wang, J.; Geoff Wang, G.; Li, R.; Sun, Y.; Yu, L.; Zhu, J.; Yan, Q. Cascading effect of source limitation on the granivore-mediated seed dispersal of Korean pine (*Pinus koraiensis*) in secondary forest ecosystems. *Ecol. Process.* **2022**, *11*, 1. [CrossRef]
15. Zhou, Y. *Vegetation in the Small Khingan Mountains of China*; Science Press: Beijing, China, 1994.
16. Yu, Z.; Zhao, S.; Wang, Q.; Zhai, Y. Simulation of polar-birch forest succession in broad-leaved-Korean pine forest zone in Mt. Changbai. *Chin. J. Appl. Ecol.* **1997**, *8*, 455–458.
17. He, H.S.; Hao, Z.; Mladenoff, D.J.; Shao, G.; Hu, Y.; Chang, Y. Simulating forest ecosystem response to climate warming incorporating spatial effects in north-eastern China. *J. Biogeogr.* **2005**, *32*, 20432056. [CrossRef]
18. He, H.S.; Gustafson, E.J.; Lischke, H. Modeling forest landscapes in a changing climate: Theory and application. *Landscape Ecol.* **2017**, *32*, 1299–1305. [CrossRef]

19. Shifley, S.R.; He, H.S.; Lischke, H.; Wang, W.J.; Jin, W.; Gustafson, E.J.; Thompson, J.R.; Thompson, F.R.; Dijak, W.D.; Yang, J. The past and future of modeling forest dynamics: From growth and yield curves to forest landscape models. *Landsc. Ecol.* **2017**, *32*, 1307–1325. [CrossRef]
20. Liu, K.; He, H.; Xu, W.; Du, H.; Zong, S.; Huang, C.; Wu, M.; Tan, X.; Cong, Y. Responses of korean pine to proactive managements under climate change. *Forests* **2020**, *11*, 263. [CrossRef]
21. Wang, W.J.; He, H.S.; Spetich, M.A.; Shifley, S.R.; Thompson, F.R., III; Dijak, W.D.; Wang, Q. A framework for evaluating forest landscape model predictions using empirical data and knowledge. *Environ. Modell. Softw.* **2014**, *62*, 230–239. [CrossRef]
22. He, H.S.; Mladenoff, D.J. The effects of seed dispersal on the simulation of long-term forest landscape change. *Ecosystems* **1999**, *2*, 308–319. [CrossRef]
23. Yoda, K. Self-thinning in overcrowded pure stands under cultivated and natural conditions (Intraspecific competition among higher plants. XI). *J. Inst. Polytech. Osaka City Univ. Ser. D.* **1963**, *14*, 107–129.
24. Wang, W.J.; He, H.S.; Fraser, F.R.T.J.S., III; Hanberry, B.B.; Dijak, W.D. Importance of succession, harvest, and climate change in determining future composition in US Central Hardwood Forests. *Ecosphere* **2015**, *6*, 277. [CrossRef]
25. Xiao, J.; Liang, Y.; He, H.S.; Thompson, J.R.; Wang, W.J.; Fraser, J.S.; Wu, Z. The formulations of site-scale processes affect landscape-scale forest change predictions: A comparison between LANDIS PRO and LANDIS-II forest landscape models. *Landsc. Ecol.* **2017**, *32*, 1347–1363. [CrossRef]
26. Shannon, C.E. A mathematical theory of communication. *Bell Syst. Tech. J.* **1948**, *27*, 379–423. [CrossRef]
27. Faith, D.P.; Minchin, P.R.; Belbin, L. Compositional dissimilarity as a robust measure of ecological distance. *Vegetatio* **1987**, *69*, 57–68. [CrossRef]
28. Team, R.C. *R: A Language and Environment for Statistical Computing (Version 3.2.2)*; R Foundation for Statistical Computing: Vienna, Austria, 2015.
29. Zhao, H.; Kang, X.; Guo, Z.; Yang, H.; Xu, M. Species interactions in spruce–fir mixed stands and implications for enrichment planting in the Changbai Mountains, China. *Mt. Res. Dev.* **2012**, *32*, 187–196. [CrossRef]
30. Shao, G.; Schall, P.; Weishampel, J.F. Dynamic simulations of mixed broadleaved-*Pinus koraiensis* forests in the Changbaishan Biosphere Reserve of China. *For. Ecol. Manag.* **1994**, *70*, 169–181. [CrossRef]
31. De Oliveira Wadt, L.H.; Faustino, C.L.; Staudhammer, C.L.; Kainer, K.A.; Evangelista, J.S. Primary and secondary dispersal of Bertholletia excelsa: Implications for sustainable harvests. *For. Ecol. Manag.* **2018**, *415*, 98–105. [CrossRef]
32. Wang, J.; Wang, G.G.; Zhang, T.; Yuan, J.; Yu, L.; Zhu, J.; Yan, Q. Use of direct seeding and seedling planting to restore Korean pine (*Pinus koraiensis* Sieb. Et Zucc.) in secondary forests of Northeast China. *For. Ecol. Manag.* **2021**, *493*, 119243. [CrossRef]
33. Freitas, M.G.; Rodrigues, S.B.; Campos-Filho, E.M.; do Carmo, G.H.P.; da Veiga, J.M.; Junqueira, R.G.P.; Vieira, D.L.M. Evaluating the success of direct seeding for tropical forest restoration over ten years. *For. Ecol. Manag.* **2019**, *438*, 224–232. [CrossRef]
34. Raupp, P.P.; Ferreira, M.C.; Alves, M.; Campos-Filho, E.M.; Sartorelli, P.A.R.; Consolaro, H.N.; Vieira, D.L.M. Direct seeding reduces the costs of tree planting for forest and savanna restoration. *Ecol. Eng.* **2020**, *148*, 105788. [CrossRef]
35. Atondo-Bueno, E.J.; Bonilla-Moheno, M.; López-Barrera, F. Cost-efficiency analysis of seedling introduction vs. direct seeding of Oreomunnea mexicana for secondary forest enrichment. *For. Ecol. Manag.* **2018**, *409*, 399–406. [CrossRef]

forests

MDPI

Article

Rainfall Modifies the Disturbance Effects on Regulating Ecosystem Services in Tropical Forests of Bangladesh

Fahmida Sultana [1], Mohammed A. S. Arfin-Khan [1,*], Md Rezaul Karim [2] and Sharif A. Mukul [3,4,*]

1 Department of Forestry and Environmental Science, School of Agriculture and Mineral Sciences, Shahjalal University of Science and Technology, Sylhet 3114, Bangladesh
2 Institute of Forestry and Conservation, University of Toronto, 33 Willcocks Street, Toronto, ON M5S 3B3, Canada
3 Tropical Forests and People Research Centre, University of the Sunshine Coast, Maroochydore DC, QLD 4556, Australia
4 Department of Earth and Environment, College of Arts, Sciences and Education, Florida International University, Miami, FL 33199, USA
* Correspondence: khan-for@sust.edu (M.A.S.A.-K.); smukul@usc.edu.au (S.A.M.)

Abstract: Tropical forests are critical for ecosystem functioning and are the major provider of global ecosystem services. The combined effects of climate change and anthropogenic disturbances in the tropics, however, have caused a mass decline in major ecosystem services in recent decades. Nevertheless, most previous studies focused entirely on these disturbances' effects on forest ecosystem services without taking into consideration the combined effects of climate change and disturbances on key ecosystem services' indicators. This is more pronounced in Bangladesh, which is one of the countries most adversely affected by global climate change. We investigated anthropogenic disturbances' effects on regulating ecosystem services under different rainfall regimes in Khadimnagar National Park (KNP) and Satchari National Park (SNP), two major forested, protected areas in Northeast Bangladesh. We considered twelve indicators for five aboveground ecosystem services (i.e., primary production, carbon storage, biodiversity, wildlife habitat, and timber production) and four indicators for four belowground ecosystem services (i.e., organic carbon storage, soil aggregate stability, soil physical support, and soil water regulation) in our study. The effects of disturbances on the indicators of four aboveground ecosystem services (i.e., primary production, tree carbon storage, biodiversity conservation, and timber production) and one belowground ecosystem service (soil physical support) differed in terms of rainfall regimes. Disturbances negatively influenced indicators of primary production ($p = 0.03$, $r = -0.28$), tree carbon storage ($p = 0.02$, $r = -0.29$), and timber production ($p = 0.002$, $r = -0.39$), and positively impacted soil organic carbon storage ($p = 0.04$, $r = 0.27$) only under a medium-rainfall regime. Interestingly, a significant positive impact of disturbances on soil physical support ($p = 0.001$, $r = 0.39$) was observed under heavy rainfall in our study. In addition, disturbances were significantly associated with biodiversity conservation ($p < 0.05$) and wildlife habitat quality ($p < 0.05$) under both rainfall classes. Our findings provide critical insights into the synergistic role of anthropogenic perturbations in regulating ecosystem services with rainfall pattern as a key climatic parameter, which can be useful in restoring large tracts of degraded tropical forests in the United Nations (UN) Decade on Ecosystem Restoration.

Keywords: ecosystem functions; regulating ecosystem services; disturbance; rainfall pattern; tropical forest

1. Introduction

Tropical forests are a major ecosystem services provider through a wide range of critical ecosystem functions [1,2]. One-fourth of Earth's terrestrial carbon (C) is stored in aboveground biomass (AGB) in tropical forests, which is also one of the major forest ecosystem services with implications for climate change mitigation [3,4]. The increasing frequency of anthropogenic disturbances in the tropics, along with global climate change, has significantly altered forest health and its capacity to provide key ecosystem functions and services [3,5]. A limited number of studies have assessed the effects of disturbances [6] and climate change [3] on tropical forest ecosystem services. The combined effects of these two major factors have rarely been studied. Our study, therefore, aims to assess the combined effects of disturbances and climatic patterns on tropical forest regulating ecosystem services in Bangladesh, with a focus on rainfall regimes.

Most tropical forests historically have supported low-income forest-resource-dependent communities by providing fodder, firewood, and timber [7,8]. Disturbing small portions of forested areas or the removal of plant biomass is known as an 'anthropogenic disturbance' [9]. The effects of such disturbances on the aboveground and belowground biodiversity of tropical forests and the provision of ecosystem services have been investigated globally in recent years [10,11]. For anthropogenic disturbances, we refer to the following: decreases in plant population and densities [12], illegal tree felling/removal in forested areas and disrupted forest regeneration, the decrease in tree species richness and tree abundance [13], alterations in forest structure, the composition of tree species [12], and finally the decrease in tree functional diversity [10].

Global ongoing climate change also alters tropical forest biodiversity and health through imposing strong environmental filters within this unique ecosystem by climate extremes [14,15]. Tropical forests will face warmer temperatures by 2100 [16], leading to extreme drought events which can reduce overall forest ecosystem services by lessening net primary production [17]. Future climate change can also alter annual population growth [18] and enhance the chance of species elimination [19]. In fact, less precipitation can reduce trees' natural regeneration and survival rate in dry forests, hypothetically changing the structure and composition of tropical ecosystems to new rigid states, such as shrubland or grassland [20]. Changing rainfall regimes also have a significant impact on tropical forests' soil ecosystem services [21].

There has been growing evidence that anthropogenic disturbances and climate change can have complex (sometimes synergistic) effects on biological diversity [22]. It is likely that hot and arid environments are at a higher risk [23]. Yet, to our knowledge, the isolated and interactive effects of climate (mostly the degree of local rainfall) and anthropogenic disturbances on tropical forest ecosystem services have not fully been explored. To address this research gap, we examined two protected areas in Northeast Bangladesh, located in the most diverse and wettest region of the country [24,25]. The region has historically been subject to anthropogenic disturbances and is highly prone to global warming [11,26–28]. Our study assessed the combined effects of local rainfall regimes and anthropogenic disturbances on both the above and belowground ecosystem services in the area. We believe our study will improve our understanding of the separate and combined effects of rainfall patterns and anthropogenic disturbances on major forest ecosystem services, thus allowing us to predict (and possibly mitigate) forthcoming changes in forest vegetation and to restore this biodiverse forest ecosystem in the recent UN Decade on Ecosystem Restoration [29].

2. Materials and Methods

2.1. The Study Area

The present study was conducted in two forested, protected areas, namely Khadimnagar National Park (KNP) and Satchari National Park (SNP). These two forested, protected areas are located in the northeastern region of Bangladesh (Figure 1). Ecologically, both areas lie in tropical evergreen and semi-evergreen biogeographic zone. The Forest Department declared KNP as a national park in 2009 and SNP in 2005 [24]. The total areas of the KNP and SNP are 679 ha and 243 ha, respectively [30]. Geographically, the KNP lies between 24°56′ and 24°58′ N latitude and 91°55′ and 91°59′ E longitude. The SNP lies between 24°5′ and 24°10′ N latitude and 91°25′ and 91°30′ E longitude. The maximum and minimum mean temperatures of KNP are 30.7 °C and 18.9 °C, whereas the maximum and minimum mean temperatures of SNP are 34.8 °C and 9.6 °C, respectively. The total annual precipitation is 3931 mm/year^{-1} with July being the wettest month, having an average of about 1250 mm of rain [11]. In SNP, the average annual rainfall is 4162 mm [1]. The soil of Khadimnagar hills is slightly fertile compared to that of the SNP. The soil textures of both forested areas are sandy loams [11].

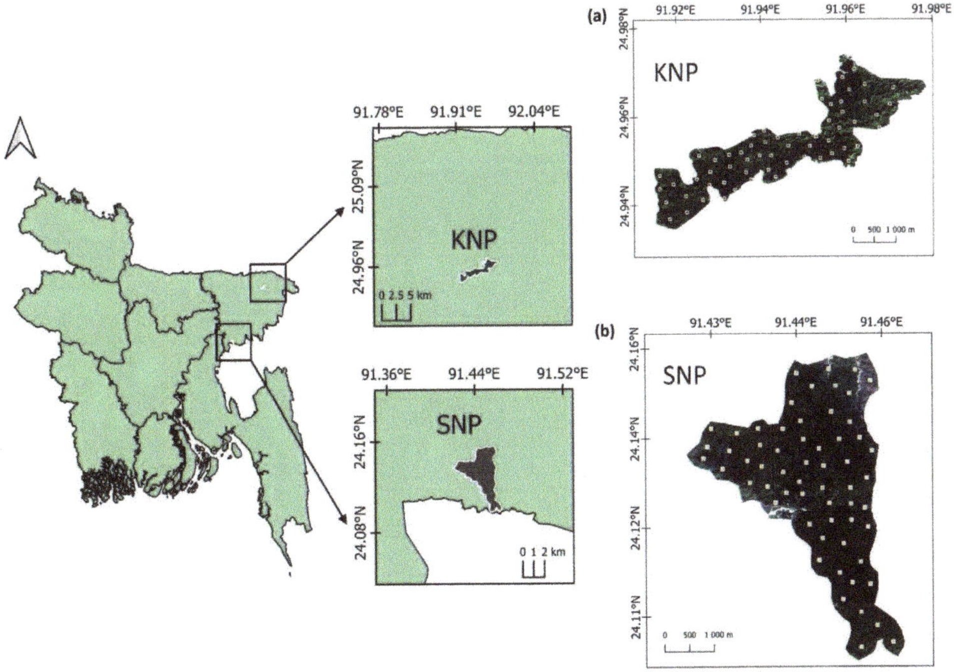

Figure 1. Location map of two studied protected areas (**a**) Khadimnagar National Park (KNP) and (**b**) Satchari National Park (SNP) in Northeast Bangladesh. The yellow dots indicate the plot location within each of the study areas.

2.2. Sampling Design and Plot Selection

From each study area, 60 main plots were selected for tree data collection. The plot size was 20 × 20 m^2. Within each main plot, 3 sampling sub-plots of 2 × 2 m^2 were chosen randomly for shrub and herb species. Altogether, data were collected from 120 main plots and 180 sub-plots. Main plots were randomly generated through ArcGIS Desktop (Version 10.5.1, Redlands, California, USA) geoprocessing tool with surrounding areas. When a randomly generated plot location was unreachable, plots were carefully chosen from the adjacent forested areas with similar biophysical conditions. The boundaries of each of the main plots and sub-plots were pegged and marked before data collection.

2.3. Indicators of Forest Ecosystem Services

We measured effects of disturbances on nine forest ecosystem services based on 16 response parameters (Table 1).

Table 1. List of responses associated with key ecosystem services in our study based on previous research.

	Ecosystem Service	Response	Source
Aboveground	Primary production	Tree biomass	Richmond et al., 2007 [31]
	Carbon storage	Tree biomass carbon	Fahey et al., 2010 [32]
	Biodiversity conservation	Tree abundance Tree richness Herb abundance Herb richness Shrub abundance Shrub richness	Gould, 2000 [33]
	Wildlife habitat	Herb–shrub cover Tree cover	Grebner et al., 2013 [34]
	Timber production	Average tree diameter Average tree height	Aishan et al., 2015 [35]
Belowground	Organic carbon storage Soil aggregate stability Soil physical support Soil water regulation	Soil organic carbon % Soil sand % Soil bulk density Soil moisture content	Fahey et al., 2010 [32] Karim et al., 2020a [11] Karim et al., 2020a [11] Doran and Parkin, 2015 [36]

2.4. *Assessment of Disturbance and Ecosystem Services*

2.4.1. Field Data Collection

Field data collection was conducted from August to October 2019. For each site, the general characteristics of each plot were recorded. Geographic coordinates and elevation were measured for topographical factors in the center of plots with a hand-held GPS (Garmin Ltd., Schaffhausen, Switzerland). Tree, shrub, and herb coverage was collected at the plot level. A mobile device with the application Compass (Xiaomi Compass 9.2.8, Beijing, China) was used to measure slopes (degree) [37]. Non-destructive approach was followed to accumulate tree biomass data. According to a previous study [27], all the stems with a diameter at breast height (DBH) $\geq$ 5 cm were considered trees, identified by species in the plot, and selected for height and DBH measurements. Diameter tape was used to measure over bark stem diameter and rangefinder was used for tree height calculation (LTI TruPulse 200, USA). The stem diameters of all stems were measured separately. Around 100 gm of tree biomass from each plant part, such as leaves, branches, twigs, bark, and roots, was collected at the species level for laboratory analysis. For dry weight determination and organic carbon analysis, the contents were put in plastic bags and the samples were carried to the laboratory. The tree canopy openness, number of cut stems, shrub and herb canopy openness, and bare ground percentage of plots were recorded as indicators of disturbances in each plot. Data were collected by a densitometer at the center of each plot to measure the tree canopy coverage. The effect of the severity of disturbances (harvesting and canopy openness due to removal of trees with stem diameter $\geq$ 5 cm) on carbon stocks was evaluated. In each sub-plots of shrubs and herbs, number of each species and individuals was counted to use in the calculation of species richness and abundance. Five soil samples were taken from each corner and the center of the plot to reduce any site homogeneity. Each of the samples was taken at 10 cm of depth using a core soil sampler. Then five soil samples from each plot were thoroughly mixed up together in the field and marked as one sample used to measure soil variables [38]. Then the samples were taken to the laboratory for further analysis.

2.4.2. Anthropogenic Disturbance Index

The magnitude of disturbance in the field was measured using the following formula [37]:

$$\text{DI } (\%) = \frac{\text{CS } (\%) + \text{TCO } (\%) + \text{SHCO } (\%) + \text{BG}(\%)}{4}$$

where DI = the disturbance index, CS (%) = percentage of cut stems, i.e., trees illegally felled and/or removed (number of cut stem/total stem of the plot), TCO = tree canopy openness, SHCO = shrub and herb canopy openness, and BG = the bare ground percentage of the plot [11].

The DI measures a comprehensive bioassay that incorporates all sorts of disturbance effects throughout recent history, such as windthrow, collection of fuelwood, illicit felling, logging, or forest fires.

2.4.3. Estimation of Tree Biomass and Carbon Stock

A non-destructive method developed by Brown et al. [39] was used to access the aboveground biomass (AGB) of trees. Previous studies suggested that this method is the most appropriate method for tropical forests [37,40,41]. Belowground biomass (BGB) was accessed by considering 15% of the AGB. Total biomass of each individual tree was calculated by adding AGB and BGB. Biomass of all species was added to obtain the plot-wise total biomass [37].

For the estimation of biomass carbon, fresh weight and oven-dried weight (48 h at 65 °C) of the collected plant samples were calculated. One (1.00) gm of oven-dried grind sample was taken in pre-weighted crucibles. The crucibles with the samples were placed in the furnace for one hour at 550 °C. Then the crucibles cooled down gradually inside the furnace. Then the weight of the crucibles with ash was measured. Afterward, the percentage of organic carbon was computed by Alamgir and Al-Amin [42].

$$\text{Ash } (\%) = \frac{\text{W3} - \text{W1}}{\text{W2} - \text{W1}} \times 100 \text{ and C } (\%) = (100 - \text{Ash } \%) \times 0.58$$

Here, W1 = crucible weight; W2 = litter weight plus crucible weight; W3 = ash weight plus crucible weight; C = organic carbon; and "0.58" = considering 58% carbon in ash-free litter material.

Tree biomass carbon was calculated by multiplying AGB with the organic carbon percentage of that particular species. Plot-wise total AGB carbon was determined by adding the AGB carbon of all trees. Plot-specific biomass carbon stocks were also calculated and then converted to tons per hectare.

2.4.4. Soil Sample Analysis and Organic Carbon Estimation

Soil bulk density and soil moisture content were used as indicators of soil physical support and soil water regulation, respectively [21]. Relevant soil analysis was performed to quantify the bulk density (gm/cm^3) and moisture content (in percent) of soil. Oven-drying method at 105 °C was used to measure soil bulk density in gm/cm^3 [43]. Soil moisture content was also measured using the oven-drying method at 105 °C [44]. Soil aggregate stability was indicated by the soil sand % in this study [21]. Hydrometer method was used to access the soil sand % [45]. The corresponding carbon content of soils was calculated using a conversion coefficient of 0.58 multiplied by the soil organic matter [46]. Loss of ignition procedure was used to measure soil organic matter content of our study areas [21].

2.5. Statistical Analysis

All data were analyzed using R (Version 3.6.1, Vienna, Austria) and figures were produced using the packages "ggplot2" [47], "corrplot" for correlogram [48], "nlme" for analysis of variance, and raster package for bio-variable extraction. Before statistical analysis, data were transformed if conditions of normality were not met or to improve the homogeneity of variance. The relationships between disturbances and ecosystem services under two precipitation regimes were determined using regression analysis. All statistical significances were decided at 5% significance level.

3. Results

3.1. Aboveground Ecosystem Services

3.1.1. Primary Production

Tree biomass (t/ha) was used as an indicator of primary production in our study (Table 1). The disturbances had significant effects on primary production ($p = 0.02$, Table 2). Disturbances were negatively correlated with ($p = 0.03$, $r = -0.28$) the primary production of the study area under a medium amount of rainfall (Figure 2a). However, the association between disturbances and primary production was not significant ($p = 0.3$, $r = -0.14$) under a high-rainfall regime in our study (Figure 2a).

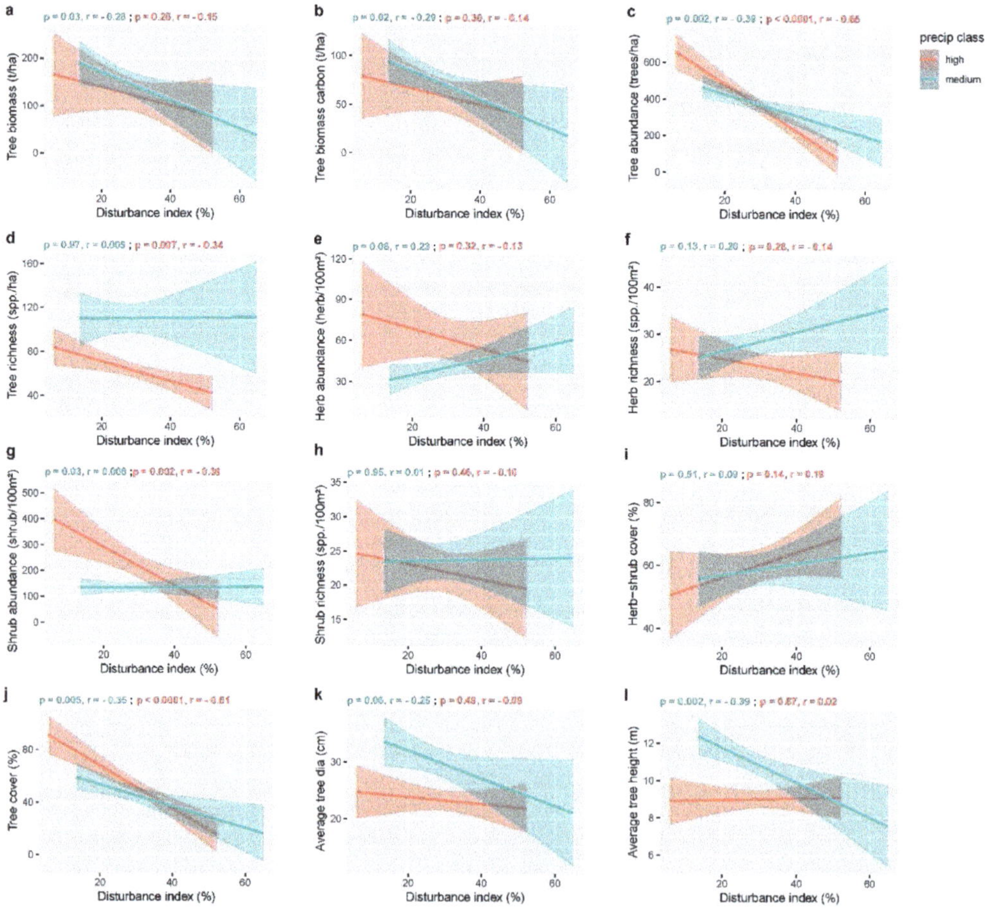

Figure 2. Relationships between disturbance and aboveground ecosystem services under high- and medium-rainfall regimes in our study sites in Northeast Bangladesh.

Table 2. *p* value and slope extent of precipitation and disturbance effects on ecosystem services under different rainfall regimes in our study sites in Northeast Bangladesh. Significant codes: 0.0001 = '***'; 0.001 = '**'; 0.05 = '*'.

	Ecosystem Service Indicators	*p* Value			Slope	
		Disturbance × Rainfall Class	Disturbance	Rainfall Class	Medium Rainfall	High Rainfall
Aboveground	Tree biomass	0.62	0.02 *	0.30	−2.9466	−1.9078
	Tree biomass carbon	0.56	0.02 *	0.25	−1.484	−0.8722
	Tree abundance	0.008 **	<0.0001 ***	0.80	−5.749	−12.617
	Tree richness	0.22	0.07	<0.0001 ***	0.0281	−0.8968
	Herb abundance	0.11	0.96	0.009 **	0.5768	−0.7458
	Herb richness	0.06	0.98	0.01 *	0.2036	−0.1459
	Shrub abundance	0.003 **	0.009 **	0.0007 ***	0.0582	−7.4966
	Shrub richness	0.53	0.55	0.44	0.0115	−0.1119
	Herb shrub cover	0.54	0.10	0.72	0.1794	0.4021
	Tree coverage	0.04*	<0.0001 ***	0.08	−0.8223	−1.6351
	Average tree diameter	0.23	0.02*	<0.0001 ***	−0.2436	−0.0616
	Average tree height	0.01*	0.004 **	<0.0001 ***	−0.0937	0.0041
Belowground	Soil organic carbon	0.27	0.002 **	<0.0001 ***	0.2766	0.0852
	Soil sand %	0.28	0.003 **	<0.0001 ***	0.0022	−0.0577
	Soil bulk density	0.42	0.0007 ***	<0.0001 ***	0.0033	0.0058
	Soil moisture	0.02 *	0.94	0.04*	0.1023	−0.1145

3.1.2. Tree Carbon Storage

Tree carbon storage was represented by tree biomass carbon (t/ha) in our study (Table 1). Disturbances had significant effects on tree carbon storage ($p = 0.02$, Table 2). Aboveground carbon storage significantly decreased ($p = 0.02$, $r = -0.29$) along the disturbance gradient under a medium amount of rainfall, but it was not significantly ($p = 0.3$, $r = -0.14$) affected in the high-rainfall regime (Figure 2b).

3.1.3. Biodiversity Conservation

We considered tree abundance (tree/ha), tree species richness (spp./ha), herb abundance (herbs/100 m^2), herb species richness (spp./100 m^2), shrub abundance (shrubs/100 m^2), and shrub species richness (spp./100 m^2) as indicators of biodiversity conservation, one of the major regulating ecosystem services (Table 1). The interactive effects of disturbances and rainfall class were significant on tree abundance ($p = 0.008$, Table 2). Disturbances had a significant negative correlation with tree abundance under both rainfall classes ($p = 0.002$, $r = -0.39$ under a medium amount of rainfall; $p < 0.0001$, $r = -0.66$ under a high amount of rainfall), though the correlation was stronger in the high-rainfall class (Figure 2c). Rainfall class had significant effects on tree richness ($p < 0.0001$), herb abundance ($p = 0.009$), and herb richness ($p = 0.01$, Table 2). Disturbances had a significant negative correlation ($p = 0.007$, $r = -0.34$) with tree richness in the high-rainfall class, whereas no significant correlation ($p = 0.97$, $r = 0.005$) was found in the medium-rainfall class (Figure 2d). Herb and shrub communities are also good indicators of forest biodiversity. Herb abundance ($p = 0.08$, $r = 0.23$ in the medium-rainfall class; $p = 0.32$, $r = -0.13$ in the high-rainfall class) and richness ($p = 0.13$, $r = 0.20$ in the medium-rainfall class; $p = 0.28$, $r = -0.14$ in the high-rainfall class) had no significant correlation with disturbances in any rainfall class in our study (Figure 2e,f). The combined effects of disturbances and rainfall were significant on shrub abundance ($p = 0.003$, Table 2). Shrub abundance was also significantly correlated with a disturbance gradient in both rainfall classes ($p = 0.03$, $r = 0.008$ for the medium-rainfall class and $p = 0.002$, $r = -0.39$ for the high-rainfall class, Figure 2g).

3.1.4. Wildlife Habitat

Vegetation coverage was the main indicator of wildlife habitat provision, another important forest ecosystem service in our study (Table 1). We did not find any significant correlation between herb–shrub coverage and disturbance gradients in the high- ($p = 0.5$, $r = 0.09$) or medium ($p = 0.1$, $r = 0.19$)-rainfall classes in the study sites (Figure 2i). On the contrary, the combined effects of disturbances and rainfall were significant on tree canopy coverage ($p = 0.04$, Table 2). Moreover, tree canopy coverage had a significant negative association with disturbance gradients in both medium- ($p = 0.005$, $r = -0.35$) and high ($p < 0.0001$, $r = -0.61$)-rainfall classes (Figure 2j).

3.1.5. Timber Production

Average tree height (m) and average tree diameter at breast height (cm) were considered as indicators for timber production in our study (Table 1). The correlation analysis showed that the disturbances had no significant association with the average tree diameter under the high- ($p = 0.06$, $r = -0.25$) or medium ($p = 0.49$, $r = -0.09$)-rainfall classes (Figure 2k). However, the interactive effects of disturbances and rainfall class were significant on tree height ($p = 0.01$, Table 2). The average tree height showed a significant negative correlation with a disturbance gradient in the medium-rainfall class ($p = 0.002$, $r = -0.39$), whereas no significant ($p = 0.87$, $r = 0.02$) relation was found in the high-rainfall class (Figure 2k,I).

3.2. *Belowground Ecosystem Services*

3.2.1. Soil Organic Carbon Storage

Soil organic carbon (%) was the indicator of soil organic carbon storage in our study (Table 1). Rainfall class had significant effects on soil organic carbon storage ($p < 0.0001$, Table 2). Belowground organic carbon storage was positively correlated ($r = 0.27$) with disturbance gradients and it was significant ($p = 0.04$) in the medium-rainfall class. On the other hand, the correlation between disturbances and belowground carbon storage was not significant ($p = 0.47$, $r = -0.10$) in our study in the high-rainfall class (Figure 3a).

3.2.2. Soil Aggregate Stability

Soil aggregate stability is an important belowground soil ecosystem function that shows the ability of soil aggregates to resist degradation when exposed to external forces. We considered soil sand (%) as an indicator of soil aggregate stability (Table 1). A higher sand percentage lowers the aggregate stability of the soil. No significant correlation was observed between the disturbance gradient and soil sand (%) in the medium- ($p = 0.9$, $r = 0.02$) and high ($p = 0.27$, $r = -0.14$)-rainfall classes (Figure 3b) in our study.

3.2.3. Soil Physical Support

Soil physical support was indicated by soil bulk density (g/cm^3) in our study (Table 1). A higher bulk density ensures stable sedimented soil in belowground forest ecosystems. Rainfall class had significant effects on soil bulk density ($p < 0.0001$, Table 2). We observed a significant positive ($p = 0.002$, $r = 0.39$) correlation between disturbances and soil bulk density in the high-rainfall class, whereas no significant correlation ($p = 0.18$, $r = 0.17$) was observed in the medium-rainfall class (Figure 3c).

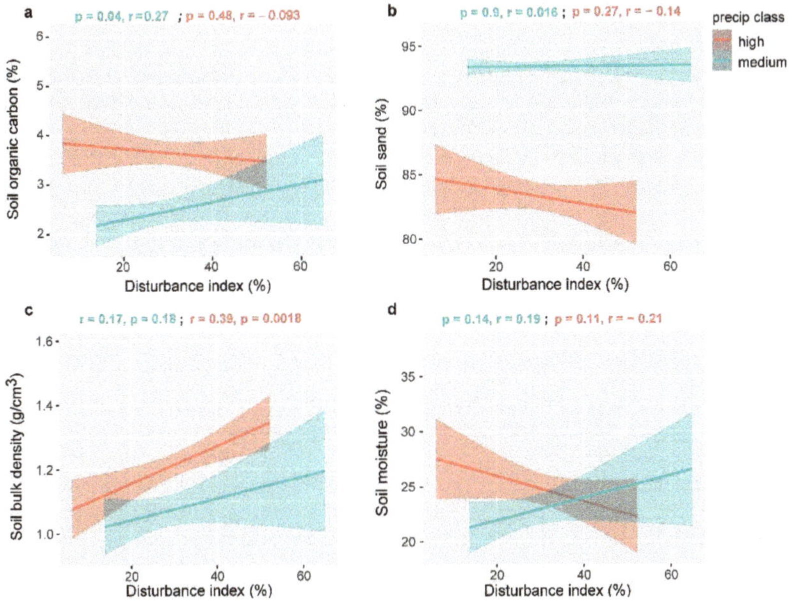

Figure 3. Relationships between disturbance and belowground ecosystem services under high- and medium-rainfall regimes in our study sites in Northeast Bangladesh.

3.2.4. Soil Water Regulation

Soil moisture content (%) was the main indicator of soil water regulation in our study (Table 1). The combined effects of disturbances and rainfall were significant on soil moisture (p = 0.02, Table 2). Although the post hoc analysis showed no significant association between disturbances and soil moisture in any rainfall class, there was a positive correlation under a medium amount of rainfall and a negative correlation under a high amount of rainfall between disturbances and soil moisture in our study (Figure 3d).

4. Discussion

4.1. Aboveground Ecosystem Services

4.1.1. Primary Production

A significant decrease in primary production was observed with an increasing disturbance gradient in our study under a medium amount of rainfall. However, this decrease was not significant under a high amount of precipitation. Previous studies reported a decrease in rainfall and water availability and an increase in the physiological stress of plants, limiting biomass production and fitness, potentially leading to the death of stress-sensitive species [49]. Yan et al. [50] reported an increase in tree biomass with increased precipitation, which might be the reason why there were limited effects of disturbances on primary production in the present study.

4.1.2. Tree Carbon Storage

In our study, tree carbon storage significantly decreased with the increasing disturbance gradient under a medium-rainfall regime, whereas disturbances had no significant impact on tree carbon storage in high-rainfall areas. A significant decrease in tree biomass lowers the tree carbon storage under less rainfall, which was also observed by Fekete et al. [51] in their study.

4.1.3. Biodiversity

Our study observed a combined association of disturbances and rainfall class with tree abundance. Greater disturbances decreased the tree abundance in both rainfall classes and mostly in the high-rainfall class. We also found disturbances significantly lower the tree richness in the high-rainfall class, but this correlation was not significant in medium-rainfall areas. The previous study showed that anthropogenic disturbances reduced the tree species richness [52]. We observed this negative effect of disturbances on tree species abundance and richness under heavy rainfall. We found no significant changes in herb richness and abundance with a disturbance gradient in both the high- and medium-precipitation classes. However, the combined effects of disturbances and rainfall class were significant on shrub abundance. In the high-rainfall class, disturbances significantly lowered the shrub abundance, but in the medium-precipitation region, this correlation was almost linear.

4.1.4. Wildlife Habitat

The interaction effects of rainfall and disturbance class were significant on tree canopy coverage in our study. We observed disturbances lower the canopy coverage in high-rainfall areas rather than medium-rainfall regions. Pfeifer et al. [53] reported the canopy coverage changes under different precipitation patterns with the disturbances fully dependent on the forest's adaptation capability under a deficit of water. Moreover, we found no interaction between herb and shrub coverage with the disturbance gradients in both medium- and high-rainfall regions. Due to the high restoring capacity of tropical understories, vegetation disturbances cannot alter their coverage significantly [54].

4.1.5. Timber Production

We observed the interaction of rainfall class and disturbances had a marginal correlation with average tree height, and in a specific trend analysis, we observed that highly disturbed areas contained species with a significantly lower average height in medium-rainfall regions, whereas no interaction was observed in high-rainfall areas. Disturbances are one of the main drivers of reduced tree density, diversity, and ecosystem production [52,55]. The present study revealed that disturbances can reduce tree height under a medium-precipitation regime. However, no significant interaction was observed in our study between the average diameter of trees and the disturbance gradients in both rainfall classes.

4.2. Belowground Ecosystem Services

4.2.1. Organic Carbon Storage

We observed a significant rise in soil organic carbon with disturbances in the medium-precipitation region, whereas in the high-rainfall region, this correlation was not significant. This is because anthropogenic disturbance can increase the illegal logging of tree species, which ultimately increases the tree debris and soil organic matter on the forest floor [38,52]. However, this positive correlation between disturbances and soil organic carbon was not observed under heavy rainfall because a higher amount of precipitation might decrease the soil organic carbon content on the forest floor due to a higher incidence of leaching or washout [56].

4.2.2. Soil Aggregate Stability

Disturbances had no significant correlation with soil aggregate stability in our study in both rainfall classes. The larger portion of sand content was deposited in our study sites due to the presence of numerous streams in the rainy season, which might be the main driving factor of soil aggregate stability, more so than the disturbance gradients.

4.2.3. Soil Physical Support

We found a significant positive association between disturbance and soil bulk density in our study in the medium-rainfall class. Our study's findings are in accordance with those of a previous study which reported an increase in soil bulk density with forest disturbances. However, under the high-rainfall class, no significant correlation was evident in our study, as also reported by Rab et al. [57].

4.2.4. Soil Water Regulation

The interaction of disturbances and rainfall class had a significant effect on soil water regulation. However, disturbances had no significant association with soil water regulation function in both the medium- and high-rainfall classes in our study.

5. Conclusions

The combined effects of climate change and anthropogenic disturbances have caused a substantial decline in the supply of ecosystem services from tropical forested regions in recent decades. Bangladesh is one of the countries most adversely affected by climate change with a major variability in rainfall patterns in recent decades. Our study found that disturbances negatively influenced indicators of primary production, tree carbon storage, and timber production, and may positively influence soil organic carbon storage under different rainfall regimes. There are, however, some indicators that we used in our study as response factors that could also be used as causal factors (e.g., soil fertility), and thus our results need careful interpretation. We believe our study will contribute to our understanding of the combined effects of rainfall regimes and disturbances on forest ecosystem services with potential implications for forest ecosystem restoration.

Author Contributions: Conceptualization, F.S., M.A.S.A.-K. and S.A.M.; methodology, M.A.S.A.-K., S.A.M.; formal analysis, F.S. and M.R.K.; investigation, F.S. and M.R.K.; resources, M.A.S.A.-K.; data curation, F.S. and M.R.K.; writing—original draft preparation, F.S. and M.R.K.; writing—review and editing, M.A.S.A.-K. and S.A.M.; supervision, M.A.S.A.-K.; project administration, M.A.S.A.-K. All authors have read and agreed to the published version of the manuscript.

Funding: This research received no external funding.

Institutional Review Board Statement: Not applicable.

Informed Consent Statement: Not applicable.

Data Availability Statement: Relevant data used in this study will be made available upon request.

Acknowledgments: We acknowledge the lab and other support provided by the Department of Forestry and Environmental Science (FES), Shahjalal University of Science and Technology (SUST), Bangladesh, which have made this study possible. We also acknowledge the funding of SUST Research center (Project ID: FES/2019/2/01; FES/2020/01/03; FES/2021/2/04).

Conflicts of Interest: The authors declare no conflict of interest.

References

1. Mukul, S.A.; Sohel, M.S.I.; Herbohn, J.; Inostroza, L.; König, H. Integrating ecosystem services supply potential from future land-use in protected area management: A Bangladesh case study. *Ecosyst. Serv.* **2017**, *26*, 355–364. [CrossRef]
2. Ninan, K.N.; Kontoleon, A. Valuing forest ecosystem services and disservices—Case study of a protected area in India. *Ecosyst. Serv.* **2016**, *20*, 1–14. [CrossRef]
3. Hofhansl, F.; Chacón-Madrigal, E.; Fuchslueger, L.; Jenking, D.; Morera-Beita, A.; Plutzar, C.; Silla, F.; Andersen, K.M.; Buchs, D.M.; Dullinger, S.; et al. Climatic and edaphic controls over tropical forest diversity and vegetation carbon storage. *Sci. Rep.* **2020**, *10*, 5066. [CrossRef] [PubMed]
4. Mukul, S.A.; Halim, M.A.; Herbohn, J. Forest Carbon Stock and Fluxes: Distribution, Biogeochemical Cycles, and Measurement Techniques. In *Life on Land. Encyclopedia of the UN Sustainable Development Goals*; Leal Filho, W., Azul, A.M., Brandli, L., Lange Salvia, A., Wall, T., Eds.; Springer: Cham, Switzerland, 2021; pp. 376–385. [CrossRef]
5. Bullock, E.L.; Woodcock, C.E.; Olofsson, P. Monitoring tropical forest degradation using spectral unmixing and Landsat time series analysis. *Remote Sens. Environ.* **2020**, *238*, 110968. [CrossRef]
6. Morris, R.J. Anthropogenic impacts on tropical forest biodiversity: A network structure and ecosystem functioning perspective. *Philos. Trans. R. Soc. B Biol. Sci.* **2010**, *365*, 3709–3718. [CrossRef]

7. Arfin-Khan, M.A.S.; Saimun, M.S.R. Forest Dependency: Status, Assessment Tools, and Influencing Factors. In *Life on Land. Encyclopedia of the UN Sustainable Development Goals*; Leal Filho, W., Azul, A.M., Brandli, L., Lange Salvia, A., Wall, T., Eds.; Springer: Cham, Switzerland, 2021. [CrossRef]
8. Nerfa, L.; Rhemtulla, J.M.; Zerriffi, H. Forest dependence is more than forest income: Development of a new index of forest product collection and livelihood resources. *World Dev.* **2020**, *125*, 104689. [CrossRef]
9. Antongiovanni, M.; Venticinque, E.M.; Matsumoto, M.; Fonseca, C.R. Chronic anthropogenic disturbance on Caatinga dry forest fragments. *J. Appl. Ecol.* **2020**, *57*, 2064–2074. [CrossRef]
10. Barlow, J.; Lennox, G.D.; Ferreira, J.; Berenguer, E.; Lees, A.C.; Nally, R.M.; Thomson, J.R.; Ferraz, S.F.B.; Louzada, J.; Oliveira, V.H.F.; et al. Anthropogenic disturbance in tropical forests can double biodiversity loss from deforestation. *Nature* **2016**, *535*, 144–147. [CrossRef]
11. Karim, M.R.; Sultana, F.; Saimun, M.S.R.; Mukul, S.A.; Arfin-Khan, M.A.S. Plant diversity and local rainfall regime mediate soil ecosystem functions in tropical forests of north-east Bangladesh. *Environ. Adv.* **2020**, *2*, 100022. [CrossRef]
12. de Oliveira Roque, F.; Menezes, J.F.S.; Northfield, T.; Ochoa-Quintero, J.M.; Campbell, M.J.; Laurance, W.F. Warning signals of biodiversity collapse across gradients of tropical forest loss. *Sci. Rep.* **2018**, *8*, 1622. [CrossRef]
13. Ribeiroa, M.C.; Metzger, J.P.; Martensen, A.C.; Ponzoni, F.J.; Hirota, M.M. The Brazilian Atlantic Forest: How much is left, and how is the remaining forest distributed? Implications for conservation. *Biol. Conserv.* **2009**, *142*, 1141–1153. [CrossRef]
14. Karim, M.R.; Mukul, S.A.; Zahira, R.B.; Saimun, M.S.R.; Arfin-Khan, M.A.S. The role of protected areas co-management in enhancing resistance and resilience of deciduous forest ecosystem to extreme climatic events in Bangladesh. *J. Environ. Manag.* **2023**, *326*, 116800. [CrossRef]
15. Moro, M.F.; Silva, I.A.; Araújo, F.S.; Nic Lughadha, E.; Meagher, T.R.; Martins, F.R. The role of edaphic environment and climate in structuring phylogenetic pattern in seasonally dry tropical plant communities. *PLoS ONE* **2015**, *10*, e0119166. [CrossRef]
16. Smith, M.N.; Taylor, T.C.; van Haren, J.; Rosolem, R.; Restrepo-Coupe, N.; Adams, J.; Wu, J.; de Oliveira, R.C.; da Silva, R.; de Araujo, A.C.; et al. Empirical evidence for resilience of tropical forest photosynthesis in a warmer world. *Nat. Plants* **2020**, *6*, 1225–1230. [CrossRef]
17. Zhao, M.; Running, S.W. Drought-induced reduction in global terrestrial net primary production from 2000 through 2009. *Science* **2010**, *329*, 940. [CrossRef]
18. Ureta, C.; Martorell, C.; Hortal, J.; Fornoni, J. Assessing extinction risks under the combined effects of climate change and human disturbance through the analysis of life-history plasticity. *Perspect. Plant Ecol. Evol. Syst.* **2012**, *14*, 393–401. [CrossRef]
19. Deb, J.C.; Phinn, S.; Butt, N.; McAlpine, C.A. Climate change impacts on tropical forests: Identifying risks for tropical Asia. *J. Trop. For. Sci.* **2018**, *30*, 182–194. [CrossRef]
20. Anderson-Teixeira, K.J.; Miller, A.D.; Mohan, J.E.; Hudiburg, T.W.; Duval, B.D.; DeLucia, E.H. Altered dynamics of forest recovery under a changing climate. *Glob. Chang. Biol.* **2013**, *19*, 2001–2021. [CrossRef]
21. Karim, M.R.; Halim, M.A.; Gale, N.V.; Thomas, S.C. Biochar effects on soil physiochemical properties in degraded managed ecosystems in northeastern Bangladesh. *Soil Syst.* **2020**, *4*, 69. [CrossRef]
22. García-Valdés, R.; Svenning, J.C.; Zavala, M.A.; Purves, D.W.; Araújo, M.B. Evaluating the combined effects of climate and land-use change on tree species distributions. *J. Appl. Ecol.* **2015**, *52*, 902–912. [CrossRef]
23. Gibb, H.; Sanders, N.J.; Dunn, R.R.; Watson, S.; Photakis, M.; Abril, S.; Andersen, A.N.; Angulo, E.; Armbrecht, I.; Arnan, X.; et al. Climate mediates the effects of disturbance on ant assemblage structure. *Proc. R. Soc. B Biol. Sci.* **2015**, *282*, 20150418. [CrossRef] [PubMed]
24. Mukul, S.A.; Rashid, A.Z.M.M.; Khan, N.A. Forest protected area systems and biodiversity conservation in Bangladesh. In *Protected Areas: Policies, Management and Future Directions*; Mukul, S.A., Rashid, A.Z.M.M., Eds.; Nova Science Publishers: Hauppauge, NY, USA, 2017; pp. 157–177.
25. Tantipisanuh, N.; Savini, T.; Cutter, P.; Gale, G.A. Biodiversity gap analysis of the protected area system of the Indo-Burma Hotspot and priorities for increasing biodiversity representation. *Biol. Conserv.* **2016**, *195*, 203–213. [CrossRef]
26. Langner, A.; Miettinen, J.; Kukkonen, M.; Vancutsem, C.; Simonetti, D.; Vieilledent, G.; Verhegghen, A.; Gallego, J.; Stibig, H. Towards operational monitoring of forest canopy disturbance in evergreen rain forests: A test case in continental Southeast Asia. *Remote Sens.* **2018**, *10*, 544. [CrossRef]
27. Mukul, S.A.; Saha, N. Conservation benefits of tropical multifunctional landscapes in and around a forest protected area of Bangladesh. *Land* **2017**, *6*, 2. [CrossRef]
28. Corbera, E.; Hunsberger, C.; Vaddhanaphuti, C. Climate change policies, land grabbing and conflict: Perspectives from Southeast Asia. *Can. J. Dev. Stud./Rev. Can. D'études Du Développement* **2017**, *38*, 297–304. [CrossRef]
29. Ota, L.; Chazdon, R.L.; Herbohn, J.; Gregorio, N.; Mukul, S.A.; Wilson, S.J. Achieving quality forest and landscape restoration in the tropics. *Forests* **2020**, *11*, 820. [CrossRef]
30. Mukul, S.A.; Herbohn, J.; Rashid, A.Z.M.M.; Uddin, M.B. Comparing the effectiveness of forest law enforcement and economic incentive to prevent illegal logging in Bangladesh. *Int. For. Rev.* **2014**, *16*, 363–375. [CrossRef]
31. Richmond, A.; Kaufmann, R.K.; Myneni, R.B. Valuing ecosystem services: A shadow price for net primary production. *Ecol. Econ.* **2007**, *64*, 454–462. [CrossRef]
32. Fahey, T.J.; Woodbury, P.B.; Battles, J.J.; Goodale, C.L.; Hamburg, S.P.; Ollinger, S.V.; Woodall, C.W. Forest carbon storage: Ecology, management, and policy. *Front. Ecol. Environ.* **2010**, *8*, 245–252. [CrossRef]

33. Gould, W. Remote sensing of vegetation, plant species richness, and regional biodiversity hotspots. *Ecol. Appl.* **2000**, *10*, 1861–1870. [CrossRef]
34. Grebner, D.L.; Bettinger, P.; Siry, J.P. Wildlife habitat relationships. In *Introduction to Forestry and Natural Resources*; Grebner, D.L., Bettinger, P., Siry, J.P., Eds.; Academic Press: San Diego, CA, USA, 2013; pp. 125–146.
35. Aishan, T.; Halik, Ü.; Betz, F.; Tiyip, T.; Ding, J.; Nuermaimaiti, Y. Stand structure and height-diameter relationship of a degraded Populus euphratica forest in the lower reaches of the Tarim River, Northwest China. *J. Arid. Land* **2015**, *7*, 544–554. [CrossRef]
36. Doran, J.W.; Parkin, T.B. Quantitative Indicators of Soil Quality: A Minimum Data Set. *Methods Assess. Soil Qual.* **2015**, *49*, 25–37.
37. Saimun, M.S.R.; Karim, M.R.; Sultana, F.; Arfin-Khan, M.A.S. Multiple drivers of tree and soil carbon stock in the tropical forest ecosystems of Bangladesh. *Trees For. People* **2021**, *5*, 100108. [CrossRef]
38. Mukul, S.A.; Herbohn, J.; Ferraen, A.; Congdon, R. Limited role of shifting cultivation in soil carbon and nutrients recovery in regenerating tropical secondary forests. *Front. Environ. Sci.* **2022**, *10*, 1076506. [CrossRef]
39. Brown, S.; Gillespie, A.; Lugo, A.E. Biomass Estimation Methods for Tropical Forests with Applications to Forest Inventory Data. *For. Sci.* **1989**, *35*, 881–902. [CrossRef]
40. Alves, D.S.; Soares, J.V.S.; Amaral, E.M.K.; Mello, S.A.S.; Almeida, O.; Fernandes, S.A.; Silveira, A.M. Biomass of primary and secondary vegetation in Rondonia, western Brazilian Amazon. *Glob. Chang. Biol.* **1997**, *3*, 451–462. [CrossRef]
41. FAO. *Estimating Biomass and Biomass Change of Tropical Forests—A Primer*; FAO: Rome, Italy, 1997; Available online: http://www.fao.org/docrep/w4095e/w4095e00.htm (accessed on 25 December 2022).
42. Alamgir, M.; Al-Amin, M. Allometric models to estimate biomass organic carbon stock in forest vegetation. *J. For. Res.* **2008**, *19*, 101–106. [CrossRef]
43. Blake, G.R.; Hartge, K.H. Bulk Density. In *Methods of Soil Analysis*, 2nd ed.; Klute, A., Ed.; Wiley & Sons: Hoboken, NJ, USA, 1986; pp. 363–375.
44. O'Kelly, B.C. Accurate determination of moisture content of organic soils using the oven drying method. *Dry. Technol.* **2004**, *22*, 1767–1776. [CrossRef]
45. Bouyoucos, G.J. Hydrometer Method Improved for Making Particle Size Analyses of Soils. *Agron. J.* **1962**, *54*, 464–465. [CrossRef]
46. Wang, S.; Huang, Y. Determinants of soil organic carbon sequestration and its contribution to ecosystem carbon sinks of planted forests. *Glob. Chang. Biol.* **2020**, *26*, 3163–3173. [CrossRef]
47. Wickham, H. *ggplot2: Elegant Graphics for Data Analysis*; Springer: New York, NY, USA, 2016.
48. Wei, T.; Simko, V. R Package 'Corrplot': Visualization of a Correlation Matrix. (Version 0.92). 2021. Available online: https://github.com/taiyun/corrplot (accessed on 3 January 2023).
49. Figueiredo, K.V.; Oliveira, M.T.; Arruda, E.C.P.; Silva, B.C.F.; Santos, M.G. Changes in leaf epicuticular wax, gas exchange and biochemistry metabolism between Jatropha mollissima and Jatropha curcas under semi-arid conditions. *Acta Physiol. Plant.* **2015**, *37*, 108. [CrossRef]
50. Yan, H.; Liang, C.; Li, Z.; Liu, Z.; Miao, B.; He, C.; Sheng, L. Impact of Precipitation Patterns on Biomass and Species Richness of Annuals in a Dry Steppe. *PLoS ONE* **2015**, *10*, e0125300. [CrossRef]
51. Fekete, I.; Lajtha, K.; Kotroczó, Z.; Várbíró, G.; Varga, C.; Tóth, J.A.; Demeter, I.; Veperdi, G.; Berki, I. Long-term effects of climate change on carbon storage and tree species composition in a dry deciduous forest. *Glob. Chang. Biol.* **2017**, *23*, 3154–3168. [CrossRef]
52. Bentsi-Enchill, F.; Damptey, F.G.; Pappoe, A.N.M.; Ekumah, B.; Akotoye, H.K. Impact of anthropogenic disturbance on tree species diversity, vegetation structure and carbon storage potential in an upland evergreen forest of Ghana, West Africa. *Trees For. People* **2022**, *8*, 100238. [CrossRef]
53. Pfeifer, M.; Gonsamo, A.; Woodgate, W.; Cayuela, L.; Marshall, A.R.; Ledo, A.; Paine, T.C.; Marchant, R.; Burt, A.; Calders, K.; et al. Tropical forest canopies and their relationships with climate and disturbance: Results from a global dataset of consistent field-based measurements. *For. Ecosyst.* **2018**, *5*, 7. [CrossRef]
54. Kutnar, L.; Nagel, T.A.; Kermavnar, J. Effects of Disturbance on Understory Vegetation across Slovenian Forest Ecosystems. *Forests* **2019**, *10*, 1048. [CrossRef]
55. Gautam, T.P.; Mandal, T.N. Effect of disturbance on biomass, production and carbon dynamics in moist tropical forest of eastern Nepal. *For. Ecosyst.* **2016**, *3*, 11. [CrossRef]
56. Chen, X.; Zhang, D.; Liang, G.; Qiu, Q.; Liu, J.; Zhou, G.; Liu, S.; Chu, G.; Yan, J. Effects of precipitation on soil organic carbon fractions in three subtropical forests in southern China. *J. Plant Ecol.* **2016**, *9*, 10–19. [CrossRef]
57. Rab, A.; Bradshaw, J.; Campbell, R.; Murphy, S. *Review of Factors Affecting Disturbance, Compaction and Trafficability of Soils with Particular Reference to Timber Harvesting in the Forests of South-West Western Australia*; University of Melbourne: Parkville, Australia, 2005.

forests

MDPI

Article

Variations in Ecosystem Service Value and Its Driving Factors in the Nanjing Metropolitan Area of China

Shulin Chen *, Xiaotong Liu, Li Yang and Zhenghao Zhu

College of Economics and Management, Nanjing Forestry University, Nanjing 210037, China
* Correspondence: slchen@njfu.edu.cn

Abstract: More than 60% of the world's ecosystem services have deteriorated over the past few decades. Studying the spatio-temporal fluctuations in ecosystem service value and its influencing factors is important for identifying regional ecosystem service value issues, upholding regional ecological harmony, and encouraging regionally healthy and coordinated sustainable development. Ecosystem service value has so far been studied primarily in relation to the effect of socioeconomic and physical–geographical variables. However, the trade-offs and synergies among ecosystem service values also drive the spatio-temporal variations in ecosystem service value. Few studies have been conducted to date to investigate the trade-offs and synergies between ecosystem service values and their impact on ecosystem service value. Therefore, this paper used sensitivity analysis, correlation analysis, trade-offs and synergies analysis, and a Geodetector to examine changes in ecosystem service value and their influencing factors within the Nanjing metropolitan region. The ecosystem service value decreased somewhat overall between 2000 and 2020, with a decline rate of 2.19 million CNY/year. In comparison to the north of the Nanjing metropolitan region, the ecosystem service value was relatively higher in the south. The water bodies had the highest total ecosystem service value, followed by forest land, cultivated land, and grassland, with construction land and unused land having the lowest ecosystem service values overall. The main socioeconomic factor influencing the spatial variations in ecosystem service value was population density, while the main physical–geographical factors were the digital elevation model, the normalized difference vegetation index, and precipitation. As a result, the Nanjing metropolitan area should tighten its grip on excessive population growth. In contrast to the expository strength of a single factor on the ecosystem service value, the influence of all individual elements on the ecosystem service value under interaction was significantly increased, and the interaction among the normalized difference vegetation index and gross economic product had the most obvious effect on the ecosystem service value. The spatial variation in the ecosystem service value was also influenced by trade-offs and synergies between the value of supply services, regulation services, support services, and cultural services. Therefore, trade-offs and synergies among ecosystem services also need to be considered in land-use decisions.

Keywords: ecosystem service value; influencing factors; trade-offs and synergies; spatial and temporal variations; Geodetector method

1. Introduction

Global climate change and anthropological disturbances are affecting the stability and biodiversity of ecosystems, which in turn leads to the degradation of ecosystem services. The Millennium Ecosystem Assessment (MEA) found that in the last 50 years, approximately 60% of the world's ecosystem services have become worse [1]. Among the 18 categories of ecosystem services assessed by the Intergovernmental Science-Policy Platform on Biodiversity and Ecosystem Services (IPBES), 14 categories of ecosystem services have declined since 1970 [2]. According to the New Nature Economy (NNE) report, published by the World Economic Forum (WEF) in partnership with the consultancy firm

Price Waterhouse Coopers (PwC), an analysis of 163 industry sectors and their supply chains found that more than half of the global GDP is moderately or highly dependent on natural ecosystems and their services [3]. Therefore, the degradation or loss of ecosystem services is a development issue that directly threatens the health of human society, sustainable development, and global ecological security. Ecosystem service value (ESV) is a monetary evaluation of the various products and services ecosystems provide for human welfare and long-term economic and social development, including the values of supply services, regulation services, support services, and cultural services provided by ecosystems. It is critical to investigate the changes in ecosystem service value (ESV) and its influencing factors in order to correctly clarify regional ecosystem service issues, uphold regional ecological harmony, and encourage regionally healthy and coordinated sustainable development [4].

Numerous studies have shown that the ESV is influenced by both physical–geographical factors and socioeconomic factors, and the influencing factors vary within different regions. According to some studies, land use change is responsible for the global ESV loss state [5]. The primary cause of the reduction of terrestrial ESV, notably throughout Asia, Africa, and South America, is the spread of cultivated land in tropical forests. Additionally, the impact of urban growth contributes to the loss of ESV across Europe [6]. In China, the reconstruction and industrialization of agricultural land in recent decades have also been important reasons for the fluctuating downward trend in ESV. The ESV has a distinct spatial pattern of being high in the west and south regions of China and low in the east and north regions of China. The ESV in northeastern and northern China decreased significantly, while in Fujian and western Xinjiang, the ESV increased significantly [7,8]. Numerous studies have concluded that in ecologically fragile areas, ESV variations are more influenced by physical–geographical factors than by socioeconomic factors [9], while in areas with better hydrothermal conditions, ESV variations are more influenced by socioeconomic factors [10–12].

Ecosystem services are not totally independent of each other at different spatial and temporal scales, and there is a complex non-linear variation among them, where the growth or decline of one ecosystem service impacts the growth or decline of another service, resulting in trade-offs and synergies between ecosystem services [13–15]. Earlier studies have generally concentrated on straightforward trade-offs and synergistic interactions between ecosystem services but have neglected to explore the drivers and mechanisms of these relationships. Despite this, some research has indicated that trade-offs and synergies across ecosystem services also drive spatial and temporal variability in ecosystem services [16]. The analysis of trade-offs and synergies is important for global ecological and environmental governance, as well as providing a theoretical framework for the wise exploitation of natural resources, given the increasing rise of the global economy, population, and resource scarcity. Therefore, clarifying the impact of trade-offs and synergies on ESV can help us to eliminate the negative effects of trade-offs on ESV and achieve sustainable socio-ecological system development goals. Currently, one of the most pressing issues confronting ecologists is the variation in ESV and its response to trade-offs or synergistic relationships.

So far, the unit area value equivalent factor method developed by Costanza et al. [17] has been widely employed to account for regional ESVs. A new version of the unit area value equivalent factor approach and an equivalent factor table for the ESV of terrestrial ecosystems in China were produced by Xie et al. [18] and are extensively utilized in numerous studies in China [9,12]. However, because ESV is estimated using unit area value coefficients, this method has a few disadvantages. Firstly, the determination of unit area value coefficients is to some extent subjective [19,20]. Secondly, whether the unit area value coefficients correspond to the current condition in the research area has a direct impact on the ESV estimation accuracy [21]. As a result, the unit area value coefficients must be updated to reflect the current condition in the research region [22,23].

Studies have shown that spatial variations in ESV are caused by the combination of physical–geographical and socioeconomic factors, that is, any two elements have a bigger impact on the ESV than any one factor alone [9,12]. Currently, researchers focus on the influencing factors and their interactions with ESV [24]. The relationship between ESV and the influencing factors is not linear and shows significant spatial phenotypic variation [21]. The majority of earlier research on the spatial variations in ESV and their underlying factors has relied on qualitative and correlation analysis methods, such as analysis of spatial autocorrelation, logistic regression, and grey correlation [25]. However, the internal coupling effect is not taken into consideration in the majority of research, which also overlooks the spatial connection between the driving elements. The Geodetector method can be used to identify the driving factors in ESV variation [26,27]. It can identify both the influence of a single factor and the combined effect of several factors on the ESV [28].

In summary, the research on ESV has progressed from a conceptual definition and methodological exploration to a dynamic assessment and spatio-temporal variations to influence factors and their driving mechanisms, which has achieved relatively fruitful results. However, the spatial and temporal dynamics of ESVs need to be deepened, the interrelationships between different ecological functions and their trade-offs and synergies are yet to be clarified, and the driving causes of ESV spatial variation need to be discovered. As the first metropolitan area to be officially approved by the Chinese National Development and Reform Commission, the struggle between economic expansion and environmental preservation in the Nanjing metropolitan region is gradually intensifying. Behind the rapid economic and social advancement is a slew of ecological difficulties, including the degradation of water quality, the destruction of wetlands, the yearly decline of forest area, and the growth of industrial land constantly occupying ecological territory [29]. Assessing the ESV and its influencing factors in the Nanjing metropolitan region is, therefore, one of the urgent issues faced by the government and ecologists. However, only a few researchers have analyzed the spatial variations in ESV and its underlying forces in the Nanjing metropolitan area. The spatio-temporal variations in the ESV and its driving factors in the Nanjing metropolitan area from 2000 to 2020 were explored in this study. The results can be helpful for the coordinated urban sustainability in the Nanjing metropolitan region as well as for the cooperative conservation and management of the environmental ecology. The following are the objectives of this paper: (1) to build a regional ESV estimation model and an index system of driving factors; (2) to measure the spatio-temporal dynamics in ESVs, as well as the links between each ecosystem service's trade-offs and synergies; and (3) to explore the driving forces of the ESV variation using Geodetectors.

2. Materials and Methods

2.1. Study Area

The Nanjing metropolitan area (29°57′ N–34°06′ N, 117°09′ E–119°59′ E) is located in eastern China and is the core area of the urban zone along the middle and lower reaches of the Yangtze River (see Figure 1). It is the first inter-provincial metropolitan area to be officially approved in China, and the cities in it are Nanjing, Zhenjiang, Yangzhou, Huai'an, Ma'anshan, Chuzhou, Wuhu, Xuancheng, Liyang, and Jintan. The area of Nanjing metropolitan area is about 6.6 million hm^2. As of 2020, the resident population is about 35.2 million, and the regional GDP is 4198.2 billion CNY. It generates an economic output of 4.1% of the total in China, with a land area of 0.7% of China and a resident population of 2.5%. The climate in the region is mostly subtropical monsoon, with an average annual temperature of approximately 17 °C and an annual precipitation of around 977 mm. The DEM rises in the south and falls in the north [30].

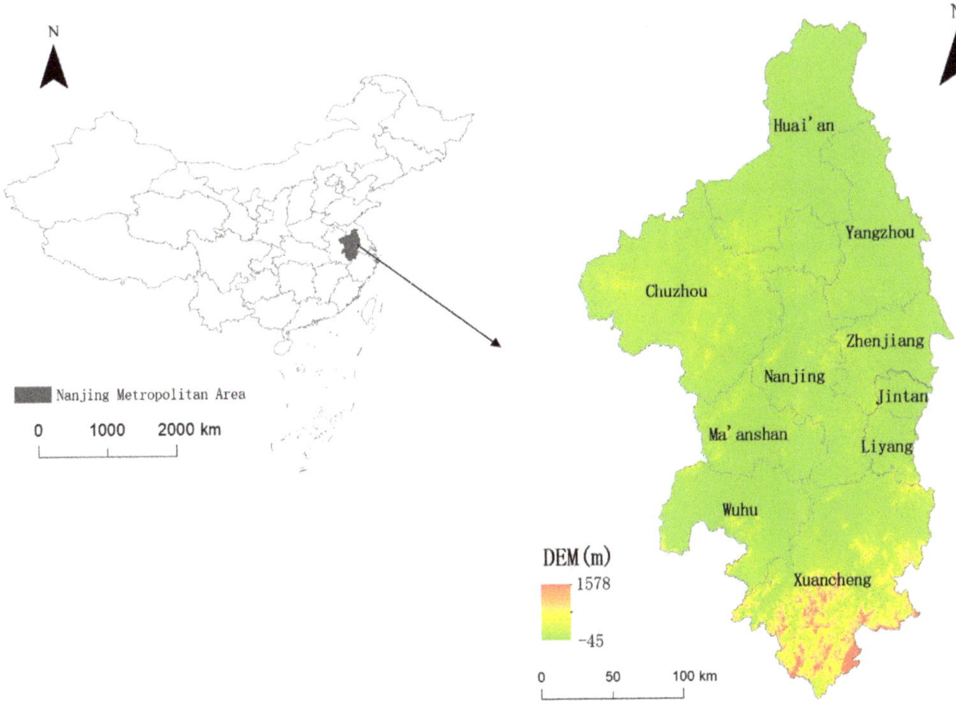

Figure 1. Location of the Nanjing metropolitan area.

2.2. Data Sources and Processing

Seven driving factors were selected in this paper, including (1) physical–geographical factors: temperature (TEM), precipitation (PRE), soil organic matter (SOM), the normalized difference vegetation index (NDVI), a digital elevation model (DEM), and land use maps and (2) socioeconomic factors: the gross economic product (GDP) and population density (POP).

The datasets, including temperature, precipitation, NDVI, land use maps, DEM, GDP, and population density figures, were obtained from the Chinese Academy of Sciences' Resource and Environmental Science and Data Center (http://www.resdc.cn/Default.aspx, accessed on 10 August 2022). Based on the continuous time series of SPOT/VEGETATION NDVI remote sensing data, the monthly NDVI dataset was obtained using the maximum synthesis method. Six groups were utilized to categorize the different types of land use: cultivated land, forest land, grassland, water body, construction land, and unused land (see Figure 2). Unused land indicates saline land, marsh land, bare land, and bare rocky land. The DEM dataset was derived from the Shuttle Radar Topography Mission (SRTM) payload onboard the Space Shuttle Endeavour. The dataset for soil organic matter (SOM) was obtained from the National Tibetan Plateau Data Center (http://data.tpdc.ac.cn/en/, accessed on 10 August 2022) [31]. These datasets were resampled to fit a regional resolution of 1 km. The datasets, including grain production, sown area, and prices, were gathered from the China Yearbook of Agricultural Price Survey and the Statistical Yearbooks produced by the Bureau of Statistics.

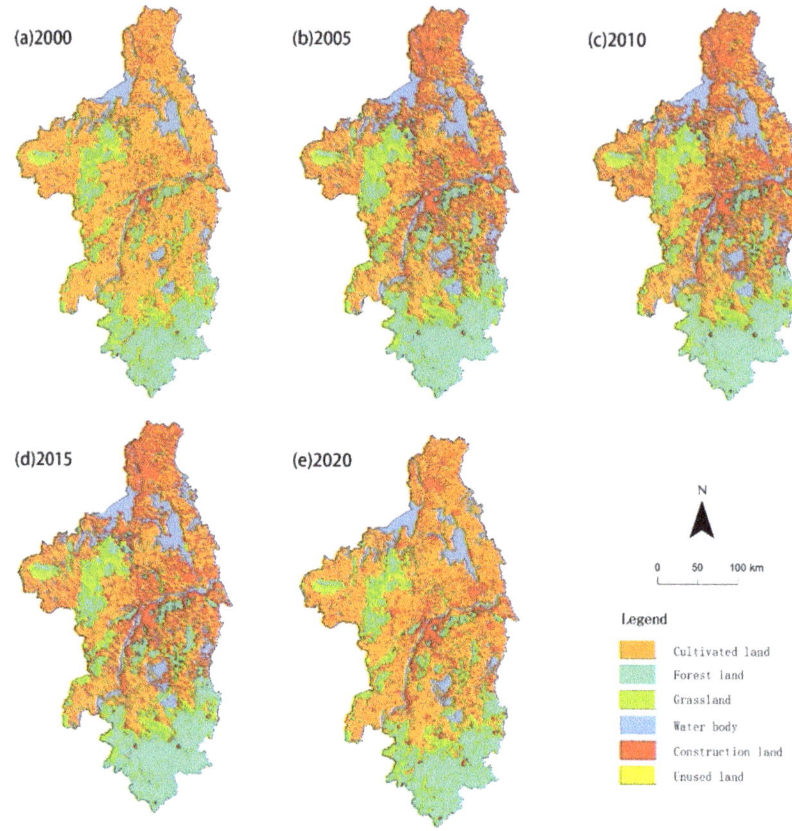

Figure 2. Maps of land use in the Nanjing metropolitan area from 2000 to 2020.

2.3. ESV Evaluation

Starting with the world's and China's ESV equivalent factor tables established by Costanza et al. [17] and Xie et al. [18,32], this article updated the ESV factor table from 2000 to 2020 based on grain production and prices inside the Nanjing metropolitan area. The ESV equivalent factor coefficient, whose value is around one-seventh of the average market value of grain produced in the research area, is determined using the comparative donation of ecosystem services that various ecosystems may supply [18]. The mean grain production in the Nanjing metropolitan area during 2000–2020 was 6427.88 kg/hm^2, and the national mean minimum purchase price of rice and wheat in 2020 was 2.45 CNY/kg, so the study area's updated ESV unit factor value was 2249.76 CNY/hm^2, and the ESV coefficients per unit area were calculated and displayed in Table 1. Previous studies have discovered that the ESV in construction land is extremely low [9]; therefore, the ESV in the construction land was not considered and assigned a value of zero in this article. The ESV in the Nanjing metropolitan area can be calculated as:

$$\begin{aligned} ESV &= \sum(A_k \times VC_k) \\ ESV_f &= \sum\left(A_k \times VC_{fk}\right) \end{aligned} \quad (1)$$

where ESV is the value of all ecosystem services (CNY), A_k is the area of the research area's land use type k (hm^2), VC_k is the ESV coefficient of land use type k (CNY/hm^2/year), ESV_f

is the value of the ecosystem service function f (CNY), and VC_{fk} is the ESV coefficient of the ecosystem service function f in the land use type k (CNY/hm^2/year).

Table 1. The ESV coefficients per unit area in the Nanjing metropolitan area.

Class One Type	Class Two Type	Unit Area Value Coefficient by Land Use Type/(CNY/hm^2/year)				
		Cultivated Land	Forest Land	Grassland	Water	Unused Land
Supply services	Food production	2249.76	497.90	424.14	737.63	0.00
	Raw material production	313.49	1161.76	626.98	258.17	0.00
Regulation services	Gas regulation	1844.07	3817.22	2231.32	885.15	36.88
	Climate regulation	958.91	11,433.21	5882.57	2618.57	0.00
	Environment purification	276.61	3319.32	1936.27	5274.03	184.41
	Hydrological regulation	3872.54	7118.09	4315.11	83,960.32	55.32
Support services	Soil conservation	479.46	4647.05	2710.78	866.71	36.88
	nutrient cycle maintenance	313.49	350.37	202.85	73.76	0.00
	Biodiversity	350.37	4241.35	2471.05	2360.40	36.88
Cultural services	Aesthetic landscape	147.53	1862.51	1088.00	1825.62	18.44

2.4. Sensitivity Test

A sensitivity test was performed to assess the reasonableness of the revised ESV coefficients in the Nanjing metropolitan region. In order to determine the sensitivity coefficient in this study, 50% of the ESV coefficient for each ecosystem service function was moved up and down [33], and the formula is as follows:

$$CS = \left| \frac{(ESV_j - ESV_i)/ESV_i}{\left(VC_{jk} - VC_{ik}\right)/VC_{ik}} \right| \quad (2)$$

where CS is the traditional sensitivity coefficient, ESV_i and ESV_j are the overall value of ecosystem services prior to and following modification, respectively, and VC_{ik} and VC_{jk} are the ESV coefficients prior to and following modification for the land use type k, respectively. If $CS \leq 1$, it means that the revised ESV coefficients are reliable, while if $CS \geq 1$, it means that the revised ESV coefficients are unreliable.

2.5. Ecosystem Service Trade-Offs and Synergies

Relationships between ecosystem services can be positive (synergies), where two services are provided more frequently or less frequently at the same time, or negative (trade-offs), where one service is provided more frequently while another is provided less frequently. Ecosystem service trade-offs and synergies can be determined using the Pearson correlation coefficient. A positive result implies a synergistic relationship between the two ecosystem service functions, whereas a negative result shows a trade-off relationship [15]. The trade-offs and synergies among four categories of ecosystem services, namely supply services, regulation services, support services, and cultural services, were calculated in this paper as follows:

$$R_{xy} = \frac{\sum_{i=1}^{n}[(x_i - \overline{x})(y_i - \overline{y})]}{\sqrt{\sum_{i=1}^{n}(x_i - \overline{x})^2 \sum_{i=1}^{n}(y_i - \overline{y})^2}} \quad (3)$$

where R_{xy} is the correlation coefficient, x_i and y_i are the values of two types of ecosystem services, and $\overline{x}$ and $\overline{y}$ are the average values of x and y, respectively.

2.6. ESV Spatial Analysis

The worldwide spatial autocorrelation analysis in this study was evaluated using Moran's I index. Moran's I has a range of values within (−1, 1). If Moran's I is greater than zero, it means that the spatial distribution of ESV is positively relevant, and the higher the index, the more significant the spatial aggregation. If the value of Moran's I is negative, it means that the spatial distribution of ESV is negatively relevant, and the lower the index, the more significant the spatial dispersion. The spatial distribution of the regional ESV hotspots and cold spots in the Nanjing metropolitan area between 2000 and 2020 was also obtained using the Hotspot Detection Tool (Getis-Ord Gi*) in the ArcGIS software (Version 10.5, Redlands, CA, USA) [34].

2.7. Driving Factors Analysis

The Geodetector method is a statistical approach created by Wang and Xu [26] that detects spatial heterogeneity and identifies the underlying variables. The method determines the impact of multi-factor interactions on the ESV in addition to detecting the influence of a single component. The Geodetector method includes four detectors: factor detector, interaction detector, risk detector, and ecological detector [26]. The factor detector and interaction detector were used in this paper to investigate the underlying factors of ESV.

The factor detector may determine how well a single factor x explains the spatial variance of attribute y using a q-value, whose model is stated as:

$$q = 1 - \frac{\sum_{h=1}^{L} N_h \sigma_h^2}{N\sigma^2} \tag{4}$$

where q is the strength of an explanation for a driving factor on the spatially divergent characteristics of ESV, which takes the range (0,1), and a larger number denotes a stronger ability of the independent variable x to explain the attribute y, and vice versa, that is, the greater explanatory power of the independent variable x on the attribute y, and vice versa; h is the partition of the attribute y or variable x; N_h and N are the numbers of cells in partition h and the number of partitions, respectively; and σ_h^2 and σ^2 are the variances of y in partition h and the variances of y in the study area, respectively.

The explanatory power of the attribute y is evaluated using the interaction detector to determine whether the interaction between factors $x1$ and $x2$ increases or decreases that power. The calculation process is as follows:

Firstly, separately compute the q-values of the two factors $x1$ and $x2$ on y: $q(x1)$ and $q(x2)$.

Secondly, compute the q-value of the interaction between the two variables $x1$ and $x2$: $q(x1 \cap x2)$.

Finally, compare $q(x1)$, $q(x2)$, and $q(x1 \cap x2)$. The relationship between the two factors can be classified into five categories: two-factor enhancement, linear enhancement, nonlinear attenuation, one-factor nonlinear attenuation, and independence.

3. Results and Analysis

3.1. ESV Variation and Sensitivity Analysis

The ESV coefficients for the Nanjing metropolitan region (see Table 1) and land use maps for 2000 to 2020 were utilized to calculate the ESV within each land use category (Table 2). The overall ESV rose significantly between 2000 and 2005, with the increasing value of 27,784 billion CNY. After 2005, the total ESV decreased significantly, with the decreasing value of 28,222 billion CNY. High ESV ecosystems, such as forests, grasslands, wetlands, and water bodies, have an important influence in the increase and reduction of regional ESV. Although there was considerable growth in construction land in the early stages of the study, the area of cultivated land reduced significantly as a result of the strategy of returning cropland to forestland and grassland, which led to an increase rather than a decrease in ESV. After 2005, with each region's rapid economic growth, forms of land usage such as rivers or streams and grasslands with high ESV coefficients were encroached upon by forms of land usage such as agricultural lands and industrial lands with low ESV coefficients. This caused forest and grassland to be degraded, and then the ecological quality of the region declined. The largest total ESV appeared in water bodies; forest land, cultivated land, and grassland came next, with unused land having the smallest total ESV.

Among them, the ESV of cultivated land remained in decline during 2000–2015 and then increased after 2015. The ESV of water bodies, grassland, and forest land trended in the same direction as the overall ESV, first increasing and then declining. The ESV of unused land fluctuated from year to year but showed a generally increasing trend.

Table 2. Total ESV of the Nanjing metropolitan area.

Land Type	ESV/×10^8 CNY				
	2000	2005	2010	2015	2020
Cultivated land	433.092	285.706	281.124	278.293	398.620
Forest land	412.863	506.255	502.180	500.911	409.979
Grassland	60.786	91.584	91.387	91.146	59.538
Water	621.239	922.268	917.424	913.964	655.444
Unused land	0.004	0.011	0.010	0.014	0.022
Total	1527.984	1805.824	1792.125	1784.328	1523.603

The regulation services had a larger total ESV than the support and supply services, while the cultural services had the least overall ESV, as seen in Table 3. The slight increase in ESV of the hydrological regulation function and environmental purification function in the research area was driven by the expansion of the water body surface area at the time of the study, as the ESV of these two ecosystem service functions was mainly influenced by the area of water bodies. In contrast, other ecosystem service functions saw a decline in ESV, with the largest declines in the food production function and nutrient cycle maintenance function, which were caused by the decrease in the area of cultivated land and forest land during the research period because the area of either cultivated land or forest land had a significant impact on the ESV of various ecosystem service functions.

Table 3. Changes of individual ESV in the Nanjing metropolitan area.

Class One Type	Class Two Type	ESV/×10^8 CNY					ESV Change Rate/%				
		2000	2005	2010	2015	2020	2000–2005	2005–2010	2010–2015	2015–2020	2000–2020
Supply services	Food production	101.33	74.69	73.65	73.01	94.34	−26.94	−1.39	−0.87	29.22	−6.90
	Raw material production	28.40	28.62	28.34	28.21	27.37	0.77	−0.98	−0.46	−2.98	−3.63
Regulation services	Gas regulation	126.65	116.61	115.36	114.70	120.67	−7.93	1.07	−0.57	5.20	−4.72
	Climate regulation	195.25	226.80	224.99	224.20	191.97	16.16	−0.80	−0.35	−14.38	−1.68
	Environment purification	85.30	108.33	107.58	107.20	85.84	27.00	−0.69	−0.35	−19.93	0.63
	Hydrological regulation	771.23	997.43	990.88	986.65	787.15	29.22	−0.66	−0.43	−20.22	2.06
Support services	Soil conservation	82.09	93.29	92.53	92.19	80.39	13.64	−0.81	−0.38	−12.80	−2.07
	nutrient cycle maintenance	17.35	14.44	14.26	14.16	16.34	−16.77	−1.25	−0.70	15.40	−5.82
	Biodiversity	81.28	97.47	96.73	96.39	80.52	19.92	−0.76	−0.35	−16.46	−0.94
Cultural services	Aesthetic landscape	40.41	50.01	49.65	49.47	40.37	23.76	−0.72	−0.36	−18.39	−0.10

The accuracy of an ESV assessment greatly depends on the validity of the revised ESV coefficients. Therefore, a sensitivity analysis on the revised ESV coefficients was carried out in this paper (see Table 4), and the findings indicate that for every land use category, the sensitivity coefficients of the revised ESV coefficients were below 1. This indicates that the ESV for the Nanjing metropolitan region was trustworthy and stable.

Table 4. Sensitivity test of the revised ESV coefficients in the Nanjing metropolitan area.

Land Type	2000	2005	2010	2015	2020
Cultivated land	0.283	0.158	0.157	0.156	0.262
Forest land	0.270	0.280	0.280	0.281	0.269
Grassland	0.040	0.051	0.051	0.051	0.039
Water	0.407	0.511	0.512	0.512	0.430
Unused land	0.000	0.000	0.000	0.000	0.000

3.2. Spatial Pattern in ESV

The overall ESV of the Nanjing metropolitan region displayed a geographical trend of being high in the south and low in the north in the last 20 years (see Figure 3). The reason for this is that cultivated land dominated the area north of the Nanjing metropolitan region, while forest land was extensively spread in the south of the research area (see Figure 2), and the ESV of cultivated land was lower than that of forest land (see Table 2). The lowest ESV (0–0.04 million CNY) appeared in the construction land and unused land, while the highest value (3.9–10 million CNY) appeared in the rivers, lakes, and other water bodies. The forest land also had a high ESV (2.2–3.9 million CNY), which was dispersed towards the Nanjing metropolitan area's southern region. The ESV of the cultivated land was low, with the values ranging from 0.04 to 1.1 million CNY. The majority of the grassland was situated in the western portion of the Nanjing metropolitan region, and its ESV ranged from 1.1 to 2.2 million CNY.

Moran's I index was computed in order to assess the global spatial autocorrelation study (Table 5). The results indicate that the Moran's I were higher than 0.5, revealing that the geographic distribution of ESV in the Nanjing metropolitan region was dramatically spatially positively correlated and had a high degree of spatial clustering, with high values converging and low values close to each other. In general, lower levels of ecological space fragmentation are associated with more concentrated geographic distributions of ESV. The higher positive agglomeration is helpful for the division of ecological space and the realization of the co-protection and co-administration of the natural ecosystem in the Nanjing metropolitan area.

Table 5. The analysis of spatial autocorrelation in ESV.

Year	Moran's I	Z Value	p Value
2000	0.511	256.022	<0.001
2005	0.506	253.504	<0.001
2010	0.505	252.939	<0.001
2015	0.503	252.056	<0.001
2020	0.507	254.019	<0.001

Regional ESV hot spots and cold spots were examined using the Hotspot Detection Tool to determine their spatial distribution, and the results show that the ESV hot spots were located around the Yangtze River and its environs, or in the lakes in Huai'an, northern Yangzhou, and southern Nanjing (see Figure 4 and Table 6). The ESV cold spots were mainly distributed in the construction land in Huai'an, Nanjing, Yangzhou and Zhenjiang, and the low ESV clustered significantly, indicating that the ESV in those regions was low. The number of cold spots displayed a pattern of initial increase and subsequent decrease, with the maximum number appearing in 2015. The number of hot spots fluctuated from year to year but showed a generally increasing trend. The interannual fluctuations in the number of cold and hot spots were generally similar with the regional ESV fluctuations.

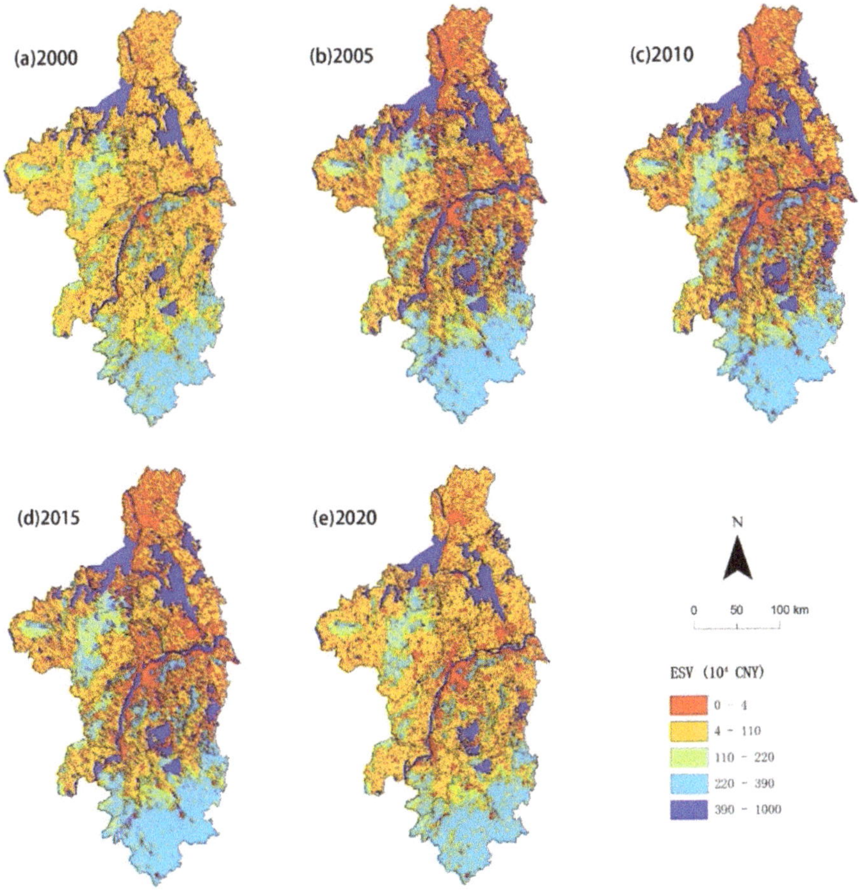

Figure 3. Spatial distribution of ESV per unit area during 2000–2020.

Table 6. Changes in the number of ESV cold spots and hot spots during 2000–2020.

Type	2000	2005	2010	2015	2020
Cold spots	4147	17,239	17,561	17,643	7033
Hot spots	8967	9125	9116	9048	9113

3.3. Spatial Variation in ESV

Figure 5 shows that during the period 2000–2020, the ESV had primarily demonstrated a downward trend in the Nanjing metropolitan area's northwest, central, and southern regions, while in the water bodies and their surrounding areas, the ESV had mainly shown an increasing trend. The regions with decreasing ESV accounted for 33.69% within the Nanjing metropolitan area. Among them, there was approximately 17.67% of the research area where the ESV decline rate was greater than 2, and they were primarily spread in the research area's central and northern regions. Within the Nanjing metropolitan area, regions with increasing ESV accounted for 20.40% of the total. Among them, there was approximately 13.37% of the study area where the ESV growth rate was greater than 2, and they were primarily spread in southwestern Nanjing and northern Xuancheng.

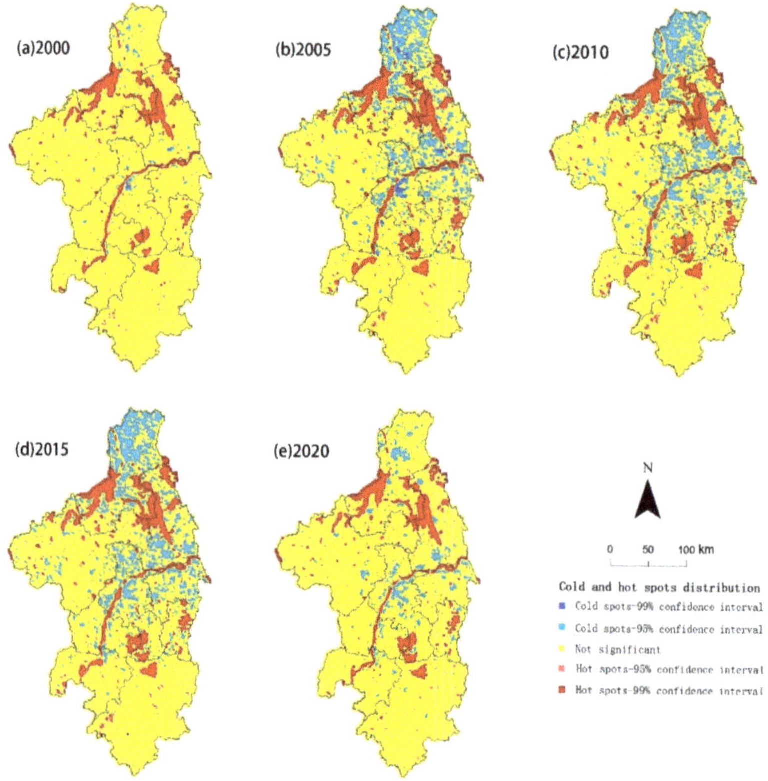

Figure 4. ESV cold spots and hot spots distribution during 2000–2020.

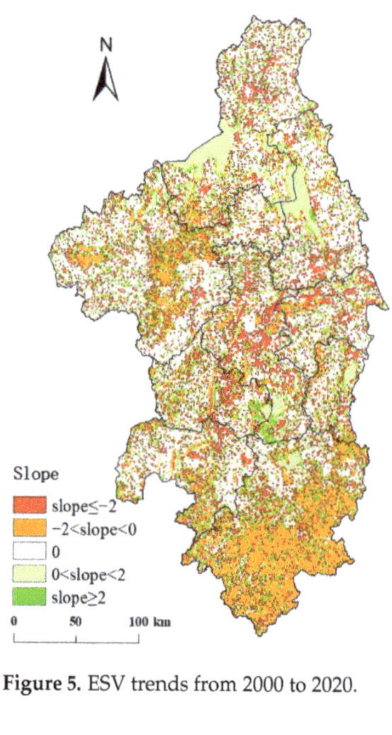

Figure 5. ESV trends from 2000 to 2020.

3.4. Factors Affecting ESV

The ESV in the Nanjing metropolitan area was affected by both physical–geographical factors and socioeconomic factors. Using the factor detector, it was possible to assess the explanatory power of the variables influencing the geographic variations in ESV (see Figure 6), and the results show that the explanatory strength of factors in descending order was population density (17.5%) > DEM (14.7%) > NDVI (13.0%) > precipitation (11.7%) > GDP (8.3%) > soil organic matter (7.2%) > temperature (4.3%). The population density was the main socioeconomic factor influencing ESV spatial variation, while the major physical–geographical factors were the DEM, NDVI, and precipitation. The impact of temperature on ESV was the lowest.

factor	DEM	GDP	NDVI	POP	PRE	SOM	TEM
DEM	0.1471						
GDP	0.2190	0.0830					
NDVI	0.2367	0.3171	0.1302				
POP	0.2268	0.2597	0.3127	0.1749			
PRE	0.1979	0.2790	0.2847	0.2608	0.1173		
SOM	0.1970	0.2715	0.3096	0.2881	0.2745	0.0716	
TEM	0.2019	0.2688	0.3155	0.2958	0.2544	0.2313	0.0434

q value: High – Low

Figure 6. The impact of single-factor and two-factor interaction on the ESV variation. DEM is the digital elevation model, GDP is the gross economic product, NDVI is the normalized difference vegetation index, POP is population density, PRE is precipitation, SOM is soil organic matter, and TEM is temperature.

The effect of two-factor interaction on the ESV variation was also investigated using the interaction detector (see Figure 6). The results show that, in contrast to the explanatory strength of a single element on the ESV, the impact of any individual element on the ESV under interaction is significantly increased. After interacting with physical–geographical factors, the impact of the GDP and population density on ESV was brought to light. It implies that socioeconomic factors had a significant interaction in combination with physical–geographical factors. The interplay between the NDVI and GDP showed the most obvious effect on the ESV, reaching 31.71% of the explanatory strength for ESV variation.

3.5. Trade-Offs and Synergies Affecting ESV

Figure 7 depicts the trade-offs and synergies between supply service value, regulation service value, support service value, and cultural service value. The results show that during the period 2000–2020, the trade-offs and synergies between the ESVs were mainly synergies. However, there were still trade-offs among the ESVs. In the Nanjing metropolitan region, the trade-offs between supply and cultural service values made up 17.85% of the total area, while those between supply and regulation service values made up 17.15%. These areas were distributed spatially in a largely uniform manner and were most spread in the central part of Xuancheng, Wuhu, Ma'anshan, Liyang, Chuzhou, and southern Huai'an. The areas with trade-offs between supply service value and support service value made up 9.28% of the Nanjing metropolitan area, primarily in the eastern Ma'anshan, southern Huai'an, and northern Wuhu. In the Nanjing metropolitan region, the trade-offs between regulation service value and support service value made up 7.85% of the total area, while those between support service value and cultural service value made up 7.27%. These areas were distributed spatially in a largely uniform manner and were mainly distributed in northern Wuhu, southeastern Ma'anshan, northern Yangzhou, western Chuzhou, and southern Huai'an. The areas with trade-offs between regulation service value and cultural service value only made up 0.42% of the Nanjing metropolitan area.

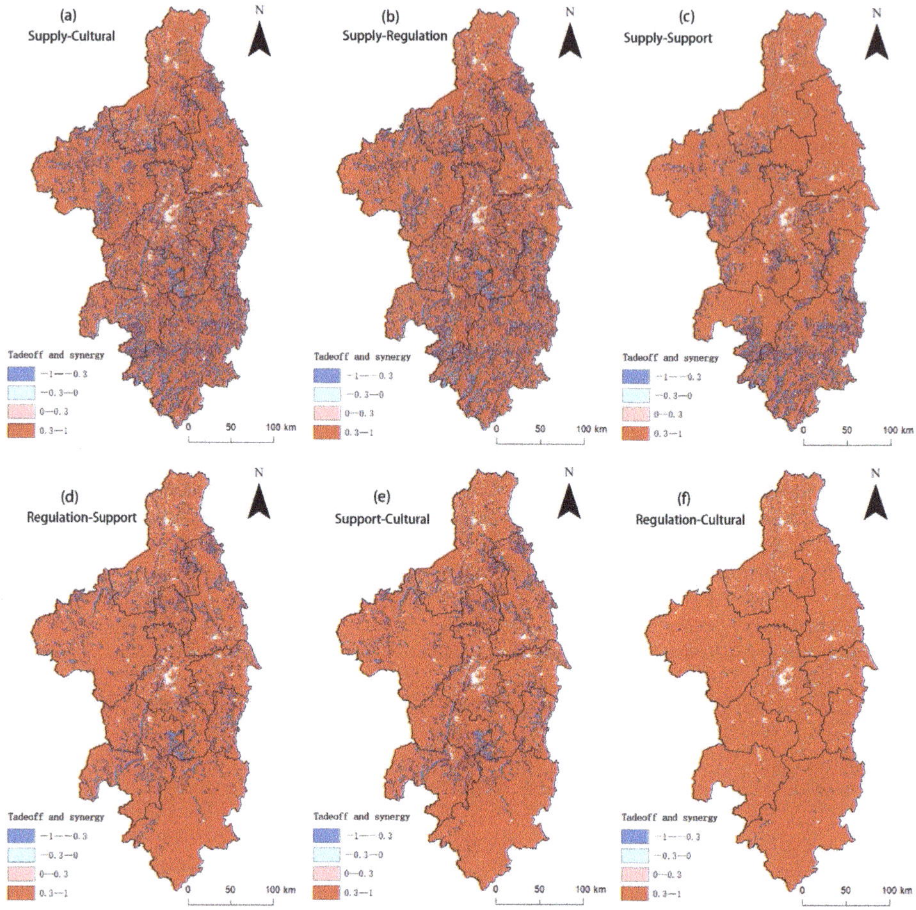

Figure 7. The trade-offs and synergies between (**a**) supply service value and cultural service value, (**b**) supply service value and regulation service value, (**c**) supply service value and support service value, (**d**) regulation service value and support service value, (**e**) support service value and cultural service value, and (**f**) regulation service value and cultural service value.

According to Figures 5 and 7, the results show that the trade-offs among the ESVs affected the spatial variation in ESV. The areas with trade-offs between supply-cultural service value, supply-regulation service value, supply-support service value, and regulation-support service value accounted for 26.81%, 21.66%, 16.33% and 14.97% of the area where the ESV was declining, respectively.

3.6. Factors Affecting Trade-Offs and Synergies

The physical–geographical factors and socioeconomic factors affected the trade-offs and synergies (TOSs) among supply service value, regulation service value, support service value, and cultural service value. Figure 8 shows that, in contrast to the limited strength of a single factor on the TOSs, the influence of all single factors on the TOSs under interaction is greatly increased. The interaction between SOM and GDP had the most obvious effect on the TOSs between supply service value and support service value, while the effect of the interaction between SOM and precipitation was more substantial for the other five TOSs.

regulation-cultural	GDP	NDVI	POP	PRE	SOM	TEM	DEM
GDP	0.0161						
NDVI	0.1189	0.0216					
POP	0.0819	0.1187	0.0360				
PRE	0.1352	0.1448	0.1251	0.0431			
SOM	0.1511	0.1584	0.1310	0.1595	0.0147		
TEM	0.1289	0.1528	0.1325	0.1187	0.1501	0.0151	
DEM	0.0729	0.0712	0.0680	0.0840	0.0795	0.0789	0.0222

regulation-support	GDP	NDVI	POP	PRE	SOM	TEM	DEM
GDP	0.0079						
NDVI	0.0886	0.0081					
POP	0.0496	0.0707	0.0094				
PRE	0.0935	0.1078	0.0760	0.0128			
SOM	0.1164	0.1352	0.0964	0.1468	0.0095		
TEM	0.1217	0.1201	0.1050	0.1022	0.1412	0.0100	
DEM	0.0833	0.0643	0.0557	0.0818	0.1267	0.0905	0.0129

supply-cultural	GDP	NDVI	POP	PRE	SOM	TEM	DEM
GDP	0.0175						
NDVI	0.0946	0.0089					
POP	0.0557	0.0709	0.0104				
PRE	0.1217	0.1171	0.0970	0.0222			
SOM	0.1384	0.1390	0.1032	0.1492	0.0111		
TEM	0.1269	0.1337	0.1044	0.1107	0.1412	0.0163	
DEM	0.0821	0.0635	0.0496	0.0837	0.1030	0.0778	0.0157

supply-regulation	GDP	NDVI	POP	PRE	SOM	TEM	DEM
GDP	0.0160						
NDVI	0.0921	0.0093					
POP	0.0506	0.0685	0.0089				
PRE	0.1170	0.1144	0.0898	0.0195			
SOM	0.1340	0.1390	0.0991	0.1466	0.0105		
TEM	0.1215	0.1283	0.1012	0.1049	0.1389	0.0137	
DEM	0.0808	0.0622	0.0480	0.0818	0.1037	0.0768	0.0149

supply-support	GDP	NDVI	POP	PRE	SOM	TEM	DEM
GDP	0.0210						
NDVI	0.1047	0.0133					
POP	0.0768	0.0932	0.0245				
PRE	0.1410	0.1288	0.1257	0.0396			
SOM	0.1544	0.1502	0.1224	0.1544	0.0157		
TEM	0.1364	0.1491	0.1209	0.1226	0.1512	0.0213	
DEM	0.0737	0.0598	0.0532	0.0851	0.0822	0.0752	0.0201

support-cultural	GDP	NDVI	POP	PRE	SOM	TEM	DEM
GDP	0.0089						
NDVI	0.0908	0.0090					
POP	0.0558	0.0728	0.0095				
PRE	0.0950	0.1073	0.0805	0.0131			
SOM	0.1171	0.1346	0.1022	0.1506	0.0093		
TEM	0.1231	0.1218	0.1021	0.1051	0.1482	0.0117	
DEM	0.0826	0.0600	0.0567	0.0848	0.1277	0.0971	0.0138

q value: High – Low

Figure 8. The impact of single-factor and two-factor interaction on the trade-offs and synergies.

The TOSs for supply service value, regulation service value, support service value, and cultural service value were also impacted by land use change. The change in land usage from high-ESV type to low-ESV type resulted in trade-off relationships among the ESVs. According to Figure 7a–f, the regions where land usage patterns had transformed from high–ESV type to low–ESV type made up 50.27%, 46.52%, 49.91%, 43.95%, 43.23%, and 40.24% of the trade-off regions, respectively.

4. Discussion

4.1. Comparison with Previous Work

In this paper, the ESV in the Nanjing metropolitan area during 2000–2020 was calculated based on the world's and China's ESV equivalent factor tables established by Costanza et al. [5] and Xie et al. [32]. The computed ESVs agreed with the findings of Cao et al. [35]. In the Nanjing metropolitan area, the key factors impacting the spatial variations in ESV were the POP, DEM, and NDVI, with socioeconomic factors having a stronger influence on the spatial variations in ESV. This is in line with the results found in the Yangtze River Economic Belt reported by Liu et al. [12] and Long et al. [11]. However, our result is different from those found in northeastern China reported by Yi et al. [9]. It is possible that this difference is due to the region's unique climatic circumstances. The Nanjing metropolitan region is situated in eastern China, with suitable hydrothermal conditions for vegetation growth. Therefore, the ESV variations are more likely to be influenced by socioeconomic factors. This is in line with the results of He et al. [36] in Sichuan Province and Wang et al. [37] in Xinjiang.

4.2. Policy Implications

At the root of the urban ecological catastrophe is the degradation and eradication of urban ecological services [38]. While ecological land, along with forest land, grassland, and garden land, has been severely encroached upon, urbanization in the Nanjing metropolitan region has propelled the construction of construction land. Policies that are pertinent to ecological protection have an excellent effect on the growth of ESV. We need to take targeted measures to mitigate the conflict between ESV and urbanization needs based on research findings [38]. For example, the high-ESV regions and the regions with severe ESV degradation should be protected. In addition, with the significant impact of population density on ESV, the massive pressure that population increase puts on the ecosystem must be relieved in order to better control excessive and rapid population growth.

In land-use decisions, TOSs among ESVs also need to be taken into account [39]. For example, land development that guarantees food security necessarily leads to diminished preservation of soil and water and hydrological regulation. As a result, it is crucial to develop scientific, rational, and practice-oriented decisions to avoid negative trade-offs and promote positive synergies. It is possible to break through administrative boundaries in the future management of ESVs, implement integrated management of ESVs in the Nanjing metropolitan area, restrict agricultural development around the Yangtze River and lakes, and protect their regulation services and support services. At the same time, in the surrounding regions of cities and towns where cultivated land is the primary land use type, consideration should be given to protecting food production functions and strictly limiting the takeover of farming areas by the extension of industrial land.

4.3. Shortcomings and Prospects

Geographers and ecologists have devoted a lot of attention to the study of ESV since Costanza first introduced the concept of ESV [40]. Research on ESV has gradually shifted from the early days of assessing their value to examining the distribution of ESVs in various ecosystem types and their driving variables. However, the exploration of spatio-temporal patterns of change and their driving mechanisms is a step-by-step task due to the complexity of ecosystems. The Geodetector method can better reveal the geographical scale driving mechanisms of the regional ESV spatial divergence phenomenon, but the motivating factors still need to be further explored at the temporal scale. In the future, the existing index system can be expanded to examine the mechanisms underlying regional ESV evolution at multiple scales across time and space.

5. Conclusions

Using the method of updated unit area value equivalent factor, this paper estimated the ecosystem service value in the Nanjing metropolitan area and analyzed the ecosystem service value fluctuations in space and time and the factors that affect it. The findings indicate that:

(1) The overall value of ecosystem services decreased between 2000 and 2020. Moreover, the ecosystem service value in the southern section of the Nanjing metropolitan region was higher than in the northern portion. The spatial distribution of the value of ecosystem services was significantly spatially positively correlated and showed a high level of spatial clustering.
(2) The highest overall ecosystem service value appeared in the water bodies, followed by forest land, cultivated land, and grassland, with construction land and unused land having the smallest total ecosystem service value.
(3) The population density was the primary socioeconomic factor influencing the spatial variation of ecosystem service value, while the digital elevation model, normalized difference vegetation index, and precipitation were the primary physical–geographical factors.
(4) The impact of the gross economic product and population density on the ecosystem service value was highlighted after interacting with physical–geographical factors. The interaction between the normalized difference vegetation index and gross economic product had the most obvious effect on the ecosystem service value.
(5) The synergies and trade-offs that exist between the value of supply service, regulation service, support service, and cultural service affect the spatial variations in ecosystem service value.

Author Contributions: Data curation, X.L.; Funding acquisition, S.C.; Investigation, Z.Z.; Methodology, X.L. and L.Y.; Project administration, S.C.; Writing—original draft, S.C. and X.L. All authors have read and agreed to the published version of the manuscript.

Funding: This research was funded by the Major Project of the Philosophy and Social Science Foundation of the Jiangsu Higher Education Institutions of China (Grant number 2021SJZDA130).

Institutional Review Board Statement: Not applicable.

Informed Consent Statement: Not applicable.

Data Availability Statement: Not applicable.

Conflicts of Interest: The authors declare no conflict of interest.

References

1. Millennium Ecosystem Assessment Board. *Millennium Ecosystem Assessment: Frameworks*; World Resources Institute: Washington, DC, USA, 2005.
2. Brondizio, E.S.; Settele, J.; Díaz, S.; Ngo, H.T. *Global Assessment Report on Biodiversity and Ecosystem Services of the Intergovernmental Science-Policy Platform on Biodiversity and Ecosystem Services*; IPBES Secretariat: Bonn, Germany, 2019.
3. World Economic Forum. Nature Risk Rising: Why the Crisis Engulfing Nature Matters for Business and the Economy. In *Collaboration with PwC*; New Nature Economy Series; WEF: Geneve, Switzerland, 2020.
4. Liu, Z.; Wang, S.J.; Fang, C. Spatiotemporal evolution and influencing mechanism of ecosystem service value in the Guangdong-Hong Kong-Macao Greater Bay Area. *J. Geogr. Sci.* **2021**, *76*, 2797–2813.
5. Costanza, R.; de Groot, R.; Sutton, P.; van der Ploeg, S.; Anderson, S.J.; Kubiszewski, I.; Farber, S.; Turner, R.K. Changes in the global value of ecosystem services. *Glob. Environ. Chang.-Hum. Policy Dimens.* **2014**, *26*, 152–158. [CrossRef]
6. Li, Y.; Tan, M.; Hao, H. The impact of global cropland changes on terrestrial ecosystem services value, 1992–2015. *J. Geogr. Sci.* **2019**, *29*, 323–333. [CrossRef]
7. Song, W.; Deng, X. Land-use/land-cover change and ecosystem service provision in China. *Sci. Total Environ.* **2017**, *576*, 705–719. [CrossRef] [PubMed]
8. Yang, R.F.; Ren, F.; Xu, W.X.; Ma, X.Y.; Zhang, H.W.; He, W.W. China's ecosystem service value in 1992–2018: Pattern and anthropogenic driving factors detection using Bayesian spatiotemporal hierarchy model. *J. Environ. Manag.* **2022**, *302*, 114089. [CrossRef]
9. Shang, Y.; Wang, D.; Liu, S.; Li, H. Spatial-Temporal Variation and Mechanisms Causing Spatial Differentiation of Ecosystem Services in Ecologically Fragile Regions Based on Value Evaluation: A Case Study of Western Jilin, China. *Land* **2022**, *11*, 629. [CrossRef]
10. Song, F.; Su, F.L.; Mi, C.X.; Sun, D. Analysis of driving forces on wetland ecosystem services value change: A case in Northeast China. *Sci. Total Environ.* **2021**, *751*, 141778. [CrossRef]
11. Long, X.; Lin, H.; An, X.; Chen, S.; Qi, S.; Zhang, M. Evaluation and analysis of ecosystem service value based on land use/cover change in Dongting Lake wetland. *Ecol. Indic.* **2022**, *136*, 108619. [CrossRef]
12. Yang, L.; Jiao, H. Spatiotemporal Changes in Ecosystem Services Value and Its Driving Factors in the Karst Region of China. *Sustainability* **2022**, *14*, 6695. [CrossRef]
13. Ma, X.; Zhu, J.; Zhang, H.; Yan, W.; Zhao, C. Trade-offs and synergies in ecosystem service values of inland lake wetlands in Central Asia under land use/cover change: A case study on Ebinur Lake, China. *Glob. Ecol. Conserv.* **2020**, *24*, e01253. [CrossRef]
14. Liu, H.; Zheng, L.; Wu, J.; Liao, Y. Past and future ecosystem service trade-offs in Poyang Lake Basin under different land use policy scenarios. *Arab. J. Geosci.* **2020**, *13*, 46. [CrossRef]
15. Liu, H.; Wu, J.; Liao, M. Ecosystem service trade-offs upstream and downstream of a dam: A case study of the Danjiangkou dam, China. *Arab. J. Geosci.* **2019**, *12*, 17. [CrossRef]
16. Zuo, L.; Jiang, Y.; Gao, J.; Du, F.; Zhang, Y. Quantitative separation of multi-dimensional driving forces of ecosystem services in the ecological conservation red line area. *J. Geogr. Sci.* **2022**, *77*, 2174–2188.
17. Costanza, R.; d'Arge, R.; de Groot, R.; Farber, S.; Grasso, M.; Hannon, B.; Limburg, K.; Naeem, S.; O'Neill, R.V.; Paruelo, J.; et al. The value of the world's ecosystem services and natural capital. *Nature* **1997**, *387*, 253–260. [CrossRef]
18. Xie, G.; Zhen, L.; Lu, C.; Xiao, Y.; Chen, C. Expert Knowledge Based Valuation Method of Ecosystem Services in China. *J. Nat. Resour.* **2008**, *23*, 911–919.
19. Wang, Y.; Pan, J. Building ecological security patterns based on ecosystem services value reconstruction in an arid inland basin: A case study in Ganzhou District, NW China. *J. Clean. Prod.* **2019**, *241*, 118337. [CrossRef]
20. Xiao, R.; Lin, M.; Fei, X.; Li, Y.; Zhang, Z.; Meng, Q. Exploring the interactive coercing relationship between urbanization and ecosystem service value in the Shanghai-Hangzhou Bay Metropolitan Region. *J. Clean. Prod.* **2020**, *253*, 119803. [CrossRef]
21. Wu, C.; Chen, B.; Huang, X.; Wei, Y.H.D. Effect of land-use change and optimization on the ecosystem service values of Jiangsu province, China. *Ecol. Indic.* **2020**, *117*, 106507. [CrossRef]
22. Rao, Y.; Zhou, M.; Ou, G.; Dai, D.; Zhang, L.; Zhang, Z.; Nie, X.; Yang, C. Integrating ecosystem services value for sustainable land-use management in semi-arid region. *J. Clean. Prod.* **2018**, *186*, 662–672. [CrossRef]
23. Shi, L.; Halik, U.; Mamat, Z.; Aishan, T.; Abliz, A.; Welp, M. Spatiotemporal investigation of the interactive coercing relationship between urbanization and ecosystem services in arid northwestern China. *Land Degrad. Dev.* **2021**, *32*, 4105–4120. [CrossRef]
24. Li, S.; Yang, H.; Liu, J.; Lei, G. Towards Ecological-Economic Integrity in the Jing-Jin-Ji Regional Development in China. *Water* **2018**, *10*, 1653. [CrossRef]
25. Chen, M.; Lu, Y.; Ling, L.; Wan, Y.; Luo, Z.; Huang, H. Drivers of changes in ecosystem service values in Ganjiang upstream watershed. *Land Use Policy* **2015**, *47*, 247–252. [CrossRef]
26. Wang, J.; Xu, C. Geodetector: Principle and prospective. *J. Geogr. Sci.* **2017**, *72*, 116–134.

27. Bai, L.; Jiang, L.; Yang, D.Y.; Liu, Y.B. Quantifying the spatial heterogeneity influences of natural and socioeconomic factors and their interactions on air pollution using the geographical detector method: A case study of the Yangtze River Economic Belt, China. *J. Clean. Prod.* **2019**, *232*, 692–704. [CrossRef]
28. Liu, X.; Wang, H.; Wang, X.; Bai, M.; He, D. Driving factors and their interactions of carabid beetle distribution based on the geographical detector method. *Ecol. Indic.* **2021**, *133*, 108393. [CrossRef]
29. Liu, W.; Zhan, J.; Zhao, F.; Wang, C.; Zhang, F.; Teng, Y.; Chu, X.; Kumi, M.A. Spatio-temporal variations of ecosystem services and their drivers in the Pearl River Delta, China. *J. Clean. Prod.* **2022**, *337*, 130466. [CrossRef]
30. Liu, W.; Xu, H.; Zhang, X.; Jiang, W. Green Infrastructure Network Identification at a Regional Scale: The Case of Nanjing Metropolitan Area, China. *Forests* **2022**, *13*, 735. [CrossRef]
31. Shangguan, W.; Dai, Y.J.; Liu, B.Y.; Zhu, A.X.; Duan, Q.Y.; Wu, L.Z.; Ji, D.Y.; Ye, A.Z.; Yuan, H.; Zhang, Q. A China data set of soil properties for land surface modeling (EI). *J. Adv. Model. Earth Syst.* **2013**, *5*, 212–224. [CrossRef]
32. Xie, G.; Zhang, C.; Zhang, L.; Chen, W.; Li, S. Improvement of the Evaluation Method for Ecosystem Service Value Based on Per Unit Area. *J. Nat. Resour.* **2015**, *30*, 1243–1254.
33. Pan, N.; Guan, Q.; Wang, Q.; Sun, Y.; Li, H.; Ma, Y. Spatial Differentiation and Driving Mechanisms in Ecosystem Service Value of Arid Region: A case study in the middle and lower reaches of Shule River Basin, NW China. *J. Clean. Prod.* **2021**, *319*, 128718. [CrossRef]
34. Li, Z.; Sun, Z.; Tian, Y.; Zhong, J.; Yang, W. Impact of Land Use/Cover Change on Yangtze River Delta Urban Agglomeration Ecosystem Services Value: Temporal-Spatial Patterns and Cold/Hot Spots Ecosystem Services Value Change Brought by Urbanization. *Int. J. Environ. Res. Public Health* **2019**, *16*, 123. [CrossRef]
35. Cao, J.; Wang, L.; Cao, W. Research on Ecological Zoning Based on the Differential Pattern of Ecosystem Service Supply and Denmand—Taking Nanjing Metropolitan Area as an Example. *Resour. Dev. Mark.* **2022**, *38*, 520–528.
36. He, C.; Shao, H.; Xian, W. Spatiotemporal Variation and Driving Forces Analysis of Eco-System Service Values: A Case Study of Sichuan Province, China. *Int. J. Environ. Res. Public Health* **2022**, *19*, 8595. [CrossRef]
37. Wang, Y.; Shataer, R.; Zhang, Z.; Zhen, H.; Xia, T. Evaluation and Analysis of Influencing Factors of Ecosystem Service Value Change in Xinjiang under Different Land Use Types. *Water* **2022**, *14*, 1424. [CrossRef]
38. Luo, Q.; Zhou, J.; Li, Z.; Yu, B. Spatial differences of ecosystem services and their driving factors: A comparation analysis among three urban agglomerations in China's Yangtze River Economic Belt. *Sci. Total Environ.* **2020**, *725*, 138452. [CrossRef]
39. Pan, S.; Liang, J.; Chen, W.; Li, J.; Liu, Z. Gray Forecast of Ecosystem Services Value and Its Driving Forces in Karst Areas of China: A Case Study in Guizhou Province, China. *Int. J. Environ. Res. Public Health* **2021**, *18*, 12404. [CrossRef]
40. Li, W.; Wang, L.; Yang, X.; Liang, T.; Zhang, Q.; Liao, X.; White, J.R.; Rinklebe, J. Interactive influences of meteorological and socioeconomic factors on ecosystem service values in a river basin with different geomorphic features. *Sci. Total Environ.* **2022**, *829*, 154595. [CrossRef]

forests

MDPI

Article

The Influence of Visitors' Recreation Experience and Environmental Attitude on Environmentally Responsible Behavior: A Case Study of an Urban Forest Park, China

Lisheng Weng [1], Yijin Zhu [1], Xiangting Xu [1], Jiayi Yang [2], Shuhui Zheng [1], Huifang Liu [3,4], Hairong Wang [5] and Liqun Zhu [1,*]

1 College of Humanities & Social Development, Nanjing Agricultural University, Nanjing 210095, China
2 College of Economics and Management, Nanjing Agricultural University, Nanjing 210095, China
3 College of Business Administration, Xinjiang University of Science & Technology, Korla 841000, China
4 School of Management, Shandong University, Jinan 250099, China
5 Headquarter of NAU New Campus Construction, Nanjing Agricultural University, Nanjing 210095, China
* Correspondence: zhulq@njau.edu.cn

Abstract: Nature-based tourism destinations such as national forest parks have become important places for outdoor recreation. This study empirically investigates the structural relationships among the four subdivided dimensions of recreation experience (education, aesthetics, entertainment, and escapism), environmental attitude, and environmentally responsible behavior. In addition, the mediating effect of environmental attitude and the moderating effect of gender are also examined. A field survey was conducted in an urban forest park to collect data, and 468 usable questionnaires were collected. Results indicate that, in addition to the educational experience, the entertainment experience, aesthetic experience, and escape experience have significant positive impacts on visitors' environmental attitudes and environmentally responsible behavior. In addition, environmental attitude partially mediates the relationship between recreation experience and environmentally responsible behavior. The results of moderating effect analysis demonstrate that men and women have differences in terms of recreation experience and environmental attitude. Theoretical and practical implications are discussed, and nature-based tourism destinations are encouraged to emphasize the four subdivided dimensions of recreation experience, especially the educational experience.

Keywords: recreation experience; environmental attitude; environmentally responsible behavior; nature-based tourism; national forest park

1. Introduction

The global outbreak of COVID-19 at the end of 2019 has had a great impact on people's travel [1–5]. The instability brought about by the pandemic has made people more inclined to choose natural places with less population density around cities [6,7]. In this context, nature-based tourism destinations such as urban forest parks have become important places for outdoor recreation experiences for tourists and urban residents [6,8]. Urban forest parks are of great help and high significance in reducing psychological stress and improving quality of life and health status for tourists and urban residents [9–11].

The recreation experience of tourists in natural destinations (e.g., urban forest parks) is one of the important fields in tourism research [12]. Some researchers suggested that outdoor recreation experience plays a significant educational role in changing tourists' attitudes towards the environment and facilitating tourists' environmentally responsible behavior (ERB) in the context of nature-based tourism [12,13]. Subsequently, several studies explored the structural relationship between recreation experience, environmental attitude (EA), and ERB. Using 1342 questionnaires, Lee and Jan (2015) examined the impact of recreation experience on the EA and ERB of nature-based tourists. They discovered that recreation experiences strengthen tourists' EA and consequently increase their participation in ERBs [12–14]. However, most of the existing research only considers the recreation experience as a whole and then examines its relationship with EA and ERB [14,15]. In fact, according to the study of Pine and Gilmore (1998), experience can be subdivided into four aspects: education, aesthetics, entertainment, and escapism [16–18]. However, whether dividing the recreation experience into four dimensions has the same effect as viewing the recreation experience as a whole remains to be further explored. Therefore, a comprehensive analysis of the relationship between the four dimensions of recreation experience, EA, and ERB is extremely necessary.

Additionally, previous studies have claimed that there are significant differences between genders in terms of tourist experiences, motivations, and behaviors [5,19–24]. Deem (1986) argued that leisure experiences and the meanings associated with the experience may be significantly different for women than they are for men [25]. Despite the fact that a number of researchers have studied gender differences as a constraint to recreation experience, few have examined the moderating role of gender in the relationships among recreation experience, EA, and ERB.

To fill the above-mentioned research gaps, the specific goals of the present study are twofold: (1) examining the structural relationships among the four components of recreation experience (education, aesthetics, entertainment, and escapism), EA, and ERB in a nature-based tourism context (urban forest park) and (2) exploring how gender plays a moderating role in the aforementioned structural relationship. An urban forest park in Nanjing City, Jiangsu Province, China, was chosen as the study site for the present study.

2. Literature Review

2.1. The Importance of Urban Forests during COVID-19

An urban forest is basically considered to be an ecological system within the urban area and is composed of forest green space, urban garden, urban greening, and the environment in which it is located [26,27]. As a green space, an urban forest is rich in natural resources. It can provide tourists and urban residents with outdoor recreation experience, make positive contributions to people's health, and become an important space for the public to have fun and enjoyment and improve mental health, especially during the COVID-19 pandemic [5,7,9,17,28]. Numerous studies have demonstrated that recreation experience in green environments such as urban forests can not only psychologically improve the negative emotional state brought about by the pandemic and improve people's sense of well-being, but also help people physically cope with many diseases and improve their quality of life and health status [29–31]. Therefore, urban forests play an increasingly essential part in the outdoor recreation experience of tourists and urban residents.

2.2. The Relationship among Forest Recreation Experience, EA, and ERB

Experience is an important element of tourism activities and is regarded as the personal thoughts, emotions, feelings, knowledge, and skills formed by participating in an activity [32]. Recreation experience is defined as a person's mental state caused by a specific activity, and scholars believe that recreation is the perception and emotion generated by an individual participating in recreational activities [33,34]. Several studies suggest that recreation experience is often considered to include four dimensions: education, aesthetics, entertainment, and escapism [16,21,35]. Education is an important factor in motivating tourists to carry out recreational activities [18]. Tourists can gain more environmental knowledge by participating in outdoor recreational activities [36]. Aesthetics refers to the aesthetic perception of the natural environment and landscape aesthetics that tourists can acquire when participating in outdoor recreational activities [37–39]. Entertainment refers to the entertainment activities that tourists participate in in the outdoor recreation experience and the entertainment perception provided by the natural environment [29,40]. The term "escapism" refers to the extent to which visitors become totally engrossed in various types of recreational activities [35,41].

Environmental attitude is defined as a psychological disposition expressed through an individual's evaluation of a specific object or environmental issue [42,43]. Previous studies have pointed out that recreation experiences have significant impacts on improving tourists' EA [44,45]. In nature-based tourism, the recreation experience can increase tourists' environmental awareness, which in turn generates positive cognitions, emotions, and attitudes [40]. For example, Collado et al. (2013) proposed that direct recreation experience can enhance students' attitudes towards the environment [46]. Duerden and Witt (2010) argued that the exposure of tourists to the natural environment via its on-site activities might increase their awareness of nature, consequently enhancing their environmental knowledge and sensitivity to environmental issues [47]. In addition, more and more scholars have confirmed that recreation experience can be regarded as an antecedent of EA [30,47]. The following four hypotheses are presented based on the preceding elaboration:

H1. *Visitors' educational experience has a significant positive impact on their environmental attitudes.*

H2. *Visitors' aesthetic experience has a significant positive impact on their environmental attitudes.*

H3. *Visitors' entertainment experience has a significant positive impact on their environmental attitudes.*

H4. *Visitors' escape experience has a significant positive impact on their environmental attitudes.*

Environmentally responsible behavior refers to behaviors that minimize negative impacts on the environment and contribute to environmentally sustainable development, reflecting individuals' environmental concerns and beliefs in reducing environmental problems [13,48–50]. Numerous studies have confirmed that recreation experience has important effects on tourists' ERB. For example, Lin and Li (2020) used the ancient trail of Alangyi and the Historic Fish Road Trail as case sites to examine the correlation between recreation experience, place attachment, EA, and responsible behavior of tourists. The analysis of 1579 questionnaires revealed that recreational experience has significant and positive effects on EA, and influenced ERB indirectly [51]. Additionally, in the case of Nansha Wetland Park, Xu et al. (2018) found that tourists' experience can be used as the predecessor of ERB [52].

Numerous prior studies have also suggested that positive EA can obviously lead to an increase in ERB [46,53–55]. When environmental attitudes are strong, more actions can be predicted to support environmental responsibility [55]. By engaging in recreational activities, visitors can improve their EA and possibly encourage ERB when they are in close contact with the natural environment. Therefore, based on the discussion above, the following five hypotheses are presented:

H5. *Visitors' educational experience has a significant positive impact on their environmentally responsible behavior.*

H6. *Visitors' aesthetic experience has a significant positive impact on their environmentally responsible behavior.*

H7. *Visitors' entertainment experience has a significant positive impact on their environmentally responsible behavior.*

H8. *Visitors' escape experience has a significant positive impact on their environmentally responsible behavior.*

H9. *Visitors' environmental attitude has a significant positive impact on their environmentally responsible behavior.*

The above discussion has shown that visitors' recreation experience has substantial positive impacts on their EA, and EA also has substantial positive impacts on their ERB. EA basically plays a moderating role in the influence of recreation experience on ERB [14,46]. Lee and Jan (2015) have pointed out that recreation experience is significantly correlated with ERB through EA, and EA acts as a mediator between recreation experience and ERB in nature-based tourism [14]. Moreover, Collado et al. (2013) also claimed that EA mediates the relationship between children's recreational experiences and general ERB [46]. Thus, the following hypothesis is proposed based on the above evidence:

H10. *Visitors' environmental attitudes play a moderating role between recreation experience and environmentally responsible behavior.*

2.3. The Moderating Effect of Gender

Existing studies have shown that there are significant gender differences in terms of tourism experience, tourism motivation, tourist behavior, etc. [5,19–24]. Weng et al. (2022) argued that there are differences in emotional experience between men and women, and fear, worry, and sadness are more commonly reported by women than by men [5]. Similarly, Dagher, Itani, and Kassar (2015) also claimed that gender affects consumers' environmental concerns and attitudes towards green purchasing behavior, and women show more intense green purchasing behavior [56]. On this basis, considering the influence of gender on the regulation of the link between recreation experience, EA, and ERB, the following hypotheses are proposed:

H11. *Gender plays a moderating role in the relationship between recreation experience and environmental attitude.*

H12. *Gender plays a moderating role in the relationship between recreation experience and environmental responsibility behavior.*

H13. *Gender plays a moderating role in the relationship between environmental attitude and environmental responsibility behavior.*

Figure 1 presents an illustration of the conceptual framework that was used for this study.

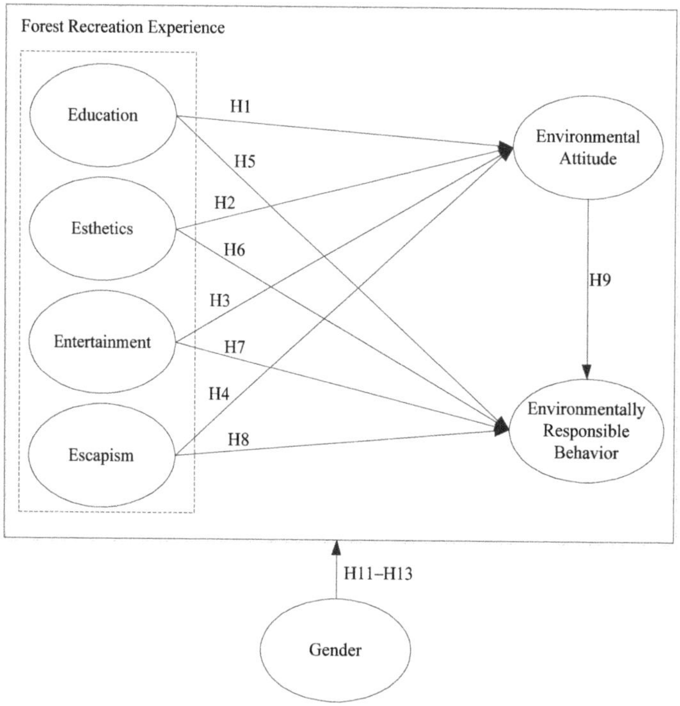

Figure 1. The conceptual framework.

3. Materials and Methods

3.1. Study Site

Zijinshan National Forest Park (hereinafter referred to as Zijinshan) is an urban forest park located in Nanjing City, Jiangsu Province (see Figure 2). The total area of Zijinshan is about 45 square kilometers, and the forest coverage rate reaches 70.2%. Zijinshan has a strong forest atmosphere, well-preserved biodiversity, and more than 200 natural and cultural landscapes. Zijinshan has rich spatial layers and numerous ravines, forming many streams and lakes, which have laid the foundation for the development of forest ecotourism. It has become one of the must-see scenic spots for tourists in Nanjing. Especially during the COVID-19 pandemic, Zijinshan has become an important destination for tourists and residents to enjoy outdoor recreation. Tourists and urban residents can carry out recreational activities such as mountain climbing, cycling, kite flying, running, swimming, camping, and barbecuing in this urban forest park. In addition, this park also organizes educational activities, such as identifying plants and insects, giving lectures on environmental protection, and watching stars and the sun at the Purple Mountain Observatory. In recent years, the overall average annual number of tourists to Zijinshan has been around 10 million.

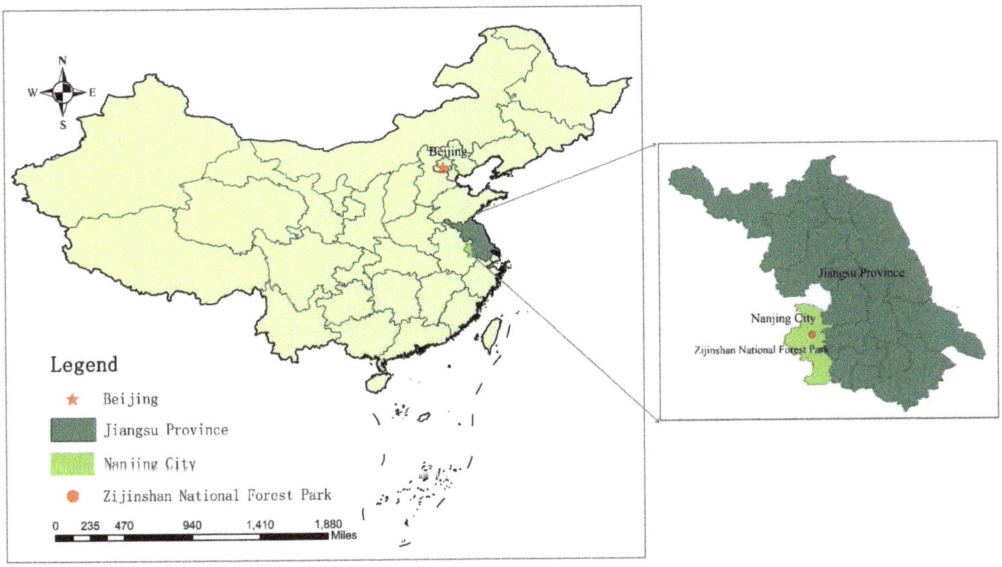

Figure 2. Location of Zijinshan National Forest Park in Jiangsu Province, China.

3.2. Questionnaire Design

To better understand the relationships among participants' forest recreation experience, EA, and ERB, a questionnaire was designed to collect data. The questionnaire consists of two sections. The evaluation of the three constructs is covered in the first section (i.e., forest recreation experience, environment attitude, and environmentally responsible behavior). The second section focuses on the demographic characteristics of the respondents, including their gender, age, education, occupation, and monthly income.

The items of each construct are based on prior research. The measurement of forest recreation experience was adapted from the research of Lee and Jan (2015) and Lee et al. (2015) and is a 4-dimensional 14-item scale [13,57]. The four dimensions are education (3 items, such as "The experience made me more knowledgeable"), entertainment (3 items, such as "The experience activities were amusing"), aesthetics (4 items; sample statement reads "The setting of this national forest park was attractive"), and escapism (4 items, such as "I completely escaped from my daily routine"). The assessment of EA was adapted from the study of Lee and Jan (2015) and Kil, Holland, and Stein (2014) [14,58]. This scale has 9 items; a sample instrument is "Humans have no right to alter the natural environment to suit their needs". ERB was measured via a 6-item scale from the study of Su et al. (2020) [59]. A sample measurement reads "I abide by the legal ways not to destroy the destination's environment".

The measurement items were assessed through a five-point Likert scale. Items that were initially designed in English were translated into Chinese. Back-translation was applied in order to guarantee the accuracy of the translation [60–62]. In addition, a pilot study with 60 participants was conducted to ensure that the questionnaire could be comprehended without ambiguity. In the pilot study, participants were required to complete questionnaires and provide feedback and suggestions. Based on the comments, a few unclear items were revised to increase clarity in relation to the context of the present study.

3.3. Field Survey

The formal field survey was conducted in Zijinshan National Forest Park in Nanjing from late April to late June 2022. Visitors were sampled at the exit of the Zijinshan Cableway. During daylight hours, data were collected using a systematic sampling technique (one out of every ten visitors was sampled) [57]. Six well-trained research assistants administered the questionnaire survey. After the screening question (visitors must be at least 18 years old), the research assistants informed each respondent of the research objectives, variables, and items prior to the completion of the questionnaires. Afterward, participants who were eligible for the study were given the opportunity to indicate the degree to which they agreed or disagreed with the questions and statements contained in the questionnaire [63]. In addition, the responders who completed the survey received small gifts as a reward for their time. Overall, 500 questionnaires were administered, 32 of which were incomplete, resulting in a total of 468 valid questionnaires.

3.4. Data Analysis

SPSS 20.0 and Amos 21.0 were utilized to analyze the data in this study. To investigate the rationality of the measurement scales, reliability and validity tests were first undertaken. Then, confirmatory factor analysis (CFA) was performed to provide the basis for the analysis of the measurement model. Finally, the research hypotheses were examined by conducting hypothesis testing and mediating and moderating analysis.

4. Results

4.1. Sample Profile

The descriptive information of the sample is outlined in Table 1. It can be found that males made up 47.6% (n = 223) of 468 respondents, while females made up 52.4% (n = 245). Nearly half of the respondents (46.8%, n = 219) are between the ages of 18 and 35. Over fifty percent of the respondents (54.7%, n = 256) have a bachelor's degree or above. In addition, more than half of the respondents (54%, n = 253) are employed as enterprise employees, self-employed, or owners. In addition, 86.3% of the respondents have a monthly salary in excess of CNY 3000 (USD 470).

4.2. Measurement Model Testing

4.2.1. Reliability Test and Confirmatory Factor Analysis

A reliability test's primary objective is to ascertain whether or not a measurement scale maintains its internal consistency. In general, a Cronbach's alpha value higher than 0.7 suggests that the scale is reliable [64]. According to the analysis results, the Cronbach's alpha coefficient of the scale has a value of 0.937. Specifically, the values of the Cronbach's coefficient for education, entertainment, aesthetics, escapism, EA, and ERB are 0.849, 0.832, 0.902, 0.909, 0.922, and 0.894 (Table 2), respectively, indicating that this study's measurements are very reliable.

Table 1. Sociodemographic information of the study sample.

		Frequency (n = 468)	Percentage (%)
Gender	Male	223	47.6
	Female	245	52.4
Age	18 to 35 years	219	46.8
	36 to 45 years	118	25.2
	46 to 55 years	59	12.6
	56 to 65 years	41	8.8
	Over 65 years	31	6.6
Education	High school or below	74	15.8
	Associate degree	138	29.5
	Bachelor's degree	187	40.0
	Master's degree or above	69	14.7
Occupation	Enterprise employee	177	37.8
	Self-employment or owner	76	16.2
	Student	49	10.5
	Government official	69	14.7
	Professional, teacher, or technical	51	10.9
	Other	46	9.8
Personal monthly income (CNY)	Less than 3000	64	13.7
	3001–6000	126	26.9
	6001–10,000	163	34.8
	10,001–15,000	69	14.7
	More than 15,000	46	9.8

heading 4 with numeration need to special approval, waitiong for AE reply to get special approval

Table 2. Reliability and confirmatory factor analysis results.

Dependent Variables	Mean (SD)	Factor Loading	CR	AVE	Cronbach's α
Education			0.849	0.652	0.849
The experience made me more knowledgeable	3.51 (0.92)	0.811			
It was a learning experience	3.43 (0.94)	0.795			
It inspired my curiosity to learn new things	3.47 (0.92)	0.816			
Entertainment			0.833	0.625	0.832
Experiential activities were amusing	3.04 (0.87)	0.804			
Experiential activities were captivating	3.16 (0.90)	0.805			
Experiential activities were entertaining	3.20 (0.90)	0.761			
Aesthetics			0.905	0.704	0.902
The setting of the forest park was attractive	3.44 (1.05)	0.860			
The forest park was set up with design details	3.38 (1.13)	0.802			
Being here was pleasant	3.39 (1.10)	0.776			
I felt real harmony	3.44 (1.10)	0.912			
Escapism			0.910	0.717	0.909
I felt like playing a different character here	3.58 (0.99)	0.875			
The experience made me imagine being someone else	3.56 (0.96)	0.773			
I escaped from my daily routine	3.55 (0.96)	0.810			
I felt I was in a different time or place	3.57 (0.99)	0.922			

Table 2. *Cont.*

Dependent Variables	Mean (SD)	Factor Loading	CR	AVE	Cronbach's α
EA			0.922	0.567	0.922
We are approaching the limit number of people the earth can support	3.87 (1.14)	0.830			
Earth is like a spaceship with limited room and resources	3.60 (1.02)	0.725			
Humans have no right to alter the natural environment to suit their needs	3.57 (0.98)	0.705			
Plants and animals have the same right to exist as humans	3.80 (1.12)	0.797			
Humans shouldn't rule over the rest of nature	3.61 (1.08)	0.753			
Human interference with nature often produces disastrous consequences	3.71 (1.02)	0.787			
Nature's balance is delicate and easily upset	3.62 (0.99)	0.703			
Humans are still subject to the laws of nature despite special abilities	3.55 (0.99)	0.729			
Humans learn enough about how nature works to control it	3.57 (1.02)	0.740			
ERB			0.895	0.587	0.894
I abide by the legal ways not to destroy the destination's environment	3.56 (1.12)	0.794			
I report environmental pollution or destruction to the destination administration	3.55 (1.06)	0.723			
I try to throw away garbage and branches when I see them	3.50 (1.05)	0.728			
I'll attend the environment cleaning activities	3.55 (1.13)	0.775			
I'd convince my travel companions to protect the natural environment	3.61 (1.02)	0.764			
During travel, I try not to disturb the fauna and flora (animals and plants)	3.53 (1.03)	0.810			

Note: Model fit indices: $\chi^2/df = 2.215$, NFI = 0.910, CFI = 0.949, GFI = 0.894, AGFI = 0.873, RMSEA = 0.051, SRMR = 0.039. All the factor loads are greater than 0.5, and the p values are significant ($p < 0.001$).

In addition, CFA was also employed to assess the measurement model. Table 2 presents the model fit indices that are satisfactory according to the predetermined cutoff thresholds ($\chi^2/df = 1.155$, NFI = 0.910, CFI = 0.987, GFI = 0.910, AGFI = 0.901, RMSEA = 0.018, SRMR = 0.031) [61,65,66], indicating that the measurement model corresponds well to the data.

4.2.2. Validity Test

Principally, convergent validity and discriminant validity are used to evaluate the validity test. The concept of convergent validity relates to the degree to which various aspects of a single variable are correlated with one another [65]. According to Hair et al. (2006), all factor loads are larger than 0.5, and the p values are significant from a statistical point of view [67]. In addition, as indicated in Table 2, the AVE is greater than 0.5, and the CR is greater than 0.6 [68]. All of these suggest that the latent variables have a high level of convergent validity.

The concept of discriminant validity relates to the capacity to differentiate between distinct variables [68]. Hu and Bentler (1999) stated that discriminant validity occurs when the square root of AVE is greater than its correlation coefficient with other variables [69,70]. According to Table 3, the correlation coefficients for each variable range between 0.108 and 0.53. The square root of AVE for each variable is bigger than its correlation coefficient with other factors, suggesting the variables have strong discriminant validity.

Table 3. Discriminant validity.

Variables	Education	Entertainment	Aesthetics	Escapism	EA	ERB
Education	**0.807**					
Entertainment	0.163 **	**0.791**				
Aesthetics	0.166 **	0.519 **	**0.839**			
Escapism	0.108 *	0.438 **	0.491 **	**0.847**		
EA	0.110 *	0.432 **	0.489 **	0.479 **	**0.753**	
ERB	0.142 **	0.475 **	0.533 **	0.482 **	0.474 **	**0.766**

Note: * < 0.05, ** < 0.01.

4.3. Structural Model and Hypotheses Testing

4.3.1. Goodness-of-Fit Test of the Structural Model

To guarantee that the data were normal, the skewness and kurtosis were assessed. The skewness and kurtosis ranged from −0.929 to 0.019 and −0.888 to 0.221, respectively, indicating that the data were regularly distributed [64]. Additionally, the goodness-of-fit test of the structural model was also conducted. The results suggested that the structural model also fits the data well (χ^2/df = 2.215, NFI = 0.910, CFI = 0.949, GFI = 0.894, AGFI = 0.873, RMSEA = 0.051, SRMR = 0.039) [64,67,69,70].

4.3.2. Hypotheses Test

The above-mentioned research hypotheses were examined using a structural equation model, and the results are presented in Table 4 and Figure 3. The standardization path coefficients of H1 and H5 are 0.001 (t = −0.008, p = 0.993) and 0.041 (t = 0.041, p = 0.410), which suggests that the educational experience provided by Zijinshan National Forest Park has no significant effect on visitors' EA and ERB. Thus, H1 and H5 were not supported. The standardization path coefficients of H2 and H6 are 0.258 (t = −4.736, p = 0.000 < 0.001) and 0.231 (t = −4.483, p = 0.000 < 0.001), indicating that the aesthetic experience provided by Zijinshan National Forest Park has a significant positive impact on visitors' EA and ERB. Therefore, H2 and H6 were both supported.

Table 4. Hypothesis testing results.

Hypothesis Paths	Estimate	S.E.	t	p	Results
H1: Education → EA	0.001	0.055	−0.008	0.993	Not support
H2: Aesthetics → EA	0.258	0.055	4.736	***	Support
H3: Entertainment → EA	0.287	0.084	3.431	***	Support
H4: Escapism → EA	0.268	0.055	4.906	***	Support
H5: Education → ERB	0.041	0.050	0.823	0.410	Not support
H6: Aesthetics → ERB	0.231	0.051	4.483	***	Support
H7: Entertainment → ERB	0.261	0.078	3.332	***	Support
H8: Escapism → ERB	0.172	0.051	3.354	***	Support
H9: EA → ERB	0.179	0.050	3.592	***	Support

Note: *** < 0.001.

In addition, the standardization path coefficients of H3 and H7 are 0.287 (t = 3.431, p = 0.000 < 0.001) and 0.261 (t = 3.332, p = 0.000 < 0.001), suggesting that the entertainment experience provided by Zijinshan National Forest Park has a significant positive impact on visitors' EA and ERB. Hence, H3 and H7 were both supported. Results of the standardization path coefficients of H4 and H8 are 0.268 (t = 4.906, p = 0.000 < 0.001) and 0.172 (t = 3.354, p = 0.000 < 0.001), which shows that the escape experience provided by Zijinshan National Forest Park has a significant positive impact on visitors' EA and ERB. Therefore, H4 and H8 were both supported. The standardization path coefficient of H9 is 0.179 (t = 3.592, p = 0.000 < 0.001), indicating that visitors' EA has a significant positive effect on their ERB. Thus, H9 was also supported.

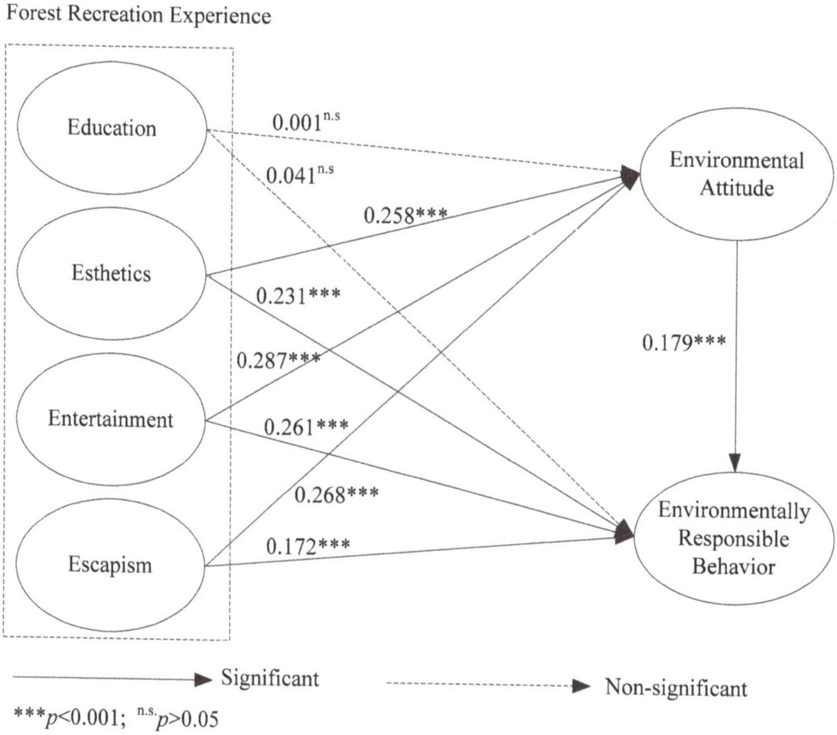

Figure 3. Standardization path coefficient and hypothesis testing results.

4.3.3. Mediating Effect Analysis

The mediating variable in this study is EA. This study employed the bootstrap approach to test the mediating effect. The results are shown in Table 5. The findings indicated that the percentile 95% CI for the mediating effect of education on ERB is −0.023 to 0.021, whereas the bias-corrected 95% CI is −0.022 to 0.021. It can be found that the bias-corrected 95% CI and the percentile 95% CI of the indirect effect include 0, demonstrating that the indirect effect does not exist.

Table 5. Result of mediating effect analysis.

Mediation Path	Path Effects	Effect Size	Bias-Corrected 95% CI		Percentile 95% CI	
			Lower	Upper	Lower	Upper
Education → EA → ERB	Total	0.041	−0.070	0.159	−0.074	0.156
	Indirect	0.000	−0.022	0.021	−0.023	0.021
	Direct	0.041	−0.066	0.151	−0.068	0.151
Aesthetics → EA → ERB	Total	0.277	0.161	0.388	0.168	0.393
	Indirect	0.046	0.019	0.087	0.017	0.083
	Direct	0.231	0.121	0.343	0.124	0.347
Entertainment → EA → ERB	Total	0.312	0.164	0.475	0.160	0.472
	Indirect	0.051	0.017	0.105	0.014	0.100
	Direct	0.261	0.113	0.421	0.110	0.419
Escapism → EA → ERB	Total	0.220	0.125	0.323	0.121	0.320
	Indirect	0.048	0.017	0.093	0.016	0.092
	Direct	0.172	0.070	0.266	0.070	0.266

Additionally, the bias-corrected 95% CI for the indirect effect (mediating effect) of aesthetics on ERB is 0.019 to 0.087, and the percentile 95% CI is 0.017 to 0.083, suggesting that the EA has a mediating effect. Similarly, the mediating role of EA also exists in the effects of entertainment and escapism on ERB.

Therefore, the environmental attitude is certified to have partial mediation effects in the model. Based on this finding, it can be concluded that H10 was partially supported.

4.4. Moderating Effect of Gender

Using a structural equation model with a multi-group sample, the moderating influence of gender was investigated. As presented in Table 6, for both male and female samples, the influence coefficients of education on EA and ERB do not reach the range of significance ($t = 0.353$, $p = 0.724 > 0.05$; $t = -0.505$, $p = 0.614 > 0.05$; $t = -0.139$, $p = 0.889 > 0.05$; $t = 1.173$, $p = 0.241 > 0.05$).

Table 6. Results of the moderating effect.

Hypothesis Paths	Male Group			Female Group		
	Estimate	*t*	*p*	Estimate	*t*	*p*
H1: Education → EA	0.027	0.353	0.724	−0.040	−0.505	0.614
H2: Aesthetics → EA	0.210	2.861	0.004	0.315	3.794	***
H3: Entertainment → EA	0.325	2.423	0.015	0.250	2.302	0.021
H4: Escapism → EA	0.275	2.885	0.004	0.264	4.018	***
H5: Education → ERB	−0.009	−0.139	0.889	0.089	1.173	0.241
H6: Aesthetics → ERB	0.155	2.438	0.015	0.305	3.636	***
H7: Entertainment → ERB	0.249	2.150	0.032	0.256	2.390	0.017
H8: Escapism → ERB	0.181	2.207	0.027	0.171	2.596	0.009
H9: EA → ERB	0.169	2.681	0.007	0.180	2.326	0.020

Note: *** < 0.001.

In addition to the above-mentioned influence relationships, it can be found that, for the male sample, the influence coefficient of aesthetics on EA is 0.210 ($t = 2.861$, $p = 0.004 < 0.01$), which is statistically significant at the 0.01 level. For the female sample, the influence coefficient of aesthetics on EA is 0.315 ($t = 3.794$, $p < 0.001$), which is also significant. The influence coefficient of the female sample is greater than that of the male sample. A similar finding can also be drawn from the influence coefficient of aesthetics on ERB (0.155, $t = 2.438$, $p = 0.015 < 0.05$ vs. 0.305, $t = 3.636$, $p < 0.001$). These results demonstrated that female groups can obtain more aesthetic experiences, which in turn will affect their EA and ERB more than male groups.

Moreover, for male groups, the influence coefficients of entertainment on EA and ERB are 0.325 ($t = 2.423$, $p = 0.015 < 0.05$) and 0.249 ($t = 2.150$, $p = 0.032 < 0.05$). The influence coefficients of escapism on EA and ERB are 0.275 ($t = 2.885$, $p = 0.004 < 0.01$) and 0.181 ($t = 2.207$, $p = 0.027 < 0.05$), respectively. For female groups, the influence coefficients of entertainment on EA and ERB are 0.250 ($t = 2.302$, $p = 0.021 < 0.05$) and 0.256 ($t = 2.390$, $p = 0.017 < 0.05$). The influence coefficients of escapism on EA and ERB are 0.264 ($t = 4.018$, $p < 0.001$) and 0.171 ($t = 2.596$, $p = 0.009 < 0.01$). These results indicated that male groups can generally obtain more entertainment and escape experiences, and this will affect their EA and ERB more than those of the female groups.

In addition, the influence coefficient of EA on ERB for female groups (0.180, $t = 2.326$, $p = 0.020 < 0.05$) is greater than that for male groups (0.169, $t = 2.681$, $p = 0.007 < 0.01$). This suggests that when the female groups obtain more environmental attitude changes, their ERB will change more than that of the male groups.

In summary, it is evident that gender acts as a partial moderator in the influence of recreational experience on EA and ERB. Therefore, H11 and H12 were partially supported and H13 was supported.

5. Discussion

5.1. Theoretical Implications

This research provides several substantial contributions to the existing body of knowledge. First, this study examines and verifies the four subdivided dimensions of tourists' recreation experience and applies this subdivision in the context of nature-based tourism. Most of the existing research usually measures tourists' recreation experience as a whole, ignoring its subdivided dimension [14,15]. This study closes this research gap and explores the four dimensions of recreation experience: education, aesthetics, entertainment, and escapism. The subdivided dimensions were examined and verified in an urban forest park. The results of the reliability test, validity test, and confirmatory factor analysis indicate that the relevant values are all in line with the statistical standard and the four constructs of recreation experience have been well verified. In the context of nature-based tourism, the natural environment affords tourists the chance to interact with and acquire environmental knowledge from nature, which in turn affects their level of satisfaction, as well as their attitude and behavior [12]. Therefore, it is necessary to apply the subdivision of recreation experience to the following research.

Second, this study is one of the first empirical studies to investigate the structural relationships between the four subdivided dimensions of recreation experience, EA, and ERB in the context of nature-based tourism. The results demonstrated that the entertainment experience, aesthetic experience, and escape experience provided by the nature-based destinations have significant positive impacts on visitors' EA and ERB. This is in line with previous studies showing that tourism experience has an impact on tourist satisfaction, attitude, behavioral intention, and loyalty [35,71,72]. However, the empirical results suggest that educational experience has no significant effect on visitors' EA and ERB, and the mediating effect of EA is partially mediated. This result is different from those of previous studies [44,45]. There are two possible reasons: On the one hand, the interpretation system of this urban forest park needs to be improved, and relevant knowledge has not been effectively passed on to tourists [14]. On the other hand, under the influence of the pandemic, tourists only wanted to take advantage of outdoor activities to fully relax physically and mentally and did not earnestly learn natural knowledge and obtain natural education.

Third, this study explores the function of gender as a moderator in the relationship between recreation experience, EA, and ERB. Existing studies have suggested that there are differences between males and females in terms of tourism experience, tourism motivation, tourist behavior, etc. [5,19–23]. For example, the research of Dagher, Itani, and Kassar (2015) found that gender affected consumers' environmental concerns and attitudes towards green purchasing behavior [56]. Similar results were discovered in this research as well. From the results of the moderating effect study, it is apparent that in addition to educational experience, women can obtain more aesthetic experience than men, which in turn affects EA and ERB more. However, men can generally obtain more entertainment and escape experiences than women. It is intended that this study will both act as a foundation for further research in this sector and motivate further research that is significantly important.

5.2. Practical Implications

The present study has the potential to provide significant practical implications for the administration of nature-based tourism destinations such as urban forest parks. First, the results of this research suggest that recreation experience is the preceding determinant of EA and ERB. To minimize environmental consequences, managers of nature-based tourism destinations should create chances for visitors to enhance their environmental knowledge through recreation experiences [14]. Particular emphasis should be placed on the four subdivided dimensions of the recreation experience. Nature-based tourism destinations should provide educational, aesthetic, entertaining, and escape recreation activities that enhance visitors' environmental attitudes. In addition, there is an urgent need to improve the interpretation system of urban forest parks [66,73]. The managers may supply systematic interpretation services, which may include interpretative logos and knowledgeable interpreters. Therefore, tourists can then change their EA and subsequently improve their ERB, eventually contributing to sustainable tourism.

Second, it was discovered that men and women have differences in the structural relationships among the four subdivided dimensions of recreation experience, EA, and ERB. According to the results of this study, women have a higher propensity than men to promote ERB if they gain more aesthetic experience and environmental attitude change. However, men are more likely than women to change their EA and improve ERB if they have access to more entertaining and escape experiences. Based on these findings, two suggestions could be offered for nature-based tourism destinations such as urban forest parks. On the one hand, policymakers in the tourism industry should take into account the differences in the needs of men and women in recreation experiences and offer different outdoor activities. On the other hand, when natural tourism destinations provide tour guide services, they should focus on interpreting different recreation experiences for male and female tourists [65]. Thus, natural interpretation information can be effectively conveyed to tourists.

Third, suggestions are also provided on how to carry out pro-environmental activities in nature-based tourism destinations such as urban forest parks. It is recommended that managers and service providers conduct regular activities (for instance, hiking, camping, insect and plant identification, and environmental education seminars) and explain environmental concerns in order to enhance visitors' environmental attitudes and environmentally responsible behaviors. Additionally, managers may urge visitors to serve as "conservation partners" rather than relying on rules and legislation to restrict and limit visitors' negative environmental behaviors during their involvement in recreational activities [74]. These diverse activities can provide visitors with good opportunities to gain more educational knowledge, improve their EA, and promote their ERB.

6. Conclusions

Nature-based tourism destinations such as urban forest parks are increasingly becoming important places for outdoor leisure and recreation. The present study employed structural equation modeling to empirically investigate the relationships among the four subdivided dimensions of recreation experience (education, aesthetics, entertainment, and escapism), EA, and ERB in an urban forest park, as well as to examine the moderating effect of gender in the aforementioned influence relationship.

The findings indicate that, in addition to the educational experience, the entertainment experience, aesthetic experience, and escape experience provided by the urban forest park have significant positive impacts on visitors' EA and ERB. Moreover, visitors' EA significantly and positively influences their ERB. In addition, the mediating effect of EA is also examined in this study. The results imply that EA provides a partial moderating effect on the relationship between recreation experience and ERB. Additionally, the role of gender as a moderator is acknowledged and explored as well, with results suggesting that women may be more likely than men to promote ERB if they gain more aesthetic experience and environmental attitude changes. However, men are more likely to change their EA and improve ERB than women if they obtain more entertaining and escape experiences. This study contributes to theory and practice in the areas of environmental education and sustainable tourism development. It also recommends that nature-based tourism destinations should pay attention to the four subdivided dimensions of recreation experience, especially educational experience. Thus, visitors' EA and ERB can be effectively enhanced.

The present study is not without limitations. First, the structural relationship in this research was examined only in one study site. Similar studies should be carried out in other case sites in the future to evaluate and validate the conclusions of this research. Second, the frequency of visits to urban forest parks should be considered in future studies so that the motivations and behaviors of first-time and repeat visitors can be compared and analyzed. Third, this study only analyzed and tested the moderator variable of gender, and future research can explore more moderator variables, such as age and location, and then examine their impact on visitors' environmentally responsible behaviors.

Author Contributions: Conceptualization, L.W. and L.Z.; methodology, L.W. and Y.Z.; software, L.W.; validation, L.W. and Y.Z.; formal analysis, L.W.; investigation, Y.Z., X.X., J.Y., H.L., S.Z. and H.W.; resources, L.Z.; data curation, L.W.; writing—original draft preparation, L.W., Y.Z. and X.X.; writing—review and editing, L.W. and L.Z.; visualization, L.W.; supervision, L.Z.; project administration, L.Z.; funding acquisition, L.Z. All authors have read and agreed to the published version of the manuscript.

Funding: This research was funded by the National Social Science Fund of China under Grant 21CH197, Jiangsu University Philosophy and Social Science Fund Project (2020SJA0052), 2020 Central University Basic Scientific Research Business Fee Humanities and Social Sciences Fund under Grants SKYC2020017 and SKYZ2020022, Jiangsu Province Education Science "Thirteenth Five-Year Plan" Project (D/2020/01/59), and The Think Tank Project of Jiangsu Province (SZKKT2022007).

Institutional Review Board Statement: Not applicable.

Informed Consent Statement: Not applicable.

Data Availability Statement: Data available on request.

Acknowledgments: The authors would like to acknowledge the helpful support received from the respondents in the field data collection.

Conflicts of Interest: The authors declare no conflict of interest.

References

1. Škare, M.; Soriano, D.R.; Porada-Rocho'n, M. Impact of COVID-19 on the travel and tourism industry. *Technol. Forecast. Soc. Chang.* **2021**, *163*, 120469. [CrossRef] [PubMed]
2. Collins-Kreiner, N.; Ram, Y. National tourism strategies during the COVID-19 pandemic. *Ann. Tour. Res.* **2020**, *89*, 103076. [CrossRef] [PubMed]
3. Yeh, S.S. Tourism recovery strategy against COVID-19 pandemic. *Tour. Recreat. Res.* **2021**, *46*, 188–194. [CrossRef]
4. McGinlay, J.; Gkoumas, V.; Holtvoeth, J.; Fuertes, R.F.A.; Bazhenova, E.; Benzoni, A.; Jones, N. The impact of COVID-19 on the management of European protected areas and policy implications. *Forests* **2020**, *11*, 1214. [CrossRef]
5. Weng, L.; Wu, Y.; Han, G.; Liu, H.; Cui, F. Emotional State, Psychological Resilience, and Travel Intention to National Forest Park during COVID-19. *Forests* **2022**, *13*, 750. [CrossRef]
6. Hsiao, C.H.; Tang, K.Y. Who captures whom–Pokémon or tourists? A perspective of the Stimulus-Organism-Response model. *Int. J. Inf. Manag.* **2021**, *61*, 102312. [CrossRef]
7. Donovan, G.H. Including public-health benefits of trees in urban-forestry decision making. *Urban For. Urban Green.* **2017**, *22*, 120–123. [CrossRef]
8. Weinbrenner, H.; Breithut, J.; Hebermehl, W.; Kaufmann, A.; Klinger, T.; Palm, T.; Wirth, K. "The Forest Has Become Our New Living Room"—The Critical Importance of Urban Forests During the COVID-19 Pandemic. *Front. For. Glob. Chang.* **2021**, *4*, 68. [CrossRef]
9. Meyer, K.; Bürger-Arndt, R. How Forests Foster Human Health—Present State of Research-Based Knowledge (in the Field of Forests and Human Health). *Int. For. Rev.* **2014**, *16*, 421–446. [CrossRef]
10. Pintilii, R.D. Forest Recreation and Landscape Protection. *Forests* **2022**, *13*, 1440. [CrossRef]
11. Musselwhite, C.; Avineri, E.; Susilo, Y. Editorial JTH 16–The Coronavirus Disease COVID-19 and implications for transport and health. *J. Transp. Health* **2020**, *16*, 100853. [CrossRef] [PubMed]
12. Ballantyne, R.; Packer, J.; Sutherland, L.A. Visitors' memories of wildlife tourism: Implications for the design of powerful interpretive experiences. *Tour. Manag.* **2011**, *32*, 770–779. [CrossRef]
13. Lee, T.H.; Jan, F.H.; Huang, G.W. The influence of recreation experiences on environmentally responsible behavior: The case of Liuqiu Island, Taiwan. *J. Sustain. Tour.* **2015**, *23*, 947–967. [CrossRef]
14. Lee, T.H.; Jan, F.H. The effects of recreation experience, environmental attitude, and biospheric value on the environmentally responsible behavior of nature-based tourists. *Environ. Manag.* **2015**, *56*, 193–208. [CrossRef]
15. Lee, T.H. How recreation involvement, place attachment and conservation commitment affect envi-ronmentally responsible behavior. *J. Sustain. Tour.* **2011**, *19*, 895–915. [CrossRef]
16. Pine, B.J.; Gilmore, J.H. Welcome to the experience economy. *Harv. Bus. Rev.* **1998**, *76*, 97–105. [PubMed]
17. Oh, H.; Fiore, A.M.; Jeoung, M. Measuring experience economy concepts: Tourism applications. *J. Travel Res.* **2007**, *46*, 119–132. [CrossRef]
18. Prentice, R.C. Tourist Motivation and Typologies. In *A Companion to Tourism*; Blackwell: Oxford, UK, 2004; pp. 261–279.

19. Shields, S.A. Gender and emotion: What we think we know, what we need to know, and why it matters. *Psychol. Women Q.* **2021**, *37*, 423–435. [CrossRef]
20. Nolen-Hoeksema, S. Emotion regulation and psychopathology: The role of gender. *Annu. Rev. Clin. Psychol.* **2012**, *8*, 161–187. [CrossRef]
21. Ekinci, Y.; Prokopaki, P.; Cobanoglu, C. Service quality in Cretan accommodations: Marketing strategies for the UK holiday market. *Int. J. Hosp. Manag.* **2003**, *22*, 47–66. [CrossRef]
22. Rajoo, K.S.; Karam, D.S.; Abdu, A.; Rosli, Z.; Gerusu, G.J. Addressing psychosocial issues caused by the COVID-19 lockdown: Can urban greeneries help? *Urban For. Urban Green.* **2021**, *65*, 127340. [CrossRef] [PubMed]
23. Schipperijn, J.; Ekholm, O.; Stigsdotter, U.K.; Toftager, M.; Bentsen, P.; Kamper-Jørgensen, F.; Randrup, T.B. Factors influencing the use of green space: Results from a Danish national representative survey. *Landsc. Urban Plan.* **2010**, *95*, 130–137. [CrossRef]
24. Lee, K. Gender differences in Hong Kong adolescent consumers' green purchasing behavior. *J. Consum. Mark.* **2009**, *26*, 87–96. [CrossRef]
25. Deem, R.; Keyes, M. *Allwork and Noplay? The Sociology of Women and Leisure*; Open University Press: London, UK, 1986.
26. WANG, C.; CAI, C.; TAO, K. The concept, range and research area of urban forest. *World For. Res.* **2004**, *17*, 23–27.
27. Tyrväinen, L.; Pauleit, S.; Seeland, K.; Vries, S.D. Benefits and uses of urban forests and trees. In *Urban Forests and Trees*; Springer: Berlin/Heidelberg, Germany, 2005; pp. 81–114.
28. Douglass, R.W. *Forest Recreation*; Elsevier: Amsterdam, The Netherlands, 1982.
29. Orsega-Smith, E.; Mowen, A.J.; Payne, L.L.; Godbey, G. The interaction of stress and park use on psycho-physiological health in older adults. *J. Leis. Res.* **2004**, *36*, 232–256. [CrossRef]
30. Lafortezza, R.; Carrus, G.; Sanesi, G.; Davies, C. Benefits and well-being perceived by people visiting green spaces in periods of heat stress. *Urban For. Urban Green.* **2009**, *8*, 97–108. [CrossRef]
31. Wolch, J.R.; Byrne, J.; Newell, J.P. Urban green space, public health, and environmental justice: The challenge of making cities 'just green enough'. *Landsc. Urban Plan.* **2014**, *125*, 234–244. [CrossRef]
32. Tynan, C.; McKechnie, S. Experience marketing: A review and reassessment. *J. Mark. Manag.* **2009**, *25*, 501–517. [CrossRef]
33. Poulsson, S.H.; Kale, S.H. The experience economy and commercial experiences. *Mark. Rev.* **2004**, *4*, 267–277. [CrossRef]
34. Joy, A.; Sherry, J.F., Jr. Speaking of art as embodied imagination: A multisensory approach to understanding aesthetic experience. *J. Consum. Res.* **2003**, *30*, 259–282. [CrossRef]
35. Hosany, S.; Witham, M. Dimensions of cruisers' experiences, satisfaction, and intention to recommend. *J. Travel Res.* **2010**, *49*, 351–364. [CrossRef]
36. Kim, J.H.; Ritchie, J.B.; McCormick, B. Development of a scale to measure memorable tourism experiences. *J. Travel Res.* **2012**, *51*, 12–25. [CrossRef]
37. Bonn, M.A.; Joseph-Mathews, S.M.; Dai, M.; Hayes, S.; Cave, J. Heritage/cultural attraction atmospherics: Creating the right environment for the heritage/cultural visitor. *J. Travel Res.* **2007**, *45*, 345–354. [CrossRef]
38. Adams, D.C.; Bwenge, A.N.; Lee, D.J.; Larkin, S.L.; Alavalapati, J.R. Public preferences for controlling upland invasive plants in state parks: Application of a choice model. *For. Policy Econ.* **2011**, *13*, 465–472. [CrossRef]
39. Gobster, P.H.; Nassauer, J.I.; Daniel, T.C.; Fry, G. The shared landscape: What does aesthetics have to do with ecology? *Landsc. Ecol.* **2007**, *22*, 959–972. [CrossRef]
40. Ballantyne, R.; Packer, J.; Falk, J. Visitors' learning for environmental sustainability: Testing short- and long-term impacts of wildlife tourism experiences. *Tour. Manag.* **2011**, *32*, 1243–1252. [CrossRef]
41. Csikszentmihalyi, M.; Csikzentmihaly, M. *Flow: The Psychology of Optimal Experience*; Harper & Row: New York, NY, USA, 1990.
42. Milfont, T.L.; Duckitt, J. The environmental attitudes inventory: A valid and reliable measure to assess the structure of environmental attitudes. *J. Environ. Psychol.* **2010**, *30*, 80–94. [CrossRef]
43. Weng, L.; Weng, S.; Luo, X. Development of petite bourgeoisie commercial landscapes in China's historic towns: A perspective from tourists' taste and distinction. *J. Tour. Cult. Chang.* **2021**, *19*, 19–37. [CrossRef]
44. Kaiser, F.G.; Wölfing, S.; Fuhrer, U. Environmental attitude and ecological behavior. *J. Environ. Psychol.* **1999**, *19*, 1–19. [CrossRef]
45. Zsóka, Á.; Szerényi, Z.M.; Széchy, A.; Kocsis, T. Greening due to environmental education? Environmental knowledge, attitudes, consumer behavior and everyday pro-environmental activities of Hungarian high school and university students. *J. Clean. Prod.* **2013**, *48*, 126–138. [CrossRef]
46. Collado, S.; Staats, H.; Corraliza, J.A. Experiencing nature in children's summer camps: Affective, cognitive and behavioral consequences. *J. Environ. Psychol.* **2013**, *33*, 37–44. [CrossRef]
47. Duerden, M.D.; Witt, P.A. The impact of direct and indirect experiences on the development of environmental knowledge, attitudes, and behavior. *J. Environ. Psychol.* **2010**, *30*, 379–392. [CrossRef]
48. Chiu, Y.T.H.; Lee, W.I.; Chen, T.H. Environmentally responsible behavior in ecotourism: Antecedents and implications. *Tour. Manag.* **2014**, *40*, 321–329. [CrossRef]
49. Ciobotaru, A.M.; Patel, N.; Pintilii, R.D. Tree Cover Loss in the Mediterranean Region—An Increasingly Serious Environmental Issue. *Forests* **2021**, *12*, 1341. [CrossRef]
50. Pintilii, R.D.; Andronache, I.; Diaconu, D.C.; Dobrea, R.C.; Zeleňáková, M.; Fensholt, R.; Ciobotaru, A.M. Using fractal analysis in modeling the dynamics of Forest areas and economic impact assessment: Maramureș County, Romania, as a case study. *Forests* **2017**, *8*, 25. [CrossRef]

51. Lin, Y.H.; Lee, T.H. How do recreation experiences affect visitors' environmentally responsible behavior? Evidence from recreationists visiting ancient trails in Taiwan. *J. Sustain. Tour.* **2020**, *28*, 705–726. [CrossRef]
52. Xu, S.; Kim, H.J.; Liang, M.; Ryu, K. Interrelationships between tourist involvement, tourist experience, and environmentally responsible behavior: A case study of Nansha Wetland Park, China. *J. Travel Tour. Mark.* **2018**, *35*, 856–868. [CrossRef]
53. Zeppel, H. Education and conservation benefits of marine wildlife tours: Developing free-choice learning experiences. *J. Environ. Educ.* **2008**, *39*, 3–18. [CrossRef]
54. Kals, E.; Schumacher, D.; Montada, L. Emotional affinity toward nature as a motivational basis to protect nature. *Environ. Behav.* **1999**, *31*, 178–202. [CrossRef]
55. Stern, P.C.; Dietz, T.; Guagnano, G.A. The new ecological paradigm in social-psychological context. *Environ. Behav.* **1995**, *27*, 723–743. [CrossRef]
56. Dagher, G.; Itani, O.; Kassar, A.N. The impact of environment concern and attitude on green purchasing behavior: Gender as the moderator. *Contem. Manag. Res.* **2015**, *11*, 179–206. [CrossRef]
57. Lee, T.H.; Jan, F.H. The influence of recreation experience and environmental attitude on the environmentally responsible behavior of community-based tourists in Taiwan. *J. Sustain. Tour.* **2015**, *23*, 1063–1094. [CrossRef]
58. Kil, N.; Holland, S.M.; Stein, T.V. Structural relationships between environmental attitudes, recreation motivations, and environmentally responsible behaviors. *J. Outdoor Recreat. Tour.* **2014**, *7*, 16–25. [CrossRef]
59. Su, L.; Hsu, M.K.; Boostrom, R.E., Jr. From recreation to responsibility: Increasing environmentally re-sponsible behavior in tourism. *J. Bus. Res.* **2020**, *109*, 557–573. [CrossRef]
60. Fischer, A.H.; Rodriguez Mosquera, P.M.; van Vianen, A.; Manstead, A.S.R. Gender and culture differences in emotion. *Emotion* **2004**, *4*, 87–94. [CrossRef] [PubMed]
61. Watson, D.; Clark, L.A.; Tellegen, A. Development and validation of brief measures of positive and negative affect: The PANAS scales. *J. Personal. Soc. Psychol.* **1988**, *54*, 1063. [CrossRef] [PubMed]
62. Connor, K.M.; Davidson, J.R. Development of a new resilience scale: The Connor-Davidson resilience scale (CD-RISC). *Depress. Anxiety* **2003**, *18*, 76–82. [CrossRef]
63. Xiaonan, Y.; Jianxin, Z. A comparison between the Chinese version of Ego-resiliency scale and Connor-Davidson resilience scale. *Psychol. Sci.* **2007**, *30*, 1169–1171.
64. Jalilvand, M.R.; Samiei, N.; Dini, B.; Manzari, P.Y. Examining the structural relationships of electronic word of mouth, destination image, tourist attitude toward destination and travel intention: An integrated approach. *J. Destin. Mark. Manag.* **2012**, *1*, 134–143. [CrossRef]
65. Weng, L.; Liang, Z.; Bao, J. The effect of tour interpretation on perceived heritage values: A comparison of tourists with and without tour guiding interpretation at a heritage destination. *J. Destin. Mark. Manag.* **2020**, *16*, 100431. [CrossRef]
66. Huang, Z.; Weng, L.; Bao, J. How do visitors respond to sustainable tourism interpretations? A further investigation into content and media format. *Tour. Manag.* **2022**, *92*, 104535. [CrossRef]
67. Hair, J.F.; Black, W.C.; Babin, B.J.; Anderson, R.E.; Tatham, R.L. *Multivariate Data Analysis*; Pearson Prentice Hall: Hoboken, NJ, USA, 2006; Volume 6.
68. Gonzalez, S.P.; Moore, E.W.G.; Newton, M.; Galli, N.A. Validity and reliability of the Connor-Davidson Resilience Scale (CD-RISC) in competitive sport. *Psychol. Sport Exerc.* **2016**, *23*, 31–39. [CrossRef]
69. O'Leary-Kelly, S.W.; Vokurka, R.J. The empirical assessment of construct validity. *J. Oper. Manag.* **1998**, *16*, 387–405. [CrossRef]
70. Hu, L.T.; Bentler, P.M. Cutoff criteria for fit indexes in covariance structure analysis: Conventional criteria versus new alternatives. *Struct. Equ. Model.* **1999**, *6*, 1–55. [CrossRef]
71. Bigné, J.E.; Mattila, A.S.; Andreu, L. The impact of experiential consumption cognitions and emotions on behavioral intentions. *J. Serv. Mark.* **2008**, *22*, 303–315. [CrossRef]
72. Wang, W.; Chen, J.S.; Fan, L.; Lu, J. Tourist experience and wetland parks: A case of Zhejiang, China. *Ann. Tour. Res.* **2012**, *39*, 1763–1778. [CrossRef]
73. Weng, L.; Huang, Z.; Bao, J. A model of tourism advertising effects. *Tour. Manag.* **2021**, *85*, 104278. [CrossRef]
74. Ballantyne, R.; Packer, J.; Hughes, K. Tourists' support for conservation messages and sustainable management practices in wildlife tourism experiences. *Tour. Manag.* **2009**, *30*, 658–664. [CrossRef]

forests

MDPI

Article

Land Use Preference for Ecosystem Services and Well-Being in Chittagong Hill Tracts of Bangladesh

Ronju Ahammad [1,*], Natasha Stacey [2], Terry Sunderland [3,4] and Kamaljit K. Sangha [2]

[1] Department of Forest Economics, Swedish University of Agricultural Sciences, 90183 Umea, Sweden
[2] Research Institute for the Environment and Livelihoods, Charles Darwin University, Darwin 0909, Australia
[3] Faculty of Forestry, University of British Columbia, Vancouver, BC V6T 1Z4, Canada
[4] Center for International Forestry Research, Bogor 16115, Indonesia
* Correspondence: ronju.ahammad@slu.se or ronj.ahammad@gmail.com

Abstract: Researchers increasingly investigate ecosystem services to assess their role in supporting livelihoods, well-being and economic value in order to inform decision-making. Many studies have explored links between ecosystem services and community-based livelihoods, with a very narrow focus on the importance of land use to well-being. We evaluated the value of ecosystem services from various land uses supporting livelihoods and the overall well-being of local communities in the Chittagong Hill Tracts (CHT) of Bangladesh. By applying a participatory habitat valuation approach with the ethnic communities from eight villages, we explored their preferences for, and perceptions of, ecosystem services and their sources in a multi-functional landscape under different land use, i.e., forest, swidden and low-land agriculture, fruit orchard and water bodies, and three land ownership contexts (state, private and mixed ownership on forest lands). Our findings revealed that community land use preference for ecosystem services supports ten different well-being needs. Among others, forests were valued land used for two-thirds of well-being needs, including the provision of shelter, nutrition, primary health care, an adequate supply of potable water, a lower level of ecological stress (i.e., protection from associated landslide soil erosion), cultural and spiritual benefits and livestock foraging. People commonly valued the food, income and nutrition contributions of all land uses. However, different forest and land ownership contexts and rights within the landscape influence people's preference for ecosystem services from land use in supporting their well-being. People with secure ownership (i.e., private and private-community) showed a broad and positive appreciation for ecosystem services to meet their well-being needs. Our study highlights that local and ethnic people's land-use preferences and ownership contexts are critical factors in assessing well-being in the context of multifunctional landscapes. We recommend that ecosystem services be considered in future decision-making related to forest and land use to support human well-being.

Keywords: valuation of ecosystem services; forest and land tenure; forest ecosystem services; landscape value; livelihoods; landscape multifunctionality

1. Introduction

Ecosystem services (ES) are the benefits people obtain from ecosystem function [1]. Over the last four decades, the ES concept has been widely used in ecological, economic and social sciences to assess the state of ecosystem states and its various values [1–6]. The concept and its application has broadened methodological dimensions to evaluate the multiple and total values of ecosystem services. Globally the assessment of ES has been an essential tool to inform and integrate the importance of natural systems in policy for facilitating conservation, poverty reduction, food security and the overall achievement of the Sustainable Development Goals (SDGs) [1,7–11]. Despite progress in research, mainly addressing the basic livelihood needs of rural people from tangible ecosystem benefits, predominantly provisioning ES (i.e., food, fibre, energy, construction materials for shelter) in developing countries are the focus of many studies today. By contrast, many intangible benefits, including those provided by regulation and cultural services that contribute significantly to regulating air and water quality, and various social and cultural aspects of people's well-being, are less understood [12].

Land use provides a wide range of well-being needs by making ecosystem goods and services available locally, regionally and globally. Land-use change can result in an improvement of one aspect of well-being, such as the economy, while causing a decline in several regulating, supporting and other ecosystem functions vital for sustaining community well-being [13]. A better understanding of how people's well-being is linked to ES is essential to sustainably using and managing different land resources [14,15]. With varying types of land use, ecosystem benefits differ even within rural populations, depending on people's knowledge, use and management of resources, and geographical location and context [16–21]. Although considerable research has been undertaken on land use change and ecosystem services, such studies have tended to focus on general valuations rather than understanding the differences in values attached to specific land uses within a landscape, e.g., [13,18,21].

People's utilisation of different land uses mainly relates to the ownership contexts in which they access natural resources and management [20]. Institutional processes, such as ownership rights and governance of land resources and usage, determine access for local communities, which can further impact the distribution of ecosystem services/benefits across spatial scales. Institutions related to land tenure imply the significant need for nature/ecosystem benefits in a specific social-ecological context and their roles in ecosystem management [20,21]. The context-specific perspective of the people and nature relationship often follows a different use of natural resources and knowledge systems within geographically and culturally defined settings [22,23]. Thereby incorporating ecosystem services within land use policies relies on people's ability to secure ownership. This is particularly critical for land ownership contexts in swidden agriculture-dominated landscapes where forest-based land tenure rights influence people's preferences for land use coupled with decisions for ensuring the sustainable provision of ES.

Due to over-population and an ever-growing demand for food production and settlement, Bangladesh has the long-standing challenge in sustainably managing natural resources, including forests [24,25]. By 2050, the population in Bangladesh will be 270 million, increasing challenges for the country to meet commitments towards SDGs [26]. Although forests and other land uses can contribute to achieving several SDGs related to poverty, health and food and nutrition, among others, unsustainable use and mismanagement of forest and tree-related agro-ecosystems pose a severe threat to food security. Over the past decades, forestland has decreased by 9%, barren land increased by 41.06%, settlement by 32.52%, shrubland by 21.76%, and standing waterbodies by 8.63% [27]. With increased competing contexts between forests, trees and other land uses, ecosystem service-based land management in particular offers a critical solution to meet people's well-being needs [25,28]. However, the disproportionate land ownership, social inequality, and environmental constraints associated with natural disasters influence the level of food production, especially for the rural population, in achieving a range of ecosystem benefits required for material and non-material well-being needs.

The south-eastern upland Chittagong Hill Tracts (CHT) contain a unique socio-ecological landscape inhabited by 12 different ethnic groups demonstrating a long tradition of forest and swidden agriculture land use [29]. Recent studies have highlighted the importance of forest ecosystem services and the contribution of traditional agriculture (i.e., swidden farming) to the livelihoods of local ethnic people as entirely or partly meeting their daily, basic material and non-material needs [18,28,29]. However, these studies report a limited recognition of ethnic people's rights to the forest, agricultural land (swidden farm and low-lying land use) or fruit orchards in the policies for the region and an undervaluation of the role of ecosystem services in sustaining people's well-being [18,28,29]. Lacking an understanding of the diverse ES values associated with land use generally can lead to conflict over natural resource management and the involvement of the local ethnic population within forest management in particular. Historically, forest clearing involved in the swidden farming process was considered the driver of deforestation in the region [28]. However, ineffective forest policies primarily driven by commercial plantations with timber trees and fruit orchards replacing native vegetation have also contributed to forest loss, often more extensively. Thus, today, forest and land tenure rights assigned to the local ethnic communities have not improved their long-term ability to access land uses or meet their livelihood and well-being needs [30]. As such, the preference of ethnic peoples for forests and other land uses concerning their well-being are currently thwarted or poorly understood.

In this study, we evaluate the perceived ES value of various land uses (mainly forest, swidden and low-lying agricultural land, fruit orchards and water bodies) for their contributions to the well-being of rural ethnic communities in the CHT region. Our study addressed two questions. Firstly, do the perceived ES values differ between land use? Secondly, is the perceived value of ES from different land uses associated with forest and land ownership? By addressing these questions, this study identifies the well-being benefits rural communities derive from land uses under different forest and land tenure contexts. We consider that assessing the value attached by local ethnic communities to ecosystem services will be essential for understanding their well-being needs and their long-term interest in improved forest and land management [31].

2. Materials and Methods

2.1. Brief Description of CHT's Land Use Context

The CHT region contains 10% of the country's total land and almost 40% of the forestlands [32]: Figure 1. Forests account for just over 30% of the land use in the region, although, in the CHT, less than a quarter is natural forest. Secondary forests, mixed with natural and planted trees, sparsely distributed trees and bamboo, are widespread in the region. Throughout the 19–20th century, planted forests with teak (*Tectona grandis*) and fast-growing species such as gamar (*Gmelina arborea*) expanded across the region driven by colonial and state policy for commercial harvesting and, later, increased plantation goals. This process of plantation impacted local ethnic people by converting the agricultural lands, mainly swidden farms, into monoculture tree plots. At least twelve ethnic groups rely on forests to meet their food needs, mainly leafy vegetables and fruits, fuel wood and construction materials. Despite this forest reliance, minimal tenure rights exist within the local ethnic population to access forestland, including planted land uses across the region.

Agriculture land use in the CHT is constrained due to hilly and medium to steep slope topography over two-thirds of the land [33]. Swidden farming along the hillslopes (also referred to as shifting cultivation) is the predominant agriculture land use practice (16% of the land use) for local people. This land use was traditionally managed for annual crop cultivation followed by a fallow period of 10–15 years, but has recently declined due to limited availability of land, lower productivity and insecure land tenure. Early assessments erroneously report that swidden agriculture land use only contributes to subsistence level production, has no significant economic benefit and causes forest loss and soil erosion [29,33,34]. However, this land use provides some critical form of provisioning services for food sources managed with traditional knowledge depending on the social and ecological contexts of the rural population [30]. Besides the swidden farming practice, small low-lying lands are cultivated with rain-fed and irrigated farming systems for paddy, sugarcane, maize, tobacco and seasonal vegetable production. Despite the low availability of low-lying agricultural land use, it is often considered productive and economically viable.

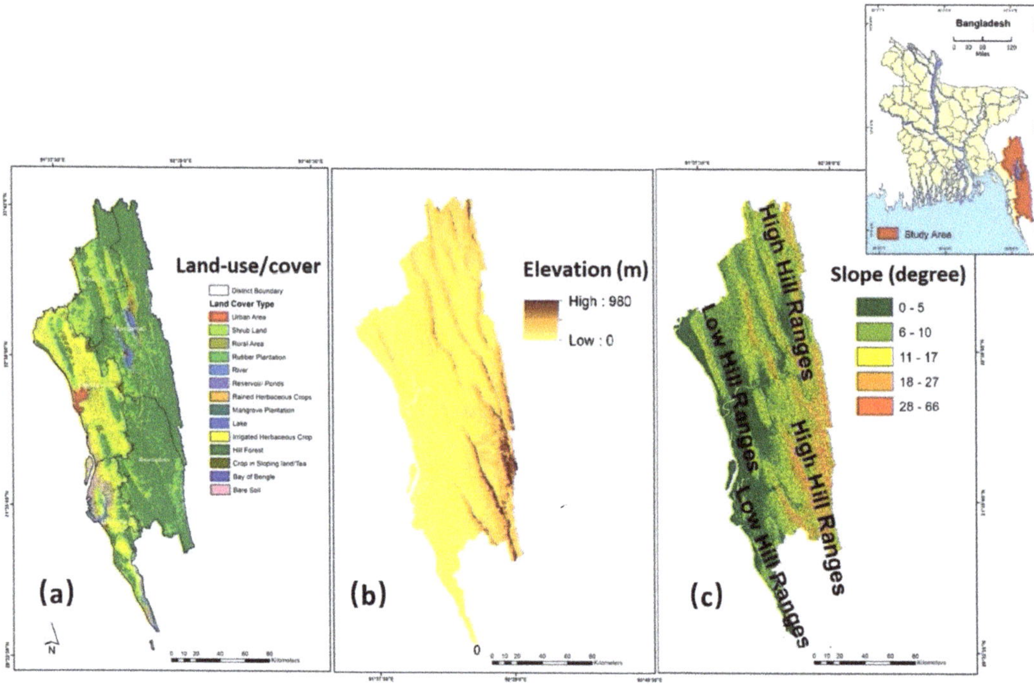

Figure 1. Land use/cover (**a**); (**b**) elevation; and (**c**) slope in the CHT.

Since the 1980s, horticulture with fruit orchard land use (i.e., tree-based agroforestry systems) has increased to improve the agriculture systems and reduce tree loss associated with swidden farming in the region. Enhancing the economic benefits, retaining tree cover on the land and limiting land degradation were the priorities of the fruit orchard based agroforestry extension. Although not extant across the region, fruit orchards with a mix of fruit trees (pineapple, orange, banana, jackfruit and mango trees) and teak plantations increased in the areas closest to markets and roads [29]. Fruit orchards with trees account for roughly 28% of the total land use [35]. However, the adoption of the fruit orchard land use is impaired by the remote locations, poor transportation, insecure market access and overall tenure situation. Notably, local ethnic people without adequate land tenure have not fully adopted fruit orchards [30,36].

Regardless of any land use for agriculture, horticulture and forestry, forest and land tenure in terms of ownership rights of the local ethnic people for natural resource management is often insecure and complicated [37]. One of the key underlying reasons is that the region's significant land (i.e., around two-thirds of the total land designated as forest, including unclassed state forests) is owned by different state departments. The local forest department regulates over 35% of the state-owned forest lands and adjacent land use with strict control on forest product collection and minimal rights for local ethnic communities (Table 1) [38]. The other 65% of forestland, especially unclassed state forests (i.e., barren lands without significant tree cover) is under the control of the local land administration. People's access to state-owned and unclassed forests is limited or partial without secure tenure rights. Customary land ownership, another form of community tenure, very often overlaps with state-designated forest-controlled areas, constraining people's access to the collection of forest and non-timber forest products, swidden farming and long-term ownership of the community-conserved regions. Only a limited extent of customary rights to lands is currently exercised, allowing people to use forest and agriculture with or without clear boundary demarcations. Customary ownership generally allows ethnic people to access swidden farming and a small area of common forest resources. Private ownership, a secure form of land ownership through land title, is minimal in the region. The secure rights through this form of private ownership of forest and planted tree land use are generally limited within CHT's poor and marginalised ethnic communities.

Table 1. Forestlands under different management authorities in three districts of the CHT [38].

District	Forest Department Managed		District Government Controlled	
	Reserved Forests (ha)	USF Lands (ha)	USF Lands (ha)	Total Forest Land (ha) by District
Bandarban	107,095	15,646	200,151	322,892
Rangamati	248,855		309,267	558,122
Khagrachari	38,800	1702	183,837	224,339
Total forest	394,750	17,348	693,255	1,105,353
% of total forest area	35.71%	1.57%	62.72%	100%

2.2. Approaches to Data Collection and Analysis

This study applied a participatory rural appraisal tool called Pebble Distribution Method to the local ethnic population in the CHT region to quantify their preferences and the perceived value they attached to important ecosystem services about their contribution to well-being [39]. This tool was used to assess people's judgement of the relative importance of various ecosystem services derived from forests and land uses for their well-being [40,41]. Value is often associated in economics with price, quality of goods and services and willingness to pay for marketed or non-marketed ecosystem services. In this study, we used ecosystem services value, not in terms of direct monetary output, but as a reflection of the local people's perceptions of the relative importance of forest and other land use benefits required to meet their well-being needs (e.g., adequate food, gain income, safe shelter, energy, etc.). The advantages of applying this approach included capturing local well-being priorities rather than the interpretations of external researchers and avoiding the complex and financial quantification of ecosystem services for which no market existed [16,39].

Although individual interviews could offer a more detailed perspective of ecosystem services, we conducted group deliberations to establish a consensus on the value of specific land use for different ecosystem services among the participants in each workshop setting. In general, the ethnic people in the region have experienced relatively similar social and ecological challenges in terms of limited availability of low-lying plain lands and insecure ownership rights to swidden farming lands or planted tree lands. However, the recent trend of the demise of swidden farming land use followed by intensive agriculture and the establishment of monoculture tree plantations has had various impacts on the livelihoods and well-being in different areas of the region. In particular, we considered that the difference in the governance of forest and land use by state, private and community management has also influenced the preference of the participants where they are located. Focus group discussions enabled us to save time in accessing and explaining the ecosystem services to the respondents, who were from socially and ecologically dispersed communities. Through this participatory approach, we were able to explore the perceived value and importance of ecosystem services to the ethnic population's livelihoods and well-being and identify what they considered to be the benefits, their sources and their linkages to well-being.

During 2015–2016, the first author conducted a scoping survey as a part of the global agrarian change study approach (led by the third author), by undertaking an initial literature review and consulting the local government, non-government organisations and local ethnic chiefs to identify representative villages in the region [29]. The first author selected 12 villages in consultation with the second and third authors that represented the dominant characteristics of current land use practices (e.g., swidden farming, forest and tree-based ecosystems, low-lying agricultural land, fruit orchards) in the region [42]. The distinct differences in the communities selected for the study villages were their locations, land use, forest and land ownership (Table 2). The communities in remote villages had a higher reliance on swidden farming and forests for subsistence and cash income without secure land ownership title [30]. In contrast, people with secure land ownership practised swidden farming to a lesser extent in intermediate and on-road locations. They were more engaged in low-lying agricultural cropping practices and planted tree areas with fruit orchards. Of the total of 12 villages, we randomly selected eight villages for focus group meetings, covering two-three villages from each location. Taking account of the characteristics (Table 1), we generally consider that the communities of villages within one location are more similar in land use practice and ownership than others.

Table 2. Salient features of the villages in three different locations of the CHT region, adapted from [29].

	Remote	Intermediate	On-Road
Name of villages	• Uluchari • Chainda • Marma • Pangkhua para	• Paglachara • Angadpara • Bijoypara • Suanglupara	• Khamadong • Kamalong • Jogeshkarbaripara • Bagmara Headmanpara
Ethnicity of inhabitants	• Tanchangya/Chakma • Pangkhua • Marma	• Tanchangya • Marma • Bawm	• Marma • Chakma
Population density (per sq. km.)	38	62	176
Land use types	• Swidden agriculture • Fallow land • Natural and/ secondary forests	• Swidden agriculture • Low-land agriculture • Secondary forests • Plantations • Fruit-tree garden	• Low-land agriculture • Plantations • Small fruit-tree garden
Land ownership	• State	• State • Customary • Private (land title)	• State • Private (land title)
Location (distance to sub-district/district market)	• 2–3 h travel by boat	• 0.5–1 h by motorbike	• 30 min by motorbike

With a small team of locally recruited research assistants, the first author organised eight focus group meetings with local communities in the CHT region to quantify their preferred (most important) ecosystem services of different land uses and their linkages to livelihood and well-being constituents (Figure 2). Each focus group consisted of six-eight individuals with various occupations, including forest users, swidden farmers, commercial farmers and politically elected community members from the group's village. The field team discussed the ecosystem services concept with participants in the local language to introduce its importance and scope in terms of more comprehensive benefits. We did not expect people to be familiar with the technical term. The field team presented the well-being concept by describing it as a combination of security, essential material for life, adequate food and nutrition, health, and social relations.

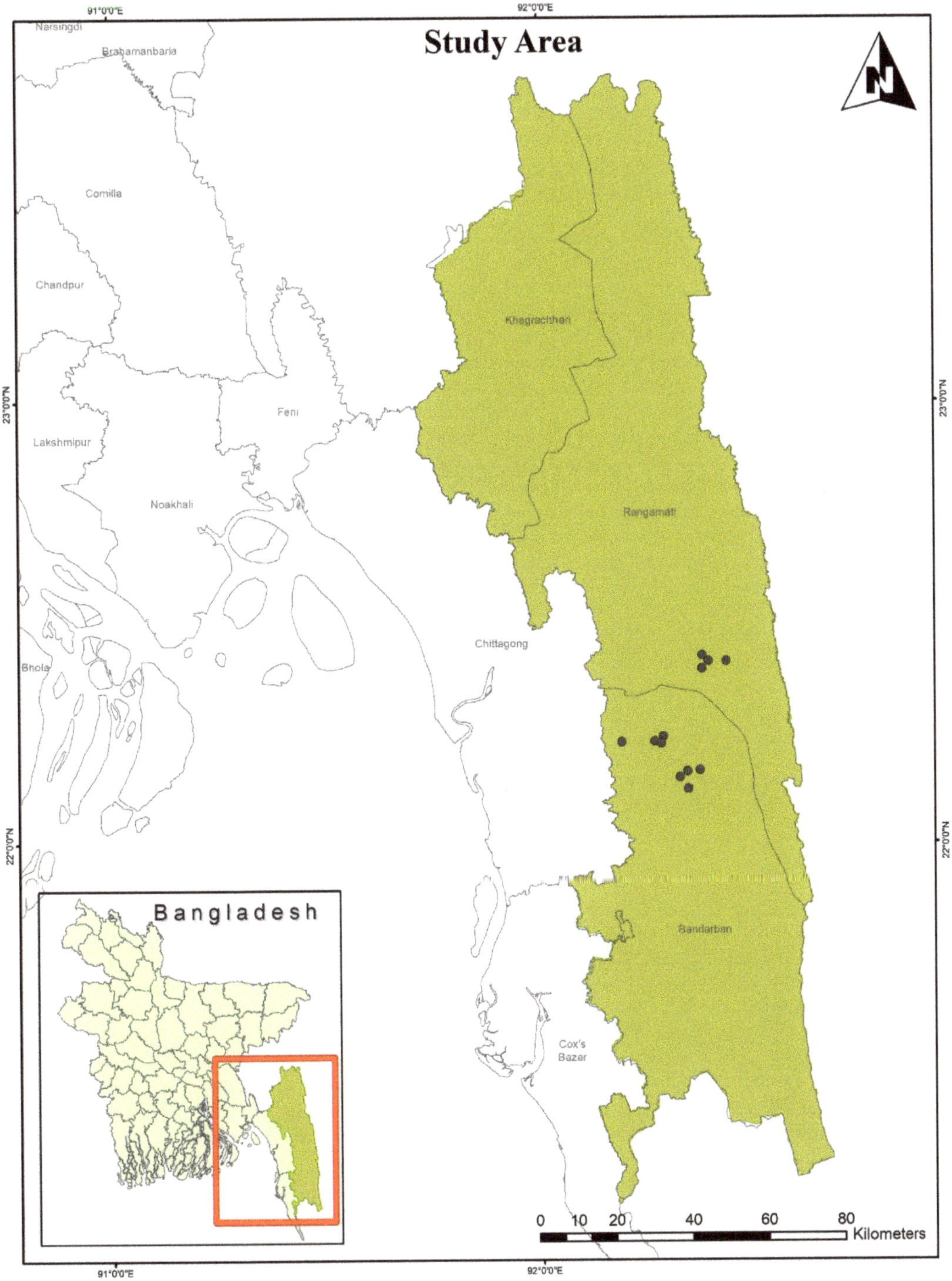

Figure 2. Study location of 8 villages in CHT region (of 12 villages) indicated on the map where focus group meeting was undertaken.

In each focus group meeting, the participants were asked two questions. The first question was: What are the most important benefits you receive from the surrounding landscape, focusing mainly on land uses? The participants were allowed to discuss the surrounding land uses and identify their valued benefits. Our study thereby identified the benefits of forests, trees and other land uses that the respondents considered most important. Occasionally, the field team introduced a list of expected benefits identified for this region related to land use [18]. Then, to determine the comparisons of land uses, we asked a second question: From where do you obtain or receive the most benefits in the landscape?

In the meeting, the participants listed the benefits they thought were important to their livelihoods and well-being. In each village, the participants discussed within the group to reach a consensus on the main forest and other landscape components/land use benefits they considered necessary for meeting their well-being. Then, the field team asked them to evaluate the benefits based on their importance and distribution across sources (land use). The participants scored the relative importance of the identified ecosystem services for the respective well-being constituents on a scale of 1–100. They evaluated the benefits more than once, according to their linkages to the livelihood and well-being requirements and sources. In so doing, they identified the main benefits of the land uses that they considered important. Once they had evaluated the important benefits for their respective well-being constituents, we discussed their reasons for choosing the benefits for the respective forest and other land uses. In this scoring part of the focus group, the participants were asked to distribute 100 counters (buttons, seeds, or pebbles) between labelled and illustrated cards in proportion to their 'importance'. We recorded the ecosystem services' values (counts) of land uses under different well-being categories in a spreadsheet.

We calculated the percentage of broad ecosystem services categories (i.e., provisioning, regulating and cultural services). Then we tested the difference among land use scores for well-being needs using Kruskal-Wallis test. We analysed the post hoc comparisons of different land-use preferred for each well-being need in R statistical programming. We also compared the well-being needs with forests and land uses and different management contexts i.e., given different forms of ownership rights of the participants in the private, private-community and state forest and land use. We obtained the ethics approval for this research through the Charles Darwin University Human Ethics Committee (ethics clearance number: H15005). We also received informed consent from all participants for our study.

2.3. Limitations of the Research

Our study was carried out via a focus group meeting rather than individual interviews or surveys. The group approach has given us a better understanding of how the majority of people in each location value ecosystem services. However, we acknowledge that the focus group may have masked individual heterogeneity on the preferences of the ecosystem services from specific land uses to meet the well-being needs, within this culturally diverse area of the country. Individual contexts of the rural ethnic population with different socio-economic characteristics, wealth conditions, ethnic backgrounds and level of land ownership may be valuable information in covering the breadth of ES value at the household level, preference of land uses and target the management interventions for improved forest and land use practices, to meet their most well-being needs.

3. Results

3.1. Preferred ES for the Well-Being

Our analysis presents reasons and people's preferences for ecosystem services regarding their importance for well-being (Table 3). Local communities identified eight of the most important ecosystem services, i.e., timber, bamboo and grass for housebuilding, protection from soil erosion, water, fuel wood, spiritual benefits, food for own uses and livestock grazing, and primary medicine (Table 3). These benefits meet their well-being needs in a number of different ways, including nutrition and income. The valued well-being needs include the ability to obtain safe shelter, adjust to ecological stress, water, secure energy, adequate food and nutrition, food for livestock, avoid disease, to meet cultural and spiritual needs. Among the specific well-being needs, people preferred and valued safe shelter (>70 points) and the ability to protect land from soil erosion (50–70 scores) (after that, this well-being worded as reduce ecological stress) as highly important services. Overall, 75% of the ecosystem services the respondents identified were provisioning services, while those remaining were regulating and cultural ecosystem services.

Table 3. Local community's identification of ES, the reason for preference and linkage to well-being.

ES Preference	Reason for Preference	Relevance for Well-Being Components (Following MA, 2005)
Timber, bamboo and grass for housebuilding/construction	Locally available raw materials, low-cost and traditional knowledge in building housing	Safe shelter
Protection from soil erosion and fertility	Reduce soil erosion for crop production and protection from landslides	Ability to reduce ecological stress such as landslides associated with soil erosion
Water	Water only sourced from springs, creeks adjacent to forest	Adequate water for use
Fuel wood	Mainly used for cooking and boiling water	Ability to secure energy use
Spiritual benefits	Worship and peaceful living	Cultural and spiritual benefits
Food	Crops cultivated; gathered from the forest and adjacent wild environment	Adequate foodAdequate nutrition
Livestock food	Availability of food for livestock grazing	Adequate food for livestock
Primary medicine	Available plant materials and traditional knowledge associated with healing practices	Ability to avoid disease
		Access to income

3.2. Land Use Preferences for ES to Support the Well-Being

The analysis presents the association of the most preferred land use for specific ecosystem services required to meet people's basic well-being (Figure 3). The five land uses preferred by local communities for ecosystem services in the landscape were forest/planted tree land, swidden/fallow land, low-lying agricultural land, fruit orchards and waterbodies. Among those land uses, the forest was the most preferred, followed by fruit orchards, swidden/fallow, low-lying land and water bodies. Forest land was preferred for almost all the ecosystem services that the local people valued for their well-being (Figure 3). Compared to all other land uses in the landscape, the forest was the single most preferred land use with diverse options for contributing to specific well-being needs through energy, safe shelter, cultural and spiritual benefits and an enhanced local people's ability to avoid disease. Aside from forests, our analysis showed the people's preference for fruit orchards and low-lying and swidden agriculture land use for meeting safe shelter and energy needs (Figure 3).

The land use preference was significantly different for ecosystem services in attaining well-being needs ($X^2 = 77.986$, $p < 0.000$). The post hoc comparisons of land use showed that forest was valued significantly higher than other land use ($p < 0.05$) for its benefit in enhancing people's ability to avoid disease, its cultural and spiritual importance, energy use, safe shelter and supporting adequate water for use. Water bodies were the most preferred land use for adequate water ($p < 0.05$) other than forest. To secure adequate food, only low-lying agricultural land was more preferred than any other land use ($p < 0.001$). No significant difference was found in the land use preference for adequate food, nutrition, income, food for livestock and ability to reducing ecological stress such as soil erosion or landslide. Forest and low-lying agricultural land were also more preferred for adequate livestock food.

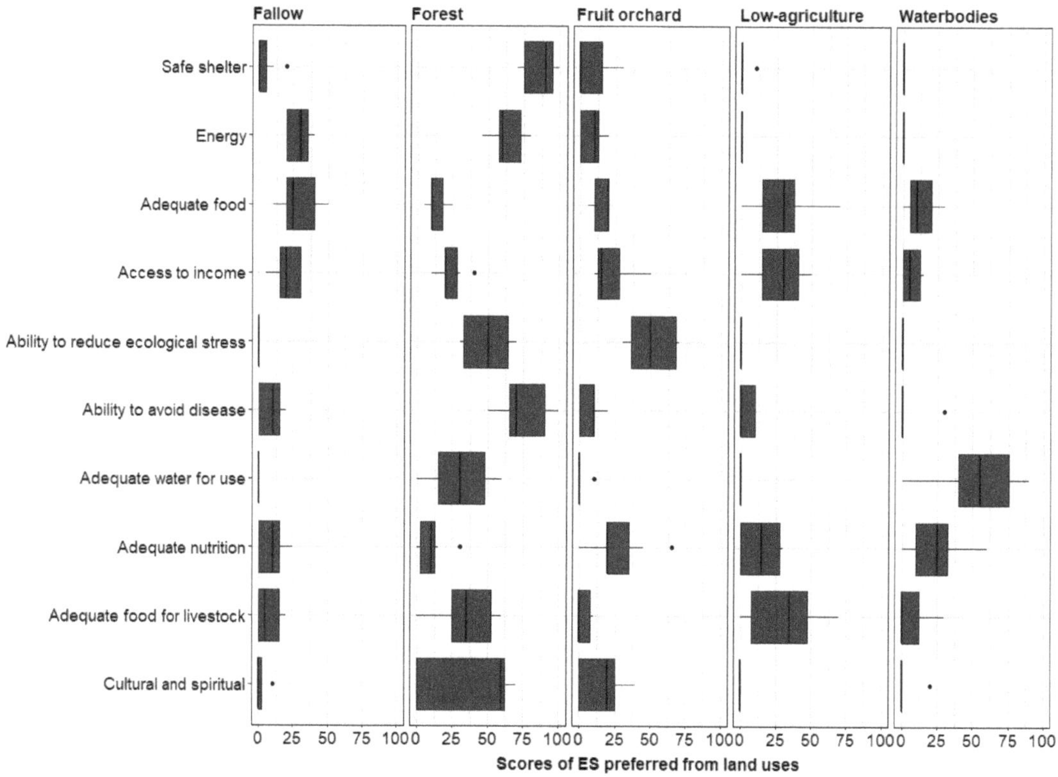

Figure 3. People's preferences for ecosystem services from different land uses to meet their well-being needs.

3.3. *Land Use Preferences for ES under Different Ownership Contexts*

Further analysis showed the influence of forest and land ownership contexts (state, private, private and community) on people's land use preference for specific well-being needs (Figure 4). Participants in state land-owned areas, reported forest over others for all well-being needs. Within the state and private-community land-owned areas, people showed more preferences for forestland use for the ability to avoid disease, energy and safe shelter, which is slightly higher than those in private ownership. Fruit orchard was the commonly preferred land use for adequate nutrition and ability to reduce ecological stress across all three ownership areas, with partly high perceived importance in private and private-community ownership areas. In private and state ownership contexts, people qualified the swidden agricultural practices in the maintenance of diverse well-being options, with slightly high value reported for this land use in state owned areas. Water bodies were highly valued land use by participants in the private and private-community ownership areas regarding water need compared to those in state-owned forest areas. People in state land ownership considered that waterbodies enhance their ability to avoid disease more than those in private and private-community villages.

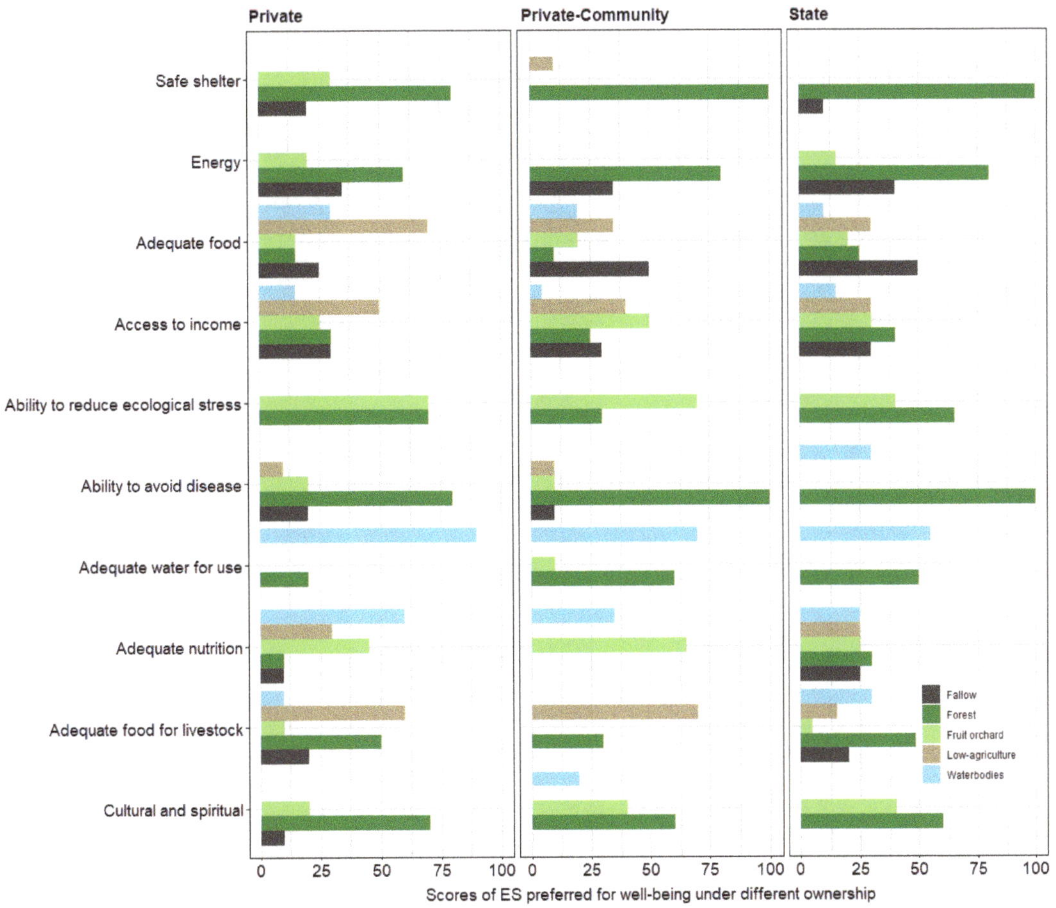

Figure 4. Variations of land use preference for ES to support well-being under different land ownership contexts. The top labels of the "x" axis show the ownership contexts. The "y" axis shows the well-being needs.

Across all the ownership contexts, participants valued swidden agricultural land as a source of adequate food and wood energy with relatively high value in a state ownership context. Among all ownership cases, the private ownership participants preferred low-lying agricultural land use. This is also observed for the swidden agricultural land use that participants generally valued for energy contribution, apart from forests, across all three ownership contexts. The participants located within private and private-ownership with secure access to land use showed a high preference for low-lying agricultural land for adequate food for livestock, compared to those in state areas.

4. Discussion

Our study revealed people's preferences for ecosystem services that supported their well-being in 10 different ways, mainly safe shelter, ecological stress reduction, disease avoidance, adequate water use, and cultural and spiritual satisfaction. Local people valued forests and land uses such as swidden farming/fallow land, low-lying agriculture, fruit orchards and waterbodies as key sources to obtain their preferred ecosystem services to support their well-being. The preferences for land use and related benefits assessed in our study reflect the different values people attached to secure their well-being [43]. The land use and associated ecosystem services, in particular, show the varied forms of well-being impacted by different management contexts within the same landscape.

4.1. Valuing ES for Well-Being

Previous studies mention individual households' perceived ecosystem services demand without explaining how this relates to people's well-being in rural contexts [19]. Our study accounted for ecosystems' highly valued contributions to basic well-being: safe shelter, the ability to reduce ecological stress, primary energy, cultural and spiritual benefits, adequate water for use and access to traditional health care. A similar set of well-being needs reported in South Africa showed a specific pattern of ecosystem services used by households [44]. In our study, people preferred safe shelter, which identified their persistent need for construction materials, including bamboo and timber provisioning ecosystem services. This finding supports previous studies showing that changes in forest areas affect rural people's well-being owing to reduced secure access to forest resources to gather construction timber and non-timber forest products, including bamboo and the inability to maintain dietary diversity because of the lack of forest-sourced wild foods [45,46].

Aside from these benefits, which are directly related to community use-values, people's preference for ecological stress reduction (i.e., protection from soil erosion or maintaining soil fertility) reveals an indirect contribution to the ecosystem itself to maintain farm productivity and support the long-term subsistence of the population, which is directly exposed to food insecurity and must rely on its surrounding ecosystem. Soil erosion and associated fertility loss resulting in a decline in land usability and further ecosystem degradation found in our study is a common concern in swidden-dominated agro-ecosystems [47]. Our participatory valuation of ecosystem services reflects people's different levels of exposure to nature, experience and effects on improving well-being [9,43]. Local farmers experiencing a decline in farm productivity showed a positive attitude toward restoring the forest landscape in the CHT region.

People's preferences for ecosystem services for well-being needs reflect a common connection to their landscape in the CHT. For instance, specific ecosystem services, such as the ability to avoid disease through traditional healing, reflects a positive attitude toward the ecosystem in maintaining health care in all the land use practices we studied. Preferences for the cultural and spiritual benefits of ecosystems also indicate better social relationships and attachment to the landscape [48,49]. These cultural benefits and the realisation of ecosystem services are strongly linked to the values and norms of Indigenous people and local communities in many countries across the world [14,50–52].

Cultural value, in terms of a sense of place, a sacred landscape, and the role of spirituality in maintaining native vegetation and conservation of watersheds, indicates the strong connection to forest conservation that many Indigenous populations have [29]. However, driven by the global focus on carbon-based forest management, national forest policy ignores local and ethnic people's cultural value given to forest conservation or their perceptions of changing landscapes [12]. Assessing local forests' social and cultural values and the customary rules and knowledge that have persisted and been co-produced over time will be a crucial area for further research to support locally adaptive forest management in maintaining its global value. In this regard, customary management can play a critical role in maintaining the traditional knowledge of community forest protection for watersheds and rule-making in sustainable use of non-timber forest provisioning ecosystem services.

4.2. ES Value Associated with Different Land Uses

The ecosystem services preferences of local people revealed that well-being was strongly associated with different land-use configurations in a landscape [16]. Forest and tree-based ecosystems, especially, were highly valued for providing most of the benefits required to achieve well-being. Forests are mainly beneficial for safe shelter, energy, cultural and spiritual needs, ecological stress reduction, primary health care or traditional healing, adequate water, and livestock food sources. Apart from forests, fruit orchard land use is valued for nutrition, adequate water, minimisation of ecological stress, safe shelter and cultural and spiritual benefits.

Previous studies have mentioned that the increased demand for fruit orchard land use is driven by government policy supporting agro-forest expansion to replace swidden agricultural practices in the CHT and other upland landscapes of Southeast Asia [34,53]. Although swidden farming land was considered to show low productivity agriculture practice in such studies, our study found that people's choice of this land use existed to support adequate food, income and energy sources. The social and cultural importance of this subsistence land use, i.e., swidden farming in our case, is more important than economic value. Interestingly low-lying agricultural land use was highly preferred for adequate food, nutrition, income and livestock grazing benefits. Low-lying agricultural land could be cultivated with commercially important crops (i.e., tobacco, potato, taro) on a rotational basis for more than two seasons in a single year, even though the proportion of this land use is relatively low due to the region's physiographic conditions. A positive attitude was also observed towards waterbodies because of the need for adequate food, nutrition, water and cultural and spiritual benefits. Therefore, it is evident that rural people have a wide range of social, economic and ecological values embedded in most land uses in their vicinity.

Agricultural expansion and the development of alternative land use with commercially valuable crops, including monoculture trees and fruit orchards, often limit the diversity of well-being options in a landscape [28]. Provisioning ecosystem services for food and income benefits showed a strong positive association with most land uses, except forests. Our study on the low valuation of forest-sourced food and nutrition contradicts the value of the forest for food of individual households recently reported by other studies in the region and elsewhere in the tropics [54]. Forest and tree-based foods were considered of low importance by community participants, who were more driven by their interest in land use to enhance food production and economic gains. In recent decades, there has been much interest in fruit orchards, monoculture crops, and planted land uses without considering people's differences in the capacity to access such land uses. This community-based valuation may contrast with that of individual households that might require forest-sourced foods to avoid food shortages and achieve adequate nutrition [28,54]. Future land management policy should consider this context concerning the well-being needs of those who may be unable to access fruit orchards and low-lying lands.

Furthermore, management contexts regarding people's different forest and land ownership rights to the landscape influence their ecosystem services and land use preferences in supporting their well-being. Our study findings supports similar research by others [40], which showed that ecosystem services use differs across people's well-being situations and partly relates to resource ownership in a social-ecological system. Land ownership status is a strong predictor of the distribution of ecosystem services with differing values for well-being needs [37,55,56]. For instance, people without secure land rights in the state ownership context preferred forest and swidden agricultural land for meeting several well-being needs (adequate food, income, nutrition, livestock foraging and water for use). Overall, people with fully secure land rights departed from relying on swidden farming lands and employed diverse forms of land use to support their well-being. It was evident that secure land ownership was attributed to a higher preference for forests, fruit orchards, low-lying agriculture and water bodies for multiple benefits. Especially private ownership influences the choice of water bodies for their protection to secure adequate water. From our study, the specific management contexts of the landscape that local people experienced in their daily lives strongly influenced their ecosystem services preferences to achieve well-being.

5. Conclusions and Future Directions

We assessed the value of ecosystem services' by eliciting community preferences for ecosystem services obtained from diverse land uses to support well-being. Our study findings demonstrate a comprehensive appreciation of local ethnic communities for their surrounding landscape, delivering a broad range of ecosystem services with diverse contributions to meet their well-being. By applying participatory habitat valuation approach, the local community's preferences revealed their shared aspiration to access a wide range of benefits, mainly safe shelter, traditional healing to avoid disease, energy, and cultural and spiritual benefits in their landscape. It is inevitable that a minimal set of ecosystem services will be necessary for the well-being of the ethnic population living in this landscape, but this may rarely be, if ever, sufficient given the patterns of land use and access. Overall, people perceived the role of land use to supply about two-thirds of their well-being needs. Although we observed a substantial variation in ecosystem services preferences from different land uses, the forest is found to be the most useful among all in delivering the maximum well-being benefits. The beneficial contributions of other land use such as swidden agriculture land use, fruit orchard and low-lying agriculture land use reflect the multifunctionality of the landscape and its roles in sustaining the ecosystem services required to meet various well-being needs [56].

Differences in land ownership influenced people's perceptions of the benefits and supply of ecosystem services from the land uses in the CHT region. The insecure land ownership affects people's reliance on the forest and swidden agricultural land use for meeting a wide range of needs. Swidden agriculture farming is currently regarded for its trade-offs in provisioning multiple ecosystem services although evidently the land use remains a critical source of food, income and energy options to a particular community. By contrast, our study found that preferences for fruit orchards, regardless of the ownership contexts, indicate its acceptance as a valuable land use. This specific preference of land use indicates an opportunity for engaging the swidden farmers in sustainable land management (i.e., agroforestry) by providing them secure land ownership. Indigenous people and local communities are likely to follow different land use pathways in accessing ecosystem services or to replace some with others to achieve well-being [11]. Thus, their existing situation and future well-being needs will be depended on a plethora of contextual social, ecological and institutional factors in accessing the most desired land use practices. Many of the well-being needs cannot be provided through access to ecosystem services without the proper ownership arrangement for land use that we found in a socio-ecological system [57].

To maintain the well-being of the local ethnic population relying on ecosystem services, it will be important to integrate their aspirations within the existing land management. Given the high value of forests, future land management approaches could consider achieving multiple benefits supported by ecosystem services-based policies. For instance, embedding ecosystem services within forest and other land-use-related strategies may avoid trade-offs in the provision of ecosystem services. The National Forest Policy (2016) in Bangladesh contains an objective for large-scale tree plantation programmes in degraded forest areas (i.e., unclassed state forests in the CHT region) and mentions the need to provide shared forest benefits for local communities [25]. However, there are no specific measures for understanding how forest and land use benefits are distributed and managed under different land ownership contexts. In the CHT region, land ownership will influence the types of forest and land management and the intended goals for ecosystem services. Without addressing ownership issues in forest and land use-related policies, there are potential challenges in securing a wide range of ecosystem services benefits to support local people's well-being.

Some broader recommendations can be drawn from this study. Firstly, forest and agricultural land management practices should consider the local community's preferences for ecosystem services from their desired land uses to realise their future well-being. Secondly, an integrated management approach may be undertaken with trees and fruit-based land use strategies to maximise ecosystem services values, as well-being is supported by more than one land use practice. Ecosystem services should be placed at the heart of forest and land use-related policies to guide local and regional natural resource management. Thirdly, a strategy to support greater land ownership for local ethnic people would allow them to a secure tenure right on forests, agriculture and fruit orchard, among others. In doing so there will be a strong possibility to maintain synergistic co-existence of the land uses and thereby sustain the provision of ES in meeting multiple well-being needs in a landscape.

The results of this study could be extended through further research to broadening understanding of the role of land use preferences in addressing the breadth of ecosystem services' contributions to well-being and decision-making. We consider specific areas worth investigation in future are: (i) integrated valuation of ecosystem services to determine the synergies and trade-offs between and within ecosystem services to inform forest and land management; (ii) individual stakeholder's land use preferences on ecosystem services and well-being; (iii) underlying social, ecological and economic drivers in mediating multiple values of ecosystem services in landscape management; (iv) the links between climatic change and land use and impacts on the provisions of ecosystem services; and (v) how to address challenges and opportunities of land ownership in operationalising ecosystem services-based forest and landscape restoration.

Author Contributions: Conceptualization, R.A., N.S. and K.K.S.; methodology, R.A. and K.K.S.; formal analysis, R.A.; writing—original draft preparation, R.A.; writing—review and editing, R.A., N.S., K.K.S. and T.S.; supervision, N.S. and T.S.; project administration, R.A.; funding acquisition, T.S. All authors have read and agreed to the published version of the manuscript.

Funding: This research was funded by the United States Agency for International Development (USAID) and the UK's Department for International Development (DFID) through grants to Center for International Forestry Research (CIFOR). This publication is an output of the CGIAR Consortium Research Program on Forests, Trees and Agroforestry. The research was also funded by an Australian Postgraduate Award, and postgraduate research funding under the Faculty of Engineering, Health, Science and the Environment of Charles Darwin University, Australia, and a Ph.D. Dissertation Fellowship of the South Asian Network for Development and Environmental Economics (SANDEE) through the Asian Centre for Development, Bangladesh.

Data Availability Statement: Not applicable.

Acknowledgments: The authors would like to express heartfelt thanks to the people who took park in numerous focus group discussions in the study villages.

Conflicts of Interest: The authors declare no conflict of interest.

References

1. Millennium Ecosystem Assessment (MA). In *Ecosystems and Human Well-Being: Synthesis*; Island Press: Washington, DC, USA, 2005; Available online: https://www.millenniumassessment.org/documents/document.356.aspx.pdf (accessed on 1 January 2020).
2. Ehrlich, P.R.; Mooney, H.A. Extinction, substitution, and ecosystem services. *BioScience* **1983**, *33*, 248–254. [CrossRef]
3. Daily, G.C. *Nature's Services: Societal Dependence on Natural Ecosystems*; Island Press: Washington, DC, USA, 1997.
4. Costanza, R.; d'Arge, R.; De Groot, R.; Farber, S.; Grasso, M.; Hannon, B.; Limburg, K.; Naeem, S.; O'neill, R.V.; Paruelo, J.; et al. The value of the world's ecosystem services and natural capital. *Nature* **1997**, *387*, 253–260. [CrossRef]
5. Pascual, U.; Muradian, R.; Brander, L.; Gómez-Baggethun, E.; Martín-López, B.; Verma, M.; Armsworth, P.; Christie, M.; Cornelissen, H.; Eppink, F.; et al. The economics of valuing ecosystem services and biodiversity. In *The Economics of Ecosystems and Biodiversity: Ecological and Economic Foundations*; Routledge: London, UK, 2010; pp. 183–256.
6. Haines-Young, R.; Potschin-Young, M. Revision of the common international classification for ecosystem services (CICES V5. 1): A policy brief. *One Ecosyst.* **2018**, *3*, e27108. [CrossRef]
7. Díaz, S.; Demissew, S.; Carabias, J.; Joly, C.; Lonsdale, M.; Ash, N.; Larigauderie, A.; Adhikari, J.R.; Arico, S.; Báldi, A.; et al. The IPBES Conceptual Framework—Connecting nature and people. *Curr. Opin. Environ. Sustain.* **2015**, *14*, 1–16. [CrossRef]
8. United Nations (UN). Transforming Our World: The 2030 Agenda for Sustainable Development. 2015. Available online: https://documents-dds-ny.un.org/doc/UNDOC/GEN/N15/291/89/PDF/N1529189.pdf?OpenElement (accessed on 1 June 2022).
9. Bratman, G.N.; Anderson, C.B.; Berman, M.G.; Cochran, B.; De Vries, S.; Flanders, J.; Folke, C.; Frumkin, H.; Gross, J.J.; Hartig, T.; et al. Nature and mental health: An ecosystem service perspective. *Sci. Adv.* **2019**, *5*, eaax0903. [PubMed]
10. High Level Panel of Experts (HLPE). Sustainable Forestry for Food Security and Nutrition. Rome, Italy. 2017. Available online: http://www.fao.org/3/i7395e/i7395e.pdf (accessed on 1 June 2022).

11. Mandle, L.; Shields-Estrada, A.; Chaplin-Kramer, R.; Mitchell, M.G.; Bremer, L.L.; Gourevitch, J.D.; Hawthorne, P.; Johnson, J.A.; Robinson, B.E.; Smith, J.R.; et al. Increasing decision relevance of ecosystem service science. *Nat. Sustain.* **2021**, *4*, 161–169. [CrossRef]
12. Plieninger, T.; Dijks, S.; Oteros-Rozas, E.; Bieling, C. Assessing, mapping, and quantifying cultural ecosystem services at community level. *Land Use Policy* **2013**, *33*, 118–129. [CrossRef]
13. Wang, X.; Dong, X.; Liu, H.; Wei, H.; Fan, W.; Lu, N.; Xu, Z.; Ren, J.; Xing, K. Linking land use change, ecosystem services and human well-being: A case study of the Manas River Basin of Xinjiang, China. *Ecosyst. Serv.* **2017**, *27*, 113–123. [CrossRef]
14. Sangha, K.K.; Preece, L.; Villarreal-Rosas, J.; Kegamba, J.J.; Paudyal, K.; Warmenhoven, T.; RamaKrishnan, P.S. An ecosystem services framework to evaluate Indigenous and local peoples' connections with nature. *Ecosyst. Serv.* **2018**, *31*, 111–125. [CrossRef]
15. Sandhu, H.; Sandhu, S. Linking ecosystem services with the constituents of human well-being for poverty alleviation in eastern Himalayas. *Ecol. Econ.* **2014**, *107*, 65–75. [CrossRef]
16. Duguma, L.A.; Hager, H. Farmers' assessment of the social and ecological values of land uses in central highland Ethiopia. *Environ. Manag.* **2011**, *47*, 969–982. [CrossRef]
17. Lau, J.D.; Hicks, C.C.; Gurney, G.G.; Cinner, J.E. Disaggregating ecosystem service values and priorities by wealth, age, and education. *Ecosyst. Serv.* **2018**, *29*, 91–98. [CrossRef]
18. Ahammad, R.; Stacey, N.; Sunderland, T.C. Use and perceived importance of forest ecosystem services in rural livelihoods of Chittagong Hill Tracts, Bangladesh. *Ecosyst. Serv.* **2019**, *35*, 87–98. [CrossRef]
19. Muhamad, D.; Okubo, S.; Harashina, K.; Gunawan, B.; Takeuchi, K. Living close to forests enhances people' s perception of ecosystem services in a forest–agricultural landscape of West Java, Indonesia. *Ecosyst. Serv.* **2014**, *8*, 197–206. [CrossRef]
20. Benra, F.; Nahuelhual, L. A trilogy of inequalities: Land ownership, forest cover and ecosystem services distribution. *Land Use Policy* **2019**, *82*, 247–257. [CrossRef]
21. Hasan, S.S.; Zhen, L.; Miah, M.G.; Ahamed, T.; Samie, A. Impact of land use change on ecosystem services: A review. *Environ. Dev.* **2020**, *34*, 100527. [CrossRef]
22. Pascual, U.; Balvanera, P.; Díaz, S.; Pataki, G.; Roth, E.; Stenseke, M.; Watson, R.T.; Dessane, E.B.; Islar, M.; Kelemen, E.; et al. Valuing nature's contributions to people: The IPBES approach. *Curr. Opin. Environ. Sustain.* **2017**, *26*, 7–16. [CrossRef]
23. Díaz, S.; Pascual, U.; Stenseke, M.; Martín-López, B.; Watson, R.T.; Molnár, Z.; Hill, R.; Chan, K.M.; Baste, I.A.; Brauman, K.A.; et al. Assessing nature's contributions to people. *Science* **2018**, *359*, 270–272. [CrossRef]
24. Adams, H.; Adger, W.N.; Ahmad, S.; Ahmed, A.; Begum, D.; Matthews, Z.; Rahman, M.M.; Nilsen, K.; Gurney, G.G.; Streatfield, P.K. Multi-dimensional well-being associated with economic dependence on ecosystem services in deltaic social-ecological systems of Bangladesh. *Reg. Environ. Chang.* **2020**, *20*, 42. [CrossRef]
25. Ahammad, R.; Stacey, N.; Sunderland, T. Analysis of forest-related policies for supporting ecosystem services-based forest management in Bangladesh. *Ecosyst. Serv.* **2021**, *48*, 101235. [CrossRef]
26. BBS 2016. Statistical Pocket Book 2016. Dhaka: Bangladesh Bureau of Statistics, Statistics and Informatics Division, Ministry of Planning, Government of the People's Republic of Bangladesh. Available online: https://www.oecd-ilibrary.org/content/publication/84971290-en (accessed on 20 August 2020).
27. Xu, X.; Shrestha, S.; Gilani, H.; Gumma, M.K.; Siddiqui, B.N.; Jain, A.K. Dynamics and drivers of land use and land cover changes in Bangladesh. *Reg. Environ. Chang.* **2020**, *20*, 54. [CrossRef]
28. Ahammad, R.; Stacey, N.; Eddy, I.M.; Tomscha, S.A.; Sunderland, T.C. Recent trends of forest cover change and ecosystem services in eastern upland region of Bangladesh. *Sci. Total Environ.* **2019**, *647*, 379–389. [CrossRef] [PubMed]
29. Ahammad, R.; Stacey, N. Forest and agrarian change in the Chittagong Hill Tracts region of Bangladesh. In *Agrarian Change in Tropical Landscapes*; CIFOR: Bogor Regency, Indonesia, 2016; pp. 191–232. [CrossRef]
30. Ahammad, R.; Stacey, N.; Sunderland, T. Assessing land use changes and livelihood outcomes of rural people in the Chittagong Hill Tracts region, Bangladesh. *Land Degrad. Dev.* **2020**, *32*, 3626–3638. [CrossRef]
31. Olander, L.P.; Johnston, R.J.; Tallis, H.; Kagan, J.; Maguire, L.A.; Polasky, S.; Urban, D.; Boyd, J.; Wainger, L.; Palmer, M. 2018. Benefit relevant indicators: Ecosystem services measures that link ecological and social outcomes. *Ecol. Indic.* **2018**, *85*, 1262–1272. [CrossRef]
32. Ahammad, R.; Stacey, N.; Sunderland, T. Determinants of forest and tree uses across households of different sites and ethnicities in Bangladesh. *Sustain. Sci. Pract. Policy* **2021**, *17*, 232–242. [CrossRef]
33. Islam, S.M.; Alam, M.; Mantel, S. Land use planning and environmental control in the Chittagong Hill Tracts. In *CHARM Project Report 3*; CHARM Project: Dhaka, Bangladesh, 2007; Available online: https://edepot.wur.nl/296235 (accessed on 22 November 2022).
34. Rahman, S.A.; Sunderland, T.; Roshetko, J.M.; Healey, J.R. Facilitating smallholder tree farming in fragmented tropical landscapes: Challenges and potentials for sustainable land management. *J. Environ. Manag.* **2017**, *198*, 110–121. [CrossRef]
35. Bala, B.K.; Majumder, S.; Altaf Hossain, S.M.; Haque, M.A.; Hossain, M.A. Exploring development strategies of agricultural systems of Hill Tracts of Chittagong in Bangladesh. *Environ. Dev. Sustain.* **2013**, *15*, 949–966. [CrossRef]
36. Nath, T.K.; Inoue, M. The upland settlement project of Bangladesh as a means of reducing land degradation and improving rural livelihoods. *Small-Scale For.* **2008**, *7*, 163–182. [CrossRef]
37. Ahammad, R.; Hossain, M.K.; Sobhan, I.; Hasan, R.; Biswas, S.R.; Mukul, S.A. Social, ecological and institutional factors affecting forest and landscape restoration in the Chittagong Hill Tracts of Bangladesh. *Land Use Policy* **2023**, *125*, 106478. [CrossRef]

38. Bangladesh Forest Department (BFD), 2016. District Wise Forest Land of Bangladesh. Government of the People's Republic of Bangladesh. Available online: http://www.bforest.gov.bd/site/page/837e6966-0fce-4274-a0d0-bcdfa49ce492/- (accessed on 2 October 2022).
39. Liswanti, N.; Basuki, I. *Guidelines for Adapted Multidisciplinary Landscape Assessment Methods for Fire Management Projects in India*; CIFOR: Bogor Regency, Indonesia, 2010.
40. Sheil, D.; Puri, R.K.; Basuki, I.; van Heist, M.; Wan, M.; Liswanti, N.; Sardjono, M.A.; Samsoedin, I.; Sidiyasa, K.; Permana, E.; et al. *Exploring Biological Diversity, Environment, and Local People's Perspectives in Forest Landscapes: Methods for a Multidisciplinary Landscape Assessment*; CIFOR: Bogor Regency, Indonesia, 2002.
41. Villamor, G.B.; Palomo, I.; Santiago, C.A.L.; Oteros-Rozas, E.; Hill, J. Assessing stakeholders' perceptions and values towards social-ecological systems using participatory methods. *Ecol. Process.* **2014**, *3*, 22. [CrossRef]
42. Sunderland, T.; Abdoulaye, R.; Ahammad, R.; Asaha, S.; Baudron, F.; Deakin, E.; Duriaux, J.-Y.; Eddy, I.; Foli, S.; Gumbo, D.; et al. A methodological approach for assessing cross-site landscape change: Understanding socio-ecological systems. *For. Policy Econ.* **2017**, *84*, 83–91. [CrossRef]
43. Dawson, N.; Martin, A. Assessing the contribution of ecosystem services to human wellbeing: A disaggregated study in western Rwanda. *Ecol. Econ.* **2015**, *117*, 62–72. [CrossRef]
44. Hamann, M.; Biggs, R.; Reyers, B. Mapping social–ecological systems: Identifying 'green-loop'and 'red-loop'dynamics based on characteristic bundles of ecosystem service use. *Glob. Environ. Chang.* **2015**, *34*, 218–226. [CrossRef]
45. Ickowitz, A.; McMullin, S.; Rosenstock, T.; Dawson, I.; Rowland, D.; Powell, B.; Mausch, K.; Djoudi, H.; Sunderland, T.; Nurhasan, M.; et al. Transforming food systems with trees and forests. *Lancet Planet. Health* **2022**, *6*, e632–e639. [CrossRef] [PubMed]
46. Ehara, M.; Hyakumura, K.; Nomura, H.; Matsuura, T.; Sokh, H.; Leng, C. Identifying characteristics of households affected by deforestation in their fuelwood and non-timber forest product collections: Case study in Kampong Thom Province, Cambodia. *Land Use Policy* **2016**, *52*, 92–102. [CrossRef]
47. van Noordwijk, M.; Ekadinata, A.; Leimona, B.; Catacutan, D.; Martini, E.; Tata, H.L.; Öborn, I.; Hairiah, K.; Wangpakapattanawong, P.; Mulia, R.; et al. Agroforestry options for degraded landscapes in Southeast Asia. In *Agroforestry for Degraded Landscapes: Recent Advances and Emerging Challenges*; Springer Nature Singapore Pte Ltd.: Singapore, 2020; pp. 307–342. [CrossRef]
48. Villamagna, A.; Giesecke, C. Adapting human well-being frameworks for ecosystem service assessments across diverse landscapes. *Ecol. Soc.* **2014**, *19*, 11. [CrossRef]
49. Leong, K.M.; Wongbusarakum, S.; Ingram, R.J.; Mawyer, A.; Poe, M.R. Improving representation of human well-being and cultural importance in conceptualizing the West Hawai 'i Ecosystem. *Front. Mar. Sci.* **2019**, *6*, 231. [CrossRef]
50. Torralba, M.; Lovrić, M.; Roux, J.L.; Budniok, M.A.; Mulier, A.S.; Winkel, G.; Plieninger, T. Examining the relevance of cultural ecosystem services in forest management in Europe. *Ecol. Soc.* **2020**, *25*, 2. [CrossRef]
51. Sangha, K.K.; Le Brocque, A.; Costanza, R.; Cadet-James, Y. Ecosystems and indigenous well-being: An integrated framework. *Glob. Ecol. Conserv.* **2015**, *4*, 197–206. [CrossRef]
52. West, S.; Haider, L.J.; Stålhammar, S.; Woroniecki, S. A relational turn for sustainability science? Relational thinking, leverage points and transformations. *Ecosyst. People* **2020**, *16*, 304–325. [CrossRef]
53. Ahammad, R. Recent Trends in Forest and Livelihood Relationships of Rural Communities in the Chittagong Hill Tracts Region, Bangladesh. Ph.D. Thesis, Charles Darwin University, Darwin, Australia, 2019. [CrossRef]
54. Rasmussen, L.V.; Fagan, M.E.; Ickowitz, A.; Wood, S.L.; Kennedy, G.; Powell, B.; Baudron, F.; Gergel, S.; Jung, S.; Smithwick, E.A.; et al. Forest pattern, not just amount, influences dietary quality in five African countries. *Glob. Food Secur.* **2020**, *25*, 100331. [CrossRef]
55. Hausner, V.H.; Brown, G.; Lægreid, E. Effects of land tenure and protected areas on ecosystem services and land use preferences in Norway. *Land Use Policy* **2015**, *49*, 446–461. [CrossRef]
56. Kumar, A.; Ekka, P.; Patra, S.; Kumar, G.; Kishore, B.S.; Kumar, R.; Saikia, P. Geospatial Perspectives of Sustainable Forest Management to Enhance Ecosystem Services and Livelihood Security. In *Advances in Remote Sensing for Forest Monitoring*; John Wiley & Sons Ltd.: Hoboken, NJ, USA, 2022; pp. 10–42.
57. Leary, J.; Grimm, K.; Aslan, C.; Mark, M.; Frey, S.; Bath-Rosenfeld, R. Landowners' Socio-Cultural Valuation of Ecosystem Services Provided by Trees in Costa Rican Agricultural Landscapes. *Environ. Manag.* **2021**, *67*, 974–987. [CrossRef] [PubMed]

forests

MDPI

Article

Disturbance Severity and Human–Nature Relationships: A New Approach to Analyze People's Well-Being along a Bark Beetle Infestation Gradient

Mareike Kortmann [1,*], Per Angelstam [2,3], Marius Mayer [4], Franz Leibl [5], Jessica Reichert [6], Christine Thorn [7,8] and Simon Thorn [1,8]

1 Field Station Fabrikschleichach, Department of Animal Ecology and Tropical Biology (Zoology III), Julius-Maximilians-University Würzburg, Glashüttenstraße 5, 96181 Rauhenebrach, Germany
2 Faculty of Applied Ecology, Agricultural Sciences and Biotechnology, Inland Norway University of Applied Sciences, 2480 Evenstad, Norway
3 School for Forest Management, Faculty of Forest Sciences, Swedish University of Agricultural Sciences (SLU), 73921 Skinnskatteberg, Sweden
4 Faculty of Tourism, Munich University of Applied Sciences, Schachenmeierstraße 35, 80636 München, Germany
5 Bavarian Forest National Park, Freyunger Str. 2, 94481 Grafenau, Germany
6 Department of Animal Ecology and Systematics, Justus Liebig University Giessen, Heinrich-Buff-Ring 26-32 (IFZ), 35392 Giessen, Germany
7 Environmental Education Centre (UBiZ), Pfarrer-Baumann-Str. 17, 97514 Oberaurach, Germany
8 Biodiversity Center, Hessian Agency for Nature Conservation, Environment and Geology, Rheingaustraße 186, 65203 Wiesbaden, Germany
* Correspondence: mareike.kortmann@protonmail.com

Abstract: Contact to nature and greenspace is important for emotional well-being and can promote human health. Forest landscapes provide such access to greenspace, especially in protected areas. However, forested protected areas are impacted by natural disturbances such as bark beetle infestations. On the one hand, such disturbances have positive impacts on ecological processes and biodiversity. On the other hand, they have allegedly negative impacts on the recreational value of a landscape. Limited knowledge about the public's perception of forests subject to natural disturbances still hampers forest management to balance ecological functions and visitors' recreational experience. Thus, our aim was to determine how attitudes towards nature influence the personal well-being in a naturally disturbed landscape. We investigated self-reported well-being and attitudes towards nature in a standardized questionnaire-based survey of 1008 German inhabitants in an experimentally adapted landscape visualization. Self-reported well-being was generally highest in landscapes with relatively few bark-beetle-killed trees. This was especially the case for people who felt included with nature and preferred an appreciative use or preservation of nature. Conversely, people who had previously visited a national park with visible bark beetle infestations rated their personal well-being highest in landscapes with larger proportions of beetle-killed trees. Our results indicate that it is necessary to analyze people's knowledge about and relations to forest landscapes as well as concepts of nature conservation, natural landscapes, and biodiversity to gain a better understanding of people's perceptions of natural disturbances.

Keywords: bark beetle disturbance; major environmental values; well-being; inclusion of nature in one's self; national park

1. Introduction

Ongoing urbanization leads to an increasing loss of human contact with nature [1,2]. However, interactions between humans and nature in both rural and urban settings are important for human well-being and can promote health [3–5]. Hence, access to forests and other greenspaces for recreational purposes is of growing importance, especially during and in the aftermath of the COVID-19 pandemic [6–8]. Forests are a huge part of European greenspace, covering over 40% of the land area [9]. Human benefits of spending time in forests span from impacts on the physical health [10–12] to positive impacts on psychological health and mental well-being [10,13]. Human well-being is a multidimensional concept ranging from material well-being, bodily well-being, and security and freedom of choice to psychological well-being [14]. While it is mostly used in the context of human development on the level of countries [15], In our study, we focus on the impacts on well-being as used in Martens et al. (2011) [13], where well-being is the current mood or mental state of an individual [16]. In this context, well-being can be described by factors like positive affect, activation, negative affect, and arousal [13]. In contrast, other studies analyzed the economic well-being ("welfare") of people in areas characterized by large-scale forest disturbances using environmental economic valuation approaches [17].

Human–nature relationships are manyfold. Hence, the influences on our perception of nature and our well-being in nature are complex, covering traditions and ethics, values, attitudes, worldviews, and lifestyle [18]. Several studies tried to find common demands and preferences for forests, exploring that dependance on tree species, the level of naturalness, and other factors whereby forest visitors perceive forests differently. For example, Martens et al. (2011) [13] showed that managed forests have more positive effects on human well-being than unmanaged forests. Furthermore, people expect forests to be natural-looking, colorful, beautiful, inviting, diverse, and blend into the landscape [19]. However, most studies that analyzed the relation between green space and people's well-being have focused on urban forests. Still, rural forests and protected forest areas are also important types of greenspace. For example, national parks are specifically set aside to not only protect ecological processes, species, and ecosystem characteristics, but also to provide scientific opportunities and recreational space for the general public [20] (for Germany, see [21]). Furthermore, the studies we described above focus on well-being in forest landscapes, but there is still a lack of knowledge on well-being in forests affected by natural disturbances.

National parks in the boreal and temperate zone include large areas of coniferous forests that are naturally prone to large-scale, stand-replacing disturbances, such as wildfires, storms, and insect outbreaks [22,23]. In most national parks, at least in their core zones, natural disturbances are allowed to proceed without human intervention [24]. Examples include wildfires in the Yellowstone National Park, USA [22], and outbreaks of the spruce bark beetle (*Ips typographus* (L.)) in the Bavarian Forest, Germany, and Białowieża National Park, Poland [25,26]. The benign neglect of disturbances in protected area management, and thus maintenance or development of spatially diverse landscapes [27], frequently lead to an increase in the diversity of different taxonomic groups [26,28,29]. For instance, higher amounts of deadwood due to unmanaged bark beetle infestations lead to a diversity of invertebrates which includes the survival of many rare and endangered species or even the rediscovering of species specialized on deadwood thought to be already extinct [30,31]. Insects are crucial parts of ecosystems as important elements of the food chain and generate decisive ecosystem services, for instance, due to their pollination effects [32]. Furthermore, invertebrates are key species for monitoring overall biodiversity decline, as exemplified by the study of Hallmann et al. (2017) [33], who reported that within 27 years, the biomass of insects in Germany has decreased by about three quarters.

However, to the best of the authors' knowledge, there are not any studies yet available which analyzed to which degree these alarming results regarding the decline of the amount and diversity of insects in Germany are recognized by the general public. Nevertheless, people obviously care about more emblematic and familiar insect species like bees, as underlined by the high level of consent to the public initiative "Biodiversity & Natural Beauty in Bavaria" („Artenvielfalt & Naturschönheit in Bayern" in German) in Bavaria in 2019, which is more well-known as "Save the bees" ("Rettet die Bienen" in German) [34]. It remains unclear, however, whether lay people link forest disturbances to such valued insect species such as bees [35]. The public outreach during the public initiative was dominated by necessary adaptations of agricultural practices or the preservation of meadow orchards while forest management practices or the (non)management of forest disturbances did not play an important role.

In contrast to these positive effects of natural forest disturbances on forest biodiversity, Flint et al. (2009) [36] reported that naturally disturbed forests are perceived as catastrophic by local residents and national park visitors, threatening safety, property, and well-being. They have also led to a deterioration of park–people relationships in the past [37–39]. Yet, it remains unclear if bark beetle infestations can diminish the positive impacts of forests on human well-being.

A balanced management strategy of protected areas should consider both the promotion of ecological processes and the preservation of the positive impacts on human well-being. Therefore, it is necessary to obtain detailed information on the extent to which trees affected by bark beetles are tolerated by the general public without diminishing their well-being derived from park visits. Recent studies on people's perceptions of natural disturbances were limited to surveys of either national park visitors or local populations adjacent to protected areas [40–44]. Still, people can also have contact to bark beetle infestations outside of protected areas, for example, in commercial forests in their neighborhood. In addition, the general public, including also nonvisitors of forests and protected areas, can have an impact on nature conservation and forest policies based on their electoral behavior (see for example [45] for Bavaria). Thus, a comprehensive assessment, covering both visitors and nonvisitors of national parks and other areas subject to large-scale bark beetle disturbances is crucial to achieve a holistic understanding of the factors contributing to self-reported well-being in landscapes affected by bark beetles.

Based on previous studies [39,45–47], it can be assumed that people generally derive their highest well-being from more green and vital forests with small amounts of dead trees. Nevertheless, personal characteristics including knowledge and previous experiences can further affect the preference for certain landscapes. For example, a strong connection with nature can promote proenvironmental behavior and environmental concern [48] (and references within). Furthermore, it also affects the perception of natural environments. Clayton (2021) [49], for example, showed that different relationships with nature can also affect the emotional well-being in certain landscapes. Being in nature can increase your commitment to the landscape and can lead to a less superficial evaluation of the respective landscape [50]. These experiences might also positively affect people's judgement of a landscape in the future. Furthermore, people who have been to a national park before might be more aware of the positive effects of natural disturbances on biodiversity, for instance, due to educational programs in the national parks on natural disturbances. Therefore, it is crucial to understand additional factors that might affect personal well-being in landscapes with different bark beetle infestation severities.

The aim of this study is to analyze the association between different infestation severities and subjectively perceived well-being based on a photographically simulated spruce forest covering an entire disturbance severity gradient. To assess the impact of bark beetle disturbance severity (i.e., the share of bark-beetle-killed trees in a landscape) on people's subjectively perceived well-being, we conducted a standardized questionnaire-based survey of 1008 German inhabitants representing an adequate sample of the German population. To control for different attitudes towards and relations with nature, we tested also inclusion of nature in one´s self (INS), major environmental values (2-MEV), and the information on prior visits to national parks characterized by significant bark beetle infestations.

We hypothesize that, when confronted with natural disturbance, the connection of people with nature will also affect their sense of well-being within this landscape. More specifically, with increasing inclusion of nature in self and more protective attitudes towards nature, people will choose more vital forests (i.e., forests with less dead trees) to comply with their personal well-being. Furthermore, we hypothesize that people who have visited a national park before and were directly confronted with disturbed forest landscapes will choose landscapes with higher amounts of dead trees for their personal well-being.

2. Materials and Methods

We conducted an online survey hosted by Norstat Deutschland GmbH, which fulfills the ICC/ ESOMAR International Code on Market, Opinion, and Social Research and Data Analytics, to achieve a representative sample of the German population. Our questionnaire assessed three groups of variables (sociodemographic parameters, proxies for people's attitudes towards nature, and preferences for different bark-beetle-affected landscapes). Sociodemographic parameters included data on age, gender, and educational level (1: no educational degree, 2: elementary school, 3: lower secondary school, 4: higher secondary school, 5: university degree). Proxies for people's attitudes towards nature included first the respondents' sense of inclusion of nature in one´s self (INS) [51], which was determined as an indicator of a self-estimated connection with nature. The INS scale is easily comprehensible and hence a suitable tool for a broad field of respondents.

In addition, we used the bidimensional scale 2-MEV (major environmental values) of Kibbe et al. (2014) [52] to assess whether or not respondents prefer an exploitative or an appreciative use of nature. In context of an exploitative use of nature, utilization represents an anthropocentric dimension including the domination and exploitation of the environment and its natural resources. An appreciative use of nature is assessed in terms of preservation, a biocentric dimension defined by a preference for environmental protection. The 2-MEV scale is a widely used and validated tool to measure environmental worldviews [53]. Whereas the New Ecological Paradigm (NEP) scale is one-dimensional and assumes that people can either be biocentric or anthropocentric, but not both, the 2-MEV scale is two-dimensional and allows respondents to have high scores for preservation and utilization at the same time [53]. For example, a person with a huge interest to protect nature can also believe that one of the main purposes of nature is to benefit humans and would hence have high scores for preservation and utilization at the same time.

Finally, we included a question on previous national park visits. We asked the respondents whether they had visited any of Germany's 16 national parks at any time in their life before, and if so, which ones. Bark-beetle-affected national parks were defined as forested national parks (excluding marine and shore parks) with significant portions of Norway spruce (*Picea abies* (L.) H. Karst) that had experienced clearly visible bark beetle infestations in the past 30 years. This is approximately the time span in which bark beetle infestations are clearly visible in a landscape. Due to natural succession, visible signs of the infestations disappear in time. Additionally, we screened the most comprehensive German online press database (www.genios.de, accessed on 25 January 2018) using the simplified search string 'National Park name + bark beetle' to quantify media coverage of bark beetle outbreaks in the respective national parks in the last 30 years. We then used this information to validate our national park selection by only choosing parks with any media coverage on bark beetle infestations (Table A1). This resulted in seven national parks where visitors could have had previous encounters with natural disturbances.

A photomontage of a forest landscape affected by the European spruce bark beetle (*I. typographus*) was used to assess the respondents' self-reported well-being. For this purpose, we used image-editing software to create an image of an artificial landscape based on a photograph of the Bavarian Forest National Park. The landscape depicted in the image ranged from entirely beetle-killed to an entirely intact and unfragmented forest canopy. On the bottom of the picture, we added a movable slider (Figure 1). We then asked the respondents the following: "Imagine you are hiking through a forested national park. The hiking trail crosses areas with different proportions of living and dead trees (0–100%). Please move the slider to the position at which your personal well-being would be the highest." This approach allowed an uncomplicated landscape-scale assessment of the proportion of beetle-killed trees associated with a personal maximum of self-reported well-being. The personal, subjectively perceived well-being is not further measured but used as an indication based on which people were asked to choose a certain proportion of dead and living trees. The possible disturbance severities ranged from zero to 100 including all integers. A German and English version of the questionnaire can be found in Supplementary File S1.

Figure 1. An artificial landscape depicting an outbreak of the European spruce bark beetle (*Ips typographus*). The landscape ranged from entirely killed (left side) to entirely intact (right side) forests. A hiking trail was inserted for visual purpose.

All data analyses were performed in R version 3.3.3 (www.r-project.org, accessed on 1 September 2022). First, multicollinearity among the predictor variables was tested through a variance inflation factor (VIF), in which values > 10 indicated highly related variables [54]. The maximum VIF detected in our variables was Preservation = 2.25. Consequently, no predictor variable was removed from the subsequent statistical analysis.

The reliability of items representing an exploitative versus an appreciative use of nature [52] was validated by means of Cronbach's α, provided by the function 'alpha' from the r-package 'psych' [55]. Cronbach's α is used to analyze the degree to which a set of items measures a single unidimensional latent construct (e.g., versus appreciative use of nature); α values < 0.65 indicate an unreliable measurement [56]. The α value of the utilization of nature was 0.77, and that of nature preservation was 0.73. The average item scores served as proxies for preservation and utilization, respectively.

Single generalized linear models were combined into one joint model using piecewise structural equation modeling (SEM). We tested the effects of age, gender and educational level, inclusion with nature, the utilization vs. preservation of nature, and previous visits to national parks (affected by bark beetle infestations) on the choice behavior of the participants (Figure 2). Different individual models were included for every response variable. Influences on the proportion of bark-beetle-killed trees associated with the highest perceived well-being were tested with a linear model in which age, gender, educational level, prior visits to national parks with bark beetle infestations, inclusion with nature, preservation, and utilization as predictors. Binomial linear models were employed to test the influence of socio-demographic parameters and attitudes on prior visits to national parks.

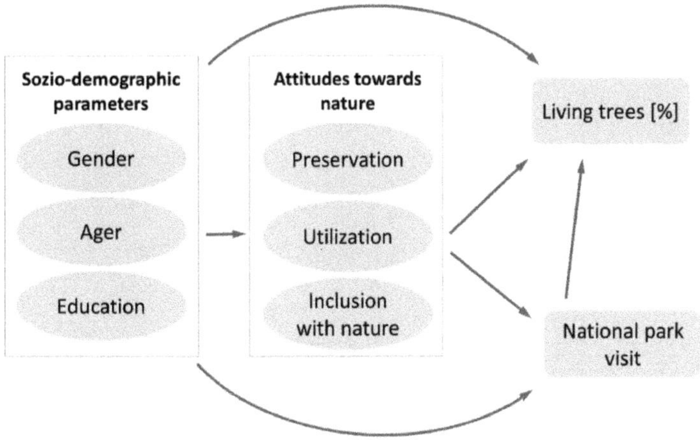

Figure 2. Variables used in the Structural Equation Modeling and the tested relations between the different variables.

Finally, different linear models were used to test the influence of age, gender, and educational level on inclusion with nature, preservation, and utilization. To account for the β-distribution of our data in the standard unit interval (0.1), the proportion of beetle-killed trees was additionally modeled via a maximum likelihood approach by means of a β-regression for proportions [57]. β-regression was used in this study to validate the results of the SEM. Therefore, the function 'betareg' was applied, with self-reported well-being as the response variable and age, gender, educational level, prior visits to national parks, inclusion with nature, preservation, and utilization as predictors. The results obtained via β-regression were very similar to those obtained with the SEM and are reported in the Appendix A (Table A2).

3. Results

The 1008 respondents who completed our questionnaire included 527 women and 481 men from all 16 German federal states. Their mean age was 44.7 (± 14.3 standard deviation) years, ranging from 18 to 70 years. According to The Federal Institute for Population Research (2019), the mean age of the German population was 44.5 in 2019. Within our respondents, 40% had a university degree (n = 287). This is above the German average of 31.3% for 25–64-year-old people [58]. An overview of the age distribution and educational levels is given in Figure 3.

A total of 41.6% (n = 416) of the respondents stated that they had visited at least one German national park with visible bark beetle infestations before. These national parks included the Bavarian Forest, Harz, Black Forest, Hunsrück-Hochwald, Saxon Switzerland, Eifel, and Berchtesgaden. On average, male respondents had visited more national parks (1.2 national parks) than female respondents (0.8 national parks) before.

The median and mean percentages of beetle-killed trees in the chosen landscape was 26% vs. 30 %, respectively. A preference for < 20 % dead trees was expressed by 41 % of the respondents (n = 409) and for < 50 % dead trees by 77 % of the respondents (n = 776) (Figure 4).

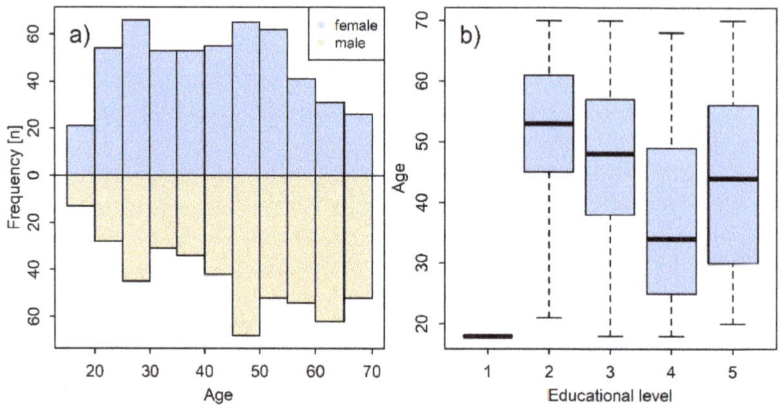

Figure 3. (**a**) Frequency of respondents age for female and male respondents. (**b**) Boxplot of respondents age per educational level (1: no educational degree, 2: elementary school, 3: lower secondary school, 4: higher secondary school, 5: university degree).

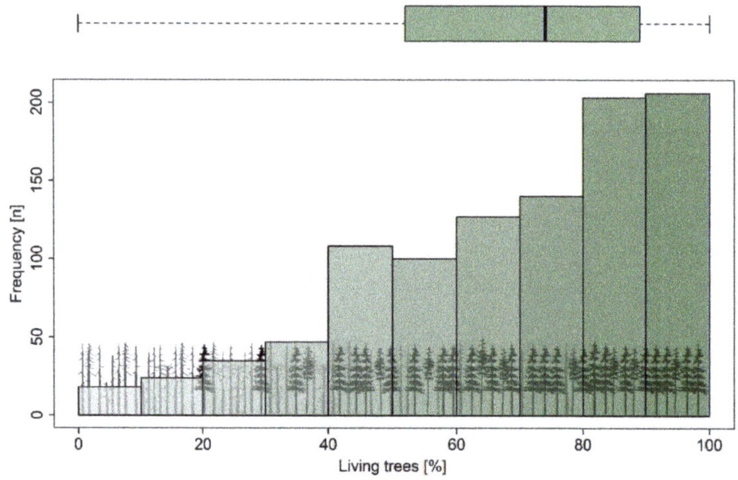

Figure 4. Frequency of respondents (n = 1008) and the percentage of living trees, where respondents rated their personal well-being highest. The boxplot shows the median self-reported highest well-being (median = 0.74).

The SEM showed a significant influence of age and gender on the psychological parameters (inclusion with nature, utilization, preservation) (Figure 5). Male respondents felt less included with nature and preferred its more exploitative use compared to women. Older respondents felt more included with nature (z = 7.41) and preferred its appreciative use (z = 9.01). Overall, respondents who felt more included with nature (z = 3.15) and placed higher value on preservation (z = 2.47) preferred landscapes with a larger proportion of living trees. Respondents with higher values for inclusion with nature and preservation were also more likely to have previously visited a national park with bark beetle infestations (z = 3.11, z = 4.65, respectively). Both male (z = 5.00) and older respondents (z = 2.74) as well as people with a higher educational level (z = 4.90) more often report previous national park visits. People with prior visits of national parks with bark beetle infestations preferred landscapes with slightly higher proportions of bark-beetle-killed trees (z = −2.30). Still, this effect was rather weak.

Table 1. Results of the structural equation model testing the influence of age, gender, education level, preservation, utilization, and inclusion with nature on prior visits and subjectively perceived well-being (n = 1008) using generalized linear models. The influence of gender, age, and education level on utilization, preservation, and inclusion with nature was tested using linear models. Boldface indicates statistical significance (significance level at 5%).

Response	Predictor	Estimate	Std. Error	z-Value
Prior visit	Gender	7.05×10^{-1}	1.41×10^{-1}	5.00
	Education	3.25×10^{-1}	6.64×10^{-2}	4.90
	Preservation	8.99×10^{-1}	1.93×10^{-1}	4.65
	Inclusion with nature	4.99×10^{-1}	1.60×10^{-1}	3.11
	Age	1.48×10^{-1}	5.39×10^{-2}	2.74
	Utilization	-1.41×10^{-2}	5.18×10^{-3}	−2.72
Perceived well-being	Inclusion with nature	7.55×10^{-2}	2.39×10^{-2}	3.15
	Preservation	2.17×10^{-1}	8.76×10^{-2}	2.47
	Prior visit	-1.51×10^{-1}	6.57×10^{-2}	−2.30
	Utilization	-5.91×10^{-2}	7.35×10^{-2}	-8.04×10^{-1}
	Gender	2.11×10^{-2}	3.07×10^{-2}	6.89×10^{-1}
	Education	4.36×10^{-2}	6.49×10^{-2}	6.72×10^{-1}
	Age	-9.37×10^{-4}	2.36×10^{-3}	-3.97×10^{-1}
Utilization	Age	-8.52×10^{-3}	1.35×10^{-3}	−6.32
	Gender	1.96×10^{-1}	3.78×10^{-2}	5.17
	Education	1.92×10^{-3}	1.81×10^{-2}	1.06×10^{-1}
Preservation	Age	1.06×10^{-2}	1.17×10^{-3}	9.01
	Gender	-1.73×10^{-1}	3.30×10^{-2}	−5.25
	Education	3.00×10^{-2}	1.58×10^{-2}	1.90
Inclusion with nature	Age	2.41×10^{-2}	3.25×10^{-3}	7.41
	Gender	-3.12×10^{-1}	9.13×10^{-2}	−3.41
	Education	6.58×10^{-2}	4.38×10^{-2}	1.50

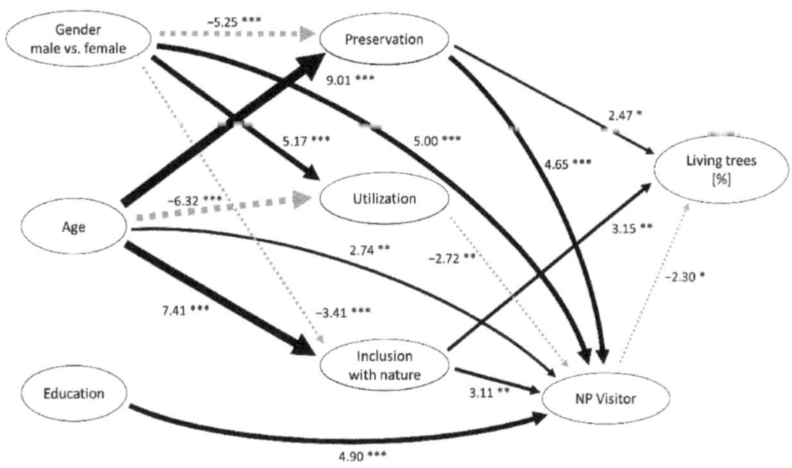

Figure 5. Structural equation model, with sociodemographic parameters on the left, preservation, utilization, and inclusion with nature in the center and previous visits of bark-beetle-affected national parks (yes/no) and the proportion of living trees on the right. Arrows are labeled with the corresponding z-values and significance levels (*** $p < 0.001$, ** $0.001 < p < 0.01$, * $0.01 < p < 0.05$). Arrow width corresponds to the z-values of the model (see Table 1 for coefficients). Black solid lines are for positive and grey dotted lines for negative effects. The figure shows only significant results.

4. Discussion

Overall, people chose landscapes with high percentages of living trees (70%) for their highest personal well-being. Our results revealed that respondents who felt most connected with nature and respondents who prioritized the preservation of nature chose landscapes with even higher portions of living trees. By contrast, respondents who had previously visited a national park with bark beetle infestations chose landscapes with larger proportions of bark-beetle-killed trees.

4.1. Environmental Perception

Female respondents in our study had higher values for preservation and inclusion with nature than male respondents, which confirms earlier findings [59,60]. While not a universal phenomenon, stronger positive environmental attitudes are generally reported more often by women than by men [61,62].

We also found a significant increase in preservation values with increasing age. Yet, there is little consensus on the effect of age on preservation values. Reported effects range from negative [63,64], to neutral [65], to positive [61]. Gifford and Nilsson (2014) [61] concluded that it is difficult to distinguish between cohort effects, i.e., impacts that more often occur in one age group than in another, era effects, which are driven by social–political changes in time, and true age effects. This might explain the ambiguous results of different studies.

4.2. Personal Well-Being Is Higher in Landscapes with Larger Proportions of Living Trees

The large majority of respondents in our study rated their personal well-being highest in landscapes with large proportions of living trees (Figure 4). This result is in line with other studies that found a preference for healthy, orderly forests [46,66] and a negative attitude towards dead wood or dying vegetation in the landscape [47,67]. As reported in previous studies, postdisturbance landscapes can be perceived as dead and desert-like and a risk for the forest ecosystem [39,43].

4.3. Impacts of Attitudes towards Nature

Our study shows that people higher values of preservation and inclusion with nature chose forests with significantly larger proportions of living trees for their personal well-being. This finding is in contrast with McFarlane et al. (2006) [40] and Müller and Job (2009) [42], where people with more proecological worldviews were less opposed to forests with bark beetle infestations and showed more positive attitudes towards bark beetles. These differences might derive from different concepts of nature conservation. Most European countries have a long history of cultural landscape management [68]. In this cultural context, persistence can be an important factor of the landscape [69]. Hence, nature conservation can be strongly linked to the preservation of the status quo. Contrary, in naturally dynamic landscapes, which are perceived more as wilderness without human interventions and management, dynamic processes might not contradict the idea of nature conservation. The difference between conservation and preservation is often discussed in the context of the measurements of human–nature relationships, which are adapted by different scientists to include new attitudes in the field of nature conservation and combine ecocentric and anthropocentric views that can change over time [70,71].

In natural forest landscapes without silvicultural use, such as national parks, bark beetle infestations and other natural disturbances might be more tolerated by the public, especially if learning about the roles of disturbances for conservation. In our data, which focuses on German citizenry with a long history of forest management, the cultural aspect as well as the preservation of the status quo might be the dominating view. Hence, people with high preservation values can perceive bark beetle infestations and dead trees as negative and incompatible with nature conservation.

4.4. National Park Visits Promote Well-Being in Disturbed Landscapes

Respondents who had previously visited a national park with bark beetle infestations located their highest personal well-being in landscape sections with larger proportion of beetle-killed trees. Hence, people presumably are able to change their perceptions of naturally disturbed forests following a national park visit. Asking national park visitors in different European national parks, Kortmann et al. (2021) [72] showed that the perceived restorativeness in disturbed landscapes was surprisingly high. In addition, after severe natural disturbances in the Bavarian Forest National Park, visitor numbers remained high in the following years [73]. Hence, visitors might learn that dead wood in a forest landscape does not reduce their recreational experience. Later on, they might even link their well-being to dead wood in the landscapes after positive experiences during a national park visit.

4.5. Methods

We used a new method to explore a well-being gradient along an ecological gradient created by natural disturbances. The method was easily applicable and allowed us to detect different choice behavior along the disturbance gradient. Due to the simpler character of our method, we were not able to determine where precisely these differences come from. Furthermore, we did not use a scale to measure peoples well-being but used it as a guiding principle for our respondents to choose a certain landscape section and disturbance severity. Hence, we cannot verify the validity of our approach. To transfer our new insights into environmental education and forest management, it would be necessary to gain additional information via further interviews and additional questions with this survey. Important additional questions would be for example what people associate with nature conservation and disturbed landscapes and how people imagine natural forest landscapes. Hence, further studies should include in-depth interviews of people with different relations to forest landscapes to explain, where the choices for certain disturbance severities stem from. These interviews might also focus on impacts of national park visits on visitors. Why do national park visitors perceive natural disturbances different after a visit than before and why does the well-being in disturbed landscapes change? Based on our results we suspect that all these additional factors and drivers are connected and need further observations.

5. Conclusions

We demonstrated that previous national park visits, attitudes towards nature, and inclusion of nature in one's self affected the subjectively perceived well-being of respondents in a naturally disturbed landscape. National park visitors seem to have generally more positive perceptions of disturbed forest landscapes. Despite the positive effects of bark beetle infestations on ecological processes and biodiversity, people do not necessarily perceive natural disturbances and nature conservation as compatible. To be able to understand the different perceptions of natural disturbances, it is necessary to analyze people's relations to forest landscapes as well as concepts of nature conservation, natural landscapes, and biodiversity. Since natural disturbances will likely increase in the future and affect more and more forest landscapes, it is necessary to understand the complex human–nature relationships in the context of forest landscapes and their management. Our research shows important aspects that need to be further analyzed and gives indications for future research projects.

Supplementary Materials: The following supporting information can be downloaded at: https://www.mdpi.com/article/10.3390/f13111954/s1.

Author Contributions: M.K., S.T., and C.T. contributed to the study conception and design. Material preparation, data collection, and analysis were performed by M.K. and S.T. Writing—review & editing: P.A., M.M., F.L., J.R. The first draft of the manuscript was written by M.K. and all authors commented on previous versions of the manuscript. All authors have read and agreed to the published version of the manuscript.

Funding: This work was supported by the scholarship program of the German Federal Environmental Foundation (20017/474).

Data Availability Statement: All data files are available from the figshare database (accession number 11662854, URL https://figshare.com/articles/questionnaires_18_xlsx/11662854, accessed on 18 November 2022).

Conflicts of Interest: The authors declare no conflict of interest. The funders had no role in the design of the study; in the collection, analyses, or interpretation of data; in the writing of the manuscript; or in the decision to publish the results.

Appendix A

Table A1. German forest-dominated national parks with bark beetle disturbances, the number of respondents who reported visiting those parks (n = 927), and the number of press releases mentioning bark beetle infestations (n = 2878).

National Park	News Releases (n)	Visitors (n)
Bavarian Forest	1649	225
Harz	398	163
Black Forest	400	157
Hunsrück-Hochwald	260	51
Saxon Switzerland	167	105
Eifel	180	127
Berchtesgaden	180	99

Table A2. Beta-regression of bark-beetle-affected forest portion, with gender, age, and education level of respondents, inclusion with nature, utilization, preservation, and prior visits as predictor variables (n = 1008). Because a beta-regression is applicable only when the dependent values are between 0 and 1, the standardization $\frac{y*(n_{obs}-1)+0.5}{n_{obs}}$ suggested by Smithson and Verkuilen [74] was applied. Boldface indicates significance.

Variable	Estimate	Std. Error	z-Value	p-Value
Intercept	-4.71×10^{-1}	5.14×10^{-1}	-9.15×10^{-1}	0.360
Gender	5.55×10^{-2}	7.17×10^{-2}	7.73×10^{-1}	0.440
Age	-5.15×10^{-3}	2.61×10^{-3}	−1.98	0.048
Education	-1.65×10^{-2}	3.40×10^{-2}	-4.85×10^{-1}	0.630
Inclusion with nature	1.03×10^{-1}	2.66×10^{-2}	3.85	<0.001
Utilization	1.61×10^{-3}	8.11×10^{-1}	2.00×10^{-2}	0.984
Preservation	2.90×10^{-1}	9.67×10^{-2}	3.00	0.003
Visits	-2.12×10^{-2}	2.51×10^{-2}	-8.44×10^{-1}	0.400

References

1. Turner, W.R.; Nakamura, T.; Dinetti, M. Global Urbanization and the Separation of Humans from Nature. *BioScience* **2004**, *54*, 585–590. [CrossRef]
2. Miller, J.R. Biodiversity conservation and the extinction of experience. *Trends Ecol. Evol.* **2005**, *20*, 430–434. [CrossRef] [PubMed]
3. Dallimer, M.; Irvine, K.N.; Skinner, A.M.J.; Davies, Z.G.; Rouquette, J.R.; Maltby, L.; Warren, P.H.; Armsworth, P.; Gaston, K.J. Biodiversity and the Feel-Good Factor: Understanding Associations between Self-Reported Human Well-being and Species Richness. *BioScience* **2012**, *62*, 47–55. [CrossRef]
4. Summers, J.K.; Smith, L.M.; Case, J.L.; Linthurst, R.A. A Review of the Elements of Human Well-Being with an Emphasis on the Contribution of Ecosystem Services. *AMBIO* **2012**, *41*, 327–340. [CrossRef] [PubMed]
5. Abraham, A.; Sommerhalder, K.; Abel, T. Landscape and well-being: A scoping study on the health-promoting impact of outdoor environments. *Int. J. Public Health* **2010**, *55*, 59–69. [CrossRef]
6. Gugerell, K.; Netsch, S. Reflection on the Austrian Newspaper Coverage of the Role and Relevance of Urban Open-and Green-Spaces in Vienna during the First COVID-19 Lockdown in 2020. *disP* **2020**, *56*, 54–63. [CrossRef]
7. Bustamante, G.; Guzman, V.; Kobayashi, L.C.; Finlay, J. Mental health and well-being in times of COVID-19: A mixed-methods study of the role of neighborhood parks, outdoor spaces, and nature among US older adults. *Health Place* **2022**, *76*, 102813. [CrossRef]

8. Jackson, S.; Stevenson, K.; Larson, L.; Peterson, M.; Seekamp, E. Outdoor Activity Participation Improves Adolescents' Mental Health and Well-Being during the COVID-19 Pandemic. *Int. J. Environ. Res. Public Health* **2021**, *18*, 2506. [CrossRef]
9. Palmero-Iniesta, M.; Pino, J.; Pesquer, L.; Espelta, J.M. Recent forest area increase in Europe: Expanding and regenerating forests differ in their regional patterns, drivers and productivity trends. *Eur. J. For. Res.* **2021**, *140*, 793–805. [CrossRef]
10. Zhang, R.; Zhang, C.-Q.; Rhodes, R.E. The pathways linking objectively-measured greenspace exposure and mental health: A systematic review of observational studies. *Environ. Res.* **2021**, *198*, 111233. [CrossRef]
11. Oh, B.; Lee, K.J.; Zaslawski, C.; Yeung, A.; Rosenthal, D.; Larkey, L.; Back, M. Health and well-being benefits of spending time in forests: Systematic review. *Environ. Health Prev. Med.* **2017**, *22*, 71. [CrossRef]
12. Sia, A.; Tam, W.W.S.; Fogel, A.; Kua, E.H.; Khoo, K.; Ho, R.C.M. Nature-based activities improve the well-being of older adults. *Sci. Rep.* **2020**, *10*, 18178. [CrossRef]
13. Martens, D.; Gutscher, H.; Bauer, N. Walking in 'wild' and 'tended' urban forests: The impact on psychological well-being. *J. Environ. Psychol.* **2011**, *31*, 36–44. [CrossRef]
14. Alkire, S. Dimensions of Human Development. *World Dev.* **2002**, *30*, 181–205. [CrossRef]
15. McGillivray, M. *Human Well-Being Concept and Measurement*; United Nations University: Tokyo, Japan, 2007.
16. Abele-Brehm, A.; Brehm, W. Zur Konzeptualisierung und Messung von Befindlichkeit. Die Entwicklung der 'Befindlichkeitsskalen' (BFS). *Diagnostica* **1986**, *32*, 209–228.
17. Holmes, T.; Koch, F. Bark Beetle epidemics, life satisfaction, and economic well-being. *Forests* **2019**, *10*, 696. [CrossRef]
18. Flint, C.; Kunze, I.; Muhar, A.; Yoshida, Y.; Penker, M. Exploring empirical typologies of human–nature relationships and linkages to the ecosystem services concept. *Landsc. Urban Plan.* **2013**, *120*, 208–217. [CrossRef]
19. Lee, T. *Perceptions, Attitudes and Preferences in Forests and Woodlands*; Forestry Commission: Edinburgh, UK, 2001.
20. Watson, J.E.M.; Dudley, N.; Segan, D.B.; Hockings, M. The performance and potential of protected areas. *Nature* **2014**, *515*, 67–73. [CrossRef]
21. Mayer, M.; Woltering, M. Assessing and valuing the recreational ecosystem services of Germany's national parks using travel cost models. *Ecosyst. Serv.* **2018**, *31*, 371–386. [CrossRef]
22. Turner, M.G.; Baker, W.L.; Peterson, C.J.; Peet, R.K. Factors influencing succession: Lessons from large, infrequent natural disturbances. *Ecosystems* **1998**, *1*, 511–523. [CrossRef]
23. Kulakowski, D.; Seidl, R.; Holeksa, J.; Kuuluvainen, T.; Nagel, T.A.; Panayotov, M.; Svoboda, M.; Thorn, S.; Vacchiano, G.; Whitlock, C.; et al. A walk on the wild side: Disturbance dynamics and the conservation and management of European mountain forest ecosystems. *For. Ecol. Manag.* **2017**, *388*, 120–131. [CrossRef] [PubMed]
24. Müller, J.; Noss, R.F.; Thorn, S.; Bässler, C.; Leverkus, A.B.; Lindenmayer, D. Increasing disturbance demands new policies to conserve intact forest. *Conserv. Lett.* **2019**, *12*, e12449. [CrossRef]
25. Mikusiński, G.; Bubnicki, J.W.; Churski, M.; Czeszczewik, D.; Walankiewicz, W.; Kuijper, D.P.J. Is the impact of loggings in the last primeval lowland forest in Europe underestimated? The conservation issues of Białowieża Forest. *Biol. Conserv.* **2018**, *227*, 266–274. [CrossRef]
26. Thorn, S.; Bässler, C.; Svoboda, M.; Müller, J. Effects of natural disturbances and salvage logging on biodiversity—Lessons from the Bohemian Forest. *For. Ecol. Manag.* **2017**, *388*, 113–119. [CrossRef]
27. Senf, C.; Seidl, R. Natural disturbances are spatially diverse but temporally synchronized across temperate forest landscapes in Europe. *Glob. Chang. Biol.* **2017**, *24*, 1201–1211. [CrossRef]
28. Müller, J.; Noss, R.F.; Bussler, H.; Brandl, R. Learning from a "benign neglect strategy" in a national park. Response of saproxylic beetles to dead wood accumulation. *Biol. Conserv.* **2010**, *143*, 2559–2569. [CrossRef]
29. Beudert, B.; Bässler, C.; Thorn, S.; Noss, R.F.; Schroder, B.; Dieffenbach-Fries, H.; Foullois, N.; Müller, J. Bark Beetles Increase Biodiversity While Maintaining Drinking Water Quality. *Conserv. Lett.* **2014**, *8*, 272–281. [CrossRef]
30. Bässler, C.; Müller, J. Importance of natural disturbance for recovery of the rare polypore Antrodiella citrinella Niemelä & Ryvarden. *Fungal Biol.* **2010**, *114*, 129–133. [CrossRef]
31. Busse, A.; Cizek, L.; Čížková, P.; Drag, L.; Dvorak, V.; Foit, J.; Müller, J. Forest dieback in a protected area triggers the return of the primeval forest specialist Peltis grossa (Coleoptera, Trogossitidae). *Conserv. Sci. Pract.* **2022**, *4*, e612. [CrossRef]
32. Vanbergen, A.J.; the Insect Pollinators Initiative. Threats to an ecosystem service: Pressures on pollinators. *Front. Ecol. Environ.* **2013**, *11*, 251–259. [CrossRef]
33. Hallmann, C.A.; Sorg, M.; Jongejans, E.; Siepel, H.; Hofland, N.; Schwan, H.; Stenmans, W.; Müller, A.; Sumser, H.; Hörren, T.; et al. More than 75 percent decline over 27 years in total flying insect biomass in protected areas. *PLoS ONE* **2017**, *12*, e0185809. [CrossRef] [PubMed]
34. Krack, K.; Oberholzer, G. » Rettet die Bienen « – Das Ergebnis des Bayerischen Volksbegehrens, zfv—Zeitschrift für Geodäsie. *Geoinf. Und Landmanagement* **2020**, *145*, 380–384.
35. Thorn, S.; Seibold, S.; Leverkus, A.B.; Michler, T.; Müller, J.; Noss, R.F.; Stork, N.; Vogel, S.; Lindenmayer, D.B. The living dead: Acknowledging life after tree death to stop forest degradation. *Front. Ecol. Environ.* **2020**, *18*, 505–512. [CrossRef]
36. Flint, C.G.; McFarlane, B.; Müller, M. Human dimensions of forest disturbance by insects: An international synthesis. *Environ. Manage.* **2009**, *43*, 1174–1186. [CrossRef] [PubMed]
37. Von Ruschkowski, E.; Mayer, M. From Conflict to Partnership? Interactions between Protected Areas, Local Communities and Operators of Tourism Enterprises in Two German National Park Regions. *J. Tour. Leis. Stud.* **2011**, *17*, 147–181.

38. Job, H.; Bittlingmaier, S.; Mayer, M.; von Ruschkowski, E.; Woltering, M. Park–People Relationships: The Socioeconomic Monitoring of National Parks in Bavaria, Germany. *Sustainability* **2021**, *13*, 8984. [CrossRef]
39. Müller, M. How natural disturbance triggers political conflict: Bark beetles and the meaning of landscape in the Bavarian. *Forest. Glob. Environ. Chang.* **2011**, *21*, 935–946. [CrossRef]
40. McFarlane, B.L.; Stumpf-Allen, R.C.G.; Watson, D.O. Public perceptions of natural disturbance in Canada's national parks: The case of the mountain pine beetle (Dendroctonus ponderosae Hopkins). *Biol. Conserv.* **2006**, *130*, 340–348. [CrossRef]
41. McFarlane, B.L.; Parkins, J.; Watson, D.O. Risk, knowledge, and trust in managing forest insect disturbance. *Can. J. For. Res.* **2012**, *42*, 710–719. [CrossRef]
42. Müller, M.; Job, H. Managing natural disturbance in protected areas: Tourists' attitude towards the bark beetle in a German national park. *Biol. Conserv.* **2009**, *142*, 375–383. [CrossRef]
43. McFarlane, B.L.; Witson, D.O.T. Perceptions of Ecological Risk Associated with Mountain Pine Beetle (Dendroctonus ponderosae) Infestations in Banff and Kootenay National Parks of Canada. *Risk Anal.* **2008**, *28*, 203–212. [CrossRef] [PubMed]
44. Parkins, J.R.; MacKendrick, N.A. Assessing community vulnerability: A study of the mountain pine beetle outbreak in British Columbia, Canada. *Glob. Environ. Chang.* **2007**, *17*, 460–471. [CrossRef]
45. Sacher, P.; Meyerhoff, J.; Mayer, M. Evidence of the association between deadwood and forest recreational site choices. *For. Policy Econ.* **2021**, *135*, 102638. [CrossRef]
46. Arnberger, A.; Ebenberger, M.; Schneider, I.E.; Cottrell, S.; Schlueter, A.C.; von Ruschkowski, E.; Venette, R.C.; Snyder, S.A.; Gobster, P.H. Visitor Preferences for Visual Changes in Bark Beetle-Impacted Forest Recreation Settings in the United States and Germany. *Environ. Manag.* **2017**, *61*, 209–223. [CrossRef] [PubMed]
47. Sheppard, S.; Picard, P. Visual-quality impacts of forest pest activity at the landscape level: A synthesis of published knowledge and research needs. *Landsc. Urban Plan.* **2006**, *77*, 321–342. [CrossRef]
48. Nisbet, E.K.; Zelenski, J.M.; Murphy, S.A. Happiness is in our Nature: Exploring Nature Relatedness as a Contributor to Subjective Well-Being. *J. Happiness Stud.* **2011**, *12*, 303–322. [CrossRef]
49. Clayton, S.D. Environment, Identity, and Response to Polluted Landscapes. *Sustainability* **2021**, *13*, 9422. [CrossRef]
50. Sumner, C.; Lockwood, J. Visitor Perceptions of Bark Beetle Impacted Forests in Rocky Mountain National Park, Colorado. *Conserv. Soc.* **2020**, *18*, 50. [CrossRef]
51. Schultz, W.P. The structure of environmental concern: Concern for self, other people, and the biosphere. *J. Environ. Psychol.* **2001**, *21*, 327–339. [CrossRef]
52. Kibbe, A.; Bogner, F.X.; Kaiser, F.G. Exploitative vs. appreciative use of nature—Two interpretations of utilization and their relevance for environmental education. *Stud. Educ. Eval.* **2014**, *41*, 106–112. [CrossRef]
53. Manoli, C.C.; Johnson, B.; Buxner, S.; Bogner, F. Measuring Environmental Perceptions Grounded on Different Theoretical Models: The 2-Major Environmental Values (2-MEV) Model in Comparison with the New Ecological Paradigm (NEP) Scale. *Sustainability* **2019**, *11*, 1286. [CrossRef]
54. Dormann, C.F.; Elith, J.; Bacher, S.; Buchmann, C.; Carl, G.; Carré, G.; Marquéz, J.R.G.; Gruber, B.; Lafourcade, B.; Leitão, P.J.; et al. Collinearity: A review of methods to deal with it and a simulation study evaluating their performance. *Ecography* **2013**, *36*, 27–46. [CrossRef]
55. Cronbach, L.J. Coefficient alpha and the internal structure of tests. *Psychometrika* **1951**, *16*, 297–334. [CrossRef]
56. Streiner, D.L. Starting at the beginning: An introduction to coefficient alpha. *J. Pers. Assess.* **2003**, *80*, 99–103. [CrossRef] [PubMed]
57. Cribari-Neto, F.; Zeileis, A. Beta Regression in R. *J. Stat. Softw.* **2010**, *34*, 1–24. [CrossRef]
58. OECD. *Education at a Glance 2021: OECD Indicators*; OECD: Paris, France, 2011.
59. Bogner, F.X.; Wiseman, M. Adolescents' attitudes towards nature and environment: Quantifying the 2-MEV model. *Environmentalist* **2006**, *26*, 247–254. [CrossRef]
60. Bogner, F.X.; Wilhelm, M.G. Environmental perspectives of pupils: The development of an attitude and behaviour scale. *Environmentalist* **1996**, *16*, 95–110. [CrossRef]
61. Gifford, R.; Nilsson, A. Personal and social factors that influence pro-environmental concern and behaviour: A review. *Int. J. Psychol.* **2014**, *49*, 141–157. [CrossRef]
62. Zelezny, L.C. Elaborating on gender differences in environmentalism. *J. Soc. Issues* **2000**, *56*, 443–457. [CrossRef]
63. Greenbaum, A. Taking Stock of Two Decades of Research on the Social Bases of Environmental Concern. In *Environmental Sociology: Theory and Practice*; Captus Press: Concord, ON, Canada, 1995; pp. 125–152.
64. Dunlap, R.E.; Van Liere, K.D.; Mertig, A.G.; Jones, R.E. Measuring endorsement of the New Ecological Paradigm: A revised NEP Scale. *J. Soc. Issues* **2000**, *56*, 425–442. [CrossRef]
65. Newman, T.P.; Fernandes, R. A re-assessment of factors associated with environmental concern and behavior using the 2010 General Social Survey. *Environ. Educ. Res.* **2015**, *22*, 153–175. [CrossRef]
66. Tyrväinen, L.; Silvennoinen, H.; Kolehmainen, O. Ecological and aesthetic values in urban forest management. *Urban For. Urban Green.* **2003**, *1*, 135–149. [CrossRef]
67. Edwards, D.; Jay, M.; Jensen, F.S.; Lucas, B.; Marzano, M.; Montagné, C.; Peace, A.; Weiss, G. Public preferences for structural attributes of forests: Towards a pan-European perspective. *For. Policy Econ.* **2011**, *19*, 12–19. [CrossRef]

68. Tieskens, K.F.; Schulp, C.J.; Levers, C.; Lieskovský, J.; Kuemmerle, T.; Plieninger, T.; Verburg, P.H. Characterizing European cultural landscapes: Accounting for structure, management intensity and value of agricultural and forest landscapes. *Land Use Policy* **2017**, *62*, 29–39. [CrossRef]
69. Lieskovský, J.; Bürgi, M. Persistence in cultural landscapes: A pan-European analysis. *Reg. Environ. Chang.* **2017**, *18*, 175–187. [CrossRef]
70. La Trobe, H.L.; Acott, T.G. A Modified NEP/DSP Environmental Attitudes Scale. *J. Environ. Educ.* **2000**, *32*, 12–20. [CrossRef]
71. Corral-Verdugo, V.; Carrus, G.; Bonnes, M.; Moser, G.; Sinha, J.B. Environmental beliefs and endorsement of sustainable development principles in water conservation: Toward a new human interdependence paradigm scale. *Environ. Behav.* **2008**, *40*, 703–725. [CrossRef]
72. Kortmann, M.; Müller, J.C.; Baier, R.; Bässler, C.; Buse, J.; Cholewińska, O.; Förschler, M.I.; Georgiev, K.B.; Hilszczański, J.; Jaroszewicz, B.; et al. Ecology versus society: Impacts of bark beetle infestations on biodiversity and restorativeness in protected areas of Central Europe. *Biol. Conserv.* **2021**, *254*, 108931. [CrossRef]
73. Müller, M.; Mayer, M.; Job, H. Totholz und Borkenkäfer im Nationalpark Bayerischer Wald aus touristischer Perspektive. In *Die Destination Nationalpark Bayerischer Wald als regionaler Wirtschaftsfaktor*; Job, H., Ed.; Nationalparkverwaltung Bayerischer Wald: Grafenau, Germany, 2008; pp. 100–116.
74. Smithson, M.; Verkuilen, J. A better lemon squeezer? Maximum-likelihood regression with beta-distributed dependent variables. *Psychol. Methods* **2006**, *11*, 54–71. [CrossRef]

forests

MDPI

Article

Through the Eyes of Forest Visitors—Perception and Scenic Preferences of Munich's Urban Proximate Woodlands

Gerd Lupp [1,*], Valerie Kantelberg [2,3], Julia Fäth [1], Theresia Hirschbeck [1], Corinna Käsbauer [1], Anna Ritter [1], Julia Schisslbauer [2] and Stephan Pauleit [1]

1 TUM School of Life Sciences, Technical University of Munich, Emil-Ramann-Str. 6, 85354 Freising, Germany
2 The Bavarian State Institute of Forestry, Hans-Carl-von-Carlowitz-Platz 1, 85354 Freising, Germany
3 Amt für Ernährung, Landwirtschaft und Forsten, Kitzingen-Würzburg, Simon-Breustr. 21, 97074 Wuerzburg, Germany
* Correspondence: gerd.lupp@tum.de; Tel.: +49-(0)8161-71-4661

Abstract: With the increasing individualization of society, perceptions and attitudes towards nature and forest management is changing. Knowledge about motives for visiting woodlands as well as the aesthetic perception of forests can support forest management and communication strategies. In Central Europe, multifunctional, close-to-nature silviculture aims to convert the still dominating conifer monocultures towards mixed, structured forests to establish forests that are better adapted to the changing climate, thus enhancing the value for nature conservation and increasing their ability to provide ecosystem services. Reflecting these management objectives, we examined the perception and scenic preferences of the resulting forest stands in Munich´s urban proximate woodlands. We applied both surveys with choice experiments using visualizations of different forest stands and conducted Photovoice walks. In the choice experiments, most interviewees preferred the visualizations of mixed forests. Deadwood was appreciated in mixed stands. Knowledge about the positive effects of deadwood influenced this choice. With the Photovoice walks, it could be shown that many forest types and even monocultures were perceived to be very attractive, such as tall Norway spruce (*Picea abies*) stands. In addition, small details such as single trees, colorful leaves, or small vegetation features contributed to a positive perception of the forest.

Keywords: forest perception; forest preferences; outdoor recreation; multifunctional forestry; urban forestry; survey; choice experiment; photos; Photovoice; Munich

1. Introduction

1.1. Forest Perception and Preferences in the Context of Ecosystem Services and Valuing Nature

Forests in urban and peri-urban areas deliver numerous ecosystem services (ES). One of the most important ES in urbanized areas is the provision of recreation opportunities for enjoying nature and the experience of nature [1–4].

Climate change will be a challenge for managing urban forests and urban proximate woodlands. Comprehensive adaptation measures to respond to the rapidly changing environmental conditions will be necessary, and their role has increased massively to mitigate the negative impacts of climate change [5]. Their importance as recreational areas to provide opportunities for citizens to leave overheated built-up areas of cities will also increase [3,6].

IPBES has examined nature's contributions and the multiple values for people. Pressure on biodiversity is increasing as a result of multiplying, diversifying, and interacting threats. These ultimately derive from the growing societal and economic demands on nature. In a world increasingly affected by climate change, comprehensive adaptation measures to respond to the rapidly changing environmental conditions will be necessary [7]. Besides protecting and restoring biodiversity, multiple benefit management and inclusive approaches through a holistic, integrated, consultative, and adaptive approach can maximize co-benefits in both conserving biodiversity, mitigating climate change, and enhancing good quality of life [8].

The predicted strong population growth in urban regions and migration to powerful economic urbanized centers implicates much more demand for recreation in urban proximate woodlands [9]. With increasing individualization and changing lifestyles based on different cultural backgrounds and socialization, economic possibilities, personal views, opinions, mentality and values, habits and the demands of forest visitors, demand for recreation opportunities, and perceptions and attitudes towards nature and forest management is changing and diversifying [3].

The IPBES highlights assessment typology with key concepts, their interrelationships to better understand the diverse values of nature, value types, and dimensions. The concept of ES is seen as an important theoretical framework linking between ecosystems, their goods, and services. Their contribution to human well-being has been globally assessed in the Millennium Ecosystem Assessment [10]. The concept of ES stresses the essential relevance of ecosystem structures and processes for human wellbeing, sociocultural values, as well as for mental and physical well-being [11]. Numerous studies attempt to assess and account sociocultural benefits also in economic terms. Recreation activities such as walking or cycling have a positive effect on health [12]. For example, in England, access to quality green space such as forests is estimated to save around £2.1 billion in health care costs for the National Health Service (NHS), and even savings of up to £1.44 billion (in health care costs) could be achieved with a 1% reduction in sedentary behavior [13].

Despite such efforts, challenges remain concerning the definition and measurement of sociocultural ES assigning values, as they are also determined by how important those services are to people and are created by people through their interactions with places individually [14,15].

Linking the recommendation of the IPBES to holistic, adaptive approaches to best maintain and maximize co-benefits in both conserving biodiversity and mitigating climate change, and enhancing good quality of life, the evaluation and application of value indicators of such management is vital.

From the perspective of a land manager, such approaches can provide valuable insights for visitor management, providing recreational infrastructure, targeted information, and adapted forest management strategies to secure the provision of multiple ecosystems and benefits, avoid conflicts, as well as increase the acceptance of forest management [16,17].

For measuring sociocultural values of such strategies and their successful integration of recreation into multifunctional forest management concepts, besides "hard data", providing visitor numbers and types of recreation activity, approaches outlined by Dickinson and Hobbs [15] that provide insights about motives for visiting the aesthetic perception of forests, are vital to evaluate and further develop adequate management strategies and concepts to minimize conflicts [18,19].

In the past, quite a large number of empirical studies on landscape preferences have been conducted. Yang et al. provide an overview and development of the respective research in their review [20]. Many of the user- or visitor-based approaches make use of choice experiments. However, comparatively few have a focus on the perception of forests and forest stands. Most published empirical studies and theories on forest preferences were conducted in Northern Europe or North America with respective forest types and silvicultural methods such as using clear-cutting and retaining seed trees like Scots Pine (*Pinus sylvestris*) in Finland [21–23]. The studies document an increasing preference for naturalness and mixed forest structures [24]. Similar tendencies have been reported from China [25–27]. In the North American context, Gobster et al. [28] postulate preferences for forests that tend towards them remaining natural as possible, which also correlates with high biodiversity.

For Central Europe, studies in the 1960s and 1970s showed preferences for coniferous forests [29,30], but preferences in Central Europe have changed. In more recent studies from the turn of the millennium, forest visitors particularly value uneven-aged, tiered stands formed by various deciduous and coniferous tree species. A proportion of very thick, large, and bizarrely shaped (supposedly) old trees are important [31,32].

Studies from Scandinavia come to similar conclusions [21–23,33], but there are also studies in which tidier, lighter forest types are preferred [22]. Some other studies contradict the aforementioned studies, claiming the importance of good recreational infrastructure and well-maintained forests, especially for urban woodlands and urban forests [34]. These studies suggest that lighter, tidier, and well-kept structures may be preferred in urban recreational areas. For example, the respondents in this study preferred the less densely forested areas in an urban park that only had a comparatively low level of biological diversity with a high visual penetrability of forests as formulated by Kaplan and Kaplan [35] in their "Information Process Theory" as a basic prerequisite for gaining high preference. However, one should be careful as some studies refer to "Urban Forests" in a broader sense also comprising wooded urban parks or urban green infrastructure and not to forests in a stricter sense. Most studies are conducted in summertime and seasonality aspects are missing.

Research on sociocultural aspects linked to forest and greenspace use has been conducted to some extent, however mainly on a more conceptual level, e.g., using household surveys or data analyses (e.g., reviewed studies in [20]). For countries with a high proportion of people who have a migration background, such as Great Britain, The Netherlands, and USA, some studies have been conducted to analyze the perceptions and desires of different societal groups and persons with a migration background or belonging to different ethnic groups [36–39]. For Central Europe, only a few examples could be found in the literature, such as the study by Jay and Schraml [40], which analyzed the forest use and perception of persons with a Russian-German and Turkish migration background.

The aim of the investigations in the greater Munich area was to examine whether the tendency towards a preference for particularly near-natural stands described in the literature applies to urban proximate forests and woodlands in Central Europe with its multifunctional management approaches. Moreover, small-scale, individual tree-related management silviculture strategies approaches are widespread in Central Europe and might be much less noticeable to forest visitors. Older studies conducted in these forests [30,41,42] help to draw comparisons with the new findings.

So far, published studies applying human-centered approaches to assess landscape and forest perceptions are quite scarce. They only tend to use single approaches, such as interviews or choice experiments. Hence, the guiding research questions for our work were:

- How are forests and forest landscapes perceived?
- What is perceived as attractive or is disliked using qualitative, user-driven approaches?
- How do the findings compare with choice experiments?

In our paper, we present findings from several studies conducted in the Greater Munich Metropolitan area. Study areas were the "Forstenrieder Park", a large forested area to the south of Munich, the "Ebersberger Forst" to the east of Munich, and in the World Forest/Weltwald near Freising, around 30 km north of Munich.

1.2. Theoretical Reflections on Landscape and Forest Perception

In order to understand people's aesthetic sensibilities, perception, and valuing of the forests and lay a theoretical foundation to embed our work, we first looked at theoretical considerations on landscape perception. Being aware of the international literature, being rooted in the German discussion on landscape perception, we took the respective German literature on landscape aesthetics [43,44] as a starting point for our reflections.

Following the definition of landscape perception according to Nohl [43], the perception of a forest therefore is not only based on physical existing elements, but also aspects such as experiences, previous knowledge about forests, attitudes, expectations, memories, desires, and needs. The current state of mind, previous experiences, knowledge (e.g., lay person, someone having a professional background, e.g., in forestry), origin (first-time or regular visitor, familiar with landscape types or not), as well as personal or collective values are important for forming the perception of a landscape or forest. Humans therefore might perceive the same landscape in a completely different way. Next, the situational influences such as seasonality, weather, the viewing point and angles from which to look at a landscape are important. The conducted activity (e.g., cycling, walking, birdwatching) plays a role in experiencing landscapes, the mood and the visitor's own feelings as well being alone or being accompanied by others. Depending on the situation, that the same forest can be perceived and valued in a completely different way by any one individual. The same person can experience the same landscape differently depending on mood or judge different during the course of life, for example, a forest can be experienced in a totally different way when being there as a child or as an adult [43,45].

According to Hoisl et al. [31] regarding the perception of a landscape, the authors distinguish between three different "layers of meaning or knowledge" (so-called levels): the perceptual layer of meaning, the symptomatic layer of meaning, and the symbolic layer of meaning. On the first level, the perceptual level, the perception of the landscape components takes place, but already some interpretations and feelings about such components belong to this layer. For example, an old large oak tree can be perceived as sublime and enchanted. On the second level, the symptomatic layer of meaning, for example, the oak could be a hint to historic land use such as a wood pasture for pig fattening or the ecology of the area, a bottomland hardwood forest. On the symbolic level, landscape elements also connect elements beyond the physical features linked to the imaginations triggered in the viewer, such as feeling at home or feeling free.

For the satisfaction of desires and needs of humans, certain landscape types and landscape features and structures are more suitable than others. Particularly noteworthy are impressions that provide a certain amount of diversity, suggest naturalness and individuality, or the uniqueness of a landscape but also interests, ideas, and knowledge thereof and ecological, functional, or historical connections play an important role for the evaluation of any landscape [31].

User-dependent procedural approaches attempt to develop and evaluate the impression that people have, resulting from their viewing and the interpretation of the perceived elements [46]. The human being is therefore at the center of the research approaches. The research mainly uses social science methods such as surveys or behavioral analyses in which mainly laypersons judge and evaluate forests and landscapes. Another form is the analysis of interactions between humans and the forest.

In the international English literature, in work and studies on landscape or forest perceptions, references or theoretical foundations are made on the basis of the work of Kaplan and Kaplan [35] in order to support, contradict, or supplement their postulated theories and findings. Despite dating back more than 30 years, nowadays, the work is still considered state-of-the art and key literature for understanding and reflecting landscape preferences.

The concept by Bourassa [47] integrates many existing theories including the ones provided from Kaplan and Kaplan [35] to better explain and cover the facets, why certain landscapes or forest types are preferred, whilst others are disliked. Bourassa postulates three strategies.

The first law postulates genetically fixed preferences for certain landscape types. These are referred to as the "prospect-and-Refuge Theory" by Appleton [48], the «Savannah Theory» by Orians [49] or a more sophisticated concept based on numerous studies by Kaplan and Kaplan [35], the «Information Process Theory». Particularly good living conditions and thus, particularly valued, would therefore be landscapes with views and cover for hunting opportunities, protection and shelter from enemies, and providing enough water. Accordingly, most preferred landscapes would be park-like with groups of trees, and open grasslands, containing rivers and lakes, but also cliffs and caves providing shelter and refuge [50]. Landscapes are structured and readable with a number of landscape elements providing some stimuli but should not overwhelm viewers with too many or unstructured features.

The second level is "Cultural Laws" [47]. It is assumed that landscape preferences are primarily acquired through human socialization, social rules and norms, learning, reflecting, collective experience, and shared knowledge [47]. Values and norms for »beauty« are formed in this way. Often, preferences shaped by this "Law" do not have universal validity. They might only be shared by small groups, some societies or in one cultural background.

As a third level, the personal development of each individual is also important and called "Personal Strategies" [47]. Every person can develop their own landscape preferences and dislikes. They are only valid for one individual. They evolve by imprinting, habituation, but are also due to traumatic experiences [47,50].

The most important human sense for perceiving landscapes and forests is sight, the second most important one, hearing [44,51]. Studies [32,52,53] show that an important and highly valued feature of the forest is "silence", meaning the absence of the noise of civilization, e.g., from car traffic. According to Newman et al. [53], the noises of civilization can significantly reduce the experience of nature. In addition to seeing and hearing, the sense of smell, taste, and touch should also be mentioned. Even if other sensory stimuli might contribute little to perception, they should not be completely neglected for the analysis of landscape preferences or their impact on human well-being [31,43,44]. A notable example was a study from Japan on the positive effect of resin smell in the forest [54].

Two methodological directions are described for capturing the quality of the landscape in order to assess landscape preferences. These are expert-based and user-dependent methods [46].

2. Materials and Methods

2.1. Methodological Approach

The methodological approach and methods used can be found in Figure 1. To get an overview, we conducted literature reviews at several points until summer 2021 collecting scientific papers from Web of Science, Scopus, and Google Scholar databases using search terms "landscape" and "forest" each in connection with one of the term "aesthetics", "perception", "preferences", "recreation" (resulting in ten combinations of terms) and the respective term combinations in the German language. It was considered important to use more than one database since search algorithms may vary across databases and potentially include some more practitioner-oriented literature gathered through Google Scholar. Resulting publications were then selected while using the PRISMA method [55]. In screening the abstracts of the articles for relevance [55], we then continued searching for additional work using a snowballing approach with author names and references in the found papers. Around 450 papers were collected and extracted this way.

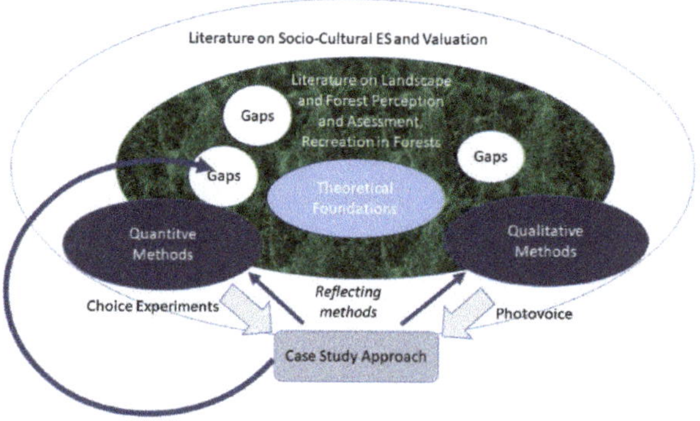

Figure 1. Methodological approach and methods used.

For our studies, we opted for the user-dependent approaches. We followed a case study approach as a flexible way for in-depth investigations at small scales that balance breadth and depth [56]. A combination of different qualitative and more quantitative approaches is seen to be the most useful in research dealing with human-nature relations [57].

Choice experiments, e.g., using a set of photos and ranking the pictures as the sole method for assessing landscape preferences are not considered sufficient in the literature [58]. To minimize the disadvantages and the limitations of different methodological approaches, combining more quantitative and more qualitative approaches can help to better determine and understand the perception and preferences for landscapes and forests. Especially the basic assumption of qualitative research approaches is that an important aspect and quality criterion is the proximity of researchers to the human being. Thus, research is supposed to be conducted as close as possible to the natural environment of the human being [59].

We chose three different forest areas in the Munich metropolitan area (Figure 2). In 2014, Munich had 1.487 million inhabitants. It is Germany´s third largest city and the most densely populated urban area in Germany [60]. Being among the top 10 most powerful economic regions in Europe [61], the Munich metropolitan area faces rapid population growth [60]. With its close vicinity to the Alps, numerous attractive green spaces in the city, nearby lakes and forests, the high recreation value is also considered to be an important driver for the rapid population growth [62].

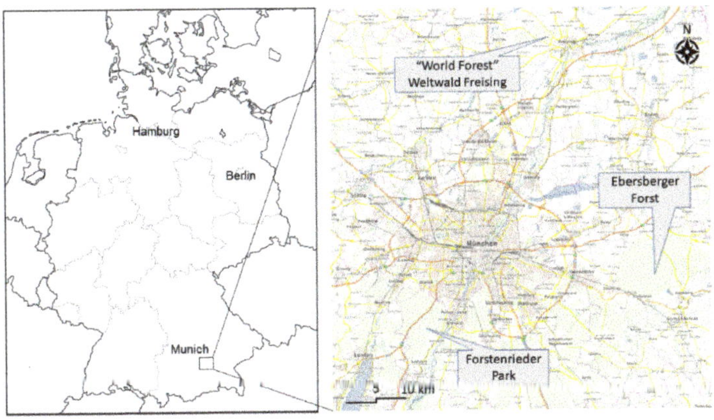

Figure 2. Map of the Study Area and selected forests in the Munich Metropolitan Area; Map Source and Geospatial Data: Bayerische Vermessungsverwaltung No. 2207-12734.

2.2. Description of the Chosen Forests Serving as Study Areas

The Forstenrieder Park together with other forests of neighboring communities form a 43 km^2 continuous forest area located to the South of Munich belonging to the state or other public bodies. It is mostly situated on the rather flat Munich gravel plain originating from the late Pleistocene glacial periods [63].

The name "Park" indicates its origin as a gated hunting area for the nobility in the past. Today, 2000 hectares have been dedicated as a game reserve with opportunities to watch deer. In the 19th century, this pasture woodland comprising oak (*Quercus robur, Quercus petrea*), beech (*Fagus sylvatica*), and other broadleaves native to the region were afforested systematically and schematically formed by a rectangular grid of forest roads using conifers, mainly Norway Spruce (*Picea abies*). After several gales, storm damages, and decimation by bark beetle (*Ips typographus*), since the early 1990s, the state forest management tries to establish mixed, structured forest stands of conifers and broadleaved tree species using methods such as establishing the next forest generation by shelterwood strategies, natural regeneration or planting broadleaves on larger empty areas resulting from calamities.

Transforming the stands will take many decades, even centuries. Therefore, forest stands currently present all phases of the transformation from pure conifer stands up to structured mixed forests. With a trail network of 43 km [63], connection to cycle routes and parking lots, and with suburban railway and metro line stations within walking distance of a few minutes, the forest is easily accessible by the general public.

The Ebersberger Forst is situated 25 km east of Munich and covers a total area of 90 km^2 [64]. While the west of the forest is located on the flat Munich gravel plain, it stretches up onto the terminal moraine range of hills to the east. Connection to cycle routes, parking lots, and public transportation in nearby villages and towns provides access to the forest. A large network of over 200 km of trails for hiking, horse riding, and cycling provide ample recreation opportunities. A forest interpretation trail, a game sanctuary, restaurants, a viewing tower, as well as ancient monuments provide additional points of interests.

Similar to the Forstenrieder Park, this forest originates from a hunting area for the nobility and forests were degraded at the end of the 18th century. In the early 19th century, the area was systematically afforested with Norway Spruce (*Picea abies*) and Scots Pine (*Pinus sylvestris*). In the 19th century, already severe damages by beetle and Black Arches (*Lymantria monacha*) were reported [64]. Forest types and management strategies are similar to the ones in the Forstenrieder Park.

The "World Forest" or "Weltwald" (in German) is a state-owned forest 30 km north of Munich. It has a size of 100 ha. The nearest town is Freising, around 4 km away to the east of the forest. Situated on tertiary Danube-Isar hilly landscape forming the northwestern end of the Munich gravel plain, it is part of a larger forest complex called the "Kranzberger Forst". The forest was established on the ground of a former village called Oberberghausen. At the end of the 19th century, the farm property had been acquired by the state and reforested also using a variety of "exotic trees" and willows (*Salix*) also for experiment purposes. The Church of St. Clemens, surrounded by a small cemetery, is the last witness to the former village [65].

Some years later, with mixed and discouraging results especially related to cultivating the willow, the forest was managed for timber production with Norway spruce (*Picea abies*) with a shift to mixed structured stands in the past three decades. With a still noticeable amount of exotic trees dating back to the reforestation in the 19th century, it was decided to host an arboretum for both the public and the life sciences faculties when the Weihenstephan university campus was further developed. Systematic planting of exotic trees started in 1987. By 2016, more than 300 different tree and shrub species had been planted in the forest matrix formed by spruces and beeches (*Fagus sylvatica*). In addition to the main routes leading through the forest, several thematic educational trails for the different tree species from different regions of the world provide access to the forest. Numerous activities, guided tours, and outdoor art exhibitions are provided to create an attractive space for visiting and recreation [65].

2.3. Overview on Methods Used

An approach using interviews with forest visitors inside the different forests in Munich and Freising was employed. Besides a set of overall questions on activities, some overall demographic data and general questions on forest preferences, choice experiments with different ranking methods of photo impressions were applied for the analysis of forest preferences including short open-ended statements as to why a picture or visualization was chosen. In addition to more quantitative approaches using ranking or selecting sets of images for preferred impressions, qualitative approaches to assessing forest images were implemented, as the choice of images or photos as well as the way of photographing or visualizing forests can influence interviewees. Besides applying such classic instruments, Photovoice forest walks were carried out in the World Forest.

In recent years, the Photovoice method has received increasing attention as a qualitative research method. A camera is handed out to a person asked to take photos of what they consider relevant in the context of the respective investigation. Then, in an interview, the persons explain the relevance of the images for them. This gives researchers a much deeper insight into the respective point of view than would be possible with structured or semi-structured interviews, since the subjects use the images to define the individual aspects of the investigation [66–68]. In addition, Photovoice can be easily combined with other research methods of qualitative social research [69]. Photovoice is originally a technique of participatory research and uses the immediacy of visual impressions to capture experiences, knowledge, and subjective perceptions of subjects [70,71].

Following the approach by Heyman [72], Photovice walks were conducted in the Freising World Forest in two studies with volunteers. In one study, persons who had lived in the region for a significant time and being German were asked to participate. A study with a similar setup focused on persons with a migration background. Participants taking the walk in the forest were given the task of taking pictures of things and impressions that they particularly like or disliked when strolling through the forest. They were accompanied by a "silent" researcher to ensure that all person took the same route and felt safe. At the end of the walk, everyone was asked to talk about the photos, explaining why they had taken them and why they represented something positive or negative.

Table 1 provides an overview of the study areas and applied methods as well as of the chosen approaches to evaluate scenic qualities of the forests.

Table 1. Overview of sites and interview approaches.

Interview Site	Season	Applied Methods	Number Participants
At frequented entry places of the Forstenrieder Park Forest Three locations at frequented entry places, 9 interview days covering a weekday weekends, interviews between 9–17	Winter without snow cover	Passers-by were asked to participate. After the interview was finished, the next person approaching was contacted. Ranking of 16 images based on the Q-Sort method [73]. From a set of 16 images, four images are selected that are like or less appreciated. The best is then selected from the pre-selection of good pictures. For the less esteemed images, the worst image is selected. The images can then be evaluated accordingly, either according to how often they were divided into certain levels or given a scale value with an allocation of values between −2 and +2. Participants were asked to give brief statements why best and worst image was chosen.	302 (out of 402 asked)
Inside Ebersberger Forst, three interview sites, 11 interview days, weekdays and weekends, time between 9–17	Autumn	Interview approach similar to Forstenrieder Park; Photos from different forest types in early autumn found in this woodland were used for a picture survey. Respondents could only pick one best and one worst image, and were asked to briefly describe their choice. How often an image was selected as the best or worst image as well the reasons given were evaluated	320 (out of 351 asked)
Inside World Forest/Weltwald Freising	Spring	Photovoice walks with persons living over a long time in the area. Recruitment of participants following sampling strategies recruiting a good cross-section of society for the study following [74,75]. Persons were selected both from persons known by a circle of acquaintances (especially for recruiting persons visiting the forest less often) and contacted on the parking lot at the entrance to the forest	26 participants, German, living in the region for a long time
Inside Weltwald Freising	late summer	See above, Photovoice with persons that migrated to Germany, who were asked to participate using contact persons in the circle of acquaintances	14 participants, originating from Canada, China, Croatia, France/Spain, Luxemburg, Norway, Peru, Poland, Syria, Tunisia/Saudi Arabia, Turkey, Ukraine USA

3. Results

3.1. Overall Demands and Wishes for the Forests

In the surveys employed in the Forstenrieder Park and the Ebersberger Forst, the respondents valued the forest's closeness to nature, its "wilderness", the recreational opportunities, and the accessibility of the forest (especially on foot, by bicycle, and by public transport). The majority of those surveyed wanted "pure nature". This included, among other things, a rather sparse supply of recreational facilities. Respondents were mostly very satisfied with the forests and the amenities found. Compared to the previous studies in the greater Munich area forests [30,33,34], the majority of those surveyed in the most recent study described the infrastructure, e.g., benches, as sufficient. In addition, around a third of those surveyed had no complaints or points of criticism. Negatives mentioned were conflicts with other user groups and races of civilization such as rubbish or noise. Especially road traffic and aircraft noise were perceived as disturbing, in particular for the World Forest as it is situated close to the approach lane of the heavily frequented Munich Airport. Disturbing traces originating from forest management such as timber harvested were only rarely mentioned.

When collecting demographic data on the two survey approaches, regarding the age of the interviewees, older age groups (from 46–55 years) and more persons with a high educational background visited the forest much more often than other groups. For young people, especially those with a migration background, if forests were visited, sporting activities were an important reason for going there, as was socializing.

3.2. Assessing Forest Landscapes Using Photos and Drawn Images

In order to obtain assessment data from a large number of respondents and to test the perception of forest landscapes resulting from different management practices, different sets of forest visualizations and photos considering the different forest types in the vicinity of Munich and seasonality were used. This was done by applying different methods for choosing, selecting, ranking, and valuing the shown impressions from the forests.

Since few studies analyze forest images in the winter, typical forest images of the local spruce stands were made as part of the investigations in the Forstenrieder Park. Most of them present themselves in different stages of transition to mixed stands that are better adapted to climate change. Impressions without snow cover and beeches (*Fagus sylvatica*) without leaves or with dried leaves were visualized. The aim was to find out whether stands with many leafless beeches might perform less well than studies of beeches with leaves.

When evaluating 16 forest images with different stands (Figure 3) and forced Q-Sort [73] ranking (Figure 4), the old-growth spruce stand of the same age was valued least. The picture of an unequal-aged, layered mixed stand of deciduous and coniferous wood with a small proportion of dead wood was particularly preferred (Figure 5). However, the statements made by those surveyed made it clear that a mixture of different stocks along the route covered in the forest is perceived as especially attractive. Even those forest pictures that came off the worst in the ranking (Figure 6) were not considered "really bad" by the participants in their feedback. The ranking and the strong gradation are more likely to be due to the fact that the task consisted of having to commit to a ranking. Studies with comparative approaches to evaluation methods, where school grades [52] were also asked for, show that forest pictures judged to be less attractive also received good school grades.

Figure 3. Set of forest images used for ranking in the Forstenrieder Park, visualization by Anna Ritter.

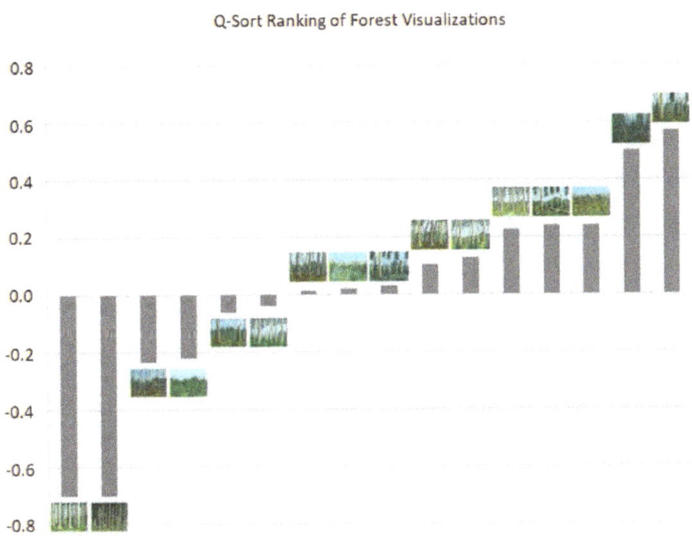

Figure 4. Results of Q-Sort Ranking from the Forstenrieder Park Study. Maximum numbers on the scale would have been −2.0 (every participant would have chosen the image as the worst one) and +2.0 (every participant would have chosen it as the best one).

Figure 5. Most preferred impression from the Forestenrieder Park Study, visualization by Anna Ritter.

Figure 6. Least preferred impression from the Forestenrieder Park Study, visualization by Anna Ritter.

From the set of photographs used in the Ebersberger Forst (Figure 7), by 31.3%, the photo of an old deciduous forest (LW_alt) was selected most frequently as the preferred forest image (Figure 8). It was considered to be a "particularly inviting" forest. Respondents explained their choice as following: the variety of colors, a "powerful" and "warm" atmosphere and the seasonality of the year can be experienced in a very good way in the shown forest type. The beech (*Fagus sylvatica*)-dominated forest with leaves covering the ground was considered to be more open, bright, warm and friendly, with the forest looking natural, and this type of forest was found to be most suitable for relaxing and taking walks. The aspect that only deciduous trees, mainly beeches, can be seen in the picture was the decisive factor for many in naming this picture as their favorite.

Figure 7. Set of pictures used for the interviews in Ebersberger Forst, photos taken by Julia Fäth. Abbreviations of the images; LW_alt, old broadleaved tree stand; NW_LN, old conifer stand with younger broadleaved trees beneath; MW, mixed forest stand in terms of species, sizes and age; NW alt, old conifer stand; Toth, old broadleaved tree stand with deadwood; LW_mi, middle aged broadleaved tree stand; LW_L, old broadleaved tree stand with initial broadleaves regeneration; NW_L, old conifer stand with initial broadleaves regeneration; LW_d, old, scattered broadleaved tree stand with several stages of regeneration; NW_d, stand with old, scattered conifers with several stages of broadleaves regeneration; NW_N, old conifer stand with young conifer regeneration; NW_mi, middle aged conifer stand.

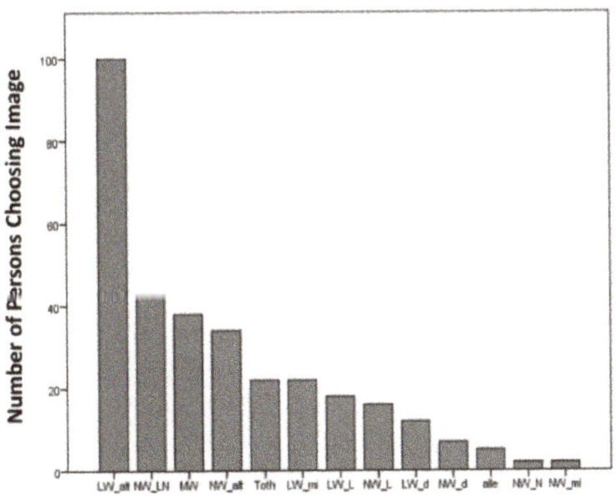

Figure 8. Diagram showing the numbers of selected most preferred pictures from the study in Ebersberger Forst. Abbreviations of the images; LW_alt, old broadleaved tree stand; NW_LN, old conifer stand with younger broadleaved trees beneath; MW, mixed forest stand in terms of species, sizes and age; NW alt, old conifer stand; Toth, old broadleaved tree stand with deadwood; LW_mi, middle aged broadleaved tree stand; LW_L, old broadleaved tree stand with initial broadleaves regeneration; NW_L, old conifer stand with initial broadleaves regeneration; LW_d, old, scattered broadleaved tree stand with several stages of regeneration; NW_d, stand with old, scattered conifers with several stages of broadleaves regeneration; NW_N, old conifer stand with young conifer regenertion; NW_mi, middle aged conifer stand; alle, persons insisted to select all pictures as their preferred ones.

On the other hand, 13.1% preferred the coniferous forest with deciduous and coniferous trees in the shelter (NW_LN). The main reason given was that it was a mixed forest, in which both coniferous and deciduous trees were present, as well as trees of different ages and sizes. The existing moss and the different colors were described by many visitors as being particularly beautiful. Of respondents, 11.9% found the mixed forest (MW) to be the preferred forest image because it appeared particularly natural due to the different tree species and tree heights; the variety and balance of the colors were also emphasized by many visitors and, according to one forest visitor, "it corresponds best to the goal of multifunctional, close to nature forest management goals". The forest stand with deadwood was chosen by 6.9% of the participants. Because of the wood left behind, the forest appears to be particularly natural, and as one of the interviewees remarked, "not man-made". The aspects of wild, restless, and disorderly were judged particularly interesting by this group. The middle-sized deciduous forest (LW_mi) was also chosen by 6.9% of the respondents as the best forest picture, since the autumn colors are particularly recognizable in this image. The blaze of color was described as particularly beautiful and bright. Of respondents, 5.6% named the deciduous forest with foliage in the undergrowth (LW_L) as their favorite. They rated the different heights and colors of the trees as positive, and especially, the large number of beeches. Everything (small, large, young, old) was present in this stand that makes a good mixed forest.

Of the interviewees, 58.4% disliked most the forest image of the middle-aged coniferous forest (NW_mi, Figure 9). The trees were considered to be too thin, unnatural, and monotonous. Furthermore, the forest stand was described as dead, with too many bare branches, dark, sick, naked, sad, not maintained, or even scary. The study participants complained about an absence of broadleaved trees and stated that only commercial aspects appear to be considered here.

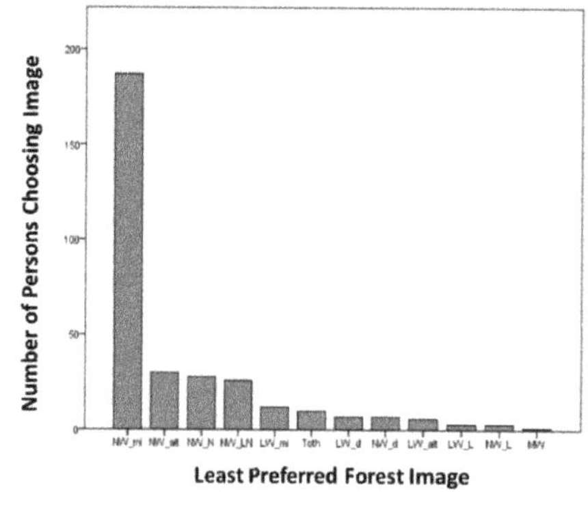

Figure 9. Diagram showing the number of selected least preferred pictures from the study in Ebersberger Forst. Abbreviations of the images; NW_mi, middle aged conifer stand; NW alt, old conifer stand; NW_N, old conifer stand with young conifer regeneration; NW_LN, old conifer stand with younger broadleaved trees beneath; LW_mi, middle aged broadleaved tree stand; Toth, old broadleaved tree stand with deadwood; LW_d, old, scattered broadleaved tree stand with several stages of regeneration; NW_d, stand with old, scattered conifers with several stages of broadleaves regeneration; LW_alt, old broadleaved tree stand; LW_L, old broadleaved tree stand with initial broadleaves regeneration; NW_L, old conifer stand with initial broadleaves regeneration; MW, mixed forest stand in terms of species, sizes and age.

Of respondents, 9.4% chose the old coniferous forest (NW_alt) as their least favorite forest image for the same reasons as for the medium-sized coniferous forest: the forest was too monotonous. The coniferous forest with conifers in the undergrowth (NW_N) was rated as the least popular forest image by 8.8%. The forest was too overgrown, monotonous, and too dense. The forest looked too uniform according to the respondents, too gloomy, and not alive enough. In addition, it was criticized that the ground could not be seen and it was therefore not possible to walk through the stand. Furthermore, 8.1% of the visitors could not name a picture that they liked the least as they considered all images beautiful.

3.3. Results from Photovoice

Almost all persons participating in the Photovoice study perceived the walk in the forest (Figure 10) as very positive and, particularly those people living for a long time in the region, took much fewer negative than positive pictures (Figures 11 and 12).

In both study groups, photos of the forest were particularly colorful showing a larger number of tree species or leaf colors. The provided benches and the signposting were mostly perceived as positive. Notably, many participants photographed tall, large trees. They were considered particularly attractive and were associated with nature, wildness, and power. Even pure Norway spruce stands (*Picea abies*) and tall non-native Douglas firs (*Pseudotsuga menziesii*) were found in a comparatively large number of pictures, as the photographed trees were considered very impressive and "mighty" due to their size. While deadwood was valued by persons living for a long time in the region, the group of migrants perceived this largely negatively. According to the justifications of the participants given, a positive perception was primarily related to (actual or assumed) knowledge about the importance of deadwood in forest ecosystems. The same diverging perspective could be observed for the insect hotels and the areas in the forest were goat grazing is practiced to maintain meadows important for nature conservation and biodiversity. While the locals photographed both as positive, the migrants perceived both as negative elements in the forest. Goat grazing received little understanding by persons originating from Mediterranean countries and was mentioned several times as the main reason for forest destruction in their countries. While the small ponds in the forest were almost always photographed by the locals as a positive element in the forest, they were only photographed once by the migrant group.

Both local persons and those originating from other countries felt disturbed by vehicles in the forest, the noise, and the rubbish, even by small pieces of litter, traces of logging such as cut trees, and the color markings on trees selected to be harvested. A clear-cut area close to the small lake was perceived very negative and foresters were blamed for this. However, this was the result of a bark beetle (*Ips typoghraphus*) invasion and had to be removed for maintaining the safety of visitors and for sanitary reasons. In other sections, the fresh lanes on an active skid trail were also perceived as disruptive by the first group of participants. Interestingly, persons who passed the same spot a few weeks later during the Photovoice walk no longer commented on the skid trail, since the lanes were leveled and had overgrown due to the growth of vegetation in the spring.

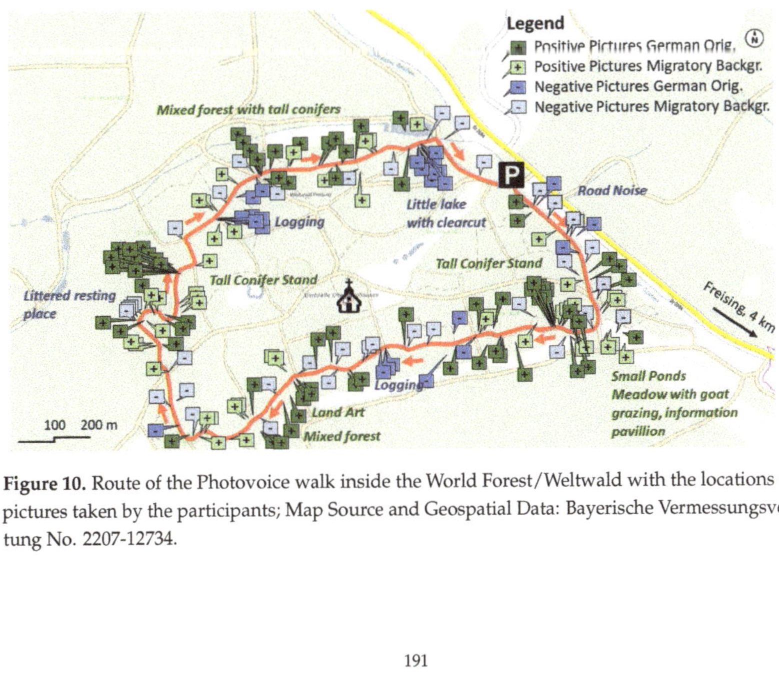

Figure 10. Route of the Photovoice walk inside the World Forest/Weltwald with the locations of the pictures taken by the participants; Map Source and Geospatial Data: Bayerische Vermessungsverwaltung No. 2207-12734.

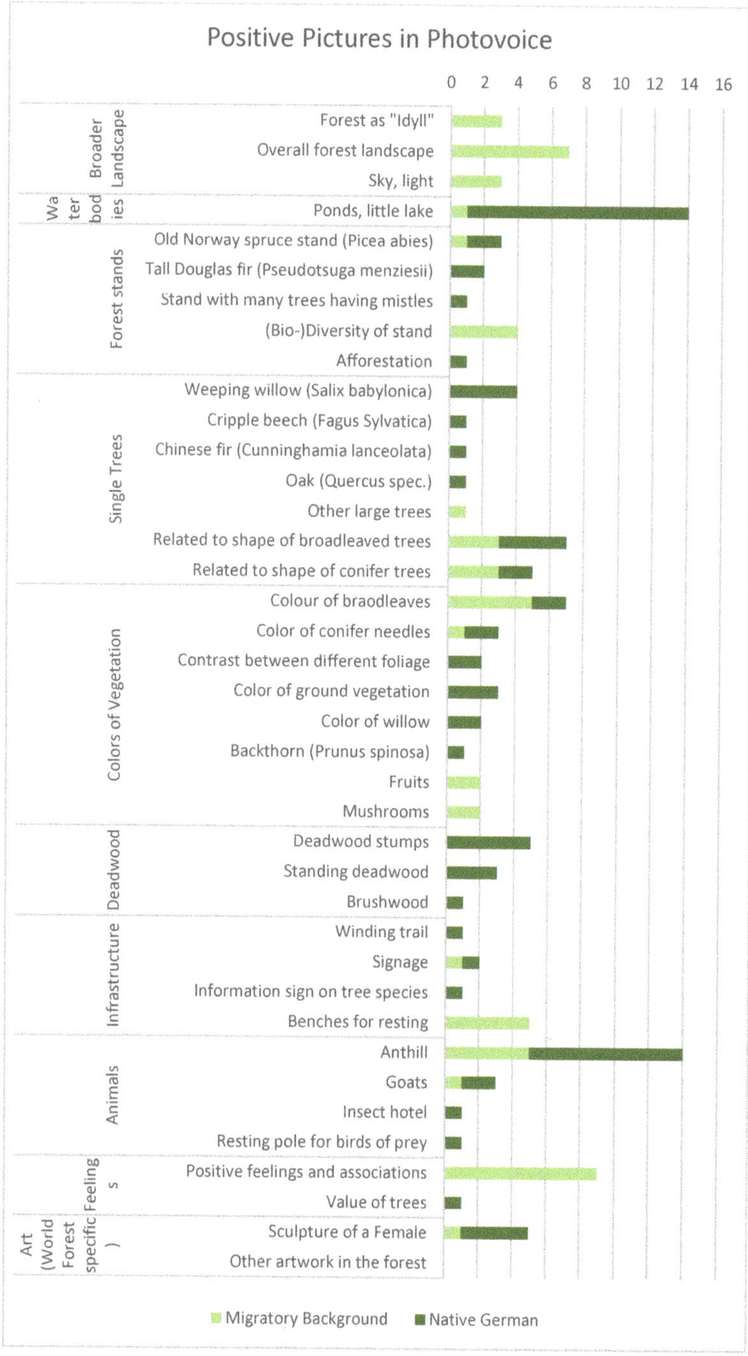

Figure 11. Picture contents chosen as positive in the Photovoice walk.

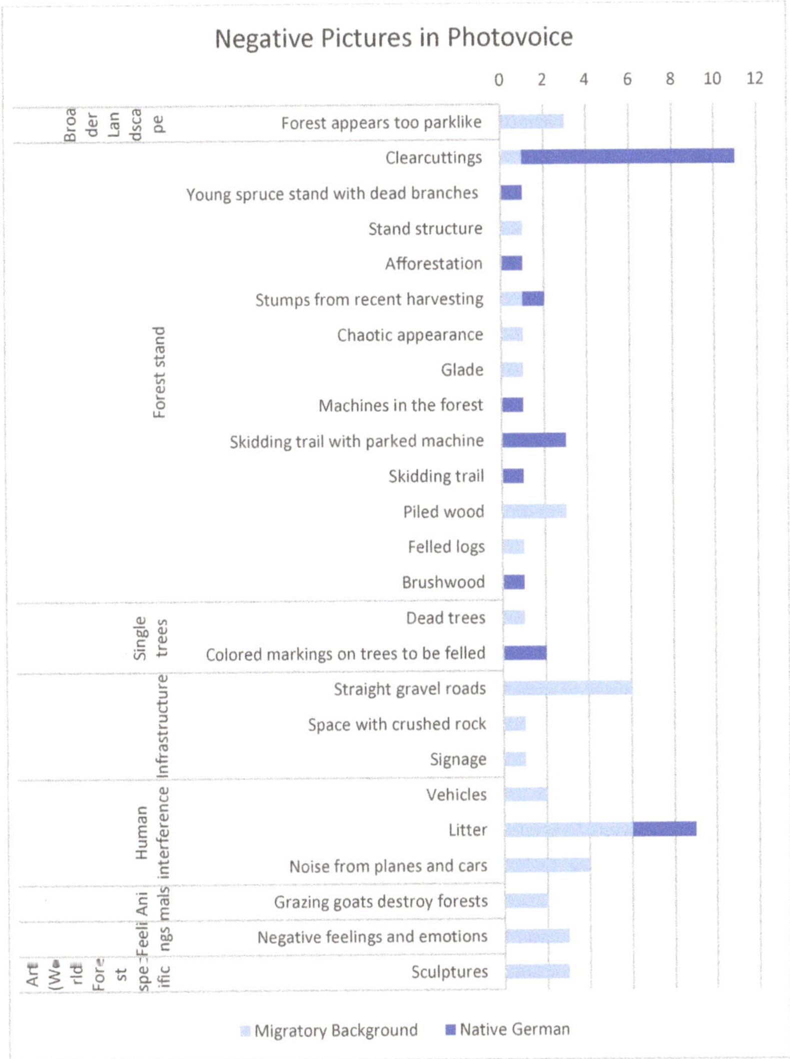

Figure 12. Picture contents chosen as negative in the Photovoice walk.

4. Discussion

4.1. Reflecting the Chosen Methods

The chosen approach covered around 300 interviewees each in the larger two surveys. This is a very small number compared to the very high visitor numbers in the forests and that were counted by a systematical approach in two forests in the Munich Metropolitan area [76]. Despite the systematical approach to collect information, e.g., selecting both weekdays and weekends, and interviews at different locations in the forest to cover a broad range of potential city dwellers and different forest visitors, the representativeness of the interviewees is minor to capture the broad range of actual forest visitors, especially capturing opinions of persons visiting the forest less frequently or recreating in other, more proximate, or more distant forests than the ones we chose for our surveys. Comparing the demographic data of our interviewees in the Forstenrieder Park and Ebersberger Forst with census data from Munich, more elderly and persons with higher vocational backgrounds were participating in the surveys.

Especially weather can influence laypersons in their perception of landscapes [77]. Using photos, even when carefully taking photos on the same day in similar light conditions, the angles, directions, or colors of photos can be slightly different and resulting bias. With drawings or computer-generated impressions, e.g., using eye-tracking methods in controlled lab settings, e.g., in a study on urban green space in Shanghai, China [78] instead of photos, bias by shade and colors can be better controlled and reduce such factors influencing the participants in their assessment [22]. Nonetheless, Hull and Stewart [79] suspected that the different assessments of the environment and the photos are related to the mood of the respondents, since the assessment of the photos in their study did not take place in a more normal situation, e.g., while hiking through the forest. Despite the option to give an open-ended answer when explaining the chosen pictures, the approach heavily depended on the pre-selection and research design developed by the researchers, selecting images and forming questions. Participants have few opportunities for stating their own, personal preferences or for expressing what they dislike(d) about the forests.

The Photovoice method, therefore, represents an interesting possibility to collect more qualitative data. While the approach aims for in-depth understanding, its ability to cover the whole range of forest visitors is even more limited. A key challenge for using Photovoice is the high and often unpredictable time needed to conduct a study, and therefore, within a given time frame, the number of participants in such studies is very limited [67]. However, with a systematic choice of representatives for certain user groups, interesting insights and understandable ways of explaining perceptions can be given. An advantage of Photovoice is that the participants can, to a large extent, determine the investigation from their perspective. This allows the researcher to see things through the visitor´s eyes at least to some extent by looking at the photos and listening to the explanations. In addition, the situation of taking a walk through the forest is different compared to an interview situation going through a survey sheet even when choosing an interview site inside the forest. Later on, Photovoice allows the test persons to speak freely and reflect about the walk. In the literature, participants are more engaged and willing to provide insights if they can co-create the investigation to some extent, take responsibility, and take photos [66,67,71]. Bethmann and Wurster [80] state that open, qualitative research methods and a change of perspective can help pick people up "from where they are" and discover what their perspective is.

While Kong et al. [67] mentioned costs for taking and evaluating the images, it can be neglected for most approaches thanks to the omnipresence of tools such as smartphones, digital cameras, laptops, or tablet computers for showing images and recording interviews. However, Berbes-Blazquez [66] stated that photos are only snapshots and are not capable of depicting dynamic processes. In the presented work, the Photovoice walks were taken in the months of April and May for the German group and in late summer for persons having a migratory background. Potentially, interviewees might have taken other pictures and expressed other things at a different time of year or in a different weather condition. More similar investigations covering the full year and different weather conditions would be feasible for this reason. With some exceptions to dynamic aspects such as short-term logging activities, at least during the covered periods, the distribution of photos taken along the trail was quite even. From the overall distribution of the images, it can be concluded that the attention to the forest remained high throughout the walk. However, this might have been a result of the tasks given to the participants of taking photos and attention to the forest at least in several parts of the walk might have been lower in a normal recreational situation.

Some sections were different with a large number of negative images showing the clear cut at the small lake and the skid trail with parked machines. For the positive impressions, accumulations can be found for the small ponds and the meadow in the southeastern edge of the trail as well as reaching the tall conifer stands when walking towards them from the mixed forest sections. Many pictures show single trees and were taken towards the canopy or showing the thick tree trunks. A large number of the images taken showed details such as colorful or nicely looking leaves, colors, single trees, fruits, or the anthill. Similarly, negative features were photographed such as litter on the ground or vehicles in the forest causing disturbance. Several participants wanted to describe non-visual effects of the forest such as noise, feelings, and emotions that they found challenging to visualize with a camera. Interestingly, only a minority of photos visualized the forest in a way that would be typically used for choice experiments to determine scenic preferences linked to the broader landscape or forest stand category.

4.2. Reflecting Studies Outcomes

In our studies, near-natural and ecologically valuable forests with only little infrastructure were preferred, even if forests were close to built-up areas and heavily frequented. Even in densely populated areas and next to built-up areas, it was said that forests should be something to contrast to the urban environment and provide "pure nature". While, visually, natural looking mixed forests were preferred; nonetheless, almost every forest type was well-suited for recreation and liked for many features. In our studies, a large share of respondents could not say something negative about the forests. Participants chose or took fewer negative images or put their negative statement into perspective, since they were asked to respond to the interviewers' questions regarding negative aspects or "had to" sort pictures according to the given tasks. Even monocultures or non-native tree species were accepted and appreciated. An important, but not decisive aspect for valuing the different scenic qualities of forests was knowledge and experiences, as well as an understanding of the relationships in the forest ecosystem. The latter was a key element in a special appreciation of "wild" forests containing deadwood. Thick, large, colorful, and bizarre-shaped trees were valued and enriched the forest experience, whereas the type of tree species was of secondary importance. Our findings are in line with and confirm theoretical (e.g., [81]) assumptions or expert-based approaches (such as [82]), as well as studies from Europe and North America [24,28].

Both Germans and persons originating from other countries with other types of ecosystems appreciated Central European forests very much. Similar elements in the World Forest create value for everyone. Depending on the origin, some specific management practices such as grassing goats for nature conservation purposes needed explanation to be understandable. Visiting a forest stimulated respondents to link, resemble, or draw comparisons to and raised memories of forests and nature providing a sense of place linked to both the visited forest and the one from the country of origin. Less frequent forest visits could be linked to mobility aspects (accessibility, e.g., public transportation), lack of knowledge, and information about the forests or awareness about recreational offers and activities as well as less familiarity with forest recreational activities such as hiking or cycling [36,38].

On closer observation, the results concerning forest preferences show that people with a higher level of formal education and long-term familiarity with respective forest ecosystems generally perceived ecologically more valuable nature as more attractive. Thus, this positive assessment is strongly related to learning and knowledge. The Finnish preference study of forest types [23] showed that unmaintained forest stands tend to be rated more negatively and monocultures more positively in terms of visual impression. However, people who had a higher level of education were more likely to accept deadwood and undergrowth and perceived pure spruce stands as very negative [21]. Quite obviously, knowledge of ecological relationships influences the perception of whether a forest stand is perceived as attractive and correlates with findings by Gobster et al. [28] from North America. The group with a higher education was therefore more aware that pure monocultures are often viewed critically from a nature conservation point of view, which is clearly reflected in the reasons given when these people had to justify their selection of negative forest images, for example in image surveys. Mixed stands, on the other hand, combine closeness to nature and biodiversity. The study by Qiu et al. [34] came to the controversial conclusion that while respondents recognize biodiversity, they do not automatically see it as particularly attractive. Picturesque, visibly cultivated landscapes were preferred to wild-looking images. However, it should be noted that in this study, more park-like structures were examined and open impressions were compared there with forest-like impressions. Hence, for this reason, one should be careful transferring findings linked to "Urban Forests" with related expectations more to a managed wooded park than to a real, larger, and less intensively managed woodlands.

Knowledge about forest management can also lead to a positive attitude towards forest management [83,84], with which our findings are in line to some extent. In a study conducted in mountain forests in Italy and Bosnia-Herzegovina, preferring standing and lying deadwood was mainly linked to the cultural background [85]. While close-to-nature-managed forest stands are perceived as positive, logging activities always cause visual disturbance for visitors to the forests and are noticed as negative. Despite the necessity to explain well and communicate the importance of wood harvesting, thereby trying to foster an understanding of the society for timber management, in heavily frequented areas, from an aesthetic standpoint, timber harvesting, wherever possible, should be as invisible, sensitive, and quickly done as possible and include a good cleaning up after logging. In many cases and when conducted with care, good practice or doing some minor restoration activities, vegetation will overgrow any such traces rapidly during the vegetation period.

Tyrväinen et al. [21] found that the preferences for a certain forest type depends not only on the level of education, but also on age, gender, frequency of forest visits, and forest activity. Thus, males, young people, and frequent and active forest visitors accepted ecological management options, abundant undergrowth, and deadwood more often than other groups. Another phenomenon from the study by Tyrväinen et al. [23] and Hauru et al. [86] was also found in our studies. In the Finnish study, for example, deadwood was viewed negatively, especially in coniferous forests, while it was hardly mentioned in mixed forests.

In the Forstenrieder Park, too, it was partly the case that deadwood in the mixed stand was hardly mentioned (or only in a positive sense as a habitat for insects), while visitors often made negative comments about deadwood in the pure deciduous forest, but also about deadwood in the pure spruce forest. The lower acceptance for deadwood in deciduous forest stands could also be related to the fact that deadwood is particularly noticeable in such forest areas in winter, the season in which the pictures were visualized.

In a study conducted both in North American and in a European context on bark beetle outbreaks with vast dying conifer stands and related impacts on scenic qualities using visualizations, visitors preferred healthy mature forest stands and disliked forests with substantial dead wood. Interestingly, the number of visitors shown in the animated visualizations influenced the expressed preferences [87].

While previous knowledge and frequent stays in nature could mean that recreational users behave in a more environmentally friendly and conscious manner when practicing sports [88], we could not find hints in our studies and a broader, evidence-based mixed-method approach would be required to verify such an assumption that is typically assessed in self-reported surveys. Variation between forest stands seem to contribute positively to recreational value, and in some cases, this may outweigh the contribution of variation within a stand according to Filyushkina et al. [89]. Taking a closer look at the results, and especially, at the Photovoice walks, it could be shown that changing stands and other forest types than the preferred mixed, uneven-aged stands are perceived to be very attractive for recreation and scenic enjoyment, such as tall spruce and Douglas fir stands. In particular, many participants photographed the old spruce stand because the size of the trees was perceived as attractive and, as well, the change between different stands was perceived in a positive way. "Mighty thick trees" drew attention and were very much appreciated, regardless of whether they were non-native trees or trees in an even-aged monoculture. With the Photovoice method, it could be shown that the view, direction, and attendance to a forest changes between a broader look at the forest stand and attention to small details such as single trees, colorful leaves, or small vegetation features. To some extent, also non-visual aspects of forest perception could be captured this way.

4.3. Implication and Further Work

With our work, it can be shown that multiple benefit management is as well highly valued, and the respective management can increase provisioning sociocultural ES. While there is a link between high biodiversity and nature conservation, from the valuation by people, a closer look shows that managed, structured, diverse, and resilient forests can provide or create very high values for humans as well and might better cope with intense recreation in urban areas. Looking at demographic and cultural backgrounds, valuing forests are similar for most visitors, including persons having more familiarity with other types of landscapes. While the forest and many services are perceived as positive, the use of timber and related actions is perceived as negative. In the eyes of most visitors, forests should be "pure nature".

For unveiling the potentials of forests, accessibility, opportunities for experiencing forests, as well as information and learning opportunities can stimulate valuing forests, forest structures, and elements. Based on the knowledge of the elements and structures especially appreciated, forest managers can develop visitor management strategies to steer or develop attractive places away from valuable, sensitive areas to more robust forests.

Often, small features and elements are highly valued and lead to a high appreciation and were reported frequently in the Photovoice walks. Qualitative, user-directed approaches in field settings can help to better understand and identify such elements assigning values or link elements with aspects that are important to create a sense of place, positive feeling, and human wellbeing. With still many gaps existing, especially to assess and value sociocultural ES, applications of in-depth, user-centered approaches can contribute to further develop a better understanding how humans appreciate and value nature.

5. Conclusions

Using in-depth approaches attempting to take the view through the eyes of forest visitors, interesting insights to the perception of forests and landscapes can be collected. Such approaches contribute to a broader understanding of preferences and satisfaction with the forest recreational experience. Making use of approaches such as Photovoice allows the participants better to determine and direct the course of the investigation. Additional insights for understanding the perception of forests and landscapes can be captured. While overall theories of landscape perceptions and approaches such as choice experiments can explain preferences in general or give feedback to management practices, it can be shown that perception of a forest goes beyond the overall visual quality of a forest stand. Many more, sometimes small visual elements such as colorful leaves or ground vegetation contribute to positive, in some cases, negative experiences, such as little pieces of litter. Even forests that might seem less attractive when looking at outcomes of choice experiments or from a literature point of view, can be perceived very attractive for recreation and scenic enjoyment by actual visitors. Qualitative, user-centered approaches provide interesting possibilities and opportunities to better understand landscape perception as well nuances that contribute to a broader understanding of landscape perception.

Author Contributions: Conceptualization, G.L. and V.K.; methodology, G.L. and V.K.; validation, G.L. and V.K.; formal analysis, J.F., T.H., C.K., A.R., J.S., G.L. and V.K.; investigation, J.F., T.H., C.K., A.R., J.S., V.K. and G.L.; resources, S.P.; writing—original draft preparation, G.L.; writing—review and editing, G.L.; visualization, G.L, A.R., J.F., T.H., J.S. and G.L.; supervision, S.P.; project administration, G.L. and S.P.; funding acquisition, S.P. All authors have read and agreed to the published version of the manuscript.

Funding: The work presented in this paper was conducted within the Project "Urban Woodlands 2050" funded by the Bavarian State Forest Authorities with grants from the Bavarian Ministry of Food, Agriculture and Forestry (Grant Number G 36).

Data Availability Statement: The data presented in this study are available on request from the corresponding author.

Acknowledgments: The authors would like to thank Stefan Huber, Thomas Stör, Herbert Rudolf, Alfred Fuchs (Freising Unit of the Bavarian State Forest Enterprise), Wilhelm Seerieder (Munich Unit of the Bavarian State Forest Enterprise), Heinz Utschig (Wasserburg Unit of the Bavarian State Forest Enterprise) and Manfred Schölch (University of Applied Sciences Weienstephan-Triesdorf) for supporting this work. We would also like to thank Elizabeth Hamzi-Schmidt at the TUM English Writing Center for polishing up the language.

Conflicts of Interest: The authors declare no conflict of interest.

Research Involving Humans: The study presented involved humans. No personal data that could be assigned to an individual were collected and anonymity of participants were maintained throughout the entire research process. Data were collected in a way so that they cannot be linked or traced back to a person. Conducting and handling of the collected data and maintaining privacy of persons follows the legal basis of the EU, REGULATION (EU) 2016/679 OF THE EUROPEAN PARLIAMENT AND OF THE COUNCIL of 27 April 2016 on the protection of natural persons with regard to the processing of personal data and on the free movement of such data, and repealing GDPA Directive 95/46/EC (General Data Protection Regulation) and corresponding country-specific regulation for the Federal Republic of Germany.

References

1. Nowak, D.J. The effects of urban forests on the physical environment. In *COST Action E12 Urban Forests and Urban Trees*; Randrup, T.B., Konijnendijk, C.C., Christophersen, T., Nilsson, K., Eds.; Office for Official Publications of the European Communities: Luxembourg, 2002; pp. 22–42.
2. Lupp, G.; Förster, B.; Kantelberg, V.; Markmann, T.; Naumann, J.; Honert, C.; Koch, M.; Pauleit, S. Assessing the Recreation Value of Urban Woodland Using the Ecosystem Service Approach in Two Forests in the Munich Metropolitan Region. *Sustainability* **2016**, *8*, 1156. [CrossRef]
3. Burkhardt, I.; Dietrich, R.; Hoffmann, H.; Leschnar, J.; Lohmann, K.; Schoder, F.; Schultz, A. *Urbane Wälder*; Bundesamt für Naturschutz: Bonn-Bad Godesberg, Naturschutz und Biologische Vielfalt: Bonn, Germany, 2009; Volume 63, 214p.
4. Fryd, O.; Pauleit, S.; Bühler, O. The role of urban green space and tress in relation to climate change. *CAB Rev. Perspect. Agric. Vet. Sci. Nutr. Nat. Resour.* **2011**, *6*, 1–18.
5. Reif, A.; Aas, G.; Essl, F. Braucht der Wald in Zeiten der Klimaänderung neue, nicht heimische Baumarten? *Nat. Landsch.* **2011**, *86*, 256–260.
6. Martens, D.; Bauer, N. Im Test: Wald als Ressource für psychisches Wohlbefinden. *Schweiz. Z. Forstwes.* **2010**, *3*, 90–96.
7. Pörtner, H.O.; Scholes, R.J.; Agard, J.; Archer, E.; Arneth, A.; Bai, X.; Barnes, D.; Burrows, M.; Chan, L.; Cheung, W.L.; et al. *IPBES-IPCC Co-Sponsored Workshop Report on Biodiversity and Climate Change*; IPBES, IPCC: Bonn, Germany; Geneva, Switzerland, 2021; 24p. [CrossRef]
8. IPBES. *Workshop Report on Biodiversity and Pandemics of the Intergovernmental Platform on Biodiversity and Ecosystem Services*; IPBES Secretariat: Bonn, Germany, 2020; 95p. [CrossRef]
9. Baur, B. Konkurrierende Raumnutzungen: Forstwirtschaft, Erholung und Naturschutz in stadtnahen Wäldern. In *Landschaftsqualitäten*; Tanner, K.M.M., Bürgi, M., Coch, T., Eds.; Haupt: Bern, Switzerland; Stuttgart, Germany; Wien, Austria, 2006; pp. 243–261.
10. Millenium Ecosystem Assessment. *Ecosystems and Human Well-Being: Current, State and Trends; Findings of the Condition and Trends Working Group*; Island Press: Washington, DC, USA, 2005; 948p.
11. Grunewald, K.; Bastian, O. (Eds.) *Ecosystem Services–Concept, Methods and Case Studies*; Springer: Berlin/Heidelberg, Germany, 2015; 312p.
12. Bell, S.; Tyrväinen, L.; Sievänen, T.; Pröbstl, U.; Simpson, M. Outdoor recreation and nature tourism: A European perspective. *Living Rev. Landsc. Res.* **2007**, *1*, 1–46. Available online: http://lrlr.landscapeonline.de/Articles/lrlr-2007-2/ (accessed on 27 July 2022). [CrossRef]
13. Forest Research. *Benefits of Green Infrastructure; Report by Forest Research*; Forest Research: Farnham, UK, 2010.
14. Bieling, C.; Plieninger, T.; Pirker, H.; Vogl, C.R. Linkages between landscapes and human well-being: An empirical exploration with short interviews. *Ecol. Econ.* **2014**, *105*, 19–30. [CrossRef]
15. Dickinson, D.C.; Hobbs, R.J. Cultural ecosystem services: Characteristics, challenges and lessons for urban green space research. *Ecosyst. Serv.* **2017**, *25*, 179–194. [CrossRef]
16. Manning, R.E. *Studies in Outdoor Recreation*, 2nd ed.; Oregon State University Press: Corvalis, OR, USA, 1999; 374p.

17. Arnberger, A.; Hinterberger, B. Visitor monitoring methods for managing public use pressures in the Danube Floodplains National Park, Austria. *J. Nat. Conserv.* **2003**, *11*, 260–267. [CrossRef]
18. Arnberger, A. Recreation use of urban forests: An inter-area comparison. *Urban For. Urban Green.* **2006**, *4*, 135–144. [CrossRef]
19. Manning, R.E.; Lime, D.W. Defining and managing the quality of wilderness recreation experiences. *USDA For. Serv. Proc.* **2000**, 13–52. Available online: https://www.fs.usda.gov/treesearch/pubs/22002 (accessed on 7 August 2022).
20. Yang, G.; Yu, Z.; Zhang, J.; Kristensen, L.S. From preference to landscape sustainability: A bibliometric review of landscape preference research from 1968 to 2019. *Ecosyst. Health Sustain.* **2021**, *7*, 1948355. [CrossRef]
21. Tyrväinen, L.; Silvennoinen, H.; Kolehmainen, O. Ecological and aesthetic values in urban forest management. *Urban For. Urban Green.* **2003**, *1*, 135–149. [CrossRef]
22. Karjalainen, E.; Tyrväinen, L. Visualization in forest landscape preference research: A Finnish perspective. *Landsc. Urban Plan.* **2002**, *59*, 15–26. [CrossRef]
23. Tyrväinen, L.; Uusitalo, M.; Silvennoinen, H.; Hasu, E. Towards sustainable growth in nature-based tourism destinations: Clients' views of land use options in Finnish Lapland. *Landsc. Urban Plan.* **2014**, *122*, 1–15. [CrossRef]
24. Lindhagen, A.; Hörnstein, L. Forest Recreation in 1977 and 1997 in Sweden: Changes in public preferences and behaviour. *Forestry* **2000**, *73*, 143–153. [CrossRef]
25. Chen, Y.; Sun, B.; Liao, S.B.; Chen, L.; Luo, S.X. Landscape perception based on personal attributes in determining the scenic beauty of in-stand natural secondary forests. *Ann. For. Res.* **2016**, *59*, 91–103. [CrossRef]
26. Xu, M.; Luo, T.; Wang, Z. Urbanization diverges residents' landscape preferences but towards a more natural landscape: Case to complement landsenses ecology from the lens of landscape perception. *Int. J. Sustain. Dev. World Ecol.* **2020**, *27*, 250–260. [CrossRef]
27. Zhang, G.; Yang, J.; Wu, G.; Hu, X. Exploring the interactive influence on landscape preference from multiple visual attributes: Openness, richness, order, and depth. *Urban For. Urban Green.* **2021**, *65*, 127363. [CrossRef]
28. Gobster, P.H.; Nassauer, J.I.; Daniel, T.C.; Fry, G. The shared landscape: What does aesthetics have to do with ecology? *Landsc. Ecol.* **2007**, *22*, 959–973. [CrossRef]
29. Wöbse, H.H. *Untersuchungen zum Nutzungs-und Bestandeswandel der Sollingwälder, die vom Urlauber bevorzugten Waldtypen und Folgerungen für die Planung*; Fakultät für Gartenbau und Landeskultur der Technischen Universität Hannover: Hannover, Germany, 1972; 205p.
30. Ammer, U.; Ammer, P. *Untersuchungen zur Erholungs-und Waldpflegeplanung in den Isarauen nördlich von München*; Lehrstuhl für Landschaftstechnik der Universität München: München, Germany, 1982.
31. Hoisl, R.; Nohl, W.; Engelhardt, P. *Naturbezogene Erholung und Landschaftsbild*; KTBL-Schrift 389; Landwirtschaftsverlag: Darmstadt, Germany, 2003; 306p.
32. Suda, M.; Schaffner, S. Der Wald in der individuellen und gesellschaftlichen Wahrnehmung. *Prowald-Mag. Des Dtsch. Forstvereins* **2006**, *3*, 4–7.
33. Lindgren, C.A. Forest Aesthetics. In *Multiple-Use Forestry in the Nordic Countries*; Hytönen, M., Ed.; METLA, The Finnish Forest Research Institute: Helsinki, Finland, 1995; pp. 279–293.
34. Qiu, L.; Lindberg, S.; Nielsen, A.B. Is biodiversity attractive?—On-site perception of recreational and biodiversity values in urban green space. *Landsc. Urban Plan.* **2013**, *119*, 136–146. [CrossRef]
35. Kaplan, R.; Kaplan, S. *The Experience of Nature—A Psychological Experience*; Cambridge University Press: Cambridge, MA, USA, 1989; 385p.
36. Gobster, P.H. Managing Urban Parks for a Racially and Ethnically Diverse Clientele. *Leis. Sci.* **2002**, *2*, 143–159. [CrossRef]
37. Comber, A.; Brunsdon, C.; Green, E. Using a GIS-based network analysis to determine urban greenspace accessibility for different ethnic and religious groups. *Landsc. Urban Plan.* **2008**, *86*, 103–114. [CrossRef]
38. Covelli Metcalfa, E.; Burns, R.C.; Graefe, A.R. Understanding non-traditional forest recreation: The role of constraints and negotiation strategies among racial and ethnic minorities. *J. Outdoor Recreat. Tour.* **2013**, *1–2*, 29–39. [CrossRef]
39. Buijs, A.E.; Elands, B.H.M.; Langers, F. No wilderness for immigrants—Cultural differences in images of nature and landscape preferences. *Landsc. Urban Plan.* **2009**, *91*, 113–123. [CrossRef]
40. Jay, M.; Schraml, U. Diversity in mind. Towards a differentiated understanding of migrants´recreational practices in urban forests. *Urban For. Urban Green.* **2014**, *13*, 38–47. [CrossRef]
41. Karameris, A. *Analyse und Prognose der Erholungsnachfrage in Wäldern als Forstlicher Beitrag zur Raumplanung*; Forstwissenschaftliche Fakultät der Universität München: München, Germany, 1982.
42. Lindenau, G. Die Erholungseignung des Auwaldes, dargestellt am Beispiel der Isaraue zwischen München und Freising. Diploma Thesis, Technical University of Munich, Munich, Germany, 1996.
43. Nohl, W. *Landschaftsplanung—Ästhetische und Rekreative Aspekte*; Patzer: Berlin/Hannover, Germany, 2001; 248p.
44. Wöbse, H.H. *Landschaftsästhetik—Über das Wesen, die Bedeutung und den Umgang mit landschaftlicher Schönheit*; Ulmer: Stuttgart, Germany, 2002; 304p.
45. Löw, G.; Michal, I. *Krajinny Raz*; Lesnicka Prace: Kostelec nad Cernymi, Czech Republic, 2003; 552p.
46. Augenstein, I. *Die Ästhetik der Landschaft—Ein Bewertungsverfahren für die Planerische Umweltvorsorge*; Weißenseeverlag: Berlin, Germany, 2002; 170p.
47. Bourassa, S.C. *The Aesthetics of Landscape*; Belhaven Press: London, UK; New York, NY, USA; Great Britain, UK, 1991; 168p.

48. Appleton, J. *The Experience of Landscape*; Wiley: New York, NY, USA, 1975; 293p.
49. Orians, G.H. Habitat Selection—A General Theory and Application to Human Behavior. In *The Evolution of Human Social Behavior*; Lockard, J.S., Ed.; Elsevier: New York, NY, USA, 1980; 336p.
50. Bürger-Arndt, R.; Reeh, T. Landschaftsästhetik—Theoretische Grundlagen. In *Handbuch Naturschutz und Landschaftspflege*; Konold, W., Böcker, R., Hampicke, U., Eds.; Ecomed: Landsberg, Germany, 2006.
51. Legewie, H.; Ehleers, W. *Knaurs Moderne Psychologie*; Droemersche Verlagsanstalt: München, Germany; Zürich, Switzerland, 1972; 312p.
52. Lupp, G. : *Landschaftswahrnehmung von Anwohnern und Besuchern des Müritz-Nationalparks und Prognose zu Erwartender Veränderungen im Landschaftsbild*; Culterra, Schriftenreihe des Instituts für Landespflege der Albert-Ludwigs-Universität Freiburg: Freiburg, Germany, 2008; 267p.
53. Newman, P.; Manning, R.; Pilcher, E.; Trevino, K.; Savidge, M. Understanding and Managing Soundscapes in National Parks: Part 1—Indicators of Quality. In *Exploring the Nature of Management, Proceedings of the Third International Conference on Monitoring and Management of Visitor Flows in Recreational and Protected Areas*; Siegrist, D., Clivaz, C., Hunziker, M., Iten, S., Eds.; University of Applied Sciences Rapperswil: Rapperswil, Switzerland, 2006; pp. 193–195.
54. Miyazaki, Y.; Morikawa, T.; Yamamoto, N. Effect of wooden odoriferous substances on humans. *J. Physiol. Anthropol. Appl. Hum. Sci.* **1999**, *18*, 189.
55. Moher, D.; Liberati, A.; Tetzlaff, J.; Altman, D.G. Reprint-Preferred Reporting Items for Systematic Reviews and Meta-Analyses: The PRISMA Statement. *Phys. Ther.* **2009**, *89*, 873–880. [CrossRef]
56. Taylor, L. Case Study Methodology. In *Key Methods in Geography*; Clifford, N., Cope, N., Gillespie, T., French, S., Eds.; Sage: London, UK; Thousand Oaks, CA, USA; New Delhi, India; Singapore; Great Britain, UK, 2016; pp. 581–595.
57. Allesch, C.; Keul, A. Analyse von Mensch-Umwelt-Beziehungen. In *Sozialwissenschaftliche Methoden. Lehr- und Handbuch für Forschung und Praxis*; Roth, E., Ed.; Oldenbourg: München, Germany; Wien, Austria, 1995; 845p.
58. Gareis-Grahmann, F.J. *Landschaftsbild und Umweltverträglichkeitsprüfung: Analyse, Prognose und Bewertung des Schutzgutes „Landschaft" nach dem UVPGM*; Erich Schmidt: Berlin, Germany, 1993; 270p.
59. Mayring, P. *Einführung in die Qualitative Sozialforschung*, 6th ed.; Beltz Deutscher Studien Verlag: Weinheim, Germany; Basel, Switzerland, 2016; 170p.
60. Statistisches Bundesamt. Einwohnerzahl der 15 Größten Städte in Deutschland. Available online: https://www.destatis.de/DE/Themen/Gesellschaft-Umwelt/Bevoelkerung/Bevoelkerungsstand/bar-chart-race.html (accessed on 27 July 2022).
61. Statistisches Bundesamt. Hamburg und Oberbayern unter den 10 wohlhabendsten EU-Regionen. Available online: https://www.destatis.de/Europa/DE/Thema/Allgemeines-Regionales/regionales-bip.html (accessed on 27 July 2022).
62. Ruppert, K.; Gräf, P.; Lintner, P. *Naherholungsverhalten im Raum München. Persistenz und Wandel freizeitorientierter Regionalstrukturen 1968/80*; Akademie für Raumforschung und Landesplanung: Hannover, Germany, 1986.
63. Verein Freunde des Forstenrieder Parks e.V. (Ed.) *Der Forstenrieder Park—Von der Jagdlandschaft zum Erholungsgebiet*; München Verlag: München, Germany, 2012; 144p.
64. Ammer, U.; Weidenbach, M.; Beer, M.; Hwang, Y.-H. Landschafts- und Erholungsplanerische Entwicklungsstudie für die Wildparke im Ebersberger Forst und im Forstenrieder Park. Available online: https://landconsult.de/markus/wildpark/ (accessed on 27 July 2022).
65. Bayerische Staatsforsten. Weltwald Freising. Available online: www.weltwald.de (accessed on 27 July 2022).
66. Berbes-Blazquez, M. A participatory assessment of ecosystem services and human wellbeing in rural Costa Rica using photo-voice. *Environ. Manag.* **2012**, *49*, 862–875. [CrossRef]
67. Kong, T.M.; Kellner, K.; Austin, D.E.; Els, Y.; Orr, B.J. Enhancing Participatory Evaluation of Land Management through Photo Elicitation and Photovoice. *Soc. Nat. Resour.* **2014**, *28*, 212–229. [CrossRef]
68. Booth, T.I.M.; Booth, W. In the Frame. Photovoice and mothers with learning difficulties. *Disabil. Soc.* **2003**, *18*, 431–442. [CrossRef]
69. Delgado, M. *Urban Youth and Photovoice. Visual Ethnography in Action*; Oxford Univ. Press: New York, NY, USA, 2015; 289p.
70. Wang, C.; Burris, M.A. Empowerment through Photo Novella. Portraits of Participation. *Health Educ. Behav.* **1994**, *21*, 171–186. [CrossRef] [PubMed]
71. Wang, C.; Burris, M.A. Photovoice. Concept, Methodology, and Use for Participatory Needs Assessment. *Health Educ. Behav.* **1997**, *24*, 369–387. [CrossRef] [PubMed]
72. Heyman, E. Analyzing recreational values and management effects in an urban forest with the visitor-employed photography method. *Urban For. Urban Green.* **2012**, *11*, 267–277. [CrossRef]
73. Stephenson, W. *The Study of Behavior—Q-Technique and Its Methodology*; The University of Chicago Press: Chicago, IL, USA; London, UK; Great Britain, UK, 1953; 376p.
74. Lamnek, S. *Qualitative Sozialforschung. Bd. 1: Methologie*, 3rd ed.; Beltz Psychologische Verlags Union: Weinheim, Germany, 1995; 308p.
75. Lamnek, S. *Qualitative Sozialforschung. Bd. 2: Methoden und Techniken*, 3rd ed.; Beltz Psychologische Verlags Union: Weinheim, Germany, 1995; 440p.
76. Lupp, G.; Kantelberg, V.; Förster, B.; Honert, C.; Naumann, J.; Markmann, T.; Pauleit, S. Visitor Counting and Monitoring in Forests Using Camera Traps: A Case Study from Bavaria (Southern Germany). *Land* **2021**, *10*, 736. [CrossRef]

77. Półrolniczak, M.; Kolendowicz, L. The influence of weather and level of observer expertise on suburban landscape perception. *Build. Environ.* **2021**, *202*, 108016. [CrossRef]
78. Li, J.; Zhang, Z.; Jing, F.; Gao, J.; Ma, J.; Shao, G.; Noel, S. An evaluation of urban green space in Shanghai, China, using eye tracking. *Urban For. Urban Green.* **2020**, *56*, 126903. [CrossRef]
79. Hull, R.; Stewart, W. *Validity of Photo-Based Scenic Beauty Judgments*; College of Architecture and Department of Recreation, Park and Tourism Services, Texas A&M University: College Station, TX, USA, 1992; pp. 112–113.
80. Bethmann, S.; Wurster, M. Zum Image der Forstwirtschaft. *AFZ-Der Wald* **2016**, *3*, 38–42.
81. Stölb, W. *Waldästhetik—über Forstwirtschaft, Naturschutz und die Menschenseele*; Kessel: Oberwinter, Germany, 2005; 400p.
82. Edwards, S.; Jay, M.; Jensen, F.S.; Beatriz Lucas, B.; Marzano, M.; Montagné, C.; Peace, A.; Weiss, G. Public preferences for structural attributes of forests: Towards a pan-European perspective. *For. Policy Econ.* **2012**, *19*, 12–19. [CrossRef]
83. Ford, R.M.; Williams, K.J.H.; Bishop, I.D.; Hickey, J.E. Effects of Information on the Social Acceptability of Alternatives to Clearfelling in Australian Wet Eucalypt Forests. *Environ. Manag.* **2009**, *44*, 1149–1162. [CrossRef]
84. Gunderson, V.; Frivold, L.H. Naturally dead and downed wood in Norwegian boreal forests: Public preferences and the effect of information. *Scand. J. For.* **2011**, *26*, 110–119. [CrossRef]
85. Pastorella, F.; Avdagić, A.; Čabaravdić, A.; Mraković, A.; Osmanović, M.; Paletto, A. Tourists' perception of deadwood in mountain forests. *Ann. For. Res.* **2016**, *59*, 311–326. [CrossRef]
86. Hauru, K.; Koskinen, S.; Kotze, D.J.; Lehvävirta, S. The effects of decaying logs on the aesthetic experience and acceptability of urban forests—Implications for forest management. *Landsc. Urban Plan.* **2014**, *123*, 114–123. [CrossRef]
87. Arnberger, A.; Ebenberger, M.; Schneider, I.E.; Cottrell, S.; Schlueter, A.C.; von Ruschkowski, E.; Venette, R.C.; Snyder, S.A.; Gobster, P.H. Visitor Preferences for Visual Changes in Bark Beetle-Impacted Forest Recreation Settings in the United States and Germany. *Environ. Manag.* **2018**, *61*, 209–223. [CrossRef] [PubMed]
88. D'Antonio, A.; Monz, C.; Newman, P.; Lawson, S.; Taff, D. The Effects of Local Ecological Knowledge, Minimum-Impact Knowledge, and Prior Experience on Visitor Perceptions of the Ecological Impacts of Backcountry Recreation. *Environ. Manag.* **2012**, *50*, 542–554. [CrossRef] [PubMed]
89. Filyushkina, A.; Agimass, F.; Lundhede, T.; Strange, N.; Bredahl Jacobsen, J. Preferences for variation in forest characteristics: Does diversity between stands matter? *Ecol. Econ.* **2017**, *140*, 22–29. [CrossRef]

forests

MDPI

Article

Ownership, Governance, Uses, and Ecosystem Services of Community Forests in the Eastern United States

Meredith Hovis [1,*], Gregory Frey [2], Kathleen McGinley [3], Frederick Cubbage [4], Xue Han [5] and Megan Lupek [4]

1 Udall Center for Studies in Public Policy, University of Arizona, Tucson, AZ 85719, USA
2 USDA Forest Service, Southern Research Station, Research Triangle Park, NC 27709, USA
3 USDA Forest Service, International Institute of Tropical Forestry, Río Piedras, PR 00926, USA
4 Department of Forestry and Environmental Resources, North Carolina State University, Raleigh, NC 27695, USA
5 Department of Management, Harbin Finance University, Harbin 150028, China
* Correspondence: hovis@arizona.edu

Abstract: Over time, community forests (CFs) have been established across the globe to meet various social, economic, and ecological needs. Benefits of CFs include conserving resilient forests and natural resources and ecosystem services, enhancing social and economic capital, and leveraging local and indigenous knowledge in forest and natural resource management and decision-making. Research on CFs in the U.S. is quite limited, and cases that have been assessed show a wide spectrum in terms of CF ownership, organizational structure, governance, property rights, and uses. Through an exploratory research approach, this study enhances the understanding of the characteristics of CFs in the U.S. and the ecosystem services and other benefits that they provide. Through online web searches, we compiled one of the first comprehensive lists of CFs in the Eastern U.S. Prior to this study, there was no publicly available comprehensive database or list of CFs in the country. Subsequently, we conducted comparative case study research, which included semi-structured in-person interviews with key stakeholders from four CFs in the Eastern U.S. to understand CF ownership, governance, uses, and benefits. CFs benefits frequently included cultural services, such as recreation and education, and regulating and supporting services, such as water quality and wildlife habitat. Much less common was a focus on provisioning services such as timber or non-timber forest products. Maintaining collaboration and funding for CF efforts in the long run without significant CF revenues remains a challenge for most forests. Overall, this research sheds lights on CF characteristics and capacities in the Eastern U.S. and identifies potential opportunities and needs for the U.S. in the future. CFs researchers, managers, and community members.

Keywords: community forests; Eastern United States; recreation; ecosystem services; governance; public participation

1. Introduction

In 2017, forests covered 766 million acres (310 ha) in the United States [1]. Although the population has increased in the U.S. substantially since 1910, the total forest land area has been relatively stable. The US forest ownership consists of 58% private forest lands (i.e., 34% private noncorporate or family forest lands, 20% corporate forest lands, 2% tribal forest lands, and 2% other private forest lands) and 42% public forest lands (i.e., 31% federal forest lands; 9% state, and 2% local government forests) [1]. The perspectives of public and private owners are too diverse to characterize narrowly, but one might surmise that industrial forest owners focus the most on commodity production such as timber, family forest lands use lands for commodities, investments, recreations, and amenities. Public lands in the U.S. may focus more on ecosystem services and recreation but retain some timber and commodity production, especially on state lands.

Despite a relatively stable land area, forests are faced with an expanding and intersecting host of issues, such as drought, wildfire, floods, and invasive species, all of which are intensified by global climate change, as well as deforestation from land-use changes and alterations [2,3]. Traditional forest policy responses to forest threats have included public ownership, regulation, education, and incentives. These focused mainly on timber and non-timber products in the past, but broader sets of ecosystem services provided by forests have become increasingly important. For example, expanding Federal government programs (e.g., Forest Legacy Program, Forest Stewardship Program) pay private forest landowners to encourage forest retention and sustainable management practices to enhance and restore ecosystem services [3].

Government ownership and management of forests and natural resources were initially implemented to protect natural resources from overexploitation. Private markets have yet to provide the socially optimal level of ecosystem goods and services from forests since there are few active markets and prices for the broad range of these goods and services. In response, various private, nongovernment, and government organizations have developed programs that provide payments for ecosystem services from forests and other natural resource systems, and voluntary environmental certification programs intended to incentivize sound resource management and protection through market access, product labeling, and price differentials.

Forests provide a large number of market commodities in the US, primarily through timber products and developed recreation, as well as a vast array of non-market commodities such as air, water, carbon storage, and biodiversity. This mix of forest goods and services requires both private markets and community management [4]. New forms of natural resources governance and management have evolved to address expanding knowledge and values of forest goods and services [5,6]. Among these new approaches are community forests (CFs), involving a broader role for local communities in forest decision-making and benefits. CFs have been put forth as an innovative prospect for enhancing forest conservation and local livelihoods in ways that traditional policy tools do not typically provide.

CFs took root in the international context in the late 1970s to meet the needs of rural populations [7,8]. In 1978, the UN's Food and Agriculture Organization (FAO) described "community forests" as a means for "provision of fuel and other goods essential to meeting the basic needs at the rural household and community level, provision of food and environmental stability necessary for continued food production, and the generation of income and employment in rural communities [9]".

Although the development and study of CFs have expanded globally since the 1970s, there is no universal definition of "community forest", because of differences in policy, history, culture, ecology, and other contexts [10–12]. In the U.S., tribal lands and New England town forests represent historical models of CFs, with more recent models including CF across a range of public and private ownerships [8,13–15]. Generally, researchers and scholars agree that CFs are defined by the following characteristics: (1) local populations have a substantive role in decision-making, (2) local values are incorporated into management and governance, and (3) CFs are managed to deliver market and nonmarket benefits to local people [11,16–19]. In the U.S. context, what makes CFs different from private or government-owned forestlands is the role that local community members play in forest decisions and outcomes, such as determining the objectives and purpose of the CF, managing and monitoring the property, and receiving social and economic benefits [8,20].

There is relatively little literature that compares the characteristics, status, and outcomes of CFs in the U.S., and examples across the U.S. portray a diverse spectrum of CF forms [15,17,21,22]. The refereed literature on CFs consists mainly of studies from the global South where marketable benefits are the community's primary focus (e.g., [23,24]). Some literature has been recently published on CFs in the Western U.S. (e.g., [25,26]), and others have focused on U.S. New England town forests (e.g., [8]).

CFs encompass organizational and governance mechanisms that may protect and provide a broad range of ecosystem services at multiple scales. The Millennium Ecosystem Assessment (MEA) (2005) developed the standard approach for classifying ecosystem services [4]. Supporting services include nutrient cycling, soil formation, and primary production. Provisioning services include commodities such as food, fresh water, wood and fiber, and fuel. Regulating services include climate regulation, flood regulation, disease regulation, and water purification. Cultural services include aesthetic, spiritual, educational, and recreational [4]. These various services are linked in multiple ways to well-being—security, the basic material for a good life, health, good social relations, and freedom of choice and action [27]. We utilized the MEA approach to ecosystem services to frame our research questions, methods, and instruments.

The purpose of this study was to explore the diversity of self-identified "community forests" in the Eastern U.S. (including Puerto Rico) in terms of: ownership, organizational structure, governance, property rights, and forest uses (i.e., ecosystem services). To better understand these topics, we explored the background, history of the establishment, current practices and activities, and community perceptions of a subset of CFs in the Eastern U.S. The results shed light on the links between these key characteristics of CFs and ecosystem services objectives and outcomes. In addition, the study results can help CF managers, local volunteers, managers, practitioners, and researchers in the U.S., and potentially across the globe, to understand the best establishment and management practices of CFs and how they benefit local communities.

2. Methods

We conducted exploratory social science research to address our research objective. First, we collected data on key characteristics of CFs in the Eastern U.S. and the range of ecosystem services that they aim to provide and protect from self-identified "community forests" from primary and secondary sources available on the internet. Second, we conducted in-depth comparative case study research to gain detailed information from four selected CFs representing a range of CF characteristics and outputs.

2.1. Compilation of Online Information

Through online web-based searches, we assembled a list of CFs in the Eastern U.S. (east of the Great Plains and including Puerto Rico). We selected the Eastern U.S. due to the lack of current refereed literature on CFs in the region, except for some research on town forests in the U.S. Northeast (e.g., [8]).

The online search provided publicly available information about the CFs and formed a basis for selecting comparative case studies for the second phase of this research. At the time of initial data collection, there was no publicly available comprehensive database or list of CFs in the country. However, several organizations had compiled shorter lists of CFs, typically where they had assisted in land acquisition or establishment of governing institutions or conducted other preliminary/exploratory work (e.g., Trust for Public Lands, U.S. Endowment for Forestry and Communities, the Ford Foundation, USDA Forest Service [28–36]). These documents and lists formed the basis for our iterative web search. As the search proceeded, other initiatives and forests came to light, which were added.

We conducted a systematic search of publicly available and accessible websites and documents. Systematic searches are common in social science research and are utilized increasingly to initiate qualitative research [37–39]. We identified key categories of variables about forest ownership, organizational structure, governance, management, uses, and benefits from the forestry, CF, ecosystem services, and nonprofit management literature (e.g., [32–36]) that we used to guide our online search and generally characterize CFs. We used keywords plus each CF's name to discover further information about each forest. Examples of the keywords included "management", "ownership", and "uses", We included an example of our keyword search in Appendix A. We reviewed each CF webpage (if available online) and documents, such as management plans, meeting minutes, new reports, and annual reports.

2.2. Comparative Case Studies

We then selected four CFs to conduct in-depth case studies involving site visits, field observations, in-person semi-structured interviews with key CF stakeholders, and archival document collection and review. The CFs selected included Río Hondo in Mayagüez, Puerto Rico; Nine Times in Pickens, South Carolina; Birch Ridge in New Durham, New Hampshire; and Page Pond in Meredith, New Hampshire. All are located in the Eastern U.S. and received funding from the U.S. Forest Service Community Forest Program (CFP). This descriptive work helps identify some prevalent CF characteristics and capacities in the Eastern U.S. This investigation of this newer category of U.S. forest ownership and management described some key elements of CFS, and identifies opportunities for future CF researchers, managers, and community members.

We selected the four CFs from our list to learn more about the interactions between CF ownership, governance, management, and ecosystem services objectives and outcomes through a comparative case study approach. The four were chosen to be both representative and diverse, thereby highlighting variability and diversity among CFs in general, and providing rich, contextual data to give insight into the CF background, history of the establishment, current practices, and community perceptions of a subset in the Eastern U.S. The case studies were diverse in terms of their ownership status, geographic location, forest uses, forest size, and management objectives from what we discovered via online searches. We selected the cases purposively to represent a range in scale, geography, tenure arrangements, and socioeconomic benefits. A multiple case study allows for meaningful comparison and exploration of differences and diversity [40–42].

Each case study employed three types of qualitative data collection modes: (1) semi-structured interviews with key informants, (2) document review of official records, financial reports, and other relevant documents to the organization, and (3) systematic web-search of each CF project's online documents and website. This approach triangulates data sources to increase the validity of findings [43–45]. From June to August 2019, we traveled to the communities, spending approximately one week in each location.

We drew from the theory and empirical evidence, themes, and concepts presented in the CF literature to develop a semi-structured interview to elicit respondents' experiences and points of view on CFs (Interview Questionnaire shown in Appendix B). We asked various questions about CF uses and ecosystem services, ownership status, governance and management activities, public input strategies, and strengths and challenges.

We interviewed volunteers, landowners, managers, community members who live near the forest, and local elected officials who handle CF duties. This mix of managers to community members to users is a standard approach in qualitative research, in order to gain different perspectives from stakeholders about the operation, merits, outcomes, and effects of a program or system [45]. We interviewed a total of 18 participants from the case study locations; eight from Río Hondo in Mayagüez, PR, USA; one from Nine Times in Pickens, SC, USA; six from Birch Ridge in New Durham, NH, USA; and three from Page Pond in Meredith, NH, USA.

We obtained a list of documents from each case study to further inform its context in terms of land ownership and protection status; history and development; governance and associated institutional structures; finances, funding, and capital assets; collaboration and partnerships; management practices; outputs, outcomes, and benefits distribution.

We developed a common coding guide, initially consisting of predetermined codes based on the CF literature, study objectives, and case study questionnaires in a deductive approach to the data analysis. As we proceeded through the analysis and coding of case study interviews and other documents, we developed additional codes inductively based on emergent topics and variables discovered from the data through the analytical process [46,47]. We organized the coding guide by six major themes and continued developing and refining the codes from the data until no new or revised codes occurred or emerged (Figure 1). Once each case study was reviewed, we compared the overall themes across the cases.

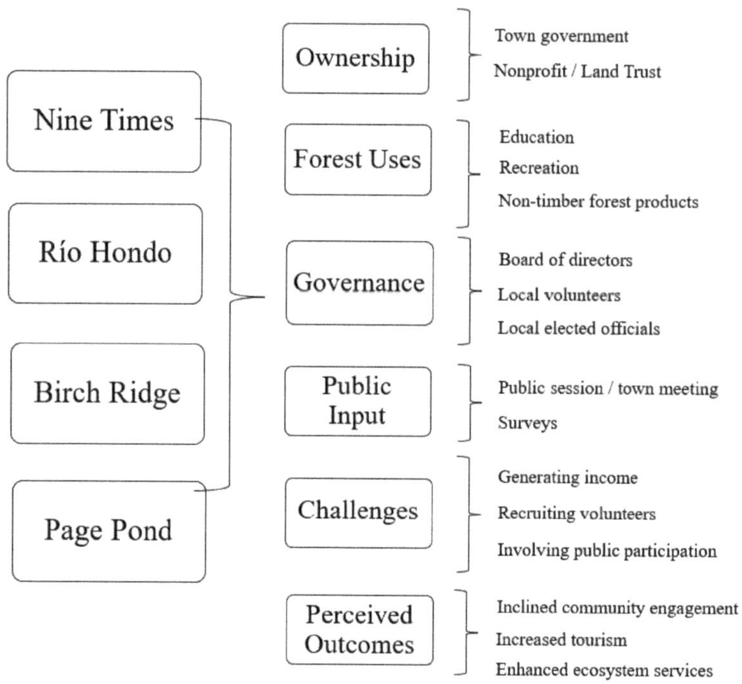

Figure 1. Qualitative data coding tree.

3. Results

3.1. Descriptive Statistics

We discovered 70 CFs in the Eastern U.S. through online searches (Appendix C). Descriptive statistics are indicative but not conclusive about CFs in the Eastern U.S. We found that 47% of the projects made their management plan available online, and 8% had bylaws accessible to the public. Fifty percent referred to working with the public or community in on-the-ground management or decision-making processes, and 80% referred to collaborating with other organizations, such as nonprofit organizations, land trusts, private entities, local businesses, or governmental agencies.

The median area of the eastern CFs was approximately 142 hectares (ha) (350 ac). The first established CF in the East was the Menominee Forest in Neopit, WI, USA, in 1854. The median year of establishment of the Eastern U.S. CFs was 2011. A number of CFs were established decades ago, particularly in the northeast, and many more CFs have been established since the 2008 Farm Bill (Food, Conservation, and Energy Act of 2008; Publ. Law 110–234), which created the U.S. Forest Service's Community Forest and Open Space Conservation Program (also referred to as the Community Forest Program (23)) [31]. The CFP grants tribal entities, local governments, and nonprofit organizations funding to acquire the land to become a CF. Sixty-three percent of eastern CFs identified through the search received funding from the CFP, with the mean amount received from CFP of USD 202,235.

CFs can be owned by various public or private entities [15,48,49]. Most (n = 35) of the eastern CFs identified are owned by local governments, and many (n = 25) of the remaining eastern CFs are owned by nonprofit organizations, such as land trusts. Seven are jointly owned, with a combination of town government and nonprofit organizations. Two of the eastern CFs in our database are owned by tribal entities. Only one is owned by the state or federal government, and none of the eastern CFs are owned by a private for-profit entity.

The majority of CF websites list that their purposes include forest access for cultural ecosystem services, including recreational purposes (93%) such as hiking, biking, snowmobiling, or horseback riding, as well as education (69%). Many projects (87%) reference on their website the importance of regulating and supporting ecosystem services that the CF provides for their community, such as clean water and air, carbon sequestration, and vital biodiversity habitat. Many of the CF's websites also reference forest provisioning ecosystem services, including harvesting and generation timber forest products (64%) and some for non-timber forest products (21%).

3.2. Qualitative Comparative Case Studies

Generally, there are three purposes for establishing CFs in the U.S.: (1) to secure local access to various goods and services, (2) to allow for public participation and input in decision-making and management processes, and (3) to protect the property from conversion to non-forest uses [8,36,50–52]. Based on the list of CFs in the eastern U.S. and the extensive online literature search, we selected our case studies that illustrate the application of these three principles to varying degrees.

The four case studies represent a range of ownership types, management objectives, and governance regimes (Table 1). We identified interview participants through managers and key players from each CF. For anonymity, interviewees are represented in the text with an acronym (RH = Rio Hondo, NT = Nine Times, BR = Birch Ridge, PP = Page Pond) and a number.

Table 1. Community Forest Case Study Sites Overview.

Community Forest	Río Hondo, Mayagüez, PR, USA	Nine Times, Pickens, SC, USA.	Birch Ridge, New Durham, NH, USA	Page Pond, Meredith, NH, USA
		Characteristics		
Size (Hectares)	28	667	815	310
Town Population	73,336	3169	2693	1495
Forest Type	Secondary subtropical; urban forest	Mixed hardwood and pine	Birch and northern hardwoods	Wetland; upland forest
Year Established	2018	2012	2019	2009
Ownership	Local entity—Town of Mayagüez	Nonprofit entity—Naturaland Trust	Nonprofit entity—Southeast Land Trust	Local entity—Town of Meredith
Administration	Río Hondo CF Board of Directors, Town of Mayagüez	Naturaland Trust Board of Directors	Southeast Land Trust	Town of Meredith Conservation Commission
Forest Manager	Río Hondo CF Board of Directors, Town of Mayagüez, local community members	Naturaland Trust Board of Directors	Southeast Land Trust, Moose Mountain Regional Greenways, Powder Mill Snowmobile Club, local community members	Town of Meredith Stewards
Forest Use	Ecological services, education, recreation, NTFPs	Recreation, timber	Ecological services, education, recreation, timber	Ecological services, education, recreation, timber

Notes: NFTPs abbreviated for non-timber forest products; Population reported from 2020 Census.

3.2.1. Bosque Comunitario de Río Hondo (Río Hondo Community Forest)

The initial vision of the Río Hondo effort was to create an eco-agritourism site in Mayagüez, Puerto Rico, in 2005. A local champion, RH4, organized the effort and recruited local individuals to join. The original idea was to create a destination for tourism with a café, a demonstration site for tobacco and sugar cane farming, a greenhouse for native plants and herbs, and a protected forested area on the property that could attract tourists and enhance the town's social and economic prosperity, and protect the forested area. At the time, a family-owned the mostly forested 28-ha (68-ac) property.

In 2005, RH4 and the recruited members promoted the idea of an eco-agritourism site to the rest of the town and created a non-profit organization, called Proyecto Comunitario Agro-Eco-Turístico Barrio Río Hondo, Inc. The organization consisted of a president and board members. With support from the local community, they pitched the idea to the municipality, who supported the idea but did not have sufficient funds to purchase the land, which was listed at USD 800,000.

In 2007, with help from professors at local universities, the organization developed a business and management plan for the site to be used for fundraising and finding investment partners. Eventually, the property owners granted a leasing contract to Proyecto Comunitario Agro-Eco-Turístico Barrio Río Hondo, Inc. (Río Hondo, Mayagüez, PR, USA) for the 10% non-forested portion of the property to provide access to the site while acquisition funds were secured. The organization installed a trailer on site for community meetings and workshops and established three greenhouses where plants and herbs were grown and sold for fundraising. For nearly a decade, the organization knocked on doors in the local community and across the island to raise money for the acquisition and eco-agritourism site without success.

Eventually, the organization learned of the USFS CFP and submitted a proposal for funding their project in 2014. Given the objectives and requirements of the CFP, development aspects of the proposal (e.g., agricultural development, café) had to be eliminated and an increased focus on sustainable forestry and natural resource management aspects had to be incorporated in the proposal to be considered for funding. Ultimately, Proyecto Comunitario Agro-Eco-Turístico Barrio Río Hondo, Inc. was granted USD 400,000 from the CFP for forest acquisition, and the municipality agreed to pay the remaining USD 400,000 to purchase the land. RH7, a member and volunteer of the Río Hondo CF recalled:

> "It was difficult to change the plans, but we did not have enough money to buy the forest without the grant's help. It was too much money for a nonprofit. The main goal was always to conserve the land. We just had to change the plan based on their restrictions".

Río Hondo is the first CF established with CFP support, owned by the municipal government, and managed by a local NGO in Puerto Rico. The property is surrounded by suburban and rural subdivisions and is approximately a five-minute drive to downtown Mayagüez. At the time of data collection, a small number of community members work daily at the property as unpaid volunteers. The 90% forested portion of the property is used for cultural services, such as recreation and education. Students from the University of Puerto Rico and Consuelo Pérez Cintrón Elementary School visit the property for research and academic purposes. RH5, a board member, states, "it is important to connect the children in the community with nature. The students can work in the forest for school credits or volunteer hours." Another volunteer, RH2, is one of the leading tour guides of the forest and notes that they "have approximately four school group visits each month". The Río Hondo Community Forest has also hosted AmeriCorps Program participants and Boy Scout groups; both helped develop trails in the forest. The CF sells trees, herbs, vegetables, other plants from the greenhouse, and sofrito (a Puerto Rican recipe made from herbs) to raise additional income.

The CF board of directors meets once a month, volunteers make and sell sofrito for additional funding, and local school groups and university students visit for educational and recreational events. The CF lands are managed by a board of director members and volunteers from the neighborhood.

Hurricane Maria struck Puerto Rico on 20 September 2017, resulting in widespread damage throughout the region. It has been cited as the costliest disaster in Puerto Rico's history and the third costliest in U.S. history [53]. Hurricane damages at the Río Hondo Community Forest included uprooted, broken, and defoliated trees throughout the forest, total destruction of the greenhouses and total loss of its produce, and damaged and destroyed trail signage and other infrastructure throughout the property.

As described by RH5:

> "We do what we can since Maria and the destruction that occurred. It has been a setback. But, this year, we will contact and reach out to more students from the university and nearby high schools and elementary schools. We plan to work on more grant proposals and complete the forest management plan".

3.2.2. Nine Times Community Forest

Nine Times Community Forest, also referred to as Nine Times Preserve per the entrance signs, is one of eight properties owned by Naturaland Trust. From the 1920s to 2012, the 667-ha (1648-ac) property was owned and logged by a local timber company. In 2007, the timber company put the property on the market. Naturaland Trust and partnering organizations were interested in purchasing the property, but they were half a million dollars short of the purchase price. Eventually, in 2012, the company dropped the asking price to about half of its 2007 listing price, making it more accessible for purchase by the organization.

In 2012, Naturaland presented the project to the CFP for acquisition funding. The proposal was approved, and these funds were leveraged to raise an additional USD 3.1 million to acquire the property later that year. Duke Energy, an electric power holding company, also holds an easement on the property. Duke's powerlines run through the middle of the forest, and they own the rights to clearing out pathways for their vehicles and spraying herbicides under the powerlines to prevent growth.

A local community member, NT1, is responsible for managing the property. As an unpaid volunteer, NT1 maintains all trails and signage and visits the CF about three times a week. Occasionally, the manager recruits individuals and local user groups to help with trail maintenance. NT1 was essentially a one-person manager and caretaker, assisted by ad hoc efforts mostly by recreation user groups.

The forested areas on the property consist of approximately 23% planted pine, 9% natural hardwood, and 68% mixed hardwood and pine forest. Since Naturaland's ownership of the property began, the organization has hired McMillian Logging Company to harvest some of the forest's timber, providing income for the CF.

The CF provides mainly cultural services, such as hiking, hunting, and rock climbing. They receive approximately USD 7000 to USD 8000 per year from Wildlife Management Areas (WMA) through hunting leases on the property. Naturaland estimates that there are 3500 visitors per year, consisting of approximately 1500 rock climbers, 1000 hunters, and 1000 hikers. The Climbing Coalition attracts the most people to the area. NT1 states, "There are a lot of social media events, mainly through Facebook, to recognize the rock-climbing club, and there are a lot of hiking clubs and online forums that visitors will post to. The word usually gets out." The property is also used for cultural services such as education from local schools, most frequently by Clemson University students. When asked about the next steps for the community forest, NT1 stated:

> "We need to attract visitors and recruit someone to help bring in volunteers. Someone needs to come and take over and organize volunteers who will help with the property. I also need to put more signs and kiosks up and make a couple more parking lots to increase visibility. We'd also like to expand the property to the north and southwest to make a wildlife corridor. We hope Naturaland will buy those properties. There is also nearby private property in Pickens that we'd like to acquire".

Despite the wealth of supporting, regulating, and cultural services provided to the local community, Nine Times had the lowest levels of community participation in forest governance and management components. During our field visits, we were unable to find anyone on adjacent properties to the forest or in the local stores who were aware that Nine

Times is a CF. There was no active board focused exclusively on the CF, but rather a loose connection to the larger organizational structure of Naturaland Trust, which had various widely dispersed properties in the region.

3.2.3. Birch Ridge Community Forest

Birch Ridge is an 820 ha (2027-acre) forest that sits on top of a hill, overlooking Merrymeeting Lake in New Durham, NH, USA. New Durham is a town that has a small population that benefits from the seasonal influx of visitors and part-time residents, by some accounts tripling in population during the summer. Birch Ridge is owned by Southeast Land Trust of New Hampshire (SELT), which owns and manages over a dozen other properties.

Before SELT, Birch Ridge was owned by a local timber company. In 2018, there were rumors that the company would sell the property to a developer, who would convert the land to condominium housing. When BR1, a local champion and business owner, heard about the proposed sale, he approached the property manager that day. The manager supported the mission to conserve the property and agreed to sell the property for USD 2 million. Still, he needed a non-refundable deposit of USD 200,000 in the next 30 days. Confidentially, BR1 shook on the deal.

Approximately 400 customers pass through BR1's local business. BR1 was able to rally together the members of the Association, individuals who passed through the marina, small businesses, others who lived in the New Durham area, and elected town officials. BR1 states:

> "This town has VERY deep pockets ... It was the perfect storm. This place that people care about was going to be destroyed. It was helpful that people who live on the lake have a lot of money".

In two days, the community raised USD 400,000. The next step was raising the rest to purchase the property. The town members reached out to SELT for additional help to raise the funding and to act as the primary owner. They created a steering committee, consisting of a representative from the following organizations and agencies: SELT, Town of New Durham Conservation Commission, New Durham Department of Recreation, Powder Mill Snowmobile Club, Lakes Regional Technical School, Camp Lions Pride, Farmington Fish, and Game Club, New England Mountain Bike Association, Merrymeeting Lake Association, Moose Mountains Regional Greenway, and New Hampshire Fish and Game Department. The group met monthly to plan for fundraising, community and public input events, and develop the management plan. Once the steering committee completes the management plan, they plan to elect and re-elect members and meet approximately twice a year. To assist with forming the management plan, the committee reached out to other New Hampshire CFs.

In approximately five months, SELT and the community raised USD 3 million; USD 500,000 granted through the CFP, and USD 25,000 donated from the town from the former timber owner's timber tax. BR2, a member of SELT, says that "this was the quickest turnaround on a project that we had ever seen".

The steering committee sent out surveys to New Durham taxpayers, asking questions about what recreation opportunities they wanted to see and participate in and the aspects and benefits of the CF they value most. Fifty-seven percent of the community members were most concerned with the property's impact on water quality.

In June 2019, the steering committee held its first workshop to gain the community member's input; the workshop attracted 80 attendees, a member of the Merrymeeting Lake Association, BR3, announced at the event, "This is YOUR forest!" In July 2019, the steering committee led a kick-off for the establishment of Birch Ridge Community Forest. The event consisted of guided hikes and a picnic lunch, and attracted more than 100 people. A second public input workshop was held in October 2019. Commenting on public input, BR2 explains:

> "This forest is different from all of the other properties SELT owns. We need this extra step for public input before we do anything else. It can be difficult, though, because we have a lot of different cooks in the kitchen and need to hear from everyone".

Additionally, referring to the decision-making processes of the CF, BR2 says, "the steering committee discusses the community member's input from the brainstorming sessions, and then they report to the SELT board. We (referring to the SELT board) have the final say, especially regarding finances and regulations".

The property is managed by members of the Powder Mill Snowmobile Club, consisting of 400 members. In return for utilizing the property for snowmobiling, approximately 40 club members maintain 138 km of trails each month. One key player in the club, BR4, also manages game cameras on the property, which capture moose, black bears, eastern coyotes, and other wildlife. BR4 states, "the cameras help monitor the property for illegal access and usages, such as ATV and camping activities. I always catch those ATV guys". The CF is also managed by Moose Mountain Regional Greenways (MMRG), a local land trust organization in New Hampshire, which holds a conservation easement on the property.

BR5, a member of the CF committee, says, "there is a need for reforestation and regrowth due to the recent clearcutting". The volunteers and SELT members plan to create signs, kiosks and install more accessible parking lots and trails for community members and visitors. In addition, SELT hopes timber harvest can occur in the future to generate income for the sustainability of the CF.

3.2.4. Page Pond Community Forest

Page Pond CF started with a 229-ha (567-ac) property owned by a religious group and operated as a youth summer camp and conference facility. In 2002, the religious group planned to enhance its facilities by adding an administration building, a dining facility, camper housing, single-family and multi-family employee housing, ball fields, and waterfront improvements, including an extensive dock and a water slide into Page Pond. However, the town Department of Environmental Services did not grant the approval.

In 2008, the town negotiated a payment to purchase the property from the religious group. Meredith's town government worked closely with Trust for Public Land (TPL) to raise the necessary funds to conserve the property permanently. PP3, a TPL employee who played a large role in the acquisition process, states:

> "Generally, towns do not have the experience to acquire large areas of forest land. We mainly help them fundraise and draft a management plan, but this is a really capable town, and this time we didn't have to do that much work. TPL usually has to start from scratch, but Meredith had a very engaged conservation commission and planning committee".

The town obtained enough funding through state and local grants and private donations to obtain the 229 ha—completing Phase I of Page Pond CF. PP2, an elected town official, notes, "This was the town's biggest project". Most town forest lands are donated from private properties, typically left in citizens' wills. PP2 discusses:

> "The concept of a community-owned forest is not uncommon to the people in New England. Town forests have been around since the 1700s. It's very traditional to the culture. In New England, counties are the main form of government. In other parts of the country, people aren't familiar with county lands or county forests. Here, everyone has heard of a town forest".

Nine years after the acquisition of Page Pond CF Phase I, the town fought to obtain the 81-ha (199-acre) tract adjacent to the Phase I property. The Phase II expansion property encompasses approximately 61 ha of undeveloped forestland, six has of abandoned agricultural fields, and 13 ha of designated wetlands.

The CF is utilized for timber, recreation, education, wildlife habitat, and other environmental benefits. Most of the forest is a mixed lowland type, with hemlock, pine, oak, and beech as dominants. In 2009, CF members conducted a timber harvest, generating over USD 50,000 in net proceeds. The forest was harvested extensively in 2011 and 2012, removing a large amount of marketable timber. The town hopes the Phase II parcels can be utilized for non-timber forest products (NTFP), such as a pumpkin patch or sunflower field, to attract tourism. The property also includes Page Pond, which PP3, a volunteer mentions, "as a local hotspot for fishing and swimming".

Page Pond is located less than one km from the Inter-Lakes School District (Inter-Lakes Junior and High School, Inter-Lakes Middle Tier School, and Inter-Lakes Elementary School), attracting many school groups for educational tours and field trips. Every year the MCC hosts several nature walks on the property, including educational walks with county foresters and other natural resource professionals. "We hope that we can build a sidewalk that connects the schools to the property for easy-to-access field trips or other after-school programs" PP2 notes.

The administration and governance of the CF are undertaken by the Meredith Conservation Commission (MCC), a town committee of elected officials who makes decisions about and protects the town's natural resources. The MCC consists of seven full members, five alternate members, and volunteer stewards who monitor town properties. The full and alternate members are elected to the local office. The members hold an annual town hall meeting for public input about the natural resources, including the forests. The volunteer stewards are the caretakers, maintaining trails for hiking, cross-country skiing, and mountain biking. The property does not contain a conservation easement but has an executory interest, meaning the town cannot sell the property, nor can it be deforested.

3.3. *Case Study Summary*

In summary, recreation is the most common use of the forests in the four case CF studies. In Birch Ridge and Page Pond CFs, snowmobiling is a widely practiced use of the forests in the winter. Two of the four cases (Page Pond and Nine Times) allow visitors to engage in consumptive services such as fishing and hunting. One of the four case studies (Río Hondo) provides non-timber forest products, such as small plants and herbs.

In addition, all four cases describe their CF as playing a crucial ecological role by offering regulating and supporting services for the community, such as clean water and air, carbon sequestration, and biodiversity. For example, interviewees at Birch Ridge indicated that the CF provides substantially healthier water quality and clarity of Merrymeeting Lake, which is the center of New Durham. In addition, members of Río Hondo recognize the value that the forest provides for the urban community, such as carbon sequestration and sheltering valuable wildlife.

The CFs involved in our case study analysis did not generate a large amount of revenue from timber or non-timber products, nor did they directly create a large number of local jobs. No direct paid jobs were reported for the four case studies that we visited. This may or may not be the case for other CFs in the Eastern U.S., and future research should examine this factor. Although Page Pond and Nine Times generate some income from timber harvests or hunting leases, timber income was not enough to sustain forest management on its own. All four case studies rely heavily on the support of their volunteers. From our literature review, we discovered this is not the case for many CFs outside of the U.S., but it occurs most commonly in developing countries where communities depend on the economic benefits that the CF provides [54,55]. Revenue generation was not the focus for many of these CFs, but rather their primary mission was to preserve land and provide cultural services. The CFs we visited all resemble a nature preserve, park, or other public green space more so than an active commercial timber-producing forest. The main difference between these CFs and traditional public parks was the ownership status and governance processes, particularly in terms of their public input and participation.

4. Discussion

CFs in the Eastern United States can be viewed as a hybrid forest governance approach that shares some ownership, governance, management, access, and benefits characteristics with both public and private forests. Consistent with the literature, we found that CFs in the Eastern U.S. display a diverse spectrum of forest ownership, governance, management authority, and uses [17,21,22,56].

4.1. Ownership

Individuals are more likely to protect forests or other natural resources when they have a sense of ownership and control [36,57,58]. CFs in the U.S. utilize several ownership models, most often varying by region. For example, we found four ownership categories represented in the Eastern U.S.: local governments, nonprofit organizations, tribal entities, and private for-profit corporations. Ownership by local governments appears to be the most common type in the East, including 50% in our search. CFs in the U.S. also may be owned by state and federal governmental agencies, though these ownerships are not known to exist in the Eastern U.S.

Town forests owned by local governments are most prevalent in the New England states, and many of these forests represent the key characteristics of CFs, including public participation and local community benefits. Town forests have highlighted the collaborative, community-based nature of conservation efforts; however, with limited focus on revenue generation and other economic-based benefits for the local communities [59]. Local governments throughout the Northeast also own CFs. CFs in the Eastern U.S. also may be under joint ownership, for example, held by local community nonprofit groups and land trusts or local government.

4.2. Governance

CFs in the U.S. span a wide range of governance and management frameworks [17,18]. In all but one of the cases in this study, we found that the landowner had the ultimate decision-making authority over the CF's management objectives, financials, and other duties of the property [36,52,60]. Geographic trends and histories can influence governance structures. For example, public input in decision-making processes is frequent in town forests in New England states, as demonstrated in our case studies, and consistent with the reports from the Community Forest Collaborative [61]. Town forests have been a part of New England's municipal history, and community members' role in decision-making processes is not new, dating to the early colonial settlements [61].

For three of our four case studies, Nine Times, Birch Ridge, and Page Pond, the landowners, being either a nonprofit organization (e.g., SELT and Naturaland Trust) or municipal government (e.g., Town of Meredith), are the final decision-makers. For Page Pond, large decisions, such as purchasing additional parcels, are voted on by the community members at town hall meetings. For minor questions, decisions are made by local elected officials of the Meredith Conservation Commission. For Birch Ridge, the CF steering committee, consisting of local citizens, incorporates public comments into the decision-making process. The committee gives their opinions and decision options to the landowner, SELT, who then has an ultimate say. The town of Mayagüez owns Río Hondo, but the nonprofit's Board of directors is the main decision-making body. Occasionally, a staff member from the municipality will join board meetings and contribute to board discussions and decisions.

4.3. Public Participation

Public participation and input in decision-making have been theorized to prevent conflict and promote the public's trust in government [62–64]. Public participation in forests may occur in decision-making and on-the-ground management. For instance, local community members partake in activities, such as trail-cleaning, non-timber and timber harvest, and property monitoring. From these CF case studies, we learned that public participation in decision-making processes is not always a priority nor explicitly included in management decisions or processes. Some CFs seemed to mostly incorporate these elements to comply with CFP requirements that mandate incorporating some form of public participation.

CF public participation and input typically occurred through engagements in informational meetings and participation in site tours hosted by the CFs (e.g., Río Hondo). The election of local community members to decision-making bodies and management roles of the CF also was quite common (e.g., Page Pond). Less common for these CFs was widespread or deep participation by a large portion of local community members. Birch Ridge CF in New Hampshire was unique in its incorporation of in-depth public input and participation in forest decision-making.

The four case study CFs fall along different points of the International Association of Public Participation's (IAP2) public participation spectrum in decision-making processes [65]. The IAP2 encompasses five levels of public participation: inform, consult, involve, collaborate, and empower [65]. Some CFs minimally engage in public participation by informing the public via forest entrance signs and kiosks, hiking trail maps, or advertising community meetings or events (e.g., inform). On the other end of the spectrum, some CFs depend heavily on volunteers and local community members as the primary leaders and decision-makers (e.g., empower). Most CFs lie in the middle of the two extremes, many leaning towards IAP2's consult and involve levels of public participation. We found that our case study CFs and many of the CFs we identified in the Eastern U.S. involve the public by obtaining comments, including community members in direct decision-making conversations, or partnering with local groups and organizations to seek advice or resources.

4.4. Uses, Benefits, and Outcomes

In addition to a sense of ownership, people are more likely to protect and manage natural resources when they have secure access [36,57,58,66]. All four case studies allow some form of public access and mainly focus on social and ecological outcomes. The case study CFs did not generate much direct income and employment from timber or non-timber (e.g., hunting leases for Nine Times, NTFPs for Río Hondo) markets. The online search also indicated that profits from commodity production were seldom mentioned. Approximately 20% of the CF's website pages reference revenues from NTFPs, and more commonly, 64% of CF's websites mentioned timber production benefits.

The principal outcomes from our cases, and perhaps from many CFs to date, are the provision of ecosystem services across the four broad MEA classes of supporting, regulating, cultural, and provisioning services of forests [4]. CFs often are fomented by desires to set aside and protect pubic and common pool goods—regulating (e.g., water, wildlife habitat, and biodiversity, air quality, carbon storage), recreational or cultural (e.g., dispersed, hiking, water sports, climbing, birding), and more broadly supporting services (e.g., nutrient and soil formation and primary production) functions. While they also produce some modest income from provisioning timber harvests and non-timber goods or services, this use alone does not appear to be the primary cause for creating or keeping CFs. Most CFs primarily aim to preserve and protect the forest from future degradation or conversion, thereby protecting common-pool resources.

The cases studied also described social benefits from their CFs, including building a sense of community and place, engaging the community to become involved in natural resource management, and bringing local community members together. In addition, social outcomes from CF included other cultural uses such as environmental education and recreational purposes.

4.5. CF Challenges

From the four case studies, three of the CFs (Río Hondo, Nine Times, and Birch Ridge) struggled to obtain sufficient funding to either support full-time or part-time employees and maintain features and functions of the CF, such as hiking and snowmobile trails, roads, signage, and parking lot installation. Financial resources are necessary to sustain a local and collaborative project for the long term [67].

Other challenges include recruiting and retaining community members to assist in public participation in decisions and management; balancing needs to attract visitors to promote community engagement with concerns for resource protection; and increasing the forest's visibility as a CF. For instance, Page Pond's grounds management crew consisted of retired and older individuals. The need for younger stewards to join the Page Pond efforts is essential for the CF's long-term viability. In addition, Río Hondo strived to increase the number of visitors to the forest. Other than local school and church groups that visited for site tours, the forest had one visitor a day on average at the time of data collection. Recruiting, motivating, and retaining volunteers and visitors is crucial for social and economic outcomes, especially for programs with limited resources [67].

4.6. Study Limitations

This exploratory research was intended to provide an initial descriptive introduction and analysis of CF efforts in the Eastern U.S. since related research and literature are extremely limited. The prospective benefits of CFs as a potentially promising model of forest ownership and management in the U.S. are large, but research information about this approach is scarce. Our lists of CFs are thorough, and the four cases selected are representative of the organizational and geographic range of CFs identified. Given the diversity of CFs nationwide and globally, our cases aim to provide a portrait of the range in types of CFs and their governance, public input processes, collaborative efforts, management plans, and recreation, commodity, and ecosystem benefits.

An apparent limitation is that we had only four detailed case studies and 18 persons interviewed. Nevertheless, the CFs visited, and the persons interviewed, were quite interesting and candid, and provided considerable insights into their experiences and prospects. A greater number of cases would be the next step to enhancing our knowledge about CFs, and we are continuing to examine this line of research in greater depth as well. The purposive cases examined here as drawn from our list of CFs, however, did serve as quite typical examples of relatively active New England CFs, where Town government owners and management have existed for decades, and of a nongovernment agency/land trust, which is increasingly common. The southern U.S. case with a land conservancy is a common approach, and the Puerto Rico case demonstrates how CFs can achieve forest protection and management as a promising alternative to development. The cases, merits, issues, and implications should be useful for guiding further research.

Last, the information obtained from online searches only represents the documentation made available online. Nevertheless, the case study information can give us an idea of which CFs are most active, what management and governing characteristics are most employed, and other general characteristics to help descriptive CFs in the Eastern U.S.

5. Conclusions

CFs reveal both direct and indirect benefits for local communities and the natural environment. In this study, we compiled a list of Eastern U.S. CFs through a systematic web-based search. We selected four CFs to conduct in-depth comparative case study research. Generally, we found that CFs benefits included better protected and more resilient biophysical resources, enhanced social capital, enabled local and indigenous knowledge in forest management and governance, and increased economic resources from timber and NTFP production. In addition, this study revealed that CFs in the Eastern U.S. have often been formed to protect forest land from development and protect associated common-pool resources. These efforts frequently bring together a group of diverse stakeholders, or at least a small number of very committed groups and individuals, to protect and manage the common-pool forest resources.

The initial efforts at CF establishment must be maintained for long-term viability and success. Maintaining and funding such efforts in the long run without significant revenues remains a challenge for most CFs. As CF initiatives increase in the Eastern U.S. and across the country, the information generated through this study can be used by CF practitioners, researchers, landowners, managers, and community members to further establish and maintain those efforts to increase ecosystem goods and services for local communities. In addition, a similar methodological framework could be adapted in other parts of the country to gain further information on the number of, types of, and characteristics of CFs in the United States.

Author Contributions: M.H., F.C., K.M. and G.F. contributed to writing the original draft, finalizing methodology protocols, and supervising the overall study. M.H., K.M. and X.H. conducted semi-structured interviews with community forest case study participants, and M.H. formally analyzed the qualitative data and curated it. All authors (M.H., F.C., K.M., G.F., X.H. and M.L.) contributed to online web search data collection, validation, and the analysis of community forests and characteristics descriptive statistics and contributed to the formation of the community forests in the Eastern United States list. All authors also contributed to reviewing and editing the manuscript. All authors have read and agreed to the published version of the manuscript.

Funding: Travel for this research was funded by the North Carolina State University Department of Forestry and Environmental Resources Zobel and Laarman Endowments.

Institutional Review Board Statement: The study was conducted in accordance with the Declaration of Helsinki, and approved by the Institutional Review Board of NORTH CAROLINA STATE UNIVERSITY (protocol code 16837, 1 April 2019).

Informed Consent Statement: Informed consent was obtained from all subjects involved in the study.

Data Availability Statement: The data are available on request from the corresponding author.

Acknowledgments: We thank the community forest case study participants for their time and participation in the interviews.

Conflicts of Interest: The authors declare no conflict of interest. The findings and conclusions in this publication are those of the authors and should not be construed to represent any official USDA or U.S. Government determination or policy.

Appendix A Keywords Search

Table A1. Keywords used in systematic search, using 13 mile woods community forest as an example.

"13 Mile Woods" + "Community forest" *
"13 Mile Woods" + "Area"
"13 Mile Woods" + "Association"
"13 Mile Woods" + "Mission"
"13 Mile Woods" + "Vision"
"13 Mile Woods" + "Purpose"
"13 Mile Woods" + "Manager"
"13 Mile Woods" + "Management"
"13 Mile Woods" + "Management plan"
"13 Mile Woods" + "Stewardship plan"
"13 Mile Woods" + "Annual report"
"13 Mile Woods" + "Owner" *
"13 Mile Woods" + "Strategic plan"

Table A1. *Cont.*

"13 Mile Woods" + "Committee"
"13 Mile Woods" + "Board"
"13 Mile Woods" + "Members*"
"13 Mile Woods" + "Partners" *
"13 Mile Woods" + "Budget"
"13 Mile Woods" + "Income"
"13 Mile Woods" + "Revenue"
"13 Mile Woods" + "Recreation"
"13 Mile Woods" + "Education"
"13 Mile Woods" + "Timber"
"13 Mile Woods" + "Timber products"
"13 Mile Woods" + "Non-timber products"
"13 Mile Woods" + "Public access"
"13 Mile Woods" + "Users"

* Represents keywords that have been truncated, meaning the web-based search include words with various word endings or phrases. For example, "community forest" could also search for phrases like "community forestry" or "community forests".

Appendix B Interview Questionnaire

Introductory questions:

- Can you discuss a bit about the history of the community forest?
- What role did you play in the establishment of the forest?
- What is your responsibility or position of the forest?
- **How has the forest changed over time?**

Optional follow-up questions:

- For example, in terms of ownership, size, organization, management, uses, productivity
- Are there any growth trends over time based on the organization's members, resources, and capacity? Can you give examples?

(Q1) What activities occur on the forest?
Optional follow-up questions:

- Does the CBF produce timber and wood products, the materials derived from a forest for commercial and personal use (but not limited to) pulpwood, firewood, and sawtimber?
- Does the CBF produce non-timber forest products (NTFPs), the materials derived from forestlands other than round wood or wood chips, including (but not limited to) fruits, nuts, honey, mushrooms, seeds? Is it a permit required to retrieve NTFPs from the CBF?
- Does the CBF provide for leisure activities such as hiking, walking, running, canoeing, kayaking, etc?
- What types of recreation activities are offered?
- What types of educational purposes to learn about forest management, conservation, wildlife, etc. for school groups and other community organizations are offered?

(Q2) Who owns the forest?
Required follow up:

- How has the ownership evolved since establishment of the forest?

Optional follow-up questions:

- Who has complete legal possession of the forestland or land rights? Out of the following ownership categories:

- ○ Land trust or conservancy
- ○ Environmental or conservation nongovernment organization (NGO)
- ○ County, city, township, or other local government entity
- ○ Sole proprietorship
- ○ Joint Venture, General Partnership Limited Partnership
- ○ Limited Liability Company+ (LLC, LLP, LLLP)
- ○ Corporation
- ○ Trust
- ○ Tribal Reservation
- ○ Other

- Who has the right to obtain resources or products from the CBF (such as timber, fishing or diverting water)?
- Who has the right to regulate, change, and implement the CBF actions and its objectives?
- Who has the right to decide who has access rights and withdrawal rights?
- Who has the right to sell products or lease management on the CBF?

(Q3) Who manages the forest?
Required follow up question:

- How has management authority evolved since establishment of the CBF?
- Who performs the management work on the forest? (e.g., staff, contractors, volunteers, CBF Members?)

Optional follow-up questions:

- What is the manager's authority, title, and responsibility? What is the process of managing the forestland?
- Is this person (or organization) a paid member or volunteer?

(Q4) How are decisions made?
Required follow-up question:

- What individual or group has the final authority to make the decisions of the forest?

Optional follow-up questions:

- How are decisions made on the strategic level?
- How are decisions made on an operational/managerial level?
- What is the balance between Board and the CEO/EVP/professional staff authority in making decisions? In implementing decisions?
- How is the budget agreed upon by the board members and managers?

(Q5) Can you discuss a little about the CBF's organizational structure?
Optional follow-up questions:

- Does the CBF have a charter, bylaws, a strategic plan, operating policies, or management and operational regulations in place?
- Does the CBF have a formal organizational hierarchy and policies or practices manual?
- Is the CBF comprised of various businesses, nonprofits, or community organizations? Are there alliances involved in the management processes? Are the organizations a part of the private or public sector?
- Does the forest monitor the lands through rangers or other staff members?
- What are processes for resolving conflicts either between board members, staff, volunteers, or volunteers?

(Q6) Who receives forest benefits?
Required follow-up question:

- What are the ecological and human benefits, both locally and globally, provided by the forested land does the CBF recognize?
    - ○ *e.g., Local: healthier water, cleaner air supply, food, wood and fiber, fuel, biodiversity habitats, education, recreation, spiritual and cultural opportunities (Millennium Ecosystem Assessment, 2005).*

- *e.g., Global: climate regulation, flood regulation, disease regulation, and water purification (Millennium Ecosystem Assessment, 2005).*

Optional follow-up questions:

- Who is eligible to become a user?
- What does membership/usership require and/or cost?

(Q7) Does the forest generate income? If yes, how so?
Optional follow-up questions:

- (Such as sale of forest products, member dues, user fees, service sales, publications, conventions and meetings, advertising, gifts and contributions, interest and dividends from investments, etc.)
- Do you receive periodic or recurring state or federal grants for the CBF? How much per year?
- What funds were used to purchase/establish the forest property?
- Does the CBF provide any indirect impact? For example, indirect income is one which is earned by way of non-forest products and services, such as income through local gas stations, restaurants, and lodging.

(Q8) What are the forest's expenses?
Optional follow-up questions:

- What are the current assets— such as cash, receivables, prepaid expenses, land, property, forests, investments, buildings, equipment, and liabilities— such as lines of credit, accounts payable, deferred membership dues, etc.? If any, what are the unrestricted and restricted net assets?
- What funds are used to manage the CBF?
- If federal or state grant funds are received, do they require specific management actions, research, or other deliverables? Are they discretionary?
- If there is net profit or income in excess of costs, how are those funds used? Is any net profit distributed to community organizations or community members?

Concluding questions:

(Q9) Do you think the property should be owned and managed as a community forest, or might another public or private management system work better?

(Q10) What suggestions do you have for other communities that might be interested in community forestry? Under what conditions do you think a community works best together?

(Q11) What are effective public engagement processes or methods for developing a community forest?

(Q12) Who are the individuals or organizations that might oppose community forest development? What roles might they play? How can they be effectively engaged?

(Q13) What are the strengths of the community forest you are associated with?

(Q14) What are possible weaknesses or biggest challenges of the community forest you are associated with?

(Q15) What other opportunities could the forest you are associated with take advantage of?

(Q16) What threats are there to the community forest you are associated with?

(Q17) Do you have any final thoughts you would like to share about your community forest or your personal experience with the community forest?

Appendix C

Table A2. List of eastern U.S. community forests identified through an online search.

Community Forest	Location
13 Mile Woods	Errol, NH, USA
Albany Town Forest	Albany, NH, USA
Albert Family Community Forest	East Nassau, New York, NY, USA
Barre Town Forest	Websterville, VT, USA
Bethel Community Forest	Bethel, ME, USA
Birch Ridge Community Forest	New Durham, NH, USA
Blackstrap Hill Community Forest and Preserve	Falmouth, ME, USA
Bobcat Woods Community Forest	Bolton, CT, USA
Browns Mill Community Forest	Atlanta, GA, USA
Brushwood Community Forest	West Fairlee, VT, USA
Buffam Brook Community Forest	Pelham, MA, USA
Canaan Community Forest	Canaan, VT, USA
Catamount Community Forest	Williston, VT, USA
Clarence Oak Savannah Community Forest	Clarence, New York, NY, USA
Cooley-Jericho Community Forest	Easton, NH, USA
Dead River Community Forest	Negaunee, MI, USA
Dorset Town Community Forest	Dorset, VT, USA
Downeast Lakes Community Forest	Grand Lake Stream, ME, USA
Falmouth Town Forest	Falmouth, ME, USA
Freedom Community Forest	Freedom, NH, USA
Gill Town Forest (also known as Blake Town Forest)	Gill, MA, USA
Glen Oakes Town Forest	Fremont, NH, USA
Gorham Town Forest	Gorham, NH, USA
Hadlock Community Forest	Falmouth, ME, USA
Hall Mountain	Franklin, NC, USA
Healing Harvest Forest Foundation	Copper Hill, VA, USA
Hidden Valley Nature Center	Jefferson, ME, USA
Hoke Community Forest	Raeford, NC, USA
Holland Community Forest (previously referred to as Sichol Family Property)	Holland, MA, USA
Jefferson Community Forest	Fairdale, KY, USA
Kirby-Ivers Town Forest	Pelham, NH, USA
Kuncanowet Community Forest	Dunbarton, NH, USA
Lakeview Federal Stewardship	Falmouth, ME, USA
Lime Lake Community Forest (Portman Nature Preserve)	Almena and Antwerp, MI, USA

Table A2. *Cont.*

Community Forest	Location
Lincoln Community Forest	Lincoln, WI, USA
Lowry Woods Community Forest	Madison, CT, USA
Machias Community Forest	Machias, ME, USA
McLaughlin Crossing Community Forest	Falmouth, ME, USA
Mendon Town Forest	Mendon, MA, USA
Menominee Forest	Neopit, WI, USA
Milan Community Forest	Milan, NH, USA
Niantic River Watershed (also referred to as East Lyme Land Trust Community Forest)	East Lyme and Waterford, CT, USA
Nine Times Community Forest	Pickens, SC, USA
North Falmouth Community Forest	Falmouth, ME, USA
North Pikes Creek Community Forest	Russell, WI, USA
Oak Ridge Town Forest	Fremont, NH, USA
Page Pond Community Forest	Meredith, NH, USA
Palmetto-Peartree Preserve	Columbia, NC, USA
Peabody Town Forest	Pelham, NH, USA
Perley Mills Community Forest	Denmark, ME, USA
Pilgrim Community Forest	Houghton, MI, USA
Plimpton Community Forest	Sturbridge, MA, USA
Poestenkill Community Forest	Poestenkill, NY, USA
Ragged Mountain Community Forest	Charlottesville, VI, USA
Raymond Community Forest	Raymond, ME, USA
Richmond Town Forest (Andrews Community Forest)	Richmond, VT, USA
Rines Community Forest	Cumberland, ME, USA
Rio Hondo Community Forest	Mayagüez, PR, USA
Sink Creek Community Forest	San Marcos, TX, USA
Springfield Bluffs	Springfield, WI, USA
Stemilt Squilchuck Community Forest	Richmond, VT, USA
Towar Woods	Meridian, MI, USA
Twin Bridges Community Forest	Otisfield, ME, USA
Tyler Forks Community Forest	Mellen, WI, USA
Virginia Stranahan Town Forest	Marshfield, VT, USA
Waitsfield Scrag Town Forest	Waitsfield, VT, USA
Weston Community Forest (previously called Weston Homestead Farm land)	Madison, ME, USA
Wildcat Falls	Haight, WI, USA
Woods Road Community Forest	Falmouth, ME, USA
Yellow Dog River Community Forest	Ispheming, MI, USA

Note: This list of community forests was documented through an online search and has not been verified on the ground. Therefore, it is likely that the list includes some initiatives that are not community forests (e.g., defunct, or never started, or do not meet the characteristics of a community forest) and does not include some community forests that exist on the ground but less online presence.

References

1. Oswalt, S.N.; Smith, W.B.; Miles, P.D.; Pugh, S.A. *Forest Resources of the United States, 2017: A Technical Document Supporting the Forest Service 2020 RPA Assessment*; Gen. Tech. Rep. WO-97; Department of Agriculture, Forest Service, Washington Office: Washington, DC, USA, 2019; 223p.
2. Tidwell, T. State of Forests and Forestry in the United States. [World Conservation Congress Presentation]. 2016. Available online: https://www.fs.usda.gov/speeches/state-forests-and-forestry-united-states-1 (accessed on 20 August 2022).
3. USDA Forest Service. U.S. Forest Resource Facts and Historical Trends. 2014. Available online: https://www.fia.fs.fed.us/library/brochures/docs/2012/ForestFacts_1952-2012_English.pdf (accessed on 20 August 2022).
4. Millennium Ecosystem Assessment (MEA). *Ecosystems and Human Well-Being: Synthesis*; Island Press: Washington, DC, USA, 2005.
5. Agrawal, A.; Hajjar, R.; Liao, C.; Rasmussen, L.V.; Watkins, C. Editorial overview: Forest governance interventions for sustainability through information, incentives, and institutions. *Curr. Opin. Environ. Sustain.* **2018**, *32*, A1–A7. [CrossRef]
6. Birch, J.C.; Thapa, I.; Balmford, A.; Bradbury, R.B.; Brown, C.; Butchart, S.H.; Gurung, H.; Hughes, F.M.; Mulligan, M.; Pandeya, B.; et al. What benefits do community forests provide, and to whom? A rapid assessment of ecosystem services from a Himalayan forest, Nepal. *Ecosyst. Serv.* **2014**, *8*, 118–127. [CrossRef]
7. International Union for the Conservation of Nature/Commission on Environmental, Economic and Social Policy (IUCN/CEESP). Bio-Cultural Diversity Conserved by Indigenous Peoples and Local Communities—Examples and Analysis. Companion Document to IUCN/CEESP Briefing Note 10. Prepared by G. Borrini-Feyerabend and Others for ICCA Consortium and Others. 2010. Available online: http://www.iccaforum.org/index.php?option=com_content&view=article&id=89&Itemid=105 (accessed on 22 February 2022).
8. Lyman, M.; McDonough, M.; Danks, C. New England's Community Forests: Comparing a Regional Model to ICCAs. *Conserv. Soc.* **2013**, *11*, 46–59. [CrossRef]
9. Food and Agriculture Organization (FAO). *Forestry for Local Community Development*; Forestry Paper 7; FAO: Rome, Italy, 1978; Available online: https://www.fao.org/3/t0692e/t0692e00.htm (accessed on 22 February 2022).
10. Bullock, R.; Lawler, J. *Community Forests Canada: Bridging Practice, Research and Advocacy*; Workshop and Symposium Report; Centre for Forest Interdisciplinary Research and Department of Environmental Studies and Sciences, University of Winnipeg: Winnipeg, MB, Canada, 2014.
11. Bullock, R.; Broad, G.; Palmer, L.; Smith, M.A. (Eds.) *Growing Community Forests: Practice, Research, and Advocacy in Canada*; University of Manitoba Press: Winnipeg, MB, Canada, 2017.
12. Teitelbaum, S. (Ed.) *Community Forestry in Canada: Drawing Lessons from Policy and Practice*; University of British Columbia Press: Vancouver, BC, Canada, 2016.
13. Belsky, J. Creating community forests. In *Forest Community Connections: Implications for Research, Management, and Governance Washington Resources*; Donoghue, E., Sturtevant, V., Eds.; Resources for the Future: Washington, DC, USA, 2008.
14. Abbott, G.; McCullough, R. The Landscape of Community: A History of Communal Forests in New England. *N. Engl. Q.* **1996**, *69*, 341–345. [CrossRef]
15. McGinley, K.A.; Charnley, S.; Cubbage, F.W.; Hajjar, R.; Frey, G.E.; Schelhas, J.; Hovis, M.; Kornhauser, K. Community forest ownership, rights, and governance regimes in the United States. In *Routledge Handbook of Community Forestry*; Routledge: London, UK, 2022; pp. 146–167. [CrossRef]
16. Bullock, R.; Hanna, K. *Community Forestry: Local Values, Conflict and Forest Governance*; Cambridge University Press: Cambridge, UK, 2012.
17. Charnley, S.; Poe, M.R. Community Forestry in Theory and Practice: Where Are We Now? *Annu. Rev. Anthropol.* **2007**, *36*, 301–336. [CrossRef]
18. McCarthy, J. Devolution in the Woods: Community Forestry as Hybrid Neoliberalism. *Environ. Plan. A Econ. Space* **2005**, *37*, 995–1014. [CrossRef]
19. Teitelbaum, S. Criteria and indicators for the assessment of community forestry outcomes: A comparative analysis from Canada. *J. Environ. Manag.* **2014**, *132*, 257–267. [CrossRef]
20. Danks, C.; Fortmann, L. Social forestry: Forest and tree tenure and ownership. In *Encyclopedia of Forest Sciences*; Elsevier Ltd.: Oxford, UK, 2004.
21. Edmunds, D.; Wollenberg, E. (Eds.) *Local Forest Management: The Impacts of Devolution Policies*; Earthscan Publications: London, UK, 2003; Volume 16, 208p, ISBN 1844070239.
22. Ribot, J.C.; Agrawal, A.; Larson, A.M. Recentralizing While Decentralizing: How National Governments Reappropriate Forest Resources. *World Dev.* **2006**, *34*, 1864–1886. [CrossRef]
23. Piabuo, S.M.; Foundjem-Tita, D.; Minang, P.A. Community forest governance in Cameroon: A review. *Ecol. Soc.* **2018**, *23*, 34. [CrossRef]
24. Rasolofoson, R.; Ferraro, P.; Jenkins, C.; Jones, J.P.G. Effectiveness of Community Forest Management at reducing deforestation in Madagascar. *Biol. Conserv.* **2015**, *184*, 271–277. [CrossRef]
25. Davis, E.J.; Hajjar, R.; Charnley, S.; Moseley, C.; Wendel, K.; Jacobson, M. Community-based forestry on federal lands in the western United States: A synthesis and call for renewed research. *For. Policy Econ.* **2020**, *111*, 102042. [CrossRef]

26. Fernandez-Gimenez, M.; Ballard, H.L.; Sturtevant, V.E. Adaptive Management and Social Learning in Collaborative and Community-Based Monitoring: A Study of Five Community-Based Forestry Organizations in the western USA. *Ecol. Soc.* **2008**, *13*, 4. [CrossRef]
27. Cubbage, F.; O'Laughlin, J.; Peterson, N. *Natural Resource Policy*; Waveland Press: Long Grove, IL, USA, 2017; p. 505.
28. Trust for Public Lands (TPL). Community Forests. 2018. Available online: https://www.tpl.org/our-work/community-forests (accessed on 20 February 2022).
29. US Endowment for Forestry and Communities. Community Forests. Available online: https://www.usendowment.org/community-forests/ (accessed on 20 February 2022).
30. USDA Forest Service. 2018 Farm Bill. 2018. Available online: https://www.fs.usda.gov/managing-land/farm-bill (accessed on 21 February 2022).
31. US Federal Register §230.2. 2011. Available online: https://www.federalregister.gov/documents/2011/10/20/2011-27117/community-forest-and-open-space-conservation-program (accessed on 21 February 2022).
32. Brown, W. Strategic management. In *The Jossey-Bass Handbook of Nonprofit Leadership and Management*; Renz, D., Ed.; Wiley: Hoboken, NJ, USA, 2016; pp. 217–237.
33. Laughlin, F.L.; Andringa, R.C. *Good Governance for Nonprofits*; American Management Association: New York, NY, USA, 2007; p. 210.
34. Ostrom, E. *Governing the Commons: The Evolution of Institutions for Collective Action*; Cambridge University Press: Cambridge, UK, 1990.
35. Renz, D.O. Leadership, governance, and the work of the board. In *The Jossey-Bass Handbook of Nonprofit Leadership and Management*; Renz, D., Ed.; Wiley: Hoboken, NJ, USA, 2016; pp. 127–132.
36. Schlager, E.; Ostrom, E. Property-Rights Regimes and Natural Resources: A Conceptual Analysis. *Land Econ.* **1992**, *68*, 249. [CrossRef]
37. Moher, D.; Liberati, A.; Tetzlaff, J.; Altman, D.G. PRISMA Group Preferred reporting items for systematic reviews and meta-analyses: The PRISMA statement. *J. Clin. Epidemiol.* **2009**, *62*, 1006–1012. [CrossRef] [PubMed]
38. Papaioannou, D.; Sutton, A.; Carroll, C.; Booth, A.; Wong, R. Literature searching for social science systematic reviews: Consideration of a range of search techniques. *Health Inf. Libr. J.* **2009**, *27*, 114–122. [CrossRef]
39. Sambunjak, D.; Straus, S.E.; Marusic, A. A Systematic Review of Qualitative Research on the Meaning and Characteristics of Mentoring in Academic Medicine. *J. Gen. Intern. Med.* **2009**, *25*, 72–78. [CrossRef]
40. Achen, C.H.; Snidal, D. Rational Deterrence Theory and Comparative Case Studies. *World Polit.* **1989**, *41*, 143–169. [CrossRef]
41. Eisenhardt, K.M. Building theories from case study research. *Acad. Manag. Rev.* **1989**, *14*, 532–550. [CrossRef]
42. Stake, R.E. *The Art of Case Study Research*; Sage Publications: Thousand Oaks, CA, USA, 1995.
43. Yin, R.K. Case study research. In *Designs and Methods*, 3rd ed.; Sage Publications: Thousand Oaks, CA, USA, 2003.
44. Yin, R.K. Case study research. In *Designs and Methods*, 4th ed.; Sage Publications: Thousand Oaks, CA, USA, 2009.
45. Yin, R.K. *Case Study Research and Applications*, 6th ed.; Sage: London, UK, 2018; p. 319.
46. Glesne, C. *Becoming Qualitative Researchers: An Introduction*, 4th ed.; Pearson: Upper Saddle River, NJ, USA, 2011.
47. Saldaña, J. *The Coding Manual for Qualitative Researchers*; Sage: Thousand Oaks, CA, USA, 2016.
48. Danks, C. Institutional arrangements in community-based forestry. In *Community and Forest Connections*; Donohue, E., Sturtevant, V., Eds.; Resources for the Future: Washington, DC, USA, 2008; pp. 185–204.
49. National Community Forestry Service Center/Northern Forest Region. *What is Community Forestry, Why Does It Matter?* St. Albans, V.T., Ed.; Yellow Wood Associates: Fairfield, VT, USA, 2000; Available online: http://www.communitiescommittee.org/pdfs/what%20comm%20forestry.pdf (accessed on 22 February 2022).
50. Christoffersen, N.; Harker, D.; Lyman, M.W.; Wyckoff, B. *The Status of Community Based Forestry in the United States: A Report to the US Endowment for Forestry and Communities*; Community Forest Consortium, 2008. Available online: https://www.usendowment.org/wp-content/uploads/2018/10/full_community_based_forestry_report_3.17.pdf (accessed on 22 September 2022).
51. Lyman, M.W. *Community Forests: A Community Investment Strategy*; Community Forest Collaborative, 2007. Available online: https://www.tpl.org/resource/community-forests-community-investment-strategy (accessed on 21 August 2022).
52. Ostrom, E.; Hess, C. *Private and Common Property Rights*; Social Science Electronic Publishing Inc.: Bloomington, IN, USA, 2007.
53. NOAA. U.S. Billion-Dollar Weather and Climate Disasters. National Centers for Environmental Information (NCEI). 2022. Available online: https://www.ncei.noaa.gov/access/billions/ (accessed on 31 May 2022). [CrossRef]
54. Cox, M.; Villamayor-Tomas, S.; Epstein, G.; Evans, L.; Ban, N.; Fleischman, F.; Nenadovic, M.; López, G.G. Synthesizing theories of natural resource management and governance. *Glob. Environ. Chang.* **2016**, *39*, 45–56. [CrossRef]
55. Hajjar, R.; Oldekop, J.; Cronkleton, P.; Etue, E.; Newton, P.; Russel, A.J.; Tjajadi, J.S.; Zhou, W.; Agrawal, A. The data not collected on community forestry. *Conserv. Biol.* **2016**, *30*, 1357–1362. [CrossRef]
56. Hajjar, R.; Molnar, A. Decentralization and Community-based Approaches. In *Forests, Business and Sustainability*; Panwar, R., Kozak, R., Hansen, E., Eds.; Earthscan: London, UK, 2016.
57. Pelletier, J.; Gélinas, N.; Skutsch, M. The Place of Community Forest Management in the REDD+ Landscape. *Forests* **2016**, *7*, 170. [CrossRef]
58. Persha, L.; Agrawal, A.; Chhatre, A. Social and Ecological Synergy: Local Rulemaking, Forest Livelihoods, and Biodiversity Conservation. *Science* **2011**, *331*, 1606–1608. [CrossRef]

59. Lyman, M.W.; Grimm, C.; Evans, J.R. Community forests as a wealth creation strategy for rural communities. *Community Dev.* **2014**, *45*, 474–489. [CrossRef]
60. Ferguson, P.F.B.; Conroy, M.J.; Chamblee, J.F.; Hepinstall-Cymerman, J. Using structured decision making with landowners to address private forest management and parcelization: Balancing multiple objectives and incorporating uncertainty. *Ecol. Soc.* **2015**, *20*, 27. [CrossRef]
61. Community Forest Collaborative. Community Forests: Needs and Resources for Creating and Managing Community Forests Report. 2011. Available online: https://vtcommunityforestry.org/sites/default/files/pictures/Resource/needs_and_resources_for_creating_and_managing_community_forests_updated_aug2011.pdf (accessed on 21 February 2022).
62. Leach, W.D. Public involvement in USDA Forest Service policymaking: A literature review. *J. For.* **2006**, *104*, 43–49. [CrossRef]
63. Mattor, K.M.; Cheng, A.S.; Kittler, B.; McDonough, M. Assessing Collaborative Governance Outcomes and Indicators across Spatial and Temporal Scales: Stewardship Contract Implementation by the United States Forest Service. *Soc. Nat. Resour.* **2019**, *33*, 484–503. [CrossRef]
64. Sixt, G.N.; Klerkx, L.; Aiken, J.D.; Griffin, T.S. Nebraska's Natural Resource District system: Collaborative approaches to adaptive groundwater quality governance. *Water Altern.* **2019**, *12*, 676–698.
65. International Association for Public Participation. IAP2 Spectrum of Public Participation. 2019. Available online: https://cdn.ymaws.com/www.iap2.org/resource/resmgr/pillars/Spectrum_8.5x11_Print.pdf (accessed on 21 February 2022).
66. Wright, G.D.; Andersson, K.P.; Gibson, C.C.; Evans, T.P. Decentralization can help reduce deforestation when user groups engage with local government. *Proc. Natl. Acad. Sci. USA* **2016**, *113*, 14958–14963. [CrossRef]
67. Hovis, M.; Cubbage, F.; Rashash, D. Designing a Citizen Science Project for Forest Landscapes: A Case from Hofmann Forest in Eastern North Carolina. *Open J. For.* **2020**, *10*, 187–203. [CrossRef]

forests MDPI

Article

The Carbon Sink of Mangrove Ecological Restoration between 1988–2020 in Qinglan Bay, Hainan Island, China

Peihong Jia [1], Weida Huang [1], Zhouyao Zhang [2,*], Jiaxuan Cheng [1] and Yulong Xiao [1]

[1] School of Public Administration, Hainan University, Haikou 570228, China
[2] School of Earth and Environmental Sciences, University of Queensland, Brisbane, QLD 4072, Australia
* Correspondence: zhouyao.zhang@uqconnect.edu.au

Abstract: As the world's largest reactive carbon reservoir, the ocean plays a critical role in global climate change. Among coastal plant ecosystems, mangroves have the highest carbon storage efficiency and are prone to the impact of anthropogenic activities. In this study, taking the mangrove wetland of Qinglan Bay as an example, we extracted information on mangrove coastal surface change from 1988 to 2020 based on multi-temporal Landsat remote sensing data through field ground surveys and laboratory analysis and used the InVEST model to calculate the spatial and temporal structure of blue carbon in the mangrove area to investigate the effects of mangrove change in an ecological restoration context. The result shows that the overall area of mangrove forests exhibited a decreasing trend from 1988 to 2020, and the area of mangroves decreased from 1559.34 ha to 737.37 ha of which 52.71% was transformed into aquiculture, construction, and farm land. Accordingly, the mangrove carbon sinks from 1988 to 2020 were significantly reduced and the carbon stock decreased at an annual tendency from 1,025,901.71 tons to 712,118.69 tons. With the implementation of mangrove restoration, the decline of mangrove forests has decreased since 2003, promoting the stabilization and enhancement of carbon sinks in the mangrove wetlands of Qinglan Bay. The results of this study provide a technical method to evaluate the carbon sink contribution of mangrove wetland restoration in Hainan Province, a scientific basis and methodological innovation to monitor the carbon sink of mangrove cover change in larger regions of China, a theoretical basis to select criteria for mangrove restoration, and a scientific and systematic management strategy for ecological and mangrove restoration on Hainan Island.

Keywords: mangrove forest; blue carbon; InVEST model; ecological restoration; Hainan Island

1. Introduction

According to the latest data from the United Nations Intergovernmental Panel on Climate Change (IPCC AR6) February 2022, global temperatures have risen by 1.09 °C over the last decades compared to 1850–1900. The Earth's climate is approaching an irreversible turning point [1]. The imbalance of the global carbon cycle has led to the occurrence of extreme climate events [2]. Source reduction is mainly achieved by means of conservation, pollution reduction, and clean energy, while sink enhancement is based on the absorption and elimination of greenhouse gases in the atmosphere [3].

The ocean carbon sink, also known as blue carbon, consists of the carbon absorbed and sequestered by oceanic and coastal zone ecosystems. This carbon is mainly stored in the form of biological biomass and sediment carbon [4]. The oceans are the largest active carbon reservoir on Earth and have an irreplaceable function in regulating climate change.

Mangroves are (i) a unique ecosystem, spanning terrestrial, marine, and wetland ecosystems, (ii) one of the four most productive natural marine ecosystems on earth, and (iii) they have the largest carbon stocks per unit area of any typical ecosystem in the coastal zone [5,6]. Mangroves are also a contiguous area between terrestrial and marine environments and are characterized by high salinity, high temperatures, strong winds and tides, muddy sediments, and anaerobic soils. In many tropical and subtropical countries, mangrove ecosystems are productive ecosystems that provide economic and environmental benefits to coastal areas. The main roles of mangroves include protecting against storms and tsunamis, regulating water systems, and providing habitat for a variety of fish and other animals. Another important function of mangrove forests is their significant role in climate mitigation activities, with a much stronger carbon sink functionality than terrestrial forests in tropical regions that store and sequester carbon.

Globally, there are approximately 152,000 km^2 of mangrove forests, accounting for 0.4% of the terrestrial forest area. The average carbon stock of tropical mangrove wetlands is as high as 1023 Mg Chm^{-2}, and the global carbon sink capacity of mangrove wetlands is 0.18–0.228 PgC-a^{-1} [7]. The total area of blue carbon ecosystem habitats in China's coastal zone is 3325.61 km^2, with a carbon sequestration rate of 559.72 Gg C-a^{-1} [8]. The mangrove forests of Hainan Island include almost all species of mangrove plants in China, and Hainan Island contains the largest number of mangrove species and preservation areas in China [9]. From the 1950s to the present, the global mangrove forest area was reduced by approximately 30% [10]. The mangrove area in China was reduced by approximately 42.5% from the 1950s to 2001, and the mangrove forest area of Hainan Island decreased by 62% [11]. Large areas of mangrove forests have been destroyed in Sanya, and the remaining mangrove forests are secondary forests or artificial protection forests [12]. From 1959 to 2002, nearly 50% of the natural mangrove forests in Dongzhai Bay were destroyed [13,14]. Historically, the main cause of mangrove destruction has been changes in land use [15]. Studies on the causes of mangrove degradation in China all point to the increased intensity of human activities among which beach polder farming and land reclamation are the main factors that encroach on natural mangrove areas and limit their development [16]. Donato et al. quantified the carbon stocks of an entire ecosystem by measuring the biomass of trees and dead wood, soil carbon content, and soil depth in 25 mangrove sites in the Indo-Pacific region and estimated that mangrove deforestation produces 0.02–0.12 Pg of carbon emissions per year, which is approximately 10% of global deforestation emissions [17]. Miteva et al. conducted the first rigorous large-scale evaluation of the effectiveness of coastal conservation areas, which can reduce mangrove loss by approximately 140 km^2 and reduce blue carbon emissions 0.013 $GtC{\cdot}a^{-1}$ as well [18]. Since 2001, ecological management has gradually gained importance in China, and the implementation of initiatives that promote the return of ponds to forestry and water areas throughout the coastal zone has led to a number of mangrove reconstruction/restoration projects being carried out each year [19]. However, since the 1980s, the overall survival rate of mangrove reconstruction/restoration projects nationwide in China has been below 66%, only reaching approximately 17% in most areas [20]. Due to the lack of systematic and scientific ecological effect monitoring and assessment methods and management systems, artificial mangrove reconstruction/restoration projects have low survival rates and low retention rates and have not achieved the desired restoration effect [21].

The total carbon stock in mangrove wetlands is composed of two parts: one part is stored inside the plants, including the above-ground part of the plant body, the underground roots, and the dead leaves; and the other part is stored in the soil [7]. Ishil and Tateda established a Leaf Area Index (LAI) measurement method based on satellite data and determined the LAI of mangrove plantations in eastern Thailand, as well as the Net Photosynthetic Production and aboveground biomass of eastern Thailand mangrove plantations [22]. Li et al. used Landsat7 TM and radar SAR imagery fusion to conduct the Remote Sensing (RS) estimation of mangrove vegetation in the Zhujiang Estuary [23]. In most research, soil carbon stocks are usually determined using direct measurements, based on field soil profiles sampled to determine the organic carbon content of each layer. Weighting is then used to calculate the organic carbon content of the entire soil profile, and finally the mapped area is used to determine the soil carbon stocks of the whole mangrove wetland [24]. Shi et al. used the Multi-Criteria Evaluation (MCE), Cell Automata (CA), Markov chain, and InVEST model based on the data extracted from the LUCC (Land

Use/Land Cover Change) dataset from 1980 to 2020 to explore the spatial and temporal evolution patterns of terrestrial ecosystem carbon stocks in the Ili Valley from 1980 to 2030 [25]. Based on the practical studies in the abovementioned literature, it can be seen that the InVEST model has the advantages of using less data and less time and can realize the spatial mapping of carbon stock spatial distribution and dynamic changes [26]. The InVEST model is different from the previous monolithic and static service value assessment, and thus it is widely adopted in various regional studies.

The mangrove forest, as an important component of the marine blue carbon sink, is the main ecological carbon pool of Hainan Island. This mangrove forest provides an important ecological support for Hainan Island and functions as a significant economic resource for future carbon trading. Since the middle of the last century, what has happened to the spatial patterns of mangrove cover as a result of anthropogenic activities and the reconstruction/restoration of mangrove forests? What has happened to the carbon cycle of mangrove ecosystems? To address these questions, in this study, the spatial and temporal patterns of mangrove blue carbon distribution on Hainan Island were studied based on the InVEST model, using field surveys and laboratory analyses, and extracting information on mangrove cover changes in the last 30 years based on multi-temporal Landsat RS data so as to investigate the changes in mangrove cover in the context of ecological restoration. The results of the study can provide a basis for the quantitative evaluation of the mangrove carbon sink in Hainan Province. In addition, the findings of this study can provide a technical approach to evaluate the carbon sink contribution of mangrove ecological restoration in Hainan Province and aid future studies on the analysis of mangrove cover change estimations. These results could provide a scientific basis for decision-making and the development of systematic management for mangrove forest restoration projects on Hainan Island.

2. Materials and Methods

2.1. Study Area

The study area, Qinglan Bay (19°34′ N, 110°45′ E), is located in Wenchang County, northeast of Hainan Island, in the location shown in Figure 1, and it is the second largest mangrove reserve in Hainan Province. The area is dominated by a tropical monsoon climate and consists of a wetland natural reserve that is approximately 3000 ha in size, with a forested area of 2732 ha and 24 species of mangroves, accounting for 85.71% of the 28 species found in China [27].

2.2. Data Collection

To reflect the spatial and temporal patterns of mangrove coastal blue carbon distribution on Hainan Island, eight-year representative RS Landsat imageries were selected to extract information on the cover change of mangrove forests in Qinglan Bay, Hainan Island, in the past 30 years. The two main data sources used in this study were LUCC raster charts and carbon density data.

2.2.1. RS Landsat Imagery Data

The RS data used in this study were downloaded from the Geospatial Data Cloud (http://www.gscloud.cn/) (accessed on 1 April 2020). To comprehensively reflect the characteristics of mangrove cover changes in Qinglan Bay, Hainan Island, in the past 30 years, based on the clarity of RS imageries and the availability of information, the data selected included eight-year phases (1988, 1993, 1998, 2003, 2008, 2013, 2018, and 2020) of Landsat imageries, with six TM, two OLI, and a resolution of 30 m. The specific date and imagery details are shown in Table 1.

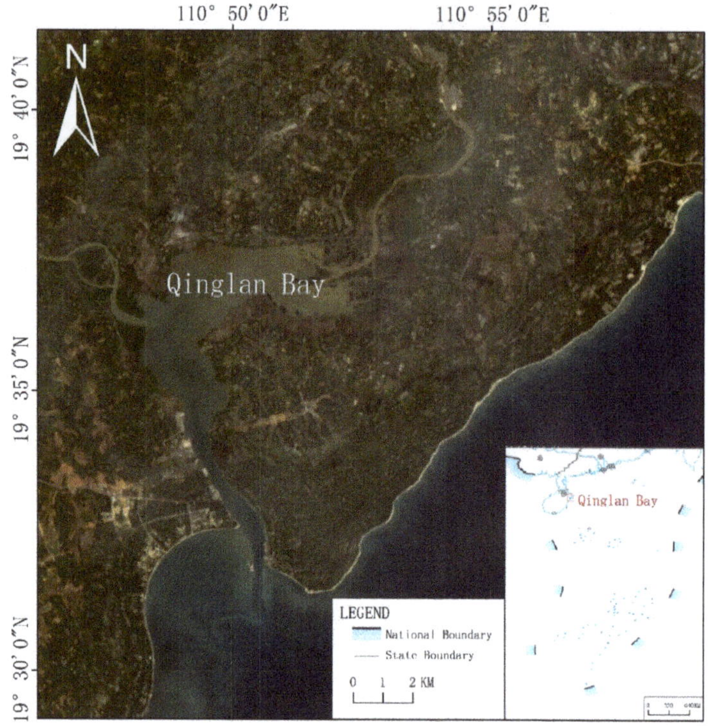

Figure 1. Study Area.

Table 1. Mangrove RS imageries in Qinglan Bay.

Sensors	Date	Band	Line No.	Resolution
TM	December 1988, April 1993, August 1998, June 2003, December 2008	46	123	30 m
OLI	December 2013, April 2018, May 2020			

2.2.2. Carbon Pool Data Extraction

This paper organized the published journal literature on mangrove blue carbon, classified the study areas by latitudes, excluded the carbon pool data with the same or similar latitudes to the distribution of mangrove areas in the study area, and replaced the missing areas with studies conducted in similar latitudes. The final combination of data was extracted from Chen [28] and Xin's [29] studies on China's coastal blue carbon and further obtained the carbon density (pool) table after calculation (Table 2).

Table 2. Carbon densities of different land use types (t/hm^2).

Land Types	Above-Ground Organisms	Underground Organisms	Soil	Dead Organic Matters
Mangrove	67.95	29.85	341.33	3.9
Forest	26.5	0	30.9	0
Aquaculture	0	0	12.2	0
Constructed land	4.8	0	23.3	0

Mangrove ecosystems consist of four main carbon pools: above-ground and below-ground biomass carbon pools, soil carbon pools, and dead organic matter carbon pools. Due to the difficulty in obtaining required timber data, the total carbon stock calculated in this study took the sum of the four major carbon pools and excluded factors, such as logging volume and the degradation rate of logged products.

2.3. Decoding and Extraction of RS Data

Taking the mangrove wetland in Qinglan Bay and its surroundings as the research object, eight periods of satellite RS imageries from 1988 to 2020 were selected, and the ENVI 5.2 software (Boulder, CO, USA) platform was used to decode and extract satellite imageries for eight periods during the last 30 years. The imagery pre-processing procedures included atmospheric radiation correction, imagery cropping, and geometric correction. The pre-processed imageries were basically consistent with the background imageries with a deviation of less than 0.5 pixels, which ensured the accuracy of this study.

2.4. The InVEST Carbon Storage and Sequestration Model

The InVEST model is a GIS (Geographic Information System) -based platform developed jointly by Stanford University, the World Wide Fund for Nature, and The Nature Conservancy for ecosystem assessment research [30]. InVEST contains several models for marine, terrestrial, and freshwater analysis among which the Carbon module, which is based on RS data, provides a fast and intuitive method for carbon estimation and is widely used in the calculation of regional carbon sinks [25]. The InVEST model links the carbon storage density to the LUCC raster by inputting the LUCC raster map, logging volume, degradation rate of logging products, and four major carbon sinks (above/below-ground biomass carbon pools, soil carbon pools, and dead organic matter carbon pools) for carbon cycle estimation and value assessment to output the present carbon stocks, the carbon stocks for a specified period of time, and the valuation of economic benefits [30].

3. Results

3.1. Extraction of RS Imageries and Land Use Data

Combined with the actual survey data, this study used the supervised classification method to extract the RS data of the current land use types in Qinglan Bay, established the land use transfer matrix, and quantitatively analyzed its spatial and temporal dynamics so as to monitor the land use dynamics of mangrove wetlands in Qinglan Bay.

The pre-processed RS imageries were loaded into the ENVI platform and color-composited with R: Band5, G: Band4, and B: Band3. According to research needs, the land in the Qinglan Bay area was classified into five categories: mangrove forest, aquaculture land, forest land, construction land, and water areas. By reviewing local historical archives and topographic maps, combined with Google Earth, different types of land features were selected as supervised classification samples. The compute ROI separability function was used on the ENVI platform, and the Jeffries–Matusita and Transformed Divergence were used to indicate the separability of the selected samples, with a sampling value range of [0, 2.0]. If the value of the sample was greater than 1.9, then it was available for use. Taking the 2008 sample data as an example, the separability values are shown in Table 3.

After selecting a neural-network-based, pattern-recognition-based vector classifier to derive supervised classification results, the sporadically distributed ground classes were organized, or clustered. The Change Detection Statistics tool was used on the ENVI platform to generate a land use type transfer matrix.

3.2. Spatial and Temporal Dynamics of LUCC in Qinglan Bay Mangrove Wetland

Eight periods of the satellite RS imageries of the mangrove wetlands in Qinglan Bay, Hainan Province were selected from the past 30 years, and the land use data were extracted using ENVI 5.2. The spatial distribution of land use and dynamic changes in the areas in the coastal zone of Qinglan Bay from 1988 to 2020 were calculated, as shown in Figure 2 and Table 4. The area of mangroves declined from 1559.34 ha to 737.37 ha in the initial phase of which 52.71% was transformed into other land categories. Most of the mangrove forests were transformed into aquaculture, while some mangroves were converted into constructed land.

Table 3. Severability of samples in 2008.

Samples	Jeffries–Matusita, Transformed Divergence
Mangrove	Forest: (1.99964123 1.99999989) Aquaculture: (1.99995880 2.00000000) Water: (2.00000000 2.00000000) Constructed Land: (1.99995791 2.00000000)
Forest	Mangrove: (1.99964123 1.99999989) Aquaculture: (2.00000000 2.00000000) Water: (2.00000000 2.00000000) Constructed Land: (1.99999665 2.00000000)
Aquaculture	Mangrove: (1.99995880 2.00000000) Forest: (2.00000000 2.00000000) Water: (1.90403571 2.00000000) Constructed Land: (1.97534302 1.99831364)
Water	Mangrove: (2.00000000 2.00000000) Forest: (2.00000000 2.00000000) Aquaculture: (1.90403571 2.00000000) Constructed Land: (1.99999579 2.00000000)
Construction Land	Mangrove: (1.99995791 2.00000000) Forest: (1.99999665 2.00000000) Aquaculture: (1.97534302 1.99831364) Water: (1.99999579 2.00000000)

From 1988 to 1993, fenced aquaculture emerged in Xiatian Village, Shatou Port, and Qingtou Village, and the aquaculture area gradually expanded. Because the tideline in the RS imagery in 1993 was lower than that in 1988, some water bodies transitioned to mangrove forests, and the number of mangrove forests also increased slightly.

From 1993 to 1998, the increasing farming activities around Lujue Village, Danchang Village, Qunjian Village, Paigang Village, Wenhu Village, Baobiao Village, Xiayang Village, and Wenjiao River caused a significant disturbance to the mangrove ecosystem, with a rapid decline of mangrove areas and expansion of the aquaculture area.

From 1998 to 2003, no new large-scale waters were added within the mangrove forest of Qinglan Bay. Compared with 1993–1998, the growth rate of the mangrove area slowed, but the growth rate of the farming area remained rapid. Mangroves far from coastal areas were transformed into construction land, and the decline of mangrove area reached its peak. Governments at all levels took various measures to strengthen mangrove protection and vigorously promote mangrove protection and restoration.

From 2003 to 2008, mangroves, water areas, and forests were transformed into expanding aquaculture areas, although this process slowed as mangrove ecological restoration projects were implemented.

From 2008 to 2013, the mangrove ecological restoration projects began to yield prominent results, with only a small increase in aquaculture area, and the decline in the area of mangroves continued to slow.

From 2013 to 2018, the scope of aquaculture waters, mainly in the Wenjiao River basin, gradually decreased, while mangrove forests increased with the implementation of the Regulations on the Protection and Development of the Coastal Zone of Hainan Special Economic Zone. The comprehensive management and restoration of coastal zones began to bring back mangrove lands.

However, from 2018 to 2020, there was a small decline in mangrove forests in Wenjiao River and Qichai River, as well as their adjacent sites.

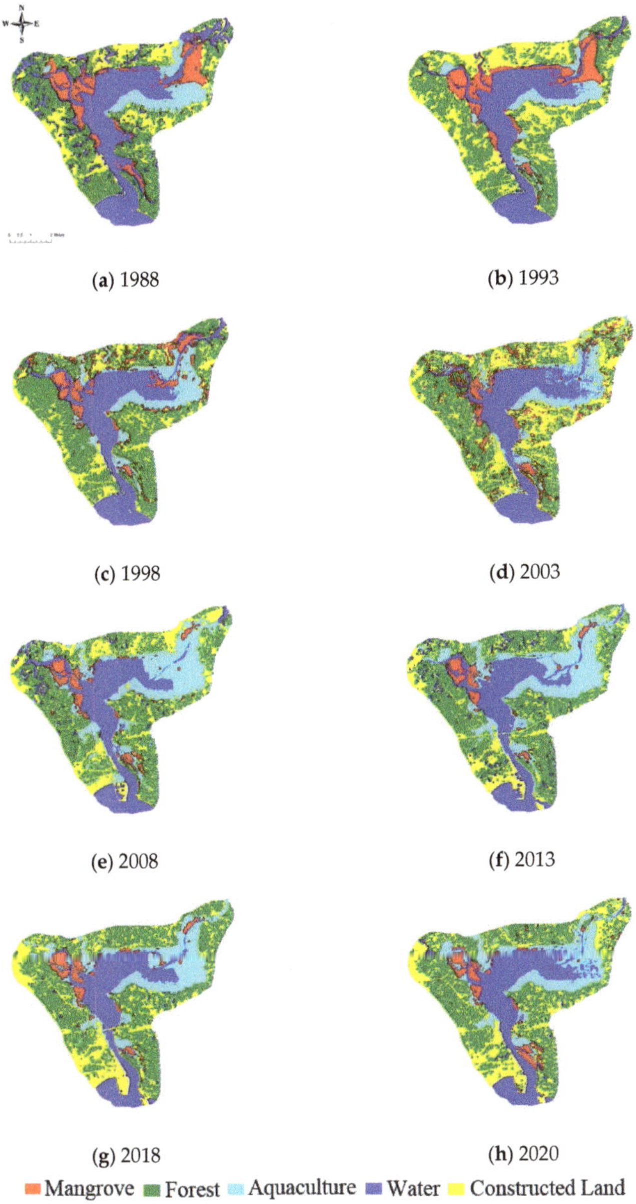

Figure 2. Spatial distribution of land use in the coastal zone of Qinglan Bay from 1988 to 2020.

Table 4. Dynamic changes in mangrove area of the Qinglan Bay from 1988 to 2020 (ha).

Land Use	1988	1993	1998	2003	2008	2013	2018	2020
Mangrove	1559.34	1619.28	1466.19	1019.61	805.5	602.01	849.33	737.37
Forest	4428.36	4461.03	5379.12	3533.67	4585.86	5049.18	5097.33	4804.65
Aquaculture	1153.35	1020.15	1747.08	1982.52	2926.98	3031.65	2474.46	2214.72
Water	4306.86	3286.35	3249.63	2928.15	3044.97	2962.53	2683.08	3127.32
Constructed Land	2377.53	3438.18	1983.15	4361.31	2461.95	2179.98	2721.15	2940.57

3.3. Estimation of Mangrove Carbon Stocks in Qinglan Bay Based on InVEST Model

The InVEST model carbon module saves the clustered data in Geotif format for analysis on the ArcGIS 10.2 platform. During this step, the processed land use data does not have an attribute table, therefore we used Data Management Tools in the toolbox and selected Build Raster Attribute Table to create a raster attribute table and to input land class names and code data to assign values to the raster map. Then, we opened the Carbon module in InVEST, inputting present and predicted data of land use and carbon pool in this step to calculate carbon stock.

3.4. Carbon Stock Distribution and Spatial Changes

In this study, the raster data of eight periods of land use types were extracted, and the net change of carbon storage over time can be calculated by the carbon module of the InVEST model, with each time period to be compared with each other. The land use type raster data and the carbon pool data were set as inputs into the InVEST model, and the spatial distribution of carbon storage in the coastal zone area of Qinglan Port, Hainan Province during 1988–2020 was acquired, as shown in Figure 3. By comparison, the differences in the spatial distribution of carbon stocks between two adjacent time phases is shown in Figure 4. As a result, the changes of the total carbon stock in eight-time phases are derived, as shown in Figure 5.

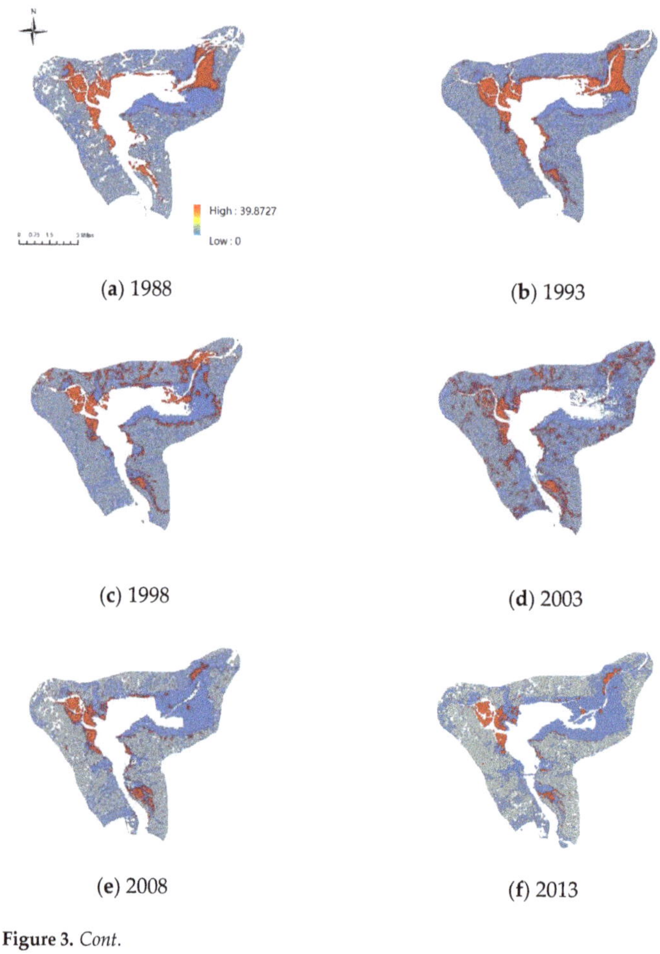

(a) 1988 (b) 1993

(c) 1998 (d) 2003

(e) 2008 (f) 2013

Figure 3. *Cont.*

(g) 2018 (h) 2020

Figure 3. Spatial distribution of carbon stock in the coastal zone in Hainan Province from 1988 to 2020.

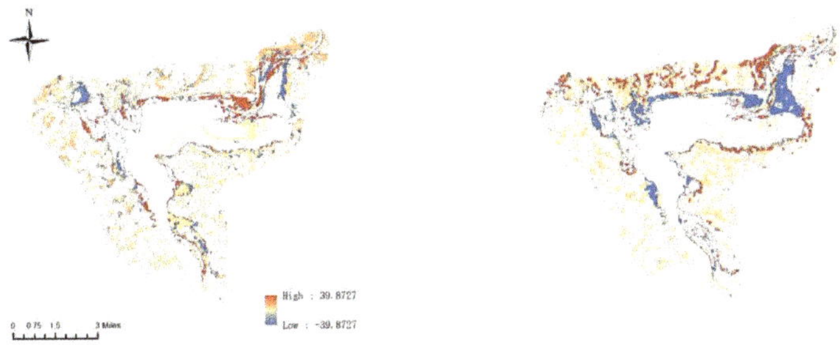

(a) 1988–1993 (b) 1993–1998

(c) 1998–2003 (d) 2003–2008

(e) 2008–2013 (f) 2013–2018

Figure 4. *Cont.*

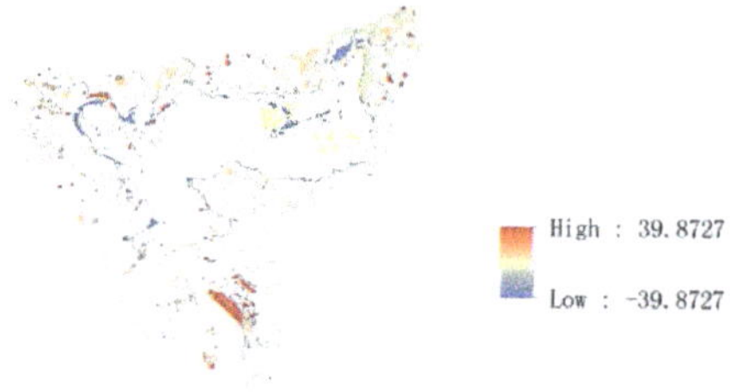

(g) 2018–2020

Figure 4. Spatial distribution map of two adjacent interphase carbon stock changes in the coastal zone of Qinglan Bay in Hainan Province from 1988 to 2020.

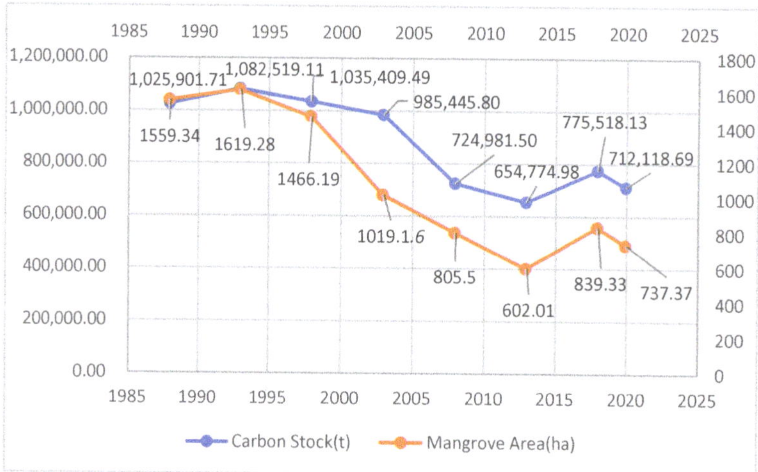

Figure 5. Total carbon stock changes with the mangrove cover change in Qinglan Port from 1988 to 2020 (ton).

The results compared by spatial analysis revealed that the curve of carbon stock change accordingly aligned with the mangrove area change. The highest distribution areas of carbon stocks were all located with the mangrove cover among the five land use types, and the mangrove vegetation distribution areas contributed the most to the carbon stocks. Through combining the spatial distribution maps of carbon stocks and land use between two adjacent time phases, the spatial and temporal evolution of carbon accumulation and carbon loss were analyzed. The results showed that carbon accumulation mainly occurred in the areas where the mangrove area increased, and carbon loss mainly occurred in the areas where the mangrove area decreased. In terms of time, the total carbon stock of the five land use types in the study area during 1988–2020 showed an overall decreasing trend from 1,025,901.71 tons to 712,118.69 tons annually, except for brief increases during 1988–1993 and 2013–2018, and the total carbon stock reached a maximum value of 1,082,519.11 tons in 1993 and a minimum of 712,118.69 tons in 2020. Since 2003, the ecological restoration project has slowed the decline of the mangrove area and the rate of carbon stock reduction in the Qinglan Bay mangrove wetland, thereby contributing to the stabilization and increase of this blue carbon sink.

4. Discussion

The InVEST model used in this study simplified the carbon cycle estimation and allowed it to be completed with relatively limited information. Nonetheless, the model also has its drawbacks in analysis. First, InVEST assumes that carbon stocks vary linearly over a given time period, or it may underestimate the carbon sink rates. Second, carbon storage estimates are influenced by land use types and carbon pool data. The more detailed the classification of land use types, the more concrete corresponding carbon pool data used and the closer the carbon storage to the actual values that would be estimated. In this study, only 8 periods of RS imagery in the past 30 years were selected. Therefore, in future studies, the accuracy of the research results can be further improved by adding more time RS imagery; increasing the land use types by function and planning; and classifying and analyzing the mangrove communities based on different plant categories and growth periods. Meanwhile, we propose that a more carbon emission monitoring system for the mangrove wetlands of Qinglan Bay can be developed in the near future by establishing RS dynamic stations for monitoring purposes in the mangrove area and adopting professional approaches.

5. Conclusions

In this study, the 3S technique was combined with the InVEST model to demonstrate the changes of blue carbon stock by investigating the changes of mangrove cover and land use and mangrove ecological restoration in Qinglan Bay mangrove forest in the past 30 years.

The results show that:

The transformation of carbon stock rate has aligned with the cover change rate of the mangrove forest in Qinglan Bay. The overall area of mangrove forests in the Qinglan Bay mangrove wetland in Hainan Province showed a decreasing trend during 1988–2020. The ecological restoration project slowed the decreasing rate of the mangrove forest area. The area of mangrove cover decreased from 1559.34 ha to 737.37 ha in its initial phase of which 52.71% was transformed into aquiculture, construction, and other land types. Most of the mangrove lands were turned into aquaculture water, and with the expansion of aquaculture water, the mangrove landscape gradually became fragmented, and some nearshore mangrove forests were transformed into constructed land.

The overall trend of the mangrove coastal carbon stock in Qinglan Bay from 1988 to 2020 is decreasing annually, from 1,025,901.71 tons to 712,118.69 tons. Among the five types of land use, mangroves have the largest contribution to carbon stock, and the highest value of carbon stock is distributed in the mangrove land type. Carbon accumulation mainly occurs in areas of increased mangroves, and carbon loss mainly occurs in areas of decreased mangroves. Since 2003, the ecological restoration project has slowed the decline of mangrove areas, which also slowed the rate of carbon stock reduction in the Qinglan Port mangrove wetland, having a positive impact on the stabilization and increase of blue carbon sink.

The results of this study provide a technical method to evaluate the carbon sink contribution of mangrove wetland ecological restoration in Qinglan Bay, Hainan Province. This method is a scientific and methodological innovation that can be used to monitor the carbon stork of mangrove cover changes in China on a large regional scale. In addition, this method provides a theoretical basis to select criteria for restoring and rebuilding mangrove habitats, and functions as an academic guidance for the development of scientific and systematic management strategies for mangrove ecological restoration on Hainan Island.

Author Contributions: Conceptualization, methodology, and writing—editing, P.J. and Z.Z.; writing—original draft and software, Z.Z., W.H., J.C. and Y.X.; funding acquisition, supervision, and project administration, P.J. All authors have read and agreed to the published version of the manuscript.

Funding: This research was funded by the National Social Science Found of China, grant number 21XGL019 and the Hainan Provincial Natural Science Foundation of China, grant number 721RC1048.

Data Availability Statement: The data presented in this study are openly available in Geospatial Data Cloud of the Computer Network Information Center, Chinese Academy of Sciences (http://www.gscloud.cn, 1 April 2020).

Acknowledgments: The authors gratefully acknowledge the two anonymous reviewers for their valuable comments and suggestions, which strengthened the quality of the paper substantially.

Conflicts of Interest: The authors declare no conflict of interest.

References

1. Keating, M. Bridging research and policy. *Nat. Clim. Chang.* **2021**, *11*, 1003.
2. Yu, G. *Global Change, Carbon Cycle and Storage in Terrestrial Ecosystem*; China Meteorological Press: Beijing, China, 2003; pp. 26–60, ISBN 7502935983.
3. Li, H. Analysis on the Object of Coastal Blue Carbon Right and Its Legal Attributes. *J. China Univ. Geosci.* **2020**, *20*, 25–38. [CrossRef]
4. Wang, X.; Zhang, H.; Han, G. Carbon Cycle and "Blue Carbon" Potential in China's Coastal Zone. *Bull. Chin. Acad. Sci.* **2016**, *31*, 1218–1225.
5. Rastogi, R.P.; Gupta, D.K.; Phulwaria, M. *Mangroves: Ecology, Biodiversity and Management*; Springer Nature Singapore Pte Ltd.: Singapore, 2021; pp. 16–52, ISBN 978981162494-0.
6. Mcleod, E.; Chmura, G.L.; Bouillon, S.; Salm, R.; Björk, M.; Duarte, C.M.; Lovelock, C.E.; Schlesinger, W.H.; Silliman, B.R. A blueprint for blue carbon: Toward an improved understanding of the role of vegetated coastal habitats in sequestering CO_2. *Front. Ecol. Environ.* **2011**, *7*, 362–370. [CrossRef]
7. Alongi, D.M. Impacts of Climate Change on Blue Carbon Stocks and Fluxes in Mangrove Forests. *Forests* **2022**, *13*, 149. [CrossRef]
8. Wang, F.; Tang, J.; Ye, S.; Liu, J. Blue Carbon Sink Function of Chinese Coastal Wetlands and Carbon Neutrality Strategy. *Bull. Chin. Acad. Sci.* **2021**, *36*, 241–251.
9. Fang, F.; Li, Z.; Gui, H. Investigation and Research on Current Situation of Mangrove in Hainan. *Trop. For.* **2022**, *50*, 42–49.
10. Zhang, Q.; Sui, S. The mangrove wetland resources and their conservation in China. *J. Nat. Resour.* **2005**, *13*, 28–36.
11. Tu, Z.; Chen, X.; Zhang, J.; Wu, J.; Chen, M.; Wang, D. Current Status of Coastal Wetland Resources in Hainan Island and Strategies for Protection. *Wetl. Sci. Manag.* **2014**, *10*, 49–52.
12. Tang, L.; Huang, S. Research on Mangrove Resources and Communities in Sanya. *Nat. Sci. J. Hainan Univ.* **2004**, *22*, 70–74. (In Chinese)
13. Liao, B.; Zhang, Q. Area, Distribution and Species Composition of Mangroves in China. *Wetland Sci.* **2014**, *12*, 435–440.
14. Huang, X.; Zhang, T.; Tan, R. The Studies on Current Status and Pre-Warning Mechanism of Mangroves in Hainan. *J. Jiangxi Norm. Univ.* **2018**, *42*, 236–241. (In Chinese)
15. Houghton, R.A.; Nassikas, A.A. Global and regional fluxes of carbon from land use and land cover change 1850–2015. *Glob. Biogeochem. Cycles* **2017**, *31*, 456–472. [CrossRef]
16. Fan, Z.; Zhang, X.; Li, J.; Yue, T.; Liu, J.; Xiang, B.; Kuang, W. Land-cover changes of national nature reserves in China. *J. Geogr. Sci.* **2013**, *23*, 258–270. [CrossRef]
17. Donato, D.C.; Kauffman, J.B.; Murdiyarso, D.; Kurnianto, S.; Stidham, M.; Kanninen, M. Mangroves among the most carbon-rich forests in the tropics. *Nat. Geosci.* **2011**, *4*, 293–297. [CrossRef]
18. Miteva, D.A.; Murray, B.C.; Pattanayak, S.K. Do protected areas reduce blue carbon emissions? A quasi-experimental evaluation of mangroves in Indonesia. *Ecol. Econ.* **2015**, *119*, 127–135. [CrossRef]
19. Zhang, Y.; Liao, B. The Research Status of Ecological Restoration Technology of Mangrove Wetlands in China. *Bull. Natl. Nat. Sci. Found. China* **2022**, *36*, 412–419. (In Chinese)
20. Jia, M.M.; Wang, Z.M.; Mao, D.H.; Huang, C. Spatial-temporal changes of China's mangrove forests over the past 50 years: An analysis towards the Sustainable Development Goals (SDGs). *Chin. Sci. Bull* **2021**, *66*, 3886–3901. (In Chinese) [CrossRef]
21. Chen, B.; Yu, W.; Chen, G.; Zheng, X.; Huang, H.; Hu, W.; Ma, Z. Coastal wetland restoration: An overview. *J. Appl. Oceanogr.* **2019**, *38*, 464–473.
22. Ishil, T.; Tateda, Y. Leaf area index and biomass estimation for mangrove plantation in Thailand. In Proceedings of the IEEE International Geoscience & Remote Sensing Symposium, Anchorage, AK, USA, 20–24 September 2004.
23. Li, X.; Y Anthony, G.; Wang, S.; Liu, K.; Liu, X.; Xian, J.; Chen, X.; He, Z.; Qin, C. Estimating Mangrove Wetland Biomass Using Radar Remote Sensing. *J. Remote Sens.* **2006**, *10*, 387–396.
24. Zhang, L.; Guo, Z.; Li, Z. Carbon storage and carbon sink of mangrove wetland: Research progress. *Chin. J. Appl. Ecol.* **2013**, *24*, 1153–1159.
25. Shi, M.; Wu, H.; Jia, H.; Zhu, L.; Dong, T.; He, P.; Yang, Q. Temporal and spatial evolution and prediction of carbon stocks in Yili Valley based on MCE-CA-Markov and InVEST models. *J. Agric. Resour. Environ.* **2021**, *38*, 1010–1019.
26. Parman, R.P.; Kamarudin, N.; Faridah Hanum, I.; Nuruddin, A.A.; Omar, H.; Abdul Wahab, Z. Geostatistical Analysis of Mangrove Ecosystem Health: Mapping and Modelling of Sampling Uncertainty Using Kriging. *Forests* **2022**, *13*, 1185. [CrossRef]
27. Jia, P.; Han, M.; Tang, D.; Zhou, X.; Huang, W.; Huang, C. The characteristics of plastic pollution of Qinglan Port along the mangrove coast to realize resources and environmental conservation. *J. Nat. Resour.* **2022**, *37*, 1061–1072. (In Chinese) [CrossRef]

28. Chen, X. Estimation of Soil Organic Carbon Reserves in Guangxi and Comparion Study with Some Provinces in China. *Sci. Geogr. Sin.* **2014**, *34*, 1247–1253, (In Chinese with English Abstract).
29. Xin, K.; Yan, K.; Li, Z.; Hu, J.; Qiu, M. Distribution of soil organic carbon in mangrove wetlands of Hainan island and its influencing factors. *Acta Pedol. Sin.* **2014**, *51*, 1078–1086. (In Chinese)
30. Hou, H.; Dai, E.; Zhang, M. A Review on InVEST Model for the Evaluation of Ecosystem Service Function. *J. Cap. Norm. Univ.* **2018**, *39*, 62–67. (In Chinese)

forests

MDPI

Article

Dynamics and Management of Restored Forests in Post-Mining Sites with Respect to Their Recreation Value: A Matrix Growth Model

Jan Melichar [1,*], Emil Cienciala [2], Jan Albert [2], Markéta Braun Kohlová [1], Vojtěch Máca [1] and Petr Pavelčík [1]

[1] Environment Centre, Charles University, José Martího 407/2, 162 00 Prague, Czech Republic
[2] IFER—Institute of Forest Ecosystem Research, Čs. armády 655, 254 01 Jílové u Prahy, Czech Republic
* Correspondence: jan.melichar@czp.cuni.cz; Tel.: +420-220-199-464

Abstract: Afforestation has been a popular restoration procedure for spoil heaps in the sites affected by coal open-cast mining in the Czech Republic. Forest replantation is a frequent restoration variant when spoil heaps are recovered, but unreclaimed sites are often left to spontaneous succession. Studies on the dynamics of such restored forests are missing, and the evidence of restored forests with respect to their recreation value is also sporadic. To study the dynamics and management of restored forests—both replanted and recovered by spontaneous succession—on spoil heaps, we used a matrix growth model, which accounts for harvest, artificial and natural regeneration, and recreation value of these forest stands. The model calibration was performed on data from 250 inventory plots distributed across the Velká Podkrušnohorská spoil heap and the Matyáš spoil heap in the Sokolov brown-coal mining area. The growth model was applied on six restored forest types to simulate—over 65 years with a 10-year cutting cycle—the effect of various management regimes of thinning on their recreation value and aboveground biomass (AGB). The results indicate that initial planting density and stand type have an effect on the dynamics of restored forest stands in the short-term horizon. Applying the thinning management resulted in an increase in recreation value for all types of restored stands, while AGB decreased.

Keywords: forest reclamation; open-cast mining restoration; discrete growth model; spontaneous succession; thinning; attractiveness for recreation

1. Introduction

Surface mining causes extensive changes in land cover and biodiversity and forms a completely new landscape in the area affected by mining. The ecosystems in the whole mining area, not just in the mine, are covered up with overburden material excavated from the mine. During the operation of open-cast coal mines, the overlying spoil material above the coal layer is removed and deposited in spoil heaps. Therefore, the rehabilitation of the post-mining landscape is crucial not only for restoring biomass production, but also for the restoration of both the ecosystem structure and functions of such areas [1,2].

Extensive areas where mining activity terminates are usually transformed into forests, agricultural land, and artificial lakes. A technical reclamation is a traditional approach to restoration, which is focused on the restoration of soil productivity and production functions of the landscape (forestry, agriculture), or in the case of hydric reclamation, to the landscaping importance of water bodies [3]. From an ecological perspective, reclamation should enable the restoration of original ecosystems, as well as their functions and services. In this way, there is a strong relation between reclamation and ecosystem services [2,4].

The concept of ecosystem services can be used as an evaluation framework and for identification of the appropriate methods of landscape restoration in order to achieve targets related to both ecosystem health and the provision of ecosystem services to society [5,6]. In recent decades, a number of studies have looked at reclamation from a broader perspective, taking into account one or more aspects of ecosystem services. In addition to restoring productivity, they also consider other functions and important ecosystem services that the restored sites are to perform, in particular soil protection, climate regulation, carbon sequestration, water retention, biodiversity, or cultural functions such as recreation [4,7–16].

Especially when post-mining areas are located in the immediate vicinity of human settlements, the recreation functions of the restored ecosystems become more important. This is the case of the post-mining area of the Sokolov brown-coal basin—the region in the west of the Czech Republic, where mining and subsequent reclamation of the affected land has been carried out since the 1960s and is going to continue as a result of the legal obligation of mining companies to recultivate the land before it is handed over for further use [17]. Because afforestation has been a popular reclamation practice for almost six decades, there are abundant reclaimed forests in the vicinity of several towns and villages in the Sokolov mining area that are particularly suitable for short-term recreation.

However, many of these restored forests were left to grow without appropriate forest management, therefore, they are little used for recreation. As found by Braun Kohlová et al. [18,19], the attractiveness of the restored forests on the Sokolov spoil heaps for recreation is significantly lower than conventionally managed forest stands. As implied by Braun Kohlová et al. [18], to increase the recreation value of restored forest ecosystems, it is necessary to implement suitable silvicultural treatment at a young and middle age of these stands.

An appropriate way to address this problem is to develop a forest growth model —and calibrate it using empirical data—that allows the study of the dynamics of restored forests and the evaluation of the thinning management alternative, including the effect of harvesting on the recreation value of these reclaimed stands. Although Frouz et al. [16,20] carried out the vegetation surveys with the basic dendrometric measurement at the largest Sokolov spoil heap—Velká Podkrušnohorská heap—in 2006 and in 2015, respectively, the small sample of plots—28 sampling plots in 2006 and 8 plots in 2015—was not sufficient for the parametrization of the growth model. In 2018, Cienciala et al. [21] conducted a representative one-time forest survey on 250 inventory plots at the Velká Podkrušnohorská heap and the adjacent Matyáš heap. As the inventory survey was stratified according to three growth stages and seven forest stand types, the data are applicable to a growth model that allows the prediction of stand characteristics structured according to its stages or sizes.

A matrix model of forest growth—firstly elaborated in the seminal works of M.B. Usher [22,23]—is a common approach when the stand-level dynamics on a size-structured population is simulated. Matrix growth models with the Usher matrix are among the empirical dynamic models that are widely used in forest growth modeling because of their ability to provide accurate and detailed simulations of forest stand dynamics structured according to tree-size distribution [24].

Compared to the stand-level and size-structured models, individual-based models are applied when the trajectories, spatial heterogeneity of individual trees, and inter-tree competition are accounted for [25]. However, individual-based models require a lot more data than stand-based models [26]. On the contrary, the shortcomings of matrix models are seen in the arbitrary division of size classes and the inability to incorporate the variability between trees of the same size class [27]. Detailed comparison and classification of growth models can be found in Liang and Picard [28] and Vanclay [29].

Many different types of matrix growth models were developed—with a wide range of forestry applications and across various types of forest ecosystems—and their linear and non-linear specifications from constant-parameter matrices to matrix models with stochastic vital rates or a geospatial dependence [30–32]. More recently, matrix growth models, due to their accuracy in predicting forest population dynamics, have been adjusted to account for climate variability, various forest management practices, and environmental disturbances (see [28]). An example of natural or anthropogenic disturbances are wild or managed fires. A matrix model was used in several applications to study forest restoration and dynamics of forest stands subject to fire [33,34]. Nevertheless, a study of the forest dynamics of restored forest stands on spoil heaps in the post-mining areas—as an example of anthropogenic disturbance—by matrix growth models is missing.

To expand the existing applications of matrix growth models to restored forest ecosystems on spoil heaps, we adopt a density-dependent matrix growth model with harvest previously developed by Buongiorno and Michie [30] and Liang et al. [31] and with its extension by artificial regeneration made by Liang [24]. More specifically, we applied the matrix growth model calibrated on forest inventory data [21] from the Velká Podkrušnohorská and Matyáš spoil heaps in order to explore the short-term dynamics of seven types of restored forests—alder, deciduous, larch, pine, spruce, and mixed coniferous-deciduous replantation, and spontaneous succession that represent predominant forest types in the Sokolov mining area. Subsequently, we used the growth model to evaluate various management options (initial planting, thinning regime) and investigate how the respective management regime affects the recreation value and aboveground biomass of restored forest. Here, we hypothesize that thinning will increase the recreation value of the forest regardless of the restored stand type and the intensity of thinning without any significant decrease in the stock of aboveground tree biomass in the short-term at the same time.

2. Materials and Methods

2.1. Study Area

The Sokolov brown-coal mining area is situated in the western part of the Czech Republic, near the German border. The areas of interest are spoil heaps situated in the neighborhood of Sokolov town, which began to be created during the 1960s by the deposition and shaping of the overburden material excavated from up to 200 m during brown-coal open-cast mining. The geographical location of the Sokolov spoil heaps in the Czech Republic is illustrated in Figure 1. So far, about 7000 ha of land on 10 main spoil heaps in the Sokolov mining area have been recultivated, which accounts for 75% of the area affected by coal mining there. The spoil heaps are different in size; the smallest Velký Riesl heap occupies the area of 16 ha, while the largest Velká Podkrušnohorská heap occupies 1900 ha [35]. The geographical distribution of the Sokolov spoil heaps can be found in Braun Kohlová et al. [19] (p. 4).

The heaps' substrate is composed of alkaline tertiary clays—so-called cypris formation—mostly consisting of clay minerals such as illite, kaolinite, and montmorillonite; non-clay minerals such as siderite and calcite are also often present.

More than 53 thousand inhabitants live in the vicinity of the spoil heaps, namely, in three towns, including the town of Sokolov with 23 thousand inhabitants and 14 villages.

Restored forests on the spoil heaps are mainly even-aged with stands represented by one dominant tree species; there is also a mosaic of deciduous and coniferous stands, and some stands are also mixed. The restored forest stands on spoil heaps are either replanted forests with one dominant tree species or some spots are unreclaimed and left for spontaneous succession. Heaps with filled overburden were usually leveled and compacted before stands were replanted, while the terrain of unreclaimed sites was left unleveled. The surface of unreclaimed sites is characterized by numerous ridges and depressions and a series of parallel waves with a 2 m height originally created by heaping machinery during the depositing overburden on the heap [17].

Inventory plot data [21] measured at two Sokolov spoil heaps—Matyáš heap and Velká Podkrušnohorská heap—were used to calibrate the growth matrix model for a subsequent simulation of forest dynamics with respect to various forest restoration managements. The Matyáš heap with an area of 167 ha is the smaller one of both soil heaps, the neighboring Velká Podkrušnohorská heap—as the largest spoil heap in the Sokolov brown-coal mining district—has an area of 1939 ha and is about 10 km long and 2.5 km wide with a height up to 200 m above the original terrain. The altitude of the study site is in the range of 409–604 m a.s.l. The mean annual precipitation and temperature are 800 mm and 7 °C, respectively [36]. The geographical distribution of the Matyáš heap and Velká Podkrušnohorská heap is shown in Figure 2.

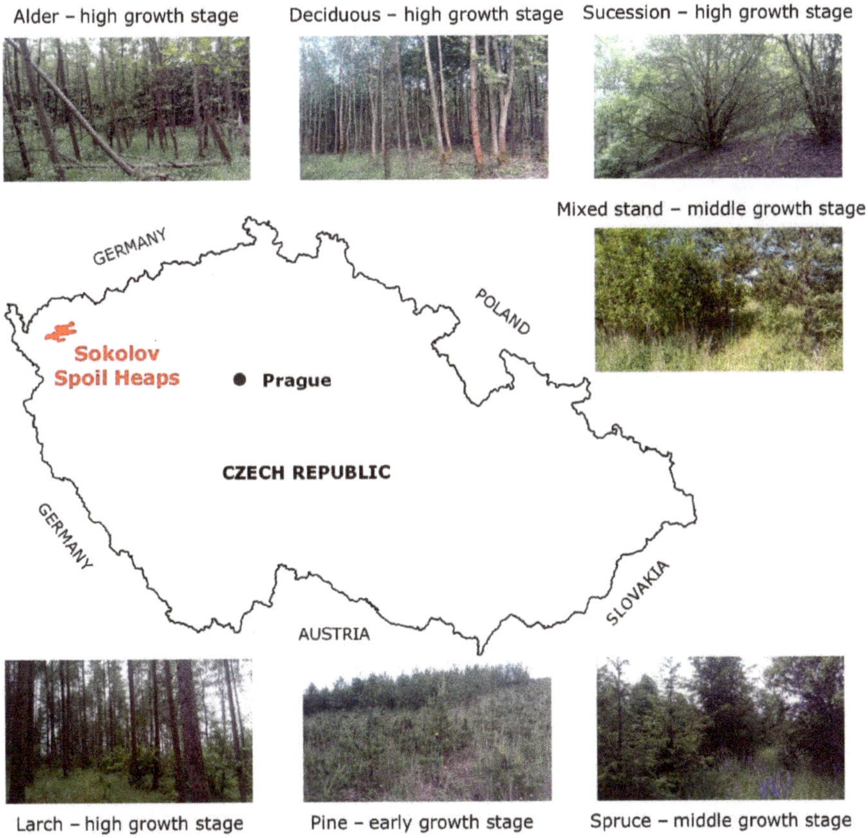

Figure 1. The geographical location of the Sokolov spoil heaps in the Czech Republic and the visualization of seven types of restored forest stands.

Replanted forest stands on the Velká Podkrušnohorská and Matyáš heap mainly consist of one tree species or two tree species of the same genus that were combined during replantation. Mostly, tree species such as Black alder (*Alnus glutinosa* (L.) Gaertn.), Norway spruce (*Picea abies* (L.) H. Karst.), European ash (*Fraxinus excelsior* L.), Scotch pine (*Pinus sylvestris* L.), European larch (*Larix decidua* Mill.), European aspen (*Populus tremula* L.), Austrian pine (*Pinus nigra* J. F. Arnold), English oak (*Quercus robur* L.), and small-leaved linden (*Tilia cordata* Mill.) have been used for replanting. Silver birch (*Betula Pendula* Roth) and goat willow (*Salix caprea* L.) are successional trees that predominate on unreclaimed sites. The overview of the relative representation of individual tree species determined from forest inventory on the investigated spoil heaps is shown in Table A1 in Appendix A.

Restored forest stands growing on the Velká Podkrušnohorská and Matyáš heaps are young and middle-aged forests. Therefore, for the forest inventory purposes, we categorized these stands according to three growth stages: early growth stage with a height of up to 2 m (10–15 years old), middle growth stage of a height from 2 to 10 m (10–30 years old), and high growth stage of a height above 10 m (30–60 years old). The geographical distribution of forest stands on the Velká Podkrušnohorská and Matyáš heap corresponding to their growth stages is illustrated in Figure 2. As prevailing stand types, stands with a dominant representation of alder, larch, pine, spruce, long-lived deciduous trees, mixed coniferous–deciduous stands, and stands with a spontaneous succession are present on the heaps (see their visualization in Figure 1). The classification of individual tree species into the prevailing stand types is given in Table A2 in Appendix A.

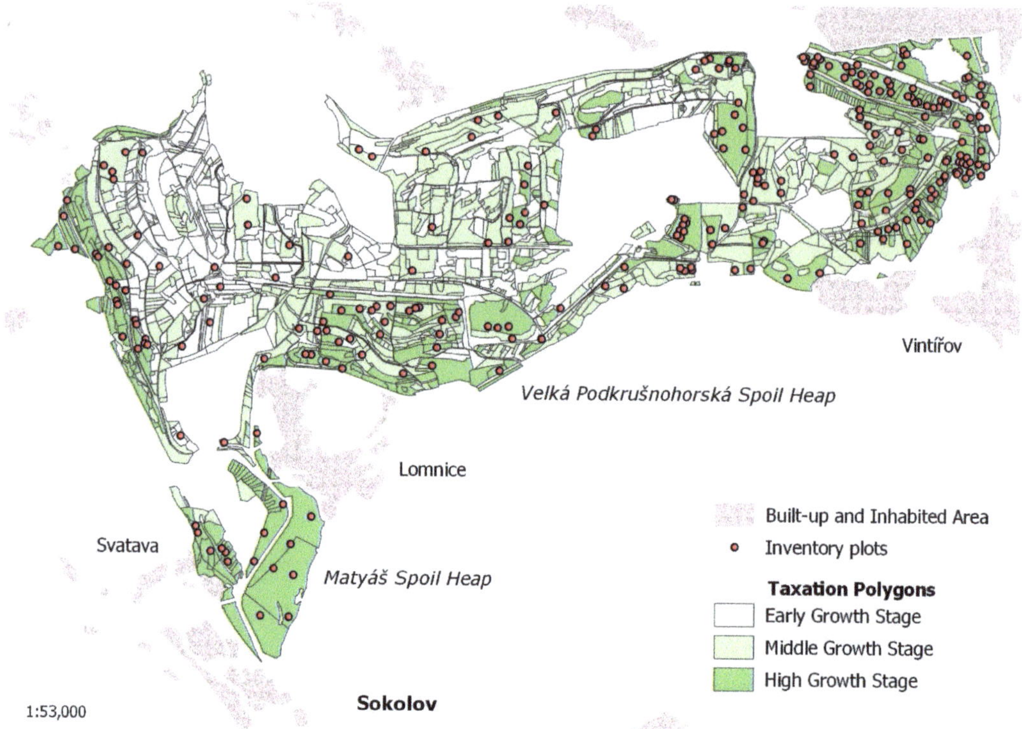

Figure 2. The geographical distribution of inventory plots and taxation polygons with corresponding growth stages of restored forest stands on the Velká Podkrušnohorská spoil heap and Matyáš spoil heap.

In addition to trees, there are also shrubs and herbaceous vegetation in the stands. Among shrub species, European black elder (*Sambucus nigra* L.), European hawthorn (*Crataegus oxyacantha* L.) and common snowberry (*Symphoricarpos albus* L.) are the most widespread species in the shrub understory of restored forest stands. Other shrub species are also present, such as wild privet (*Ligustrum vulgare* L.), alder buckthorn (*Frangula alnus* Mill.), dog rose (*Rosa canina* L.), guelder rose (*Viburnum opulus* L.), common dogwood (*Cornus sanguinea* L.), Turkish hazel (*Corylus colurna* L.), bird cherry (*Prunus padus* L.), blackthorn (*Prunus spinosa* L.), common ninebark (*Physocarpus opulifolius* Maxim.), European buckthorn (*Rhamnus cathartica*), and cornelian cherry (*Cornus mas* L.), but their occurrence is rather rare. Most plant species, and at the same time the largest coverage of the herbaceous layer (e. g. *Festuca ovina* agg., *Pastinaca sativa*, *Festuca gigantea*, *Crepis biennis*, *Epipactis helleborine*, *Melilotus albus*), are in the alder, larch, and pine stands (see [36] for more details).

2.2. Forest Inventory Data

The matrix growth model was calibrated on data from a forest inventory survey—conducted by Cienciala et al. [21]—with a sample of 250 inventory plots (see Figure 2), which were randomly distributed across the Velká Podkrušnohorská and Matyáš heap. The inventory plots included restored forests, both stands established by planting and those created by spontaneous succession. In both cases, the stands were unmanaged—left to grow without appropriate forestry management since the planting or colonization with successional trees—forests until the time of inventory.

The inventory survey was planned as a single survey, i.e., with one-time dendrometric measurement on each plot. Each inventory plot was a circular plot composed of three circular zones with the following size criteria that a tree had to meet in order to be subject to a dendrometric measurement. Trees with DBH (diameter at breast height) below 7 cm and with tree height above 25 cm were measured in the zone delimited by the inner circle (radius of 2 m, area 12.6 m^2). The trees above DBH > 7 cm were measured in the zone bounded by the middle cycle (radius of 5 m, area 78.5 m^2) and trees above DBH > 15 cm were measured in the zone delimited by the outer circle (radius of 10 m, area 314.2 m^2).

Tree and stand characteristics on the sample plots were measured throughout the year 2018. On each sample plot, the basic dendrometric measurement was taken for each individual tree that corresponded to the sampling design. Tree data collected include species, diameter, height, status (type of recruitment—natural or artificial, live, or dead standing tree), basal area, and dry matter of aboveground tree biomass. On the plot-level, stem density, natural and artificial recruitment density, total basal area, total AGB, number of tree and shrub species, total cover of shrub understory on a plot, and the age of the stand in years were derived for each plot. The list of stand characteristics and their description is provided for illustration in Table A3 in Appendix B.

Each plot was assigned to one of three growth phases: (i) early—stand up to a height of 1.5–2 m, (ii) middle—stand in the range of height of 2–10 m, and (iii) high—stand above 10 m in height (alder stand above 6 m in height). In addition to the growth stage differentiation, the plot was classified with respect to the seven forest stand types, i.e., according to the dominant tree species prevailing on the given inventory plot. The assignment of individual tree species to the appropriate forest type category is defined in Table A2 in Appendix A. Specifically, the following categories of forest types were defined—the conditions for classifying the inventory plots to the defined forest types are given in parentheses: (i) alder (alder representation of 70% and more in the stand), (ii) long-lived deciduous (≥70%), (iii) spontaneous succession (representation of succession trees >50%), (iv) larch (≥70%), (v) pine (≥70%), (vi) spruce (≥70%), and (vii) mixed coniferous–deciduous stands (if none of the above conditions are met).

While the inventory plots assigned to early and middle growth stages made up 7% and 29% of the sample, respectively, plots with trees in high growth stage accounted for 64% of the sample. The most frequent in the sample were the plots classified as mixed coniferous–deciduous, succession, and pine stand with 22%, 20%, and 15%, respectively. Plots classified as alder and long-lived deciduous stand equally accounted for 12%, plots with spruce and larch stands made up 11% and 8%, respectively.

Summary statistics of stand characteristics for plot-level variables are given in Table A4 and mean values for other stand characteristics are presented in Table A5 in Appendix B. At the same time, statistics are presented for individual forest types. At the plot-level, mean tree density was highest for spruce forest stands (6300 trees·ha^{-1}), the density of succession, pine, and larch stands was very similar. The mean stand recruitment was the highest for succession; this also applies for natural recruitment. Spruce stands had the highest mean artificial recruitment. Deciduous forest stands had on average the largest both basal area and aboveground tree biomass, while spruce stands had the lowest values of these stand characteristics. Deciduous stands were the oldest on average (34 years), while pine stands were the youngest (17 years). Spontaneous succession stands had the highest tree diversity on average, while alder stands had the lowest. The mean shrub species diversity was highest in deciduous stands and lowest in spruce stands. The coverage of shrub understory exceeding 22%, was measured in alder stands.

Summary statistics of tree characteristics for tree-level variables for all stands and according to forest type category are given in Table A6 in Appendix B. At the tree-level, succession stands had the largest mean values of DBH and aboveground biomass, while pine stands had the lowest. The average tree height was the highest for deciduous stands (14.4 m); pine stands were the lowest (8.5 m). In contrast, pine stands had the lowest mortality rate on average, while mixed coniferous–deciduous stands had the highest mortality.

Mean values of stem density per hectare based on inventory plot data differentiated according to the DBH classes—we implemented 4 DBH classes in the growth model with the first DBH class representing recruitment: (i) 0–2 cm, (ii) 2–7 cm, (iii) 7–14 cm, and (iv) 14 cm and more—and with respect to three growth stages and seven forest types are shown in the three graphs in Figure 3. Across all inventory plots, tree density decreases with diameter classes. Similarly, a decreasing tree density was observed for succession, deciduous, and spruce stands. In contrast, alder and pine stands reached the highest stem density in the second diameter class; larch stands had the highest density in the first two classes. Differentiation according to the growth stages shows that the stem density was the highest in the first diameter class for the early growth stage. Trees in the third and fourth DBH class are no longer observed for this growth stage. Plots of the middle growth stage reached maximum density in the second diameter class. The number of trees inventoried in the fourth DBH class was negligible for this growth stage. As with all forest stands, the distribution of tree density shows a decreasing trend for the high growth stage.

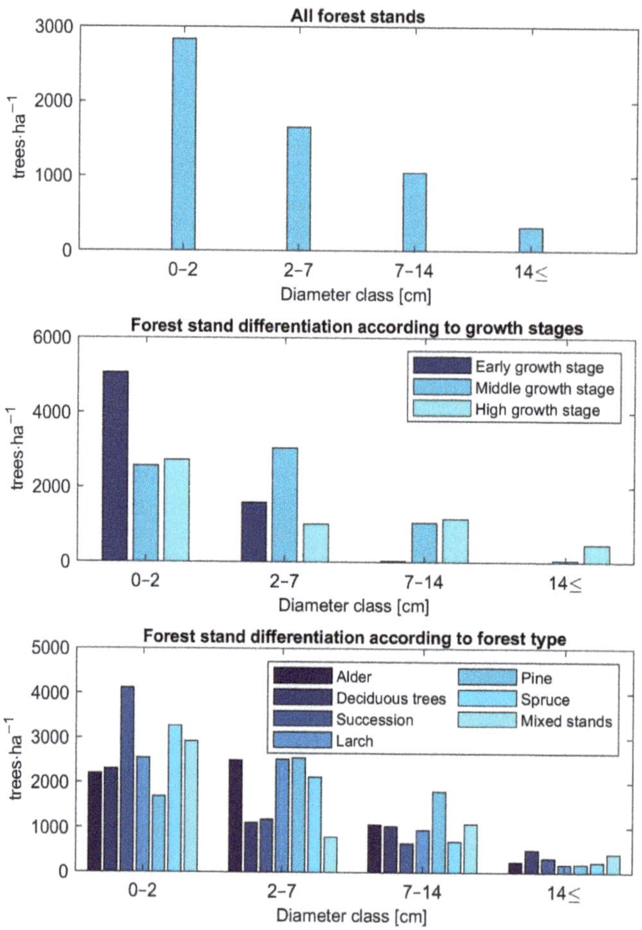

Figure 3. Tree density distribution by diameter classes and growth stages and forest types.

2.3. Data on Attractiveness for Recreation

Data on the recreation value of restored forest sites came from an environmental preference study [19], in which several representative reclaimed forest stands, including spontaneous succession stands, located on spoil heaps in the Sokolov brown-coal basin were evaluated for their recreational values. The study used a discrete choice experiment (DCE), in which the participants evaluated restored stands in response to the particular photography of a given stand as perceptual stimuli. In the DCE, the measure of recreation value was represented by a person's choice of a location with restored forests for walking in response to particular visual stimuli. The design of choice task in the DCE consisted of two hypothetical localities for recreation represented by different types of restored forest stands and a reference option of staying at home.

The survey was carried out in 2016 as an online questionnaire, which included the DCE. The resulting sample of 869 respondents consisted of the residents living in the neighborhood of spoil heaps and open-cast mines ($n = 629$) and residents from the Central Bohemian region ($n = 240$), which represented a control population [19].

The questionnaire used photographs of seven forest stands representing different reclamation practices and different growth phases. Spruce, pine, and alder replantation at the age of 35 years represented the plantation practice and stands of the middle growth stage. A 15-year-old spruce replantation was also included in the DCE as the plantation of the early growth stage. Spontaneous succession as an example of an unreclaimed forest site was another type of restored forest included in the DCE and was represented by three age and growth stage classes (15, 35, 55 years, and early, middle, high growth phase, respectively).

The mean recreation values—attractiveness for recreation—expressed as the mean probability of a person's choice of a given type of restored forest for a one-hour walk are presented in Table A5 in Appendix B, with the highest recreation value for pine replantation and sites restored with spontaneous succession. Conversely, alder stands had a very low recreation value.

2.4. Matrix Growth Model and its Structure

The model used for the simulation of the dynamics of restored forests growing at the Sokolov spoil heaps is a density-dependent matrix growth model developed by Liang [24]. This growth model is an extension of previous models [30,31] and takes into account not only harvest and natural regeneration, but also artificial recruitment. To express the dependence of the projection matrix on stand density, the tree growth, mortality, and artificial recruitment are functions of the stand basal area [37].

In the subsequent notations, t refers to time in years ($t = 0, 1, 2\ldots$), i to a type of restored forest stand, and j to a diameter class. We consider the model accounting for seven types of restored forests i ($i = 1, \ldots, 7$), as they are defined in the description of forest inventory survey. Specifically, we distinguish between alder, long-lived deciduous, spontaneous succession, larch, pine, spruce, and mixed coniferous–deciduous stands. Each stand is structured into four diameter classes j ($j = 1, \ldots, 4$) specified as follows: (i) 0–2 cm, (ii) 2–7 cm, (iii) 7–14 cm, and (iv) above 14 cm.

At the stand level, the state of forest according to its diametric structure is represented by a column vector $\mathbf{y}_t = [y_{ijt}]$ that expresses the number of live trees per hectare of restored forest type i and diameter class j at time t. The matrix growth model predicts the state of forest stand $\mathbf{y}_{t+1}$ in year $t+1$ with respect to the state of forest stand $\mathbf{y}_t$ (e.g., tree density per hectare) in year t [24,30]. The growth model [24] noticed in a matrix form is as follows:

$$\mathbf{y}_{t+1} = \mathbf{G}_t \cdot (\mathbf{y}_t - \mathbf{h}_t) + \mathbf{R}_{\mathrm{N}t} + \mathbf{R}_{\mathrm{A}t} + \varepsilon,\ \mathbf{y}_0 \geq 0,\ t = 0, 1, 2\ldots, \quad (1)$$

where $\mathbf{G}_t$ is an Usher projection matrix or also growth matrix, $\mathbf{h}_t$ is a vector of harvest, $\mathbf{R}_{\mathrm{N}t}$ is a vector of natural regeneration, $\mathbf{R}_{\mathrm{A}t}$ is a vector of artificial recruitment, and ε is a vector of random errors. The initial state of forest stand is given by a vector $\mathbf{y}_0$.

The state-dependent growth matrix $\mathbf{G}_t$ is a block matrix:

$$\mathbf{G}_t = \begin{bmatrix} \mathbf{G}_{1t} & & & \\ & \mathbf{G}_{2t} & & \\ & & \ddots & \\ & & & \mathbf{G}_{7t} \end{bmatrix} \quad (2)$$

with seven submatrices $\mathbf{G}_{it}$ on the main diagonal. Each block $\mathbf{G}_{it}$ of the growth matrix $\mathbf{G}_t$ represents the projection matrix of order 4×4 for a given type of restored forest stand i, where:

$$\mathbf{G}_{it} = \begin{bmatrix} a_{i,1,t} & & & \\ b_{i,1,t} & a_{i,2,t} & & \\ & b_{i,2,t} & a_{i,3,t} & \\ & & b_{i,3,t} & a_{i,4,t} \end{bmatrix}, (i = 1, \ldots, 7) \quad (3)$$

that describes how the trees of each restored forest type i grow or die between t and $t+1$.

The parameters a_{ijt} and b_{ijt} of $\mathbf{G}_{it}$ are vital rates representing the transition probability of ingrowth and upgrowth, respectively. The parameter of ingrowth a_{ijt} is the probability that a tree of restored forest type i and diameter class j survives and stays in the same diameter class j between t and $t+1$. The parameter of upgrowth b_{ijt} is the probability that a tree of restored forest type i and diameter class j survives and grows into the diameter class $j+1$ between t and $t+1$.

Let the parameter m_{ijt} denote the mortality rate representing the probability that a tree of forest stand i and diameter class j dies between t and $t+1$. Then, the transition parameters a_{ijt} and b_{ijt} are related as follows:

$$a_{ijt} = 1 - b_{ijt} - m_{ijt},\ (j < 4), a_{ijt} = 1 - m_{ijt},\ (j = 4). \quad (4)$$

Vector $\mathbf{R}_{\mathrm{N}t}$ represents natural recruitment between time t and $t+1$. Vector $\mathbf{R}_{\mathrm{N}it} \in \mathbb{R}^4$ is a subvector of $\mathbf{R}_{\mathrm{N}t}$ with zero elements except the first one $R_{\mathrm{N}it}$. This regeneration parameter defines the annual number of trees of restored forest type i naturally recruited in the first DBH class (0–2.0 cm) between t and $t+1$, where:

$$\mathbf{R}_{\mathrm{N}t} = \begin{bmatrix} \mathbf{R}_{\mathrm{N}1t} \\ \mathbf{R}_{\mathrm{N}2t} \\ \vdots \\ \mathbf{R}_{\mathrm{N}7t} \end{bmatrix}, \mathbf{R}_{\mathrm{N}it} = \begin{bmatrix} R_{\mathrm{N}it} \\ 0 \\ 0 \\ 0 \end{bmatrix}, (i = 1, \ldots, 7). \quad (5)$$

Vector $\mathbf{R}_{\mathrm{A}t}$ represents state-dependent artificial regeneration between time t and $t+1$. Similar to natural regeneration, vector $\mathbf{R}_{\mathrm{A}it} \in \mathbb{R}^4$ is a subvector of $\mathbf{R}_{\mathrm{A}t}$ with zero elements except the replanting parameter $R_{\mathrm{A}it}$ in the first position. This parameter expresses the annual number of trees of restored forest type i artificially recruited in the smallest diameter class between t and $t+1$, where:

$$\mathbf{R}_{\mathrm{A}t} = \begin{bmatrix} \mathbf{R}_{\mathrm{A}1t} \\ \mathbf{R}_{\mathrm{A}2t} \\ \vdots \\ \mathbf{R}_{\mathrm{A}7t} \end{bmatrix}, \mathbf{R}_{\mathrm{A}it} = \begin{bmatrix} R_{\mathrm{A}it} \\ 0 \\ 0 \\ 0 \end{bmatrix}, (i = 1, \ldots, 7). \quad (6)$$

Column vector $\mathbf{h}_t = [h_{ijt}]$ is a harvest vector formed by subvectors $\mathbf{h}_{it} \in \mathbb{R}^4$. Vector $\mathbf{h}_{it}$ consists of elements h_{ijt} representing the number of harvested alive trees per hectare for restored forest type i and diameter class j at time t:

$$\mathbf{h}_t = \begin{bmatrix} \mathbf{h}_{1t} \\ \mathbf{h}_{2t} \\ \vdots \\ \mathbf{h}_{7t} \end{bmatrix}, \ \mathbf{h}_{it} = \begin{bmatrix} h_{i1t} \\ h_{i2t} \\ h_{i3t} \\ h_{i4t} \end{bmatrix}, \ (i = 1, \ldots, 7). \tag{7}$$

2.5. Parametrization of Vital Rates and Projected Variables of Restored Stands

The matrix growth model (Equation (1)) is a density-dependent type of the model, of which vital rates—transition, mortality, and recruitment parameters—are dependent on the stand state $\mathbf{y}_t$, thus, the model is non-linear. The vital rates are hypothesized to be functions of stand density, respectively, stand basal area.

The vital rates were parametrized as empirical functions of tree- and stand-level variables from forest inventory data measured at the Sokolov spoil heaps [21]. The notation of all variables is defined in Table A3, and their summary statistics are given in Tables A4–A6 in Appendix B. Both the functions of b_{ijt}—derived from the annual diameter growth g_{ijt}, see Equation (8)—and m_{ijt} were formulated based on individual tree-level data, together with stand-level (plot-level) data from the inventory plot in which each individual tree was growing. Conversely, the natural regeneration R_{Nit} and artificial recruitment R_{Ait} were developed as empirical functions based on stand-level data.

The aboveground tree biomass AGB_{ijt} and recreation value AR_t were formulated as other empirical functions that were further used when the dynamics of restored forests were evaluated for different thinning scenarios. These functions were parametrized by stand variables—such as stand density and stand basal area—that the growth model (Equation (1)) enables us to predict. As growth rate and mortality function, the parameters of the tree biomass equation were estimated with individual tree-level data together with stand-level data. The recreation value equation was estimated with stand-level data only.

As explanatory variables, the diameter of an individual tree (D) and total stand basal area (B_t) were used in the annual tree diameter growth, mortality, and aboveground biomass equations. Stem density (N_t) was used in both recruitment equations to represent the seed abundance. Total stand basal area (B_t) was used in the artificial recruitment equation only. In addition to stem density, age of the dominant tree species (A) and total cover of shrub understory (C_S) were used in the recreation value equation. Apart from tree diameter measured at the tree-level, the other explanatory variables were defined at the stand-level.

The regression parameters of all empirical functions, except mortality function, were estimated with the maximum likelihood procedure using Matlab function 'fitnlm'. Mortality was estimated as a probit function with the generalized least squares using Matlab function 'fitglm'.

2.5.1. Growth Function

The transition probability of upgrowth b_{ijt} is computed as follows [24,31]:

$$b_{ijt} = \frac{g_{ijt}}{d_j}, \tag{8}$$

where g_{ijt} is the annual tree diameter growth [cm] of a tree of restored forest type i in diameter class j and in year t, and d_j is the width of the corresponding diameter class j [cm]. It is assumed that trees are uniformly distributed within each diameter class [24].

The diameter growth [cm·year^{-1}] is a function of tree diameter and its square (D, D^2) [cm] and stand basal area, its squared and cubed form (B_t, B_t^2, B_t^3) [m^2·ha^{-1}], with parameters α and error term η. The specification of non-linear diameter growth model is as follows:

$$g_{ijt} = \alpha_{i1} \cdot D + \alpha_{i2} \cdot D^2 + \alpha_{i3} \cdot B_t + \alpha_{i4} \cdot B_t^2 + \alpha_{i5} \cdot B_t^3 + \eta_{ij}. \quad (9)$$

According to prior biological knowledge, the tree diameter parameter α_{i1} is expected to have a positive sign and α_{i2} to have a negative sign. The stand basal area parameter α_{i3} is expected to be negative; the parameters and of its squared and cubed term, α_{i4} and α_{i5}, should be positive and negative, respectively. Therefore, the growth rate is supposed to increase with tree diameter, and the slower growth is expected in denser forests [37,38].

The total stand basal area B_t for each forest type i is defined at the stand-level as follows:

$$B_t = \sum_{j=1}^{4} B_{ij} \left(y_{ijt} - h_{ijt} \right), \quad (10)$$

where B_{ij} corresponds to the average basal area of trees growing in the stand of forest type i and diameter class j.

When predicting g_{ijt} during the simulations of the matrix model, D was replaced by D_j, the midpoint of the corresponding DBH class j, in Equation (9).

In the absence of permanent plot data in the study site, we used one-time plot data, the proposed method for the calculation of the yearly diameter growth g_{ijt} is based on the following observations and assumptions: (a) afforested forests on spoil heaps by plantation or spontaneous succession are newly formed stands on sites that previously had no tree cover; (b) the age of the stand, and thus, main trees initially planted or established by spontaneous succession, represents the time since the afforestation of the site; (c) the age of the dominant tree species that represents a given forest type i is used to determine the stand age. Table A2 in Appendix A shows the assignment of individual tree species to a given forest type. (d) The trees that played roles in the stand initiation are determined according to their diameter and growth phase of the stand as follows: (i) trees of the first and second DBH class of forest stands in the early growth phase, (ii) trees of the third and fourth DBH class of stands in the middle growth phase, and (iii) trees of the fourth DBH class of stands in the high growth phase.

Under these observations, the annual growth rate g_{ijt} of an individual tree was derived with tree-level data such as tree diameter (D) and age of the dominant tree (A) with respect to the growth phase of the inventory plot on which the tree was located. Three growth phases are classified: early, middle, and high, denoted by subindexes $s = 1$, $s = 2$, and $s = 3$, respectively. The yearly diameter growth is calculated as follows:

$$g_{ijt} = \begin{cases} \frac{D_{ij}^1}{A_{ij}^1} & \text{for } j = 1,2 \\ \frac{D_{ij}^2}{A_{ij}^2} & \text{for } 2 < j \\ \frac{D_{ij}^3}{A_{ij}^3} & \text{for } 3 < j, \end{cases} \quad (11)$$

where D_{ij}^s and A_{ij}^s are the diameter and age, respectively, of the tree of forest type i, diameter class j, and growth phase s. This means that the tree diameter growth per year is equal to DBH divided by the age of dominant tree growing on the plot. While g_{ijt} was computed for trees with DBH below 7 cm—1st and 2nd DBH class—for the inventory plots of the early growth phase, trees with DBH above 7 cm—3rd and 4th diameter class—and trees with DBH above 14 cm—4th diameter class—were considered when the growth rate was calculated for trees growing on the inventory plots of the middle and high growth phases, respectively.

2.5.2. Mortality Function

In the studies [31,32,38] with permanent plot data, the probability of annual tree mortality is derived from the inventory records representing the state whether a tree died or remained alive between two inventories. In our study with cross-section inventory data without periodic measurement, we used the individual tree records indicating the health status of a tree. We constructed a binary variable representing tree mortality M_{ijt}, which equals to 1 if a standing tree of forest type i and diameter class j is dead or a standing tree is otherwise damaged, e.g., a tree with a break, and 0 if it is a live tree without visible damage. Mean values of M_{ijt} according to forest type are presented in Table A6 in Appendix B.

We used the time interval parameter τ_{ij} with respect to the forest type i and diameter class j to convert state-based mortality M_{ijt} on the probability of annual tree mortality m_{ijt} as follows:

$$m_{ijt} = \frac{M_{ijt}}{\tau_{ij}}. \tag{12}$$

The time interval parameter τ_{ij} is set with inventory plot age (A) as:

$$\tau_{it} = \begin{cases} \Lambda_i^1 & \text{for } j = 1,2 \\ \Lambda_i^2 & \text{for } j = 3 \\ \Lambda_i^3 & \text{for } j = 4, \end{cases} \tag{13}$$

where Λ_i^s is the average age of forest stands represented by inventory plots by growth phases s and forest types i. The annual mortality rates derived from M_{ijt} according to Equations (12) and (13) are roughly similar to those mortality estimates obtained from the permanent plot data in other studies [31,38,39].

The probability of tree mortality M_{ijt} is a probit function of tree diameter D, stand basal area B_t, its squared and cubed form $\left(B_t^2, B_t^3\right)$, with parameters β and error term μ. The specification of mortality model is as follows:

$$M_{ijt} = \Phi\left(\beta_{i1} + \beta_{i2} \cdot D + \beta_{i3} \cdot B_t + \beta_{i4} \cdot B_t^2 + \beta_{i5} \cdot B_t^3\right) + \mu_{ij}, \tag{14}$$

where Φ is the standard normal cumulative function. The tree diameter parameter β_{i2} is expected to be negative. Conversely, the stand basal area parameter β_{i3} is supposed to have a positive sign, the parameters of its squared and cubed term, β_{i4} and β_{i5}, should be negative and positive, respectively. Therefore, smaller trees are supposed to have higher mortality than larger ones, and the higher mortality rate will be in denser forest stands.

Knowing b_{ijt} and m_{ijt}, the transition probability of ingrowth a_{ijt} is computed according to the relation in Equation (4).

2.5.3. Recruitment Functions

Dissimilar to growth rate and mortality, both natural regeneration and artificial recruitment, R_{Nit} and R_{Ait} [trees·ha^{-1}·year^{-1}], are developed as stand-based functions, and they are specified as linear functions of stand density, N_t [trees·ha^{-1}·year^{-1}]. Specifically, natural regeneration is a function of stand density with parameter ν and error term ϑ, and has the following form:

$$R_{Nit} = \nu_{i1} \cdot N_t + \vartheta_i. \tag{15}$$

The stand density parameter ν_{i1} is expected to have a positive sign, so that more trees increase the abundance of seeds and cause a higher rate of regeneration.

Artificial recruitment is a function of stand density and stand basal area, with parameters γ and error term θ and has the following specification:

$$R_{Ait} = \gamma_{i1} \cdot N_t + \gamma_{i2} \cdot B_t + \theta_i. \tag{16}$$

One would expect both parameters of stand density γ_{i1} and stand basal area γ_{i2} to be negative. Therefore, a smaller number of trees in the stand would require a higher intensity of artificial planting. However, the larger the stand basal area is, and thus, stand density, the higher the competition between trees will be.

Data for the recruitment, R_{Nit} and R_{Ait}, i.e., the number of trees per hectare entering yearly the smallest diameter class as natural regeneration and artificial planting, respectively, were computed from the inventory data measured on individual trees of up to 2.0 cm DBH. First, a tree from recruitment was assigned to either (1) the natural or (2) the artificial recruitment category r $(r = 1,2)$ based on the regeneration origin determined during the inventory survey. Furthermore, the average annual height increment of recruitment (δ_i) for each restored forest type i was derived with tree-level data such as tree height (h) and age of the dominant tree (A) measured on inventory plots of the early growth phase—trees up to 2 m height—as follows:

$$\delta_i = \frac{h_{i1}^s}{A_{i1}^s}, \text{ for } s = 1, \tag{17}$$

where h_{i1}^1 and A_{i1}^1 are the tree height and age, respectively, of the average tree of forest type i, diameter class $j = 1$, and growth phase $s = 1$. The annual height increment of recruitments was from 9 cm for deciduous, larch, and mixed stands to 28 cm for succession stands. The height increment of spruce, pine, and alder stands amounted to 15 cm, 18 cm, and 24 cm, respectively.

The parameter δ_i was then used to measure the annual recruitment R_{it}^r—both natural R_{Nit} and artificial recruitment R_{Ait}—on each inventory plot as follows:

$$R_{it}^r = \frac{h_{i1}^r}{\delta_i} y_{i1}^r, \ (r = 1,2), \tag{18}$$

where h_{i1}^r and y_{i1}^r are, respectively, the tree height of the average tree and the number of trees of forest type i, diameter class $j = 1$, and recruitment type r, on a given plot.

2.5.4. Function of Aboveground Tree Biomass

The volume of aboveground tree biomass of individual trees AGB_{ijt} [tonnes·tree^{-1}] is a function of tree diameter D, stand basal area B_t, its squared and cubed form $\left(B_t^2, B_t^3\right)$, with parameters σ and error term ξ. The specification of tree biomass model is as follows:

$$AGB_{ijt} = \sigma_{i1} \cdot D + \sigma_{i2} \cdot B_t + \sigma_{i3} \cdot B_t^2 + \sigma_{i4} \cdot B_t^3 + \xi_{ij}. \tag{19}$$

Since the tree volume is related to the tree diameter, the parameter σ_{i1} should be positive. The stand basal area reflects possible effect of stand density, so that tree volume should be higher in a less dense stand. Therefore, the stand basal area parameter σ_{i2} is expected to be negative, the parameters of its squared and cubed term, σ_{i3} and σ_{i4}, should be positive and negative, respectively.

The data for AGB_{ijt} in Equation (19) were determined using the allometric equations, which were predominantly derived for main tree species and different tree components in the central Europe [40–47]. Specifically, the calculation of aboveground tree biomass consisted of deriving the volume of following tree component: coarse wood ($\geq$7 cm in diameter), bark of coarse wood, small wood (<7 cm in diameter) including bar, and needles (in conifers). Individual tree inventory data such as DBH and tree height were used as explanatory variables in the component biomass functions.

2.5.5. Function of Attractiveness for Recreation

The recreation value AR_t [probability] as attractiveness of restored forest stands for recreation, is a function of the proportion of stem density (N_t) to age of the dominant tree species A, and total cover of shrub understory C_S, with parameters ϱ and error term φ. The model on recreation value is specified as linear-log model:

$$AR_t = \varrho_1 + \varrho_2 \cdot \ln(N_t / A) + \varrho_3 \cdot C_S + \varphi \tag{20}$$

The shrub understory parameter ϱ_3 is supposed to have a negative sign, so that shrubs detract from scenic beauty of forest stands, as found by Brown and Daniel [48]. The parameter ϱ_2 is assumed to be negative, since increasing stand age should have positive scenic impact, thus, increasing recreation value of the stand. The effect of tree density on recreation value is opposite. As found by Hull et al. [49], the age of stand has a positive effect on scenic beauty, but in interaction with tree density, since increasing density weakens the age effect.

The parameters of the recreation value model were estimated with forest inventory data [21] and with the estimates of recreation value derived in the environmental preference study [19]. The data for recreation value, AR_t in Equation (20), were represented by mean recreation values predicted by the DCE model for seven restored forest stands at the Sokolov spoil heaps in the environmental preference study. Forest inventory data—such as tree density, age of stand, shrub coverage—from seven inventory plots geographically corresponding to the restored forest stands, where photographs were taken in 2016 for the assessment of the attractiveness for recreation in the environmental preference study, was subsequently attached to the data on recreation value.

2.6. Verification of the Growth Model

The matrix growth model (Equation (1)) with empirical functions (Equations (9), (14)–(16), (19)) representing individual variables in the matrix model was calibrated on inventory plot data from the Velká Podkrušnohorská and Matyáš heap. Verification—accuracy of the calibrated growth model was determined based on the prediction errors, the differences between the observed values of stand characteristics—stem density, density according to the DBH classes, basal area, and AGB—and the predicted values of these simulated forest characteristics.

The observed values of stand characteristics were expressed as 5-year averages. In the given 5-year average, only those inventory plots whose age was within the given 5-year interval were included. At the same time, a 95% confidence interval was determined with the 5-year average of the observed forest characteristic.

The accuracy of the model was verified on 250 inventory plots. The growth model was used to simulate the dynamics of restored forest stands on each inventory plot in a one-year time step up to the age of 59 years. This age corresponds to the maximum age that was determined during the forest inventory survey for the dominant tree growing on the inventory plot. The initial state of forest stand $\mathbf{y}_0 \geq 0$ used for simulation corresponds to the observed state of the restored forest on the given inventory plot.

2.7. Application of the Model to Differenct Thinning Managements

The matrix growth model calibrated on inventory plot data from the Sokolov spoil heaps—Velká Podkrušnohorská and Matyáš heap—was applied to study the dynamics of six types of restored forests—alder, deciduous, successional, larch, pine, and spruce stands—with and without the management of various regimes of thinning. The simulation was mainly focused on the prediction of the recreation value and aboveground biomass under different thinning regimes, which correspond to a given type of restored forest stand. The simulation was run in a one-year time step for each thinning scenario for 65 years with a 5-year cutting cycle up to 20 years of stand age and a 10-year cutting cycle in the rest of the simulated period. The simulation period up to 65 years represents the age range of restored forests of early, middle, and high growth stage that are present in the study site. Older restored stands are absent in the Sokolov mining area, because the first forest reclamations were started here only at the beginning of the 1970s.

The parameters of thinning measures considered in the simulation are based on the framework guidelines for the management of reclaimed forests [50], which were drawn up for the Sokolov spoil heaps within this study. For each stand type and age, the guidelines propose the state of the forest stand, which corresponds to the minimum target numbers of trees per hectare after thinning in a given cutting year. The minimum target numbers of trees per hectare are given in Table A7 in Appendix C.

3. Results

3.1. Estimates of Parameters

Individual tree data were used for the estimation of the regression coefficients of the diameter growth model (Equation (9)), mortality model (Equation (14)), and aboveground tree biomass model (Equation (19)); the parameters of the recruitment models (Equations (15) and (16)) and the model of forest attractiveness for recreation (Equation (20)) were estimated with inventory plot data.

The diameter growth, mortality, recruitment, aboveground tree biomass model was compiled and estimated both as a common model for all forest types and separately for each type of forest stand. The estimates of regression parameters of the diameter growth, mortality, recruitment, AGB model, and the model on prediction of recreation value are given in Tables 1–5.

Table 1. Estimated regression parameters of the diameter growth model (Equation (9)).

Variable	All Stands	Alder	Deciduous	Succession	Larch	Pine	Spruce	Mixed Stands
D	0.044 ***	0.028 ***	0.034 ***	0.043 ***	0.045 ***	0.091 ***	0.063 ***	0.061 ***
D^2	−0.0001 ***	-	−0.00003 ***	−0.00003 ***	−0.00005 ***	−0.0002 ***	−0.0001 ***	−0.0001 ***
B	12.654 ***	17.276 ***	5.235 ***	23.215 ***	21.943 ***	−9.284 ***	8.338 ***	8.195 ***
B^2	−22.850 ***	−33.219 ***	−9.804 ***	−48.637 ***	−47.564 ***	21.326 ***	−18.396 ***	−19.135 ***
B^3	9.586 ***	15.458 ***	4.041 ***	24.036 ***	24.365 ***	−13.826 ***	8.341 ***	8.824 ***
df	4341	423	706	728	251	748	348	1108
Adj. R^2	0.372	0.457	0.521	0.488	0.598	0.621	0.525	0.361

Note: D—tree diameter [cm], B—stand basal area [$m^2 \cdot ha^{-1}$], df: degrees of freedom, Adj. R^2: adjusted index of determination, significance level: *** $p < 0.01$.

Table 2. Estimated regression parameters of the probit mortality model (Equation (14)).

Variable	All Stands	Deciduous	Succession	Mixed Stands	Joint Model (1)
Constant	−2.660 ***	−6.829 **	−18.591 **	−5.366 **	−1.291 **
D	−0.005 ***	−0.008 ***	−0.008 ***	−0.005 ***	−0.003 *
B	4.756 **	21.891 **	78.468 **	16.084 **	−2.328 **
B^2	−5.190 *	−23.646 **	−110.050 **	−17.774 **	1.942 ***
B^3	1.927 **	7.925 **	49.730 **	6.138 **	-
df	4174	685	709	1082	1684

Note: (1) Joint model includes alder, larch, pine, and spruce forest stands. D—tree diameter [cm], B—stand basal area [$m^2 \cdot ha^{-1}$], df: degrees of freedom, significance levels: * $p < 0.10$, ** $p < 0.05$, *** $p < 0.01$.

Table 3. Estimated regression parameters of the recruitment model with artificial regeneration (Equation (16)) and the recruitment model with natural regeneration (Equation (15)).

Variable	All Stands	Alder	Deciduous	Succession	Larch	Pine	Spruce	Mixed Stands
	Dependent variable: R_A—artificial recruitment							
N	0.012 ***	0.004 **	0.035 **	−0.0004	0.026 ***	0.018 ***	0.030 ***	0.010 ***
B	−1.691 **	−0.645	−4.659	1.282	−5.646 *	−2.476	−4.583 *	−1.356
df	248	27	29	49	17	36	25	53
	Dependent variable: R_N—natural recruitment							
N	0.065 ***	0.046 ***	0.076 ***	0.118 ***	0.040 ***	0.015 ***	0.034 **	0.055 ***
df	249	28	30	50	18	37	26	54

Note: N—tree density [$trees \cdot ha^{-1}$], B—stand basal area [$m^2 \cdot ha^{-1}$], df: degrees of freedom, significance levels: * $p < 0.10$, ** $p < 0.05$, *** $p < 0.01$.

Table 4. Estimated regression parameters of the aboveground tree biomass model (Equation (19)).

Variable	All Stands	Alder	Deciduous	Succession	Larch	Pine	Spruce	Mixed Stands
D	0.0014 ***	0.0010 ***	0.0013 ***	0.0020 ***	0.0013 ***	0.0006 ***	0.0011 ***	0.0013 ***
B	−0.011 ***	−0.307 ***	−0.330 ***	−1.135 ***	−0.523 ***	−0.107 ***	−0.339 ***	−0.381 ***
B^2	0.0003 ***	0.404 ***	0.322 ***	1.773 ***	0.837 ***	0.065 ***	0.399 ***	0.407 ***
B^3	−0.000003 ***	−0.172 ***	−0.105 ***	−0.822 ***	−0.411 ***	0.006	−0.142 ***	−0.138 ***
df	4317	421	701	727	251	745	341	1107
Adj. R^2	0.729	0.923	0.839	0.718	0.935	0.879	0.888	0.725

Note: D—tree diameter [cm], B—stand basal area [$m^2 \cdot ha^{-1}$], df: degrees of freedom, Adj. R^2: adjusted index of determination, significance level: *** $p < 0.01$.

Table 5. Estimated regression parameters of the recreation value model (Equation (20)).

	Constant	ln(*N*/*A*)	C_S
Coeff.	0.562 ***	−0.096 **	−0.004 **
df	4		
Adj. R^2	0.814		

Note: *N*—tree density [trees·ha^{-1}], *A*—age of the restored forest stand [years], C_S—total cover of shrub understory [%], df: degrees of freedom, Adj. R^2: adjusted index of determination, significance levels: ** $p < 0.05$, *** $p < 0.01$.

The parameter estimates of regression models for both all stands and individual forest types are statistically significant for most of the explanatory variables. The regression coefficients in the estimated models also have the expected direction of the effect on response variable and are consistent across models for individual forest stand types.

Specifically, diameter growth increases at a decreasing rate with tree diameter for all types of restored forests, more so for the pine stand. Diameter growth is positively affected by stand basal area, except the pine stands with a negative effect (Table 1). Mortality declines with tree diameter across all stand types but increases with stand basal area with the only exception of the joint model (Table 2). Both artificial and natural recruitment of each stand type increases with its tree density. However, the negative effect of tree density was expected for artificial recruitment. Moreover, artificial recruitment significantly declines with stand basal area for all stands together (Table 3). In the aboveground tree biomass model, AGB increases with increasing tree diameter and declines with stand basal area for all stand types. Basal area squared and cubed also affects AGB, the quadratic term positively and cubic term negatively, except for in the pine stand (Table 4). The attractiveness for recreation is negatively affected by the logarithm of the tree density-to-age ratio. Moreover, attractiveness declines with increasing cover of the shrub understory (Table 5).

The regression model explaining the variability of recreation value (attractiveness of restored forest stands for recreation) depending on the stem density, its age, and shrub cover was used to predict the recreation value of all inventory areas in the Velká Podkrušnohorská and Matyáš heap, based on inventory plot data. The mean values of the predicted attractiveness by growth stages and forest types with a 95% confidence interval are depicted in the following graphs in Figure 4. The difference in the mean recreation is statistically significant between the individual growth stages at the 5% level. The high growth stage has the highest recreation value; conversely, the early growth stage has the lowest. Furthermore, the mean recreation value differs between alder and deciduous stands and alder and mixed stands, with alder stands having the lowest recreation value.

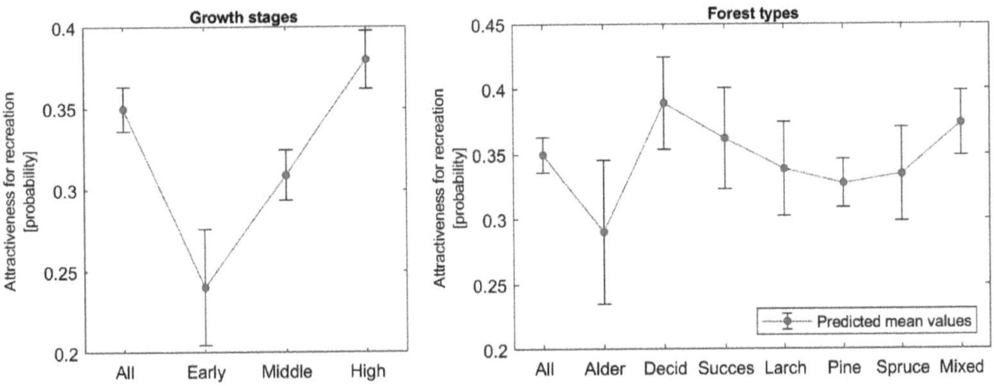

Figure 4. Mean values of the predicted recreation value (attractiveness for recreation) with a 95% confidence interval according to growth stages and forest types.

3.2. Prediction Errors

The calibrated matrix growth model was tested to be accurate between the observed values of stand characteristics and the predicted ones. The graphs in Figures 5 and 6 present the observed 5-year averages for stands of all forest types with 95% confidence intervals and predicted values by the matrix growth model (Equation (1)), parametrized by empirical functions (Equations (9), (14)–(16), (19)) of stand characteristics, such as total stem density, total basal area, total aboveground tree biomass, and stem density for each DBH class. Harvest was fixed at $\mathbf{h}_t = 0$.

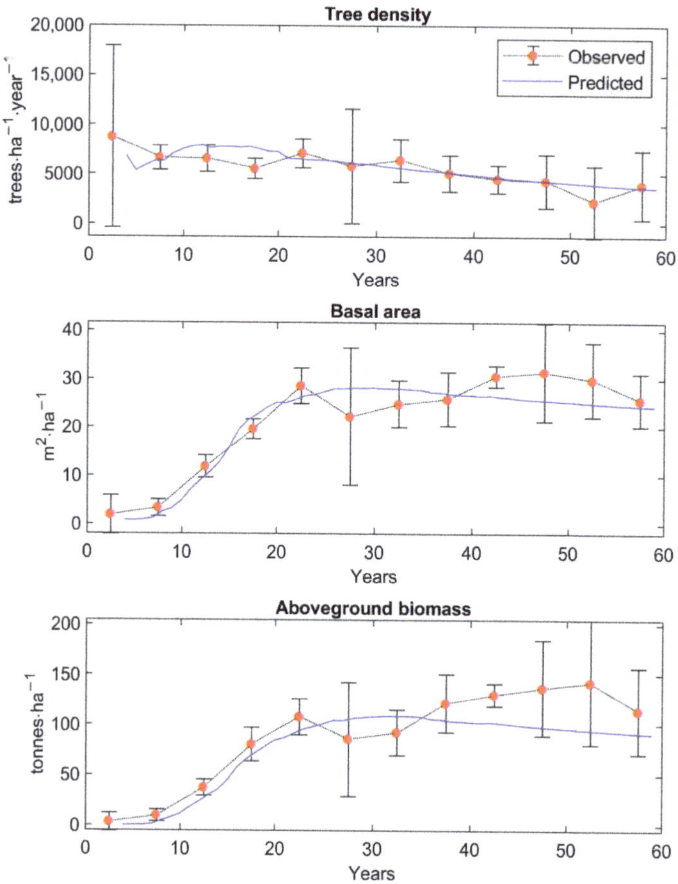

Figure 5. Verification of the model by the comparison of the observed 5-year average values with average predicted values of stem density, basal area, and AGB, with a 95% confidence interval, over 250 inventory plots.

The predicted values of stand characteristics pass through almost all 95% confidence intervals, and therefore, the predicted states of forest stands fit well the mean observed states of the 250 inventory plots.

The graphs of Figures A1–A4 in Appendix C show similarly the observed 5-year averages for individual forest types with 95% confidence intervals and predicted values of basal area (Figure A1), aboveground tree biomass (Figure A2), stem density (Figure A3), and stem density distributed according to DBH classes (Figure A4). The predicted stand states, especially for stem density, basal area, and AGB, match the observed ones.

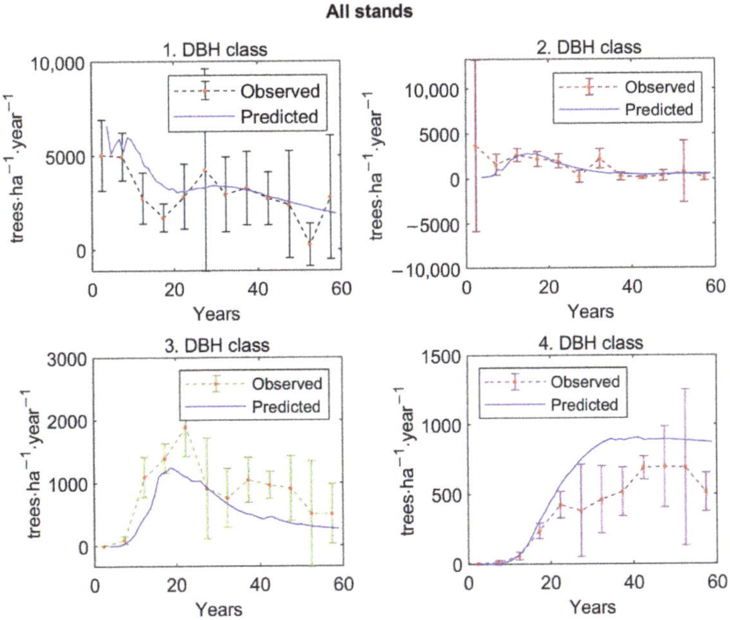

Figure 6. Verification of the model by the comparison of the observed 5-year average values with average predicted values of stem density distributed according to DBH classes, with a 95% confidence interval, over 250 inventory plots.

3.3. Application of Thinning Managements

Various thinning schemes proposed in the framework guidelines [50] for six types of restored forest stands including spontaneous succession result in different volumes of the aboveground biomass removed during 65 years. Total amount of the harvested stock is sensitive to the thinning intensities, target numbers of trees per hectare in each cutting year, the initial size of replantation, and also to the restored type of forest. Table A7 in Appendix C shows the predicted stock of aboveground biomass harvested according to different thinning managements for each stand type over 65 years.

The thinning intensity is increasing over time for all restored forests, except for the even harvesting intensity of pine and spruce in the second part of the simulated period. Alder and larch stands have the highest total volume harvested with 65.6 and 65.4 $t{\cdot}ha^{-1}$, respectively. Conversely, pine stands have the lowest AGB removed in total with 1.7 $t{\cdot}ha^{-1}$.

The graphs in Figures 7 and 8 present the simulated dynamics of aboveground tree biomass and recreation values of restored forest types with and without thinning management, respectively. The graphs of Figures A5 and A6 in Appendix D show similarly the simulated dynamics of basal area, and tree density and tree density distributed according to DBH classes of restored forest types without harvesting and with thinning, respectively.

The results indicate that initial planting density and stand type have an effect on the dynamics of restored forest stands in the short-term horizon (up to 65 years). Applying the different thinning practices resulted in an increase in recreation value for all types of restored forest stands, while aboveground tree biomass decreased. When the simulations of AGB and recreation value with and without harvest are compared at the final stand state (65-year-old stand), the highest decrease in AGB of 61% in response to the thinning management is observed for larch stands. At the same time, the recreation value of larch is increased by 26% due to the thinning harvest. The similar changes in AGB and recreation value are observed for alder replantation and spontaneous succession. The lowest decrease in AGB of 11% and the lowest increase in recreation value of 6% due to the thinning is seen for pine replantation.

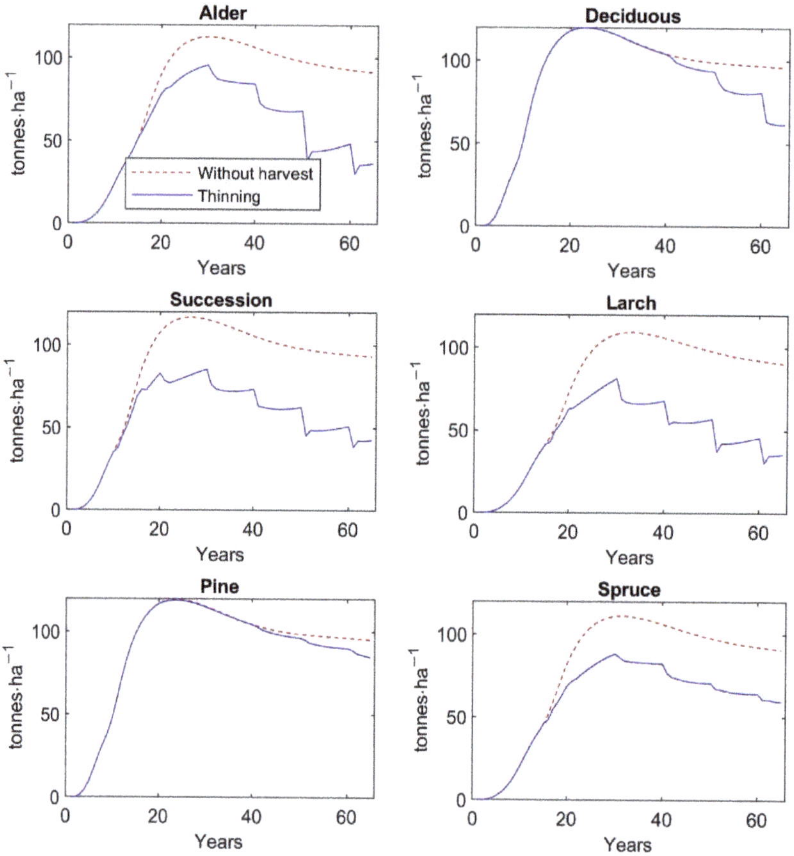

Figure 7. Predicted aboveground tree biomass over 65 years without harvest and under different thinning regimes and forest types.

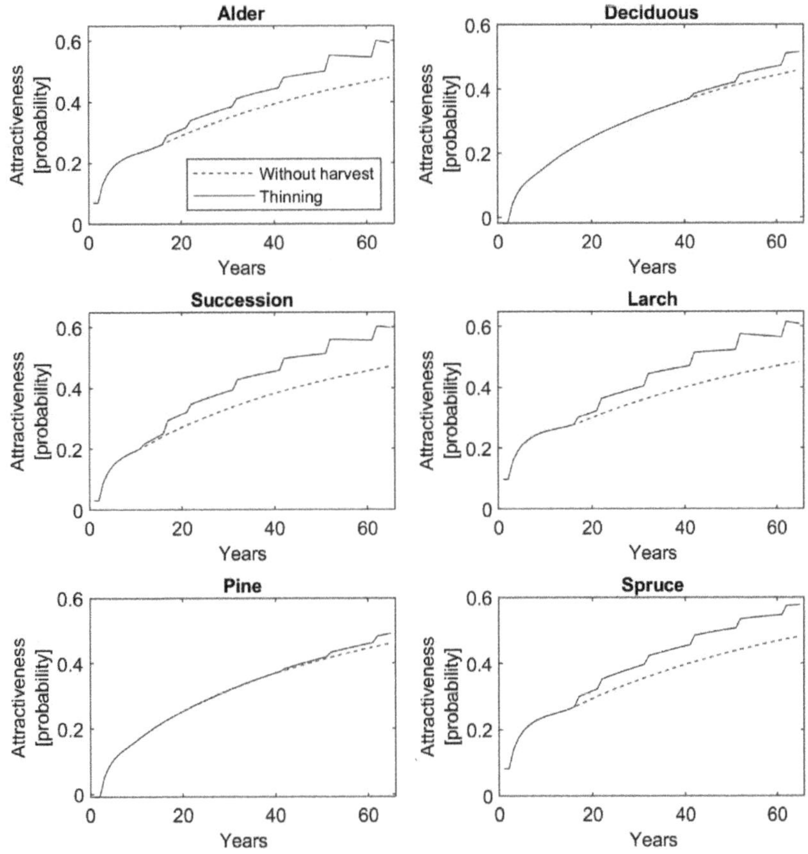

Figure 8. Predicted recreation value over 65 years without harvest and under different thinning regimes and forest types.

4. Discussion

Following the findings of previous studies [18,19], the results have shown that restored forest stands—both replanted and sites with succession—at the Velká Podkrušnohorská and Matyáš spoil heaps in the Sokolov brown-coal mining area, which are left to grow without appropriate silvicultural management, have a lower recreation value than forest stands with targeted after-care such as thinning and pruning. This is consistent with the findings of the study on perceived beauty of various restored forest stands in the North-West Bohemian brown-coal basin in the Czech Republic realized by Sklenicka and Molnarova [51], in which managed coniferous forest received higher preferences than wild, unmanaged, deciduous forest. On the other hand, the wilderness element of the forest landscape was not significant in the study by Svobodova et al. [52], in which visual preferences of different type of restored landscape were evaluated in the same brown-coal district as in the previous study.

In the matrix growth model (Equation (1)), the recreation value of the restored forests is parametrized apart from the stand age and shrub cover by tree density (Equation (20)) with its negative effect on recreation value, when the decreasing tree density increases the attractiveness of reclaimed forests for recreation. The tree density effect reflects well the stand conditions after the thinning selections in the long-term horizon when the direct remnants of logging are no longer visible. However, the model does not reflect the short-term post-thinning stand conditions, with slash piles, small diameter downed wood, and other visible harvest effects detracting from the scenic beauty and recreation value of managed stands [48].

In addition to the tree density, the results indicate that the recreation value of restored forests increases with the age of the stand. The positive effect of stand age on the attractiveness of reclaimed forests for recreation or on their perceived beauty was previously confirmed by Braun Kohlová et al. [18,19], and Sklenicka and Molnarova [51], respectively, for restored forest sites in the Sokolov and North-Bohemian mining area. This means that the restored forest stands of an early (<2 m) and middle (2–10 m) growth stage are characterized by low recreation value; this was confirmed for both successional stands [18,19] and replantations [52]. Interestingly, the significantly lower recreation value of early-stage restored stands was observed for spruce plantation compared to spontaneous succession [19]. This can be explained by a lower density of successional stands in their initial phase, while the spruce is planted in high seedling densities (10,000 seedlings per hectare).

The level of shrub and herbaceous understory and ground deformations are among other stand-level attributes that affect peoples' preferences towards recreation in restored forests. While the effect of shrubs—the presence of shrub layer detracts from visual attractivity of restored stands—was taken into account in the model on recreation value (Equation (20)), the cover of the herbaceous understory was not measured on the inventory plots, and therefore, the presence of herbaceous vegetation was not considered in the model. This may result in higher estimates of the recreation value of restored stands, especially those with a thick herbaceous vegetation. This is particularly problematic in alder stands with a high fixation of natural nitrogen that supports the spreading of grasses such as *Calamagrostis epigejos* or *Arrhenatherum elatius* [36]. It was alder stands with a large understory vegetation of *C. epigejos* that were perceived as less attractive than other restored forest stands [18].

Ground deformations caused by the conveyor belt dumping and layering of overburden material into 1–2 m high waves during the technical reclamation of the spoil heap are a suitable place for the development of spontaneously emerging ecosystems and the growth of succession tree species [10,17]. However, visible ground unevenness present at unreclaimed places left to spontaneous succession detract from the attractiveness of successional stands for recreation, as proven by Braun Kohlová et al. [18]. The highest growth phase of succession (*B. pendula*) was perceived as more attractive than the early (*B. Pendula, P. sylvestris, P. abies*) or middle (*S. caprea*) growth phases. In part, this may be influenced by less visible geomorphological irregularities in high growth stage stands due to geophysical processes. This might partially overestimate the predicted recreation value of successional forests in their early and middle growth stages.

From the long-term perspective, we can, therefore, recommend the formation of numerous ridges and depressions when a spoil heap is created, supporting the emergence and growth of successional forests. At suitable places on the spoil heap, strongly deformed terrain with spontaneous succession can be carefully combined with a phytotoxic surface. This will create future open areas with scenic landscape views and a habitat for rare plant species [10]. Deeper deformities give birth to ponds and pools suitable for amphibians. After opening these locations with spontaneous succession to visitors, we can inform the public about the importance of habitats for biodiversity and its origins. This information can take many forms, such as information boards or natural trails.

Furthermore, the findings of the current study indicate that thinning of different intensities increases the recreation value of restored forests, but at the same time the aboveground tree biomass significantly decreases, except on pine plantations. AGB and recreation value—measured as mean probability of a person's choice of a given forest type for a one-hour walk—of a 65-year-old pine stand is 84.9 t·ha^{-1} and 0.49, if thinning management is regularly realized. On the other hand, larch plantations at the same age reach 35.6 t·ha^{-1} AGB, but recreation value amounts to 0.61. In comparison to unmanaged stands, the AGB of 65-year-old larch stands due to thinning operations is decreased by

61%, while recreation value is increased by 26%. The similar values and dynamics in AGB and recreation value as for larch stands are observed for alder plantations. Deciduous and spruce stands at the age of 65 reach 61.8 and 59.2 t·ha^{-1} AGB, and their recreation value is 0.52 and 0.58, respectively. The regular thinning and, thus, subsequent decrease in AGB by 35% in both types of plantations will increase recreation value by 12% for deciduous stands and by as much as 20% for spruce stands. The highest increase in recreation value by 27% is observed for successional stands, which is accompanied by a 54% decrease in AGB. A 65-year-old successional stand reaches 42.5 t·ha^{-1} AGB with recreation value at 0.60.

From the timber production perspective of the individual forest types of restored stands, pine forests (usually *P. sylvestris* or *P. nigra*) created by plantation have an above-average wood yield, especially if the thinning interventions are regularly realized in the middle and high growth phase. This is consistent with the findings of the study by Vacek et al. [53], which proved high productivity and stand volume of *P. sylvestris* on the reclaimed forest site at the Antonín spoil heap in the Sokolov mining area, even though the pine restored stand was insufficiently thinned out. At the same time, *P. sylvestris* was found to be very adaptive towards climate change. However, pine forests have worse pedogenetic characteristics than alder or succession forests, as confirmed by Melichar et al. [54] at the Velká Podkrušnohorská and Matyáš spoil heaps.

The successional forests have an average wood yield, as was also confirmed by Vacek et al. [55] at the Antonín spoil heap. The mean stand volume of successional stands (*Populus tremula, S. caprea,* or *B. pendula*) were significantly lower than on afforested stands by *Q. robur, B. pendula,* or *Alnus glutinosa*. However, the reclaimed sites colonized with successional trees are significant for their pedogenetic process and tree and herbaceous diversity [54]. The higher species richness and total stand diversity of successional sites compared to replanted stands were also proven by Vacek et al. [55] for the Antonín heap.

Forest plantations predominantly with alder have a lower-than-average wood yield; wood volume per hectare is less than for successional forests. This was confirmed at the Velká Podkrušnohorská heap by Frouz et al. [56], when woody biomass was significantly greater on successional sites (*P. tremula* and *S. caprea*) than on replanted sites by *A. glutinosa* at their high growth stages. Nevertheless, woody biomass was greater for alder stands than for successional sites at their early and middle growth stages. The stand volume of alder plantation at the high growth stage was also lower compared to replanted stands with *Q. robur* and *B. pendula* and the successional stand with *P. tremula* at the Antonín heap [55]. However, successional forests with *S. caprea*, or *B. pendula* had a lower stand volume than the alder plantation. Nonetheless, alder planting at spoil heaps is mainly motivated by its suitability as a preparatory tree species. The pedogenetic role of alder stands is crucial, even though natural conditions on soil heaps are less favorable for alders and, thus, cause its premature aging [54]. The species diversity of alder stands is also relatively high compared to spruce, pine, long-term deciduous, or silver birch plantations [36,55].

One of the limitations of the matrix growth model is the fact that the model is calibrated on the inventory data of the restored forests from early to high growth stage up to the age of 60 years at maximum, because older forest reclamations are absent in the Sokolov mining area and also in other coal mining areas in the Czech Republic. Therefore, we have used the model to simulate the forest dynamics and evaluate the effects of the thinning interventions for a period of 65 years. Beyond this time horizon, we would not be able to evaluate the errors between the predicted and actual stand states. The possible improvement of the model would be to establish the inventory system of permanent plots capturing a wide range of restored forest types, soil, and geomorphological conditions in the study site and to extend the simulation period beyond the current time horizon of 65 years. In addition, the vital rates including the diameter growth, mortality, and both artificial and natural recruitment rates were parametrized with one-time inventory plot data. Second forest inventory with dendrometric measurements on the same inventory plots as in 2018 could greatly enhance the validity of the estimated parameters of the matrix growth model (Equation (1)).

Finally, in addition to the biomass production and recreation function, the restored forest ecosystems positively influence the properties of reclaimed soil, store carbon in AGB and in soil organic matter, and create the habitat conditions favoring herbaceous species diversity. Given the availability of data from pedological and phytocenological surveys [36,57] realized at the Sokolov spoil heaps in recent past, the opportunity exists to further extend the matrix model with the soil properties, carbon storage, tree and herbaceous diversity, and to study the stand dynamics from the perspective of the multifunctional use of restored forest ecosystems.

5. Conclusions

This study has shown that the state of restored forest stands established at the Velká Podkrušnohorská and Matyáš spoil heaps in the Sokolov brown-coal mining district is influenced, apart from forest type, by initial planting density and thinning intensity. On average for all growth phases, a higher aboveground biomass production was observed for long-lived deciduous plantations in comparison to half biomass production of spruce plantations on the forest inventory data from 2018. At the same time, larch plantations and forest stands originated by spontaneous succession showed a higher production compared to alder and pine reclamations. The simulations of the matrix growth model calibrated on the inventory data showed that the initial size of afforestation affects the size of biomass production until about 30 years of stand age, when AGB culminates. After that, the difference in tree biomass volume between stands with various seedling rate decreases. However, if we adopt thinning operations in the short-term horizon, the biomass production will decline, especially when thinning management is more intensive. On the other hand, the recreation value of restored stands is increasing with thinning interventions. Specifically, the greater the intensity of the thinning management that is carried out, the higher the increase in recreation value of restored forests is, as it was observed, for example, in larch plantations and successional stands. The possible future improvement of the matrix growth model would be to include, apart from biomass production and recreation function, the other important functions of reclaimed forest ecosystems as pedogenic function, carbon storage, or tree and herbaceous diversity restoration.

Author Contributions: Conceptualization, J.M.; methodology, J.M., E.C. and M.B.K.; validation, J.M., J.A. and V.M.; investigation, E.C., J.A., J.M. and M.B.K.; data curation, J.M., V.M., M.B.K. and J.A.; writing—original draft preparation, J.M., E.C., J.A., M.B.K., V.M. and P.P.; writing—review and editing, P.P.; visualization, J.M. and J.A.; supervision, E.C.; project administration, J.M.; funding acquisition, J.M. All authors have read and agreed to the published version of the manuscript.

Funding: This research received funding from the Ministry of Agriculture of the Czech Republic (grant number QK1710241, Optimization of management of forest restoration on sites affected by surface mining) and from the European Commission (grant number LIFE17 IPE/CZ/00005, Integrated LIFE project for the Natura 2000 network in the Czech Republic).

Acknowledgments: The authors thank their colleagues Jana Beranová, Vladimír Zatloukal, and Jan Tumajer for the coordination, work on the collection, and processing of forest inventory data.

Conflicts of Interest: The authors declare no conflict of interest. The funders had no role in the design of the study; in the collection, analyses, or interpretation of data; in the writing of the manuscript, or in the decision to publish the results.

Appendix A Forest Types and Tree Species Occurring in the Study Area

Table A1. Relative frequencies of tree species growing on the inventory plots of the Matyáš a Velká Podkrušnohorská spoil heaps.

Common Name	Scientific Name	No. of Trees	Relative Frequency [%]
Black alder	*Alnus glutinosa* (L.) Gaertn.	8180	17.81
Norway spruce	*Picea abies* (L.) H. Karst.	5681	12.37
European ash	*Fraxinus excelsior* L.	4178	9.10
Silver birch	*Betula pendula* Roth	3750	8.17
Scotch pine	*Pinus sylvestris* L.	3305	7.20
European larch	*Larix decidua* Mill.	3027	6.59
European aspen	*Populus tremula* L.	2739	5.96
Austrian pine	*Pinus nigra* J. F. Arnold	2463	5.36
English oak	*Quercus robur* L.	2158	4.70
Small-leaved linden	*Tilia cordata* Mill.	1891	4.12
Goat willow	*Salix caprea* L.	1633	3.56
Gray alder	*Alnus incana* (L.) Moench	1329	2.89
European mountain ash	*Sorbus aucuparia* L.	1150	2.50
Sycamore maple	*Acer pseudoplatanus* L.	877	1.91
Douglas fir	*Pseudotsuga menziesii* (Mirb.) Franco var. *menziesii*	804	1.75
European field elm	*Ulmus minor* Mill.	429	0.93
Serbian spruce	*Picea omorika* (Pančić) Purk.	353	0.77
Sessile oak	*Quercus petraea* (Matt.) Liebl.	352	0.77
Blue spruce	*Picea pungens* Engelm.	293	0.64
Field maple	*Acer campestre* L.	250	0.54
Wild cherry	*Prunus avium* L.	238	0.52
Box elder	*Acer negundo* L.	215	0.47
European wild pear	*Pyrus pyraster*	157	0.34
Other pines (1)	-	154	0.34
Other willows (2)	-	117	0.25
Eastern white pine	*Pinus strobus* L.	77	0.17
European beech	*Fagus sylvatica* L.	75	0.16
Norway maple	*Acer platanoides* L.	25	0.05
Black poplar	*Populus nigra* L.	15	0.03
Black locust	*Robinia pseudoacacia* L.	5	0.01
Total		45,920	100.00

Note: (1) Other pines are Lodgepole pine (*Pinus contorta* Douglas ex Loudon), Macedonian pine (*Pinus peuce* Griseb.), Ponderosa pine (*Pinus ponderosa*), Jeffrey pine (*Pinus jeffreyi*), Western white pine (*Pinus monticola*). (2) Other willows are Bay willow (*Salix pentandra* L.), Crack willow (*Salix euxina* I. V. Belyaeva), Basket willow (*Salix viminalis* L.).

Table A2. Assignment of individual tree species to the defined categories of forest types.

Forest Type Name	Tree Species
Alder	*A. glutinosa* *A. incana*
Deciduous	*Q. robur* *Q. petraea* *F. sylvatica* *A. platanoides* *A. pseudoplatanus* *A. campestre* *A. negundo* *S. aucuparia* *U. minor* *R. pseudoacacia* *P. avium* *P. pyraster* *T. cordata*
Succession	*B. pendula* *F. excelsior* *P. tremula* *P. nigra* *S. caprea* Other willows
Larch	*L. decidua*
Pine	*P. sylvestris* *P. nigra* *P. strobus* Other pines
Spruce	*P. abies* *P. pungens* *P. omorika* *P. menziesii*

Appendix B List of Variables Used in the Study and Their Descriptive Statistics

Table A3. Summary of stand variables and their definitions.

Type of Variable	Variable	Unit	Definition
Tree-level variables	D	cm	Diameter of a live tree at breast height
	G	$cm \cdot year^{-1}$	Annual diameter growth of a live tree
	M	{0,1}	Mortality: 1 = dead standing tree, 0 = live tree
	H	m	Tree height
	AGB_{tree}	$tonnes \cdot tree^{-1}$	Dry matter of aboveground tree biomass of individual tree
Plot-level variables	N	$trees \cdot ha^{-1}$	Number of trees per hectare
	R	$trees \cdot ha^{-1} \cdot year^{-1}$	Recruitment—the number of trees per hectare growing into the first DBH class in a year
	R_A	$trees \cdot ha^{-1} \cdot year^{-1}$	Artificial recruitment—the number of seedlings planted in a year
	R_N	$trees \cdot ha^{-1} \cdot year^{-1}$	Natural recruitment—the number of trees per hectare naturally regenerated in the first DBH class in a year
	B	$m^2 \cdot ha^{-1}$	Total stand basal area per hectare
	D_T	n	Number of tree species growing on a plot
	D_S	n	Number of shrub species growing on a plot
	C_S	%	Total cover of shrub understory on a plot
	AGB_{plot}	$tonnes \cdot ha^{-1}$	Total dry matter of aboveground tree biomass of all trees on a plot
	A	years	Age represented by the dominant tree species on a plot
	AR	probability	Attractiveness for recreation expressed as the mean probability of a person's choice of a given type of restored forest for a one-hour walk

Table A4. Descriptive statistics of stand characteristics for inventory plot data.

Variable	Statistics	All Stands	Alder	Deciduous	Succession	Larch	Pine	Spruce	Mixed Stands
Stem density [trees·ha^{-1}]	Mean	5847	6004	4954	6283	6192	6243	6324	5235
	S.D.	4224	5609	2631	5883	3314	2659	3435	3817
	Min	32	1210	1019	32	1783	1751	1432	382
	Max	29,316	26,261	11,077	29,316	12,955	12,350	13,528	16,329
	n	250	29	31	51	19	38	27	55
Recruitment [trees·ha^{-1}·year^{-1}]	Mean	367	320	406	648	301	175	320	289
	S.D.	663	570	605	1109	439	236	580	377
	Min	0	0	0	0	0	0	0	0
	Max	6565	2387	2653	6565	1592	1061	2785	1790
Basal area [m^2·ha^{-1}]	Mean	19.90	17.54	25.08	17.60	17.48	20.91	15.01	22.90
	S.D.	11.95	10.36	12.11	9.77	10.94	10.78	14.09	13.08
	Min	0.06	0.06	0.14	0.50	0.30	0.36	0.53	0.23
	Max	48.39	36.01	48.39	33.41	37.68	37.29	45.23	47.43

Table A5. Mean value of other stand characteristics for inventory plot data and attractiveness for recreation.

Variable	Unit	All Stands	Alder	Deciduous	Succession	Larch	Pine	Spruce	Mixed Stands
R_A	trees·ha^{-1}·year^{-1}	65	17	109	23	92	78	154	43
R_N	trees·ha^{-1}·year^{-1}	312	303	315	632	234	103	174	256
D_T	n	3.08	1.93	3.16	3.65	1.89	2.68	2.52	4.07
D_S	n	1.16	1.66	1.81	1.37	0.95	0.55	0.44	1.16
C_S	%	7.36	22.52	7.18	9.14	2.79	0.68	0.89	7.19
AGB_{plot}	t·ha^{-1}	79.40	67.52	110.18	80.74	93.27	60.90	49.86	89.58
A	years	24	24	34	25	21	17	19	28
AR	probability	0.4006	0.1681	-	0.4474	-	0.4870	0.4036	-

Table A6. Descriptive statistics of tree characteristics for individual tree data.

Variable	Statistics	All Stands	Alder	Deciduous	Succession	Larch	Pine	Spruce	Mixed Stands
Diameter [cm]	Mean	14.4	13.1	15.6	16.5	14.3	10.9	14.2	15.3
	S.D.	6.5	5.0	5.9	7.2	7.7	4.4	7.2	6.5
	Min	0.4	0.5	0.4	1.0	0.9	0.7	0.7	0.4
	Max	52.1	33.2	42.0	52.1	36.3	29.2	34.9	45.6
	n	4323	427	711	718	255	755	349	1108
Diameter growth [cm·year^{-1}]	Mean	0.53	0.45	0.42	0.65	0.53	0.57	0.52	0.52
	S.D.	0.25	0.23	0.17	0.34	0.21	0.19	0.24	0.25
	Min	0.02	0.12	0.02	0.11	0.10	0.04	0.09	0.03
	Max	2.61	1.66	1.37	2.61	1.40	1.38	1.93	1.93
	n	4323	427	711	718	255	755	349	1108
Dead tree {0,1}	Mean	0.037	0.036	0.039	0.038	0.008	0.006	0.050	0.058
	S.D.	0.188	0.187	0.194	0.190	0.091	0.075	0.218	0.234
	Min	0	0	0	0	0	0	0	0
	Max	1	1	1	1	1	1	1	1
	n	4186	415	690	720	242	710	321	1088
Tree height [m]	Mean	12.00	11.40	14.39	12.57	13.31	8.52	10.80	12.72
	S.D.	5.33	3.80	5.06	5.96	5.89	3.34	5.05	5.31
	Min	0.13	0.30	0.25	0.25	0.30	0.25	0.25	0.13
	Max	29.40	20.40	24.37	28.27	24.70	19.10	20.77	29.40
	n	4646	445	742	833	269	789	372	1196
Aboveground biomass [t·tree^{-1}]	Mean	0.080	0.063	0.097	0.115	0.085	0.028	0.074	0.089
	S.D.	0.104	0.052	0.082	0.167	0.104	0.030	0.077	0.101
	Min	3.8×10^{-6}	1.7×10^{-4}	1.6×10^{-5}	5.3×10^{-6}	5.4×10^{-5}	3.0×10^{-5}	6.5×10^{-5}	3.8×10^{-6}
	Max	1.871	0.349	0.539	1.871	0.488	0.217	0.404	1.575
	n	4403	429	713	763	256	762	350	1130

Appendix C Thinning Management Scenarios

Table A7. Target number of trees and predicted stock of aboveground biomass harvested according to different thinning managements over 65 years.

Forest Stand Type	Age of Stand	Target No. of Trees	Removed AGB
		[trees·ha^{-1}]	[t·ha^{-1}]
Alder	seedlings	4000	-
	5	4000	0.0
	10	3000	0.0
	15	2500	1.7
	20	2000	1.1
	30	1200	5.1
	40	800	10.3
	50	430	29.2
	60	310	18.1
	Total	-	65.6
Deciduous	seedlings	10,000	-
	5	9000	0.0
	10	8000	0.0
	15	7000	0.0
	20	6000	0.0
	30	2000	0.1
	40	1300	0.7
	50	1000	5.8
	60	700	16.7
	Total	-	23.3
Succession	seedlings	6000	-
	5	5000	0.0
	10	3000	1.1
	15	1500	3.0
	20	1200	6.3
	30	900	9.0
	40	700	9.6
	50	500	16.7
	60	400	12.4
	Total	-	58.0
Larch	seedlings	3000	-
	5	2500	0.0
	10	2250	0.0
	15	2000	1.4
	20	1200	3.2
	30	800	12.1
	40	600	13.5
	50	400	19.7
	60	300	15.5
	Total	-	65.4
Pine	seedlings	9000	-
	5	8000	0.0
	10	7000	0.0
	15	5500	0.0
	20	3500	0.1
	30	2500	0.0
	40	1700	0.3
	50	1500	0.5
	60	1200	0.7
	Total	-	1.7

Table A7. *Cont.*

Forest Stand Type	Age of Stand	Target No. of Trees	Removed AGB
		[trees·ha^{-1}]	[t·ha^{-1}]
Spruce	seedlings	3500	-
	5	3000	0.0
	10	2500	0.0
	15	2000	1.9
	20	1600	1.4
	30	1200	2.8
	40	900	5.1
	50	800	2.6
	60	700	2.7
	Total	-	16.3

Appendix D Verification of the Model for Investigated Stand Characteristics According to Forest Types

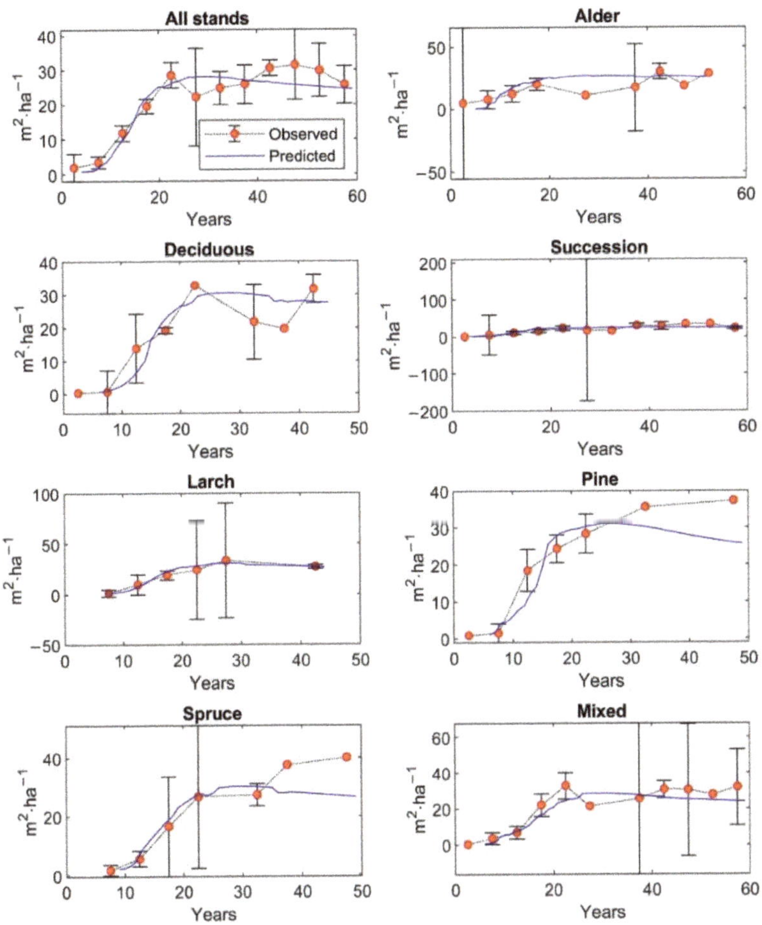

Figure A1. Verification of the model by the comparison of the observed 5-year average values with average predicted values of basal area, with a 95% confidence interval, according to individual forest types.

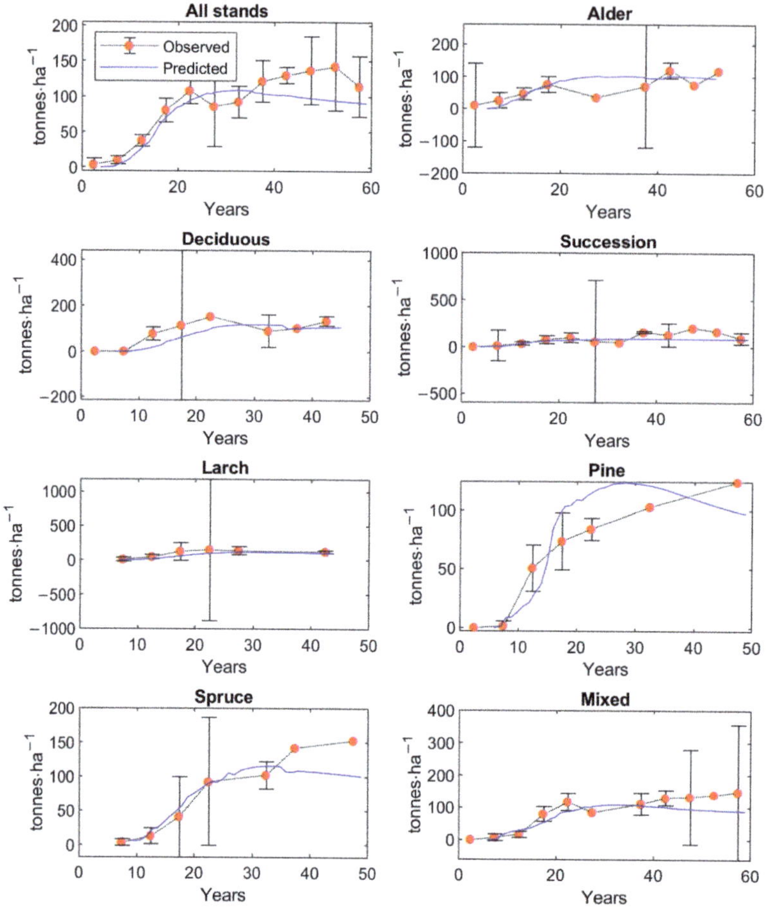

Figure A2. Verification of the model by the comparison of the observed 5-year average values with average predicted values of AGB, with a 95% confidence interval, according to individual forest types.

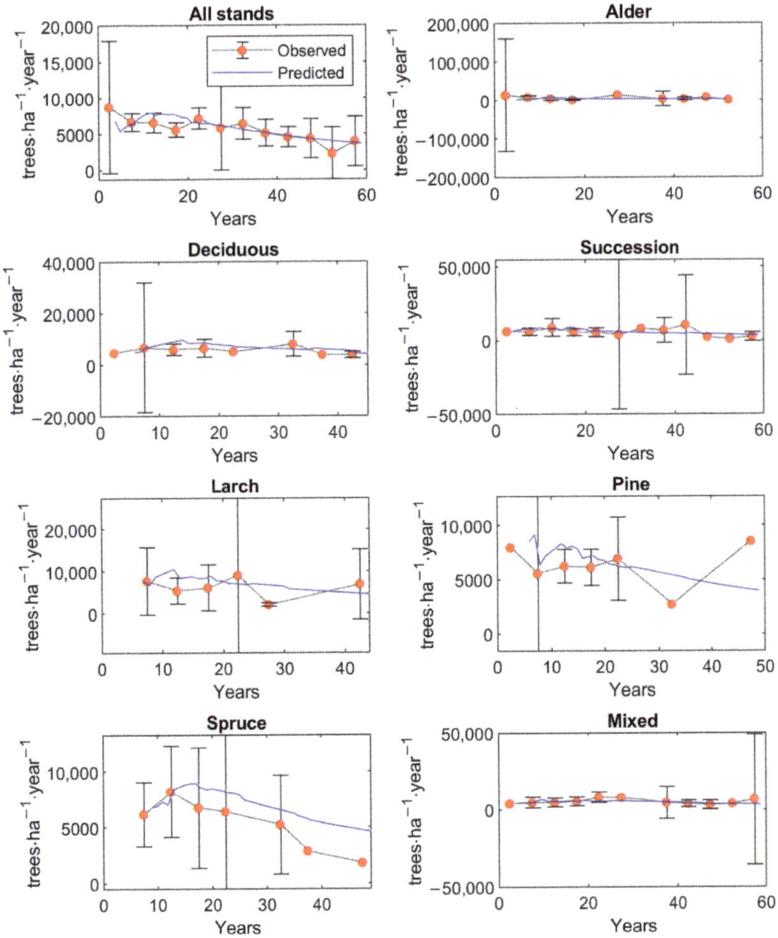

Figure A3. Verification of the model by the comparison of the observed 5-year average values with average predicted values of stem density, with a 95% confidence interval, according to individual forest types.

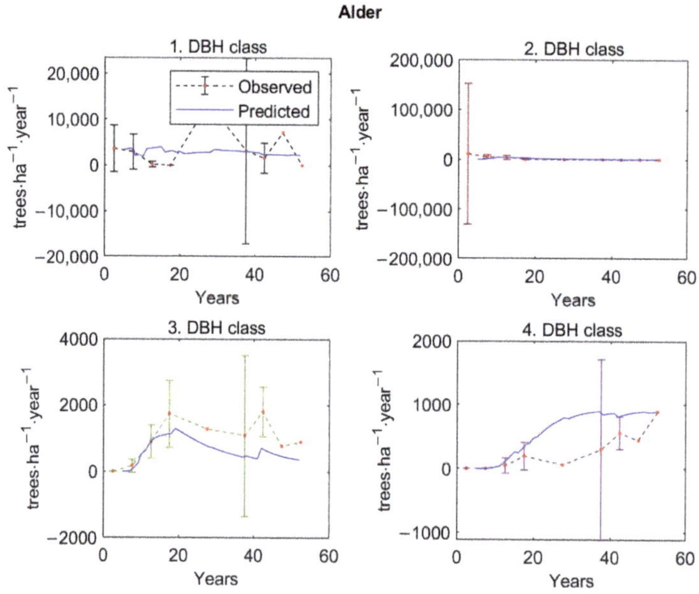

(a)

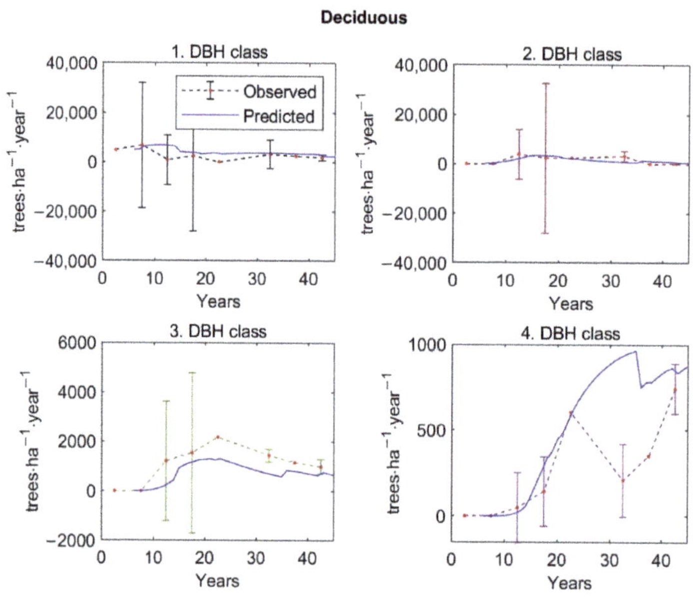

(b)

Figure A4. *Cont.*

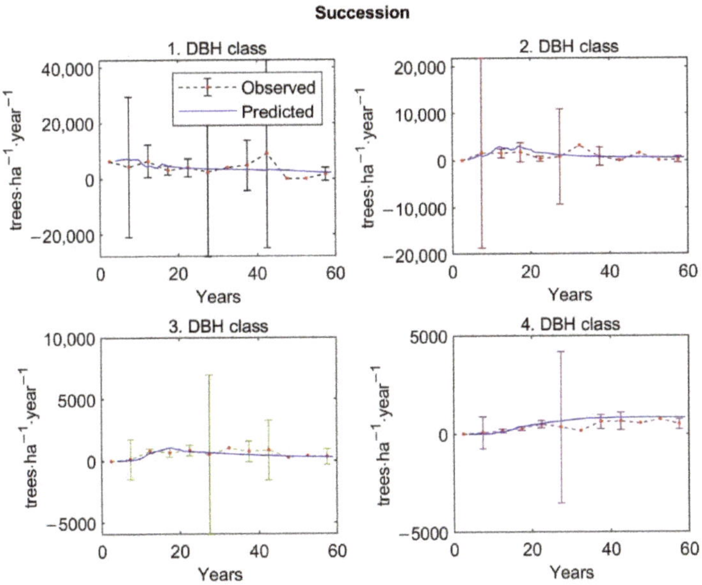

(c)

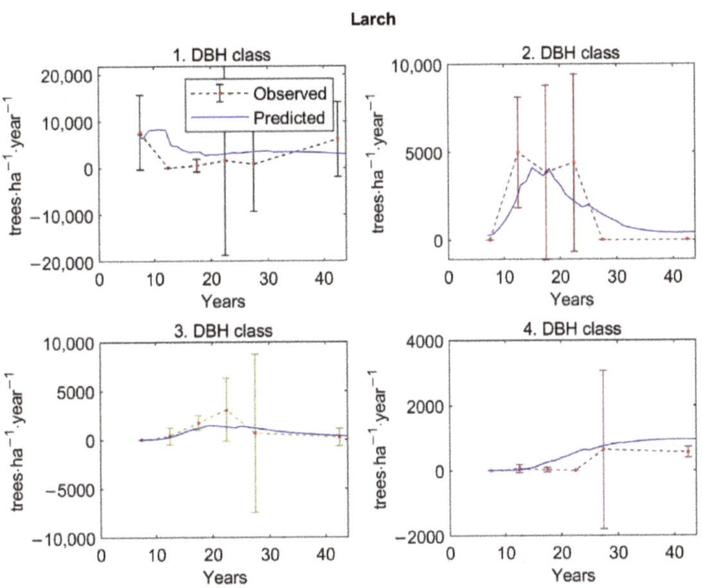

(d)

Figure A4. *Cont.*

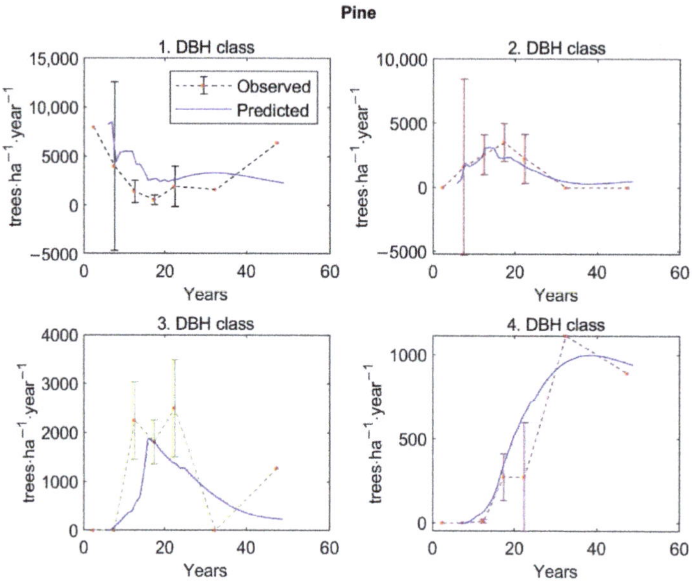

(e)

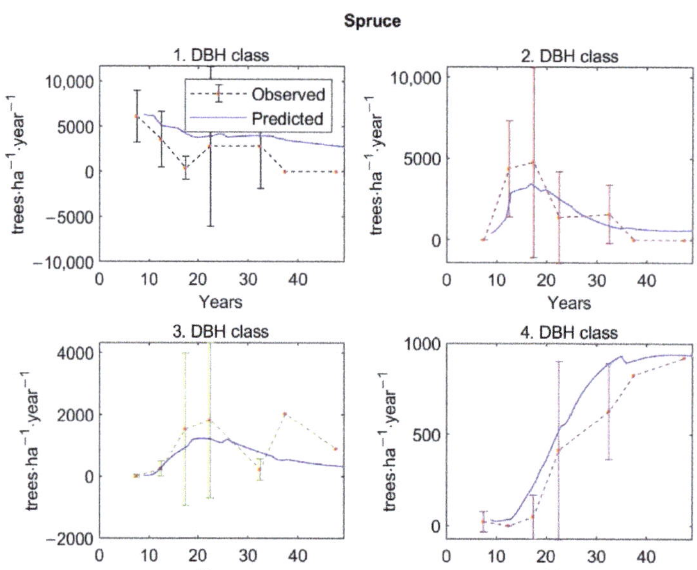

(f)

Figure A4. *Cont.*

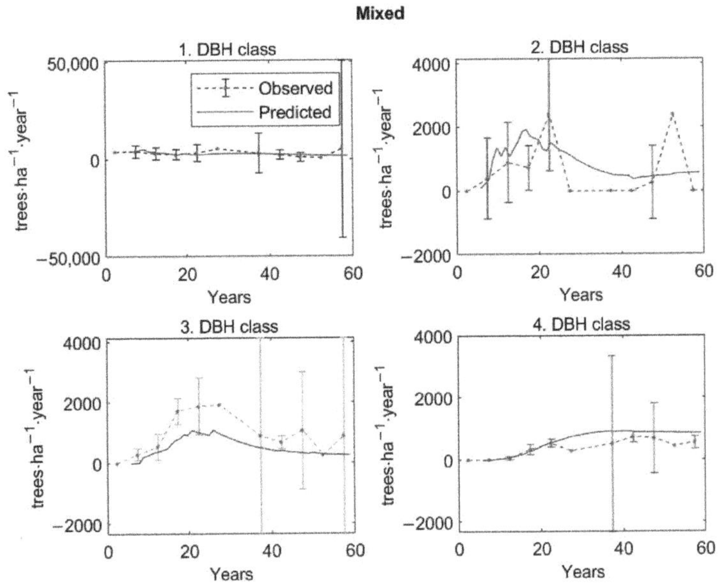

(g)

Figure A4. Verification of the model by the comparison of the observed 5-year average values with average predicted values of stem density distributed according to DBH classes, with a 95% confidence interval, according to individual forest types: (**a**) alder stand; (**b**) long-lived deciduous stand; (**c**) spontaneous succession; (**d**) larch stand; (**e**) pine stand; (**f**) spruce stand; (**g**) mixed coniferous–deciduous stand.

Appendix E Simulated Dynamics of Restored Forest Stands under Different Thinning Scenarios

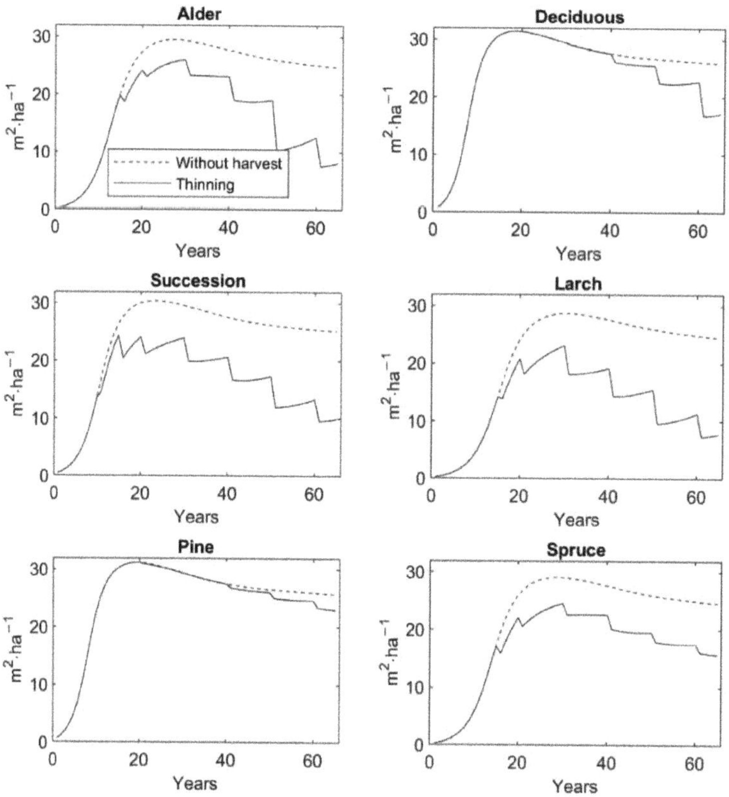

Figure A5. Predicted basal area over 65 years without harvest and under different thinning regimes and forest types.

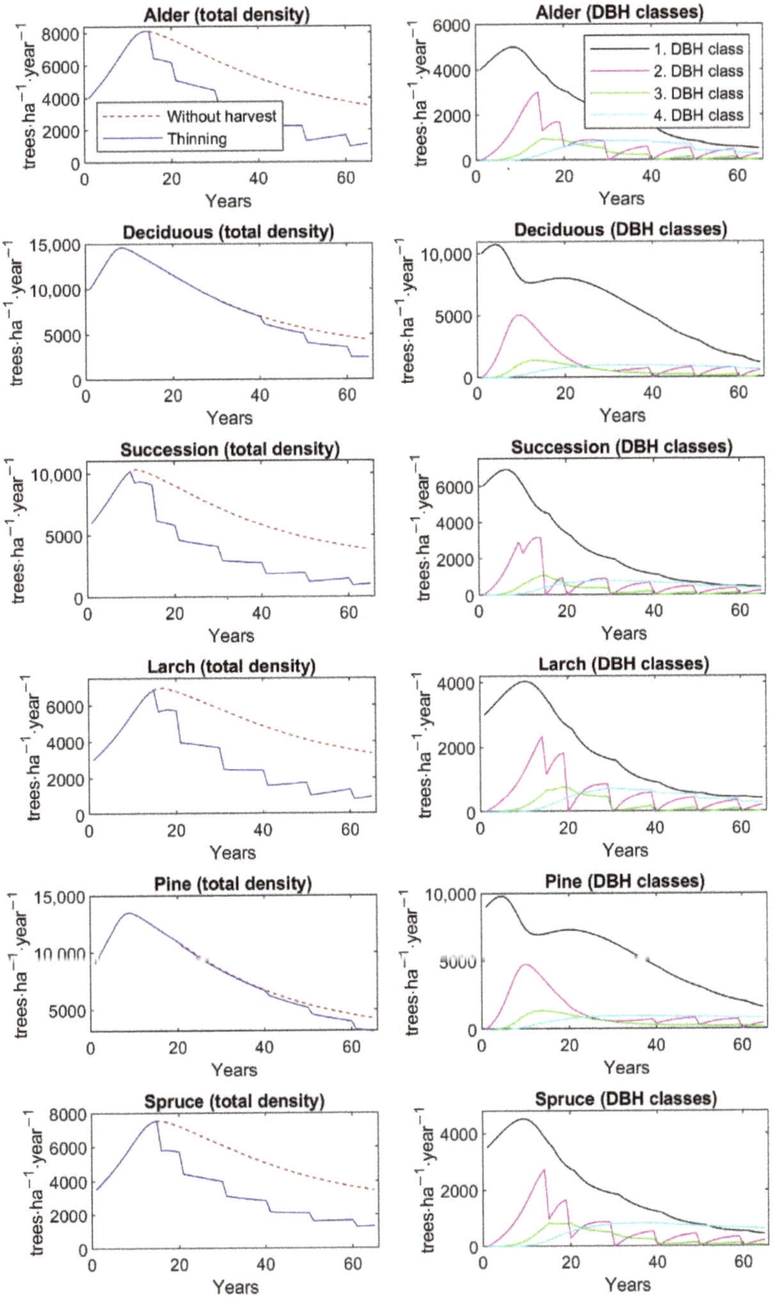

Figure A6. Predicted stem density and stem density distributed according to DBH classes over 65 years under different thinning regimes and forest types. The total stand density is also predicted without harvest.

References

1. Bradshaw, A.D. The reconstruction of ecosystems. *J. Appl. Ecol.* **1983**, *20*, 1–17. [CrossRef]

2. Macdonald, S.E.; Landhausser, S.M.; Skousen, J.; Franklin, J.; Frouz, J.; Hall, S.; Jacobs, D.; Quideau, S. Forest restoration following surface mining disturbance: Challenges and solutions. *New Forest.* **2015**, *46*, 703–732. [CrossRef]
3. Chuman, T. Restoration Practices Used on Post Mining Sites and Industrial Deposits in the Czech Republic with an Example of Natural Restoration of Granodiorite Quarries and Spoil Heaps. *J. Landsc. Ecol.* **2015**, *8*, 29–46. [CrossRef]
4. Zipper, C.E.; Burger, J.; Skousen, J.G.; Angel, P.N.; Barton, C.D.; Davis, V.; Franklin, J. Restoring forests and associated ecosystem services on appalachian coal surface mines. *Environ. Manag.* **2011**, *47*, 751–765. [CrossRef] [PubMed]
5. Alexander, S.; Aronson, J.; Whaley, O.; Lamb, D. The relationship between ecological restoration and the ecosystem services concept. *Ecol. Soc.* **2016**, *21*, 34. [CrossRef]
6. Tolvanen, A.; Aronson, J. Ecological restoration, ecosystem services, and land use: A European perspective. *Ecol. Soc.* **2016**, *21*, 47. [CrossRef]
7. Gurung, K.; Yang, J.; Fang, L. Assessing Ecosystem Services from the Forestry-Based Reclamation of Surface Mined Areas in the North Fork of the Kentucky River Watershed. *Forests* **2018**, *9*, 652. [CrossRef]
8. Larondelle, N.; Haase, D. Valuing post-mining landscapes using an ecosystem services approach—An example from Germany. *Ecol. Indic.* **2012**, *18*, 567–574. [CrossRef]
9. Maiti, S.K.; Ahirwal, J. Ecological restoration of coal mine degraded lands: Topsoil management, pedogenesis, carbon sequestration, and mine pit limnology. In *Phytomanagement of Polluted Sites;* Elsevier: Amsterdam, The Netherlands, 2019; pp. 83–111.
10. Mudrák, O.; Frouz, J.; Velichová, V. Understory vegetation in reclaimed and unreclaimed post-mining forest stands. *Ecol. Eng.* **2010**, *36*, 783–790. [CrossRef]
11. Pietrzykowski, M. Soil and plant communities development and ecological effectiveness of reclamation on a sand mine cast. *J. Forest Sci.* **2008**, *54*, 554–565. [CrossRef]
12. Walmsley, A.; Vachová, P.; Hlava, J. Tree species identity governs the soil macrofauna community composition and soil development at reclaimed post-mining sites on calcium-rich clays. *Eur. J. For. Res.* **2019**, *138*, 753–761. [CrossRef]
13. Gypser, S.; Veste, M.; Fischer, T.; Lange, P. Infiltration and water retention of biological soil crusts on reclaimed soils of former open-cast lignite mining sites in Brandenburg, north-east Germany. *J. Hydrol. Hydromech.* **2016**, *64*, 1–11. [CrossRef]
14. Pietrzykowski, M. Tree species selection and reaction to mine soil reconstructed at reforested post-mine sites: Central and Eastern European experiences. *Ecol. Eng. X* **2019**, *3*, 100012. [CrossRef]
15. Sena, K.; Agouridis, C.; Miller, J.; Barton, C. Spoil type influences soil genesis and forest development on an Appalachian surface coal mine ten years after placement. *Forests* **2018**, *9*, 780. [CrossRef]
16. Frouz, J.; Pižl, V.; Cienciala, E.; Kalčík, J. Carbon storage in post-mining forest soil, the role of tree biomass and soil bioturbation. *Biogeochemistry* **2009**, *94*, 111–121. [CrossRef]
17. Frouz, J.; Pöpperl, J.; Přikryl, I.; Štrudl, J. *Tvorba Nové Krajiny na Sokolovsku;* Sokolovská Uhelná a.s.: Sokolov, Czech Republic, 2007.
18. Braun Kohlová, M.; Nepožitková, P.; Melichar, J. How Do Observable Characteristics of Post-Mining Forests Affect Their Attractiveness for Recreation? *Land* **2021**, *10*, 910. [CrossRef]
19. Braun Kohlová, M.; Máca, V.; Melichar, J.; Pavelčík, P. How High Is the Recreation Value of Successional Forests Growing Spontaneously on Coal Mine Spoil Heaps? *Forests* **2021**, *12*, 160. [CrossRef]
20. Frouz, J.; Mudrák, O.; Reitschmiedová, E.; Walmsley, A.; Vachová, P.; Šimáčková, H.; Albrechtová, J.; Moradi, J.; Kučera, J. Rough wave-like heaped overburden promotes establishment of woody vegetation while leveling promotes grasses during unassisted post mining site development. *J. Environ. Manag.* **2018**, *205*, 50–58. [CrossRef]
21. Cienciala, E.; Tumajer, J.; Beranová, J.; Zatloukal, V.; Albert, J. *Database of Allometric Relationships;* IFER–Institute of Forest Ecosystem Research Ltd.: Jílové u Prahy, Czech Republic, 2019; Available online: https://drive.google.com/file/d/102pRwvEK_nDmfVKSbCoXHuBtaztFmGaY/view (accessed on 3 May 2022). (In Czech)
22. Usher, M.B. A Matrix Approach to the Management of Renewable Resources, with Special Reference to Selection Forests. *J. Appl. Ecol.* **1966**, *3*, 355–367. [CrossRef]
23. Usher, M.B. A Matrix Model for Forest Management. *Biometrics* **1969**, *25*, 309–315. [CrossRef]
24. Liang, J. Dynamics and management of Alaska boreal forest: An all-aged multi-species matrix growth model. *Forest Ecol. Manag.* **2010**, *260*, 491–501. [CrossRef]
25. Choi, J.; An, H. A Forest Growth Model for the Natural Broadleaved Forests in Northeastern Korea. *Forests* **2016**, *7*, 288. [CrossRef]
26. Pyy, J.; Ahtikoski, A.; Laitinen, E.; Siipilehto, J. Introducing a Non-Stationary Matrix Model for Stand-Level Optimization, an Even-Aged Pine (*Pinus sylvestris* L.) Stand in Finland. *Forests* **2017**, *8*, 163. [CrossRef]
27. Pyy, J.; Laitinen, E.; Ahtikoski, A. Demonstrating the Effect of Height Variation on Stand-Level Optimization with Diameter-Structured Matrix Model. *Forests* **2020**, *11*, 226. [CrossRef]
28. Liang, J.; Picard, N. Matrix model of forest dynamics: An overview and outlook. *Forest Sci.* **2013**, *59*, 359–378. [CrossRef]
29. Vanclay, J.K. *Modeling Forest Growth and Yield: Applications to Mixed Tropical Forests;* CAB International: Wallingford, UK, 1994.
30. Buongiorno, J.; Michie, B.R. A matrix model of uneven-aged forest management. *Forest Sci.* **1980**, *26*, 609–625.
31. Liang, J.; Buongiorno, J.; Monserud, R.A. Growth and yield of all-aged Douglas-fir–western hemlock stands: A matrix model with stand diversity effects. *Can. J. For. Res.* **2005**, *35*, 2369–2382. [CrossRef]
32. Namaalwa, J.; Eid, T.; Sankhayan, P. Amulti-species density-dependent matrix growth model for the dry woodlands of Uganda. *For. Ecol. Manag.* **2005**, *213*, 312–327. [CrossRef]

33. Cropper, W.P., Jr.; Loudermilk, E.L. The interaction of seedling density dependence and fire in a matrix population model of longleaf pine (*Pinus palustris*). *Ecol. Model.* **2006**, *198*, 487–494. [CrossRef]
34. Price, O.; Bowman, D.M.J.S. Fire-stick forestry: A matrix model in support of skilful fire management of *Callitris intratropica* R.T. Baker by north Australian Aborigines. *J. Biogeogr.* **1994**, *21*, 573–580. [CrossRef]
35. Melichar, J.; Braun Kohlová, M.; Kaprová, K.; Pavelčík, P.; Frouz, J.; Karel, J.; Krkošková, N. *Specialized Map of Production and Non-Production Functions of the Reclaimed Sites for the Mining Site of the Sokolov Coal Basin*; Charles University, ATEM–Studio of Ecological Models: Prague, Czech Republic, 2017; Available online: https://drive.google.com/file/d/1FQxHRwMh6I-puwe8ChAc7zqD9obQrMQA/view (accessed on 22 August 2022). (In Czech)
36. Vachova, P.; Vach, M.; Skalicky, M.; Walmsley, A.; Berka, M.; Kraus, K.; Hnilickova, H.; Vinduskova, O.; Mudrak, O. Reclaimed Mine Sites: Forests and Plant Diversity. *Diversity* **2022**, *14*, 13. [CrossRef]
37. Lin, C.R.; Buongiorno, J.; Vasievich, M. A multi-species, density-dependent matrix growth model to predict tree diversity and income in northern hardwood stands. *Ecol. Model.* **1996**, *91*, 193–211. [CrossRef]
38. Ralston, R.; Buongiorno, J.; Schulte, B.; Fried, J. Non-linear matrix modeling of forest growth with permanent plot data: The case of uneven-aged Douglas-fir stands. *Int. Trans. Oper. Res.* **2003**, *10*, 461–482. [CrossRef]
39. Ma, W.; Zhou, X.; Liang, J.; Zhou, M. Coastal Alaska forests under climate change: What to expect? *For. Ecol. Manag.* **2019**, *448*, 432–444. [CrossRef]
40. Vonderach, C.; Kändler, G.; Dormann, C.F. Consistent set of additive biomass functions for eight tree species in Germany fit by nonlinear seemingly unrelated regression. *Ann. For. Sci.* **2018**, *75*, 49. [CrossRef]
41. Bronisz, K.; Strub, M.; Cieszewski, C.; Bijak, S.; Bronisz, A.; Tomusiak, R.; Wojtan, R.; Zasada, M. Empirical equations for estimating aboveground biomass of *Betula pendula* growing on former farmland in central Poland. *Silva Fenn.* **2016**, *50*, 1559. [CrossRef]
42. Cienciala, E.; Černý, M.; Tatarinov, F.; Apltauer, J.; Exnerová, Z. Biomass functions applicable to Scots pine. *Trees* **2006**, *20*, 483–495. [CrossRef]
43. Cienciala, E.; Apltauer, J.; Exnerová, Z.; Tatarinov, F. Biomass functions applicable to oak trees grown in Central-European forestry. *J. For. Sci.* **2008**, *54*, 109–120. [CrossRef]
44. Čihák, T.; Hlásny, T.; Stolariková, R.; Vejpustková, M.; Marušák, R. Functions for the aboveground woody biomass in Small-leaved lime (*Tilia cordata* Mill.). *Lesn. Časopis–For. J.* **2014**, *60*, 150–158.
45. Johansson, T. Biomass equations for determining fractions of common and grey alders growing on abandoned farmland and some practical implications. *Biomass Bioenergy* **2000**, *18*, 147–159. [CrossRef]
46. Wirth, C.; Schumacher, J.; Schulze, E.-D. Generic biomass functions for Norway spruce in Central Europe—A meta-analysis approach toward prediction and uncertainty estimation. *Tree Physiol.* **2004**, *24*, 121–139. [CrossRef] [PubMed]
47. Wutzler, T.; Wirth, C.; Schumacher, J. Generic biomass functions for Common beech (*Fagus sylvatica* L.) in Central Europe–predictions and components of uncertainty. *Can. J. For. Res.* **2008**, *38*, 1661–1675. [CrossRef]
48. Brown, T.C.; Daniel, T.C. *Modeling Forest Scenic Beauty: Concepts and Application to Ponderosa Pine*; Research Paper RM-256. USDA Forest Service: Fort Collins, CO, USA, 1984.
49. Hull, R.B., IV; Buhyoff, G.J.; Cordell, H.K. Psychophysical Models: An Example with Scenic Beauty Perceptions of Roadside Pine Forests. *Landsc. J.* **1987**, *6*, 113–122. [CrossRef]
50. Cienciala, E.; Zatloukal, V.; Albert, J.; Beranová, J.; Melichar, J. *Framework Guidelines for the Management of Reclaimed Forests*; IFER–Institute of Forest Ecosystem Research Ltd.: Jílové u Prahy, Czech Republic, 2021; Available online: https://drive.google.com/file/d/10vcA8OTwOwG6yXnwB2RKOlKrRT9lIuw3/view (accessed on 10 June 2022). (In Czech)
51. Sklenicka, P.; Molnarova, K. Visual Perception of Habitats Adopted for Post-Mining Landscape Rehabilitation. *Environ. Manag.* **2010**, *46*, 424–435. [CrossRef] [PubMed]
52. Svobodova, K.; Sklenicka, P.; Molnarova, K.; Salek, M. Visual Preferences for Physical Attributes of Mining and Post-Mining Landscapes with Respect to the Sociodemographic Characteristics of Respondents. *Ecol. Eng.* **2012**, *43*, 34–44. [CrossRef]
53. Vacek, Z.; Linda, R.; Cukor, J.; Vacek, S.; Šimůnek, V.; Gallo, J.; Vančura, K. Scots pine (*Pinus sylvestris* L.), the suitable pioneer species for afforestation of reclamation sites? *For. Ecol. Manag.* **2021**, *485*, 118951. [CrossRef]
54. Melichar, J.; Walmsley, A.; Cienciala, E.; Beranová, J.; Berka, M.; Braun Kohlová, M.; Černý, M.; Holá, Š.; Nepožitková, P.; Pavelčík, P.; et al. *Optimization of Management of Forest Restoration on Sites Affected by Surface Mining*; Research Report; Project No. QK1710241; Ministry of Agriculture of the Czech Republic; Charles University, Czech University of Life Sciences Prague, IFER–Institute of Forest Ecosystem Research Ltd.: Prague, Czech Republic, 2020.
55. Vacek, Z.; Cukor, J.; Vacek, S.; Podrázský, V.; Linda, R.; Kovařík, J. Forest biodiversity and production potential of post-mining landscape: Opting for afforestation or leaving it to spontaneous development? *Cent. Eur. For. J.* **2018**, *64*, 116–126. [CrossRef]
56. Frouz, J.; Dvoršík, P.; Vávrová, A.; Doušová, O.; Kadochová, Š.; Matějíček, L. Development of canopy cover and woody vegetation biomass on reclaimed and unreclaimed post-mining sites. *Ecol. Eng.* **2015**, *84*, 233–239. [CrossRef]
57. Jačka, L.; Walmsley, A.; Kovář, M.; Frouz, J. Effects of different tree species on infiltration and preferential flow in soils developing at a clayey spoil heap. *Geoderma* **2021**, *403*, 115372. [CrossRef]

forests

Article

Management of Recreational Forests in the Romanian Carpathians

Gheorghe-Marian Tudoran [1,*], Avram Cicșa [1,2], Maria Cicșa (Boroeanu) [1] and Alexandru-Claudiu Dobre [1,2]

[1] Department of Forest Engineering, Faculty of Silviculture and Forest Engineering, "Transilvania" University, Forest Management Planning and Terrestrial Measurements, 500123 Brașov, Romania
[2] "Marin Drăcea" Romanian National Institute for Research and Development in Forestry, Department of Forest Monitoring, 128 Eroilor Blvd., 077190 Voluntari, Romania
* Correspondence: tudoran.george@unitbv.ro; Tel.: +40-26-841-8600

Abstract: Research Highlights: Forests, due to their aesthetic properties, have huge recreational potential, but their management must take into account the requirements of all parties interested in these services. Background and Objectives: We sought to determine the main indicators that characterize the structural diversity of a recreational mountain forest, with relevance for the management of these forests, given that they fulfill multiple functions. Materials and Methods: The structure of 446 stands was investigated and the Shannon (H) diversity index was applied at the level of species (Hsp), age (Hage), tree diameter (Hd_g), and tree height (Hh_g). Results: Beech occupied 49% of the forest area and fir and spruce 16% each. Generations of trees older than 100 years occupied 71% of the forest area and those older than 150 years occupied 10%. At an age of 120 years, the beech reached a diameter (d, at breast height) of 45 cm and the fir 52 cm. At the forest level, Hsp had a value of 1.63, Hd_g of 3.17, and Hh_g of 2.76. At the stand level, Hsp reached 1.54, while Hd_g and Hh_g reached 1.72. Mixed beech–coniferous stands had the greatest diversity. High values of 1.00 for Hsp were determined for 18% of the stands, for Hd_g 38%, and for Hh_g 35%. Conclusions: Stand structures are in a continuous state of change, so diversity indices can be used to monitor structural and species diversities and to evaluate the recreational potential of stands and forests. A compatibility between the aesthetic qualities of Romanian forests, which is a priority, and the other protection and production services they offer can be achieved by leading the forest stands toward a selection system.

Keywords: Shannon index; stand structure; mixed stands; forest recreation; evenness

1. Introduction

In recent decades, there has been an increasing demand for ecotourism and outdoor recreation. Society is much more concerned about health, well-being, and recreational activities [1,2] as a result of the accentuated urbanization that has created a certain distance between man and nature. The forest has become an important element due to its social and ecological value. The recreation function is especially relevant in areas close to large cities [3–5]. It has been noted that the surface area of cities is directly correlated to the need for recreation in their neighboring forests [5]. Therefore, the recreational landscape depends mainly on forest management, which aims to maintain and improve the quality of recreational resources [6]. However, recreational activities can often have negative effects. For example, trails can lead to the fragmentation of habitats, and the presence of humans can bring important changes in the behavior, and even the physiology, of wild animals [2]; all these, including land use change, are also related to natural hazards [7]. However, frequent visits to green areas would inspire the perception of the natural environment, being an important driver in improving the ecological paradigm [8].

The importance of the recreation function brings many concerns in the management of forests because more and more emphasis is placed on the integration of society's preferences in making management decisions [3]. Numerous studies have presented the role of the recreation function in forest management, and also the public's preferences on the structure of stands [3–5,9–12]. The management can improve the quality of resources and recreational opportunities for users to achieve the desired recreation experiences [6]. It needs to consider the compromise between tourism–recreation function and other ecosystem services [13]. The very wide range of land and forest ecosystems, however, provide multiple ecosystem services (ES) that do not have proper valuation indicators and, therefore, have limited use for planning and management [14]. Thus, planning decisions regarding ES could be more effective for improving overall resource quality [15].

The public's preferences are often based on cultural and regional differences; for example, in Central Europe, high-density mixed stands are preferred, whereas in Scandinavia, less-dense mixed forests are preferred [16]. The recreation location index, by comparing regional rates to a standardized reference region, also provides information for future planning decisions [17]. In general, mixed stands are preferred by the public, compared to monocultures [9,11,18–20]. However, monocultures are preferred when they are old. Visual variation is one of the factors followed by the public, and it depends on the size of the tree, the crown density of the stand, the predominant species, or the presence of undergrowth [11]. Increasing recognition of the important role that forests play in providing goods and services leads, through management practices, to the diversification of stand structure and development of mixed structures [21–23]. Stands with irregular structures, an uneven spatial distribution of the trees, and with a broad range of diameter categories are often preferred by the public [3]. Mature stands are preferred over young ones because young stands are associated with a higher density, which interferes with visibility and gives the impression of being unsafe to the public. In general, the public prefers semi-open forests, with large trees and little undergrowth [9,10,24]. Stands that have a diversified structure, in terms of tree height, are also preferred over even-aged ones with similar heights.

The species is also a factor that influences public preferences. Several studies have shown that mixed deciduous stands are more appreciated, especially by young people [3], as opposed to conifers [10]. Mixed multi-layered broadleaves and coniferous forests also have a higher dust retention effect compared to single-layered forests [25]. Evergreen mixtures in the winter season increase forest effectiveness and reduce the concentration of particulate matter [26]. Trails that contain a greater diversity of stands are more appreciated than those with a low visual diversity, which are considered monotonous [11], with the aesthetic quality of the forest being the one visually assessed by the public [27]. The public also prefers spaces as little affected by humans as possible, with a high degree of naturalness and a natural structure, over those affected by humans [12]. Thus, management plans must also be developed for the sustainability of endemic species affected by climate variability [28].

On the other hand, dead wood is one of the elements that is less appreciated by the public, even if it is very important in biodiversity conservation [9,24]. Thus, a compromise is necessary to balance the aesthetic value of forests and the conservation of forest biodiversity. Additionally, dense undergrowth or other elements associated with a high conservation value are less appreciated [24]. In Europe, the development phase of stands provides the greatest contribution to their ultimate recreational value [10].

Another factor that is taken into account by the public is the accessibility of the forest, with those that are close to pathways being preferred, but with a reduced number of tourists so as to enhance the feeling of a natural forest [3]. The necessary money or funding of an individual to cover recreation expenses, as well as the distance the forest is located [29] or the day of the week and weather conditions [30], also affects recreation participation and causes variations in recreation constraints [1].

The Carpathian Mountains have a high potential for recreation, thanks to the old-growth forests, the flora, and also the high density of cultural heritage [31]. An essential characteristic of the management, in Romania, of the forests intensely requested by tourists is the preservation of their natural, unaltered framework, produced through an uneven, or relatively uneven, structure selection system [32].

In Romania, the need for recreation has increased in recent years, and tourism has become an important element in the country's economy. This is why society's preferences concerning the structure and management of forests should represent an increasing concern, especially in the case of forests with high recreational value. Tourism can have a significant economic value, bearing in mind that a considerable increase in the number of tourists can compensate for losses caused by a reduction in, or even the absence of, silvicultural interventions. Thus, a good way forward would be the integration of nature-based tourism with traditional forestry, especially in areas where there is a high percentage of forests with recreational functions [33].

If the public expresses its preferences for forests through their structural parameters, then these parameters can be expressed as indices for evaluating and monitoring the recreational potential of forests. The considered objective of this study was to determine the values of the main indicators that characterize the structure and species diversity of mountain recreational forests, which are relevant for forest management in terms of fulfilling multiple functions.

2. Materials and Methods

2.1. Study Area

The studied forest, with an area of 2923.6 ha, was located in the Postăvarul Massif of the Romanian Carpathians, in the proximity of Brașov city (Figure 1). The forest covered altitudes of between 700 and 1400 m. The most frequently encountered relief was slope, with inclinations of between 20 and 35° and varied aspects. The average multi-annual temperature of the studied area was between 5 and 7.8 °C, with precipitation between 750 and 950 mm. The forest has been integrated into a management unit and was characterized by a management plan that has been developed since 1890.

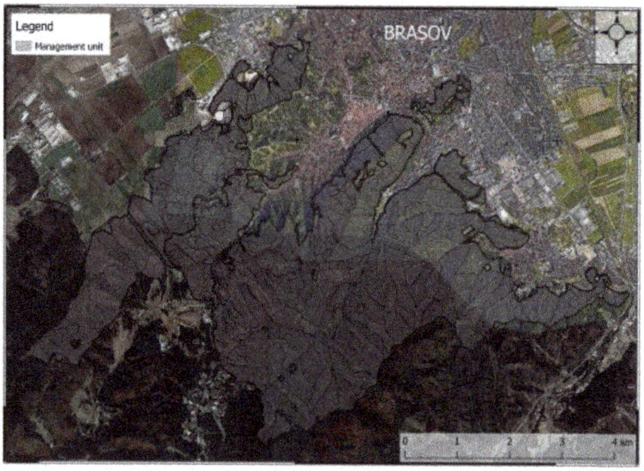

Figure 1. Location of the studied forest management unit.

2.2. Field Measurements

Data on the structure of the stands were recorded through field measurements. The measurements were carried out in the spring of 2022 and represent a sequel to the measurements carried out in this forest in 2020 [34]. Management plan maps were used to identify the forest boundaries. The terrain was traversed via perpendicular transects, following the lines of valleys and ridges, in order to describe as many stands as possible. The transects were not materialized on the ground, but the coordinates of the points in the stands where the measurements were taken were recorded. To obtain descriptions, a Bitterlich survey was placed in each of 466 stands. In the stands, the surveys were placed in representative areas, after each stand had been fully traversed so as to obtain an overview of its structure [35]. In all, 7129 trees were counted. We worked with multiplication factors 1 and 2, depending on the development stage of the stand, the diameter of the trees, and the crown density of the stand. In multi-aged stands, the measurements were carried out over several generations. In one generation, we included trees of the same age and with differences of, at most, 25 years. The age of each generation was determined by increment cores extracted from the average trees. For even-aged stands, the ages were taken from the management plan. For example, for an even-aged stand made up of three species, each of which being 80 years old, three species were considered, representing three variables that we named generations. If 120-year-old trees were identified for a species, they formed a fourth generation. Thus, for that stand, three species and four generations were considered. In the study, these representative trees were labeled "sp." or "age-sp.". To quantify the effect of age on the structure, only the age of the variables was analyzed. For the example given above, only two age classes were considered—80 and 120 years. In the study, the age class was denoted by "age". For each species or generation, the mean tree (considering the basal area) was identified and its diameter determined (i.e., the quadratic mean diameter, denoted by d_g), based on the tree diameters measured in the Bitterlich circle. The mean height (i.e., the height corresponding to the diameter d_g), denoted by h_g, was also determined. For the measurements, we used a Criterion RD 1000 device and a Vertex laser.

To characterize the structure of the stands and represent the 3D profiles, we fully inventoried stands of sessile oak–beech, beech, and mixed beech–fir formations over 4 ha. For all trees with diameters greater than 1 cm, the diameters and heights were measured and their coordinates determined [34].

We took information about the existing forest types in the area and the target compositions from inventories carried out in 2020 [34]. The forest types and target compositions we identified in the area were:

- Mixed spruce–fir forest with herbaceous mull plants: 40%–50% *Pa*, 30%–40% *Aa*, 10%–20% *Fs*, *App*, and *Ap*;
- Fir forest with herbaceous mull plants: 70%–80% *Aa*, 10%–20% *Fs*, 10% *App* and *Ap*;
- Mixed beech–fir stands: 50%–60% *Aa*, 30%–40% *Fs*, 10% *App*, *Ap*, and *Fe*;
- Mixed beech–spruce–fir forest: 30%–40% *Pa*, 30%–40% *Aa*, 30% *Fs* and *App*; and
- Beech forest with herbaceous mull plants: 70%–80% *Fs*, 10%–20% *Aa*, 10% *App*, *Ap*, *Fe* and *Pra*
- (*Aa*—*Abies alba* Mill.; *Ap*—*Acer platanoides* L.; *App*—*Acer pseudoplatanus* L.; *Fe*—*Fraxinus excelsior* L.; *Fs*—*Fagus sylvatica* L.; *Pa*—*Picea abies* Karst.; *Pra*—*Prunus avium* L.)

2.3. Data Analysis

The basal areas determined in the Bitterlich surveys were used to establish the proportion of species ("sp."), generations ("age-sp."), age classes ("age"), mean-diameter classes, d_g (i.e., classes of mean diameters with the same d_g values), and mean-height classes, h_g (i.e., average-height classes with the same h_g values). The proportions were used to determine the Shannon index (H) through the known relationship $H = -\sum (pi*\ln [pi])$, where p is the proportion of species i, in terms of the variables [36] at the stand level and the whole management unit level (i.e., the forest). This calculation resulted in H values for the: diversity of species (Hsp.); diversity of generations and species (Hage-sp.); diversity of age classes (Hage); diversity of mean diameters at the generation and species level (Hd_g); and diversity of mean heights at the generation and species level (Hh_g). The evenness ($E = H/\ln S$, where S is the number of species, in terms of the frequency of the variables) was also determined for each of the five variables (i.e., age-sp., sp., age, d_g and h_g). For fir, 26 generations (i.e., age) were determined, whereas for alder, there were only two. At the forest level, 257 generations (i.e., age-sp.), 25 species (i.e., sp.), 35 age classes (i.e., age), 41 diameter classes (i.e., d_g), and 34 height classes (i.e., h_g) were determined. These values were considered frequencies for determining the H index. Using the values of the H and E indices, maps were generated that can be used in management decision making.

3. Results

3.1. Stand Structure and Forest Composition

For the investigated forest, the recreation function was the main priority. The capacity of the forest to provide favorable conditions for recreational activities (art, relaxation, tourism, fun, and rest) can be examined through the prism of its structural characteristics. These characteristics are analyzed below.

The age of the stand was a specific characteristic of the silvicultural system age class forest. In the investigated forest, young and middle-aged stands occupied 18% of the forest area, while another 11% was occupied by stands between 85 and 100 years old. The aesthetic value of a forest increases with stand age, and stands over 100 years old had a fairly large share of the forest area at 71% (Figure 2a). Advanced age was also associated with stand density, which frequently presented values of between 0.6 and 0.8. At these densities, the trees have rich crowns, but these are transparent enough to allow the sun to create light and shadow inside the stand, which enhances the aesthetic effect. However, aesthetic properties can also be found in young stands where there is systematic intervention. Such interventions are essential for ensuring forest stability and for leading stands toward a target composition.

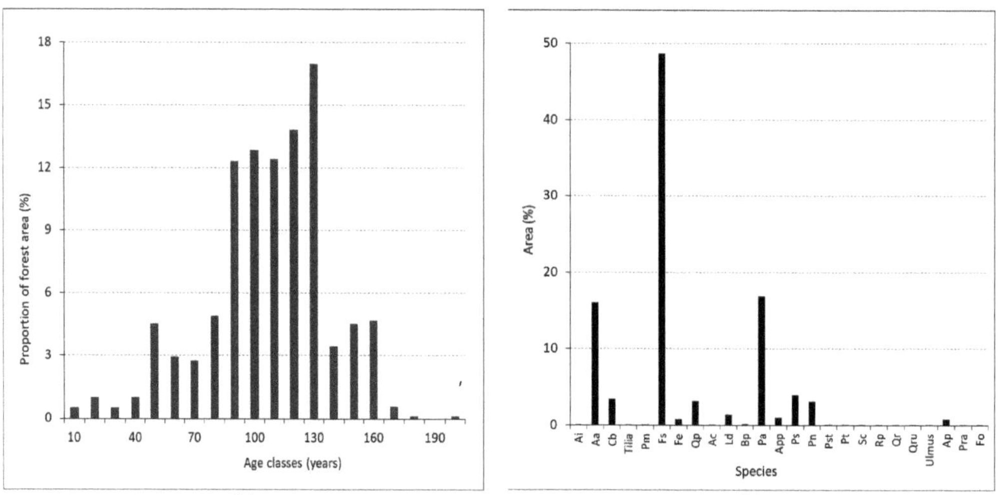

(a) (b)

Figure 2. (**a**) Forest age structure and (**b**) proportion of species (% of area) identified in the forest composition. The age class division used was based on the ages of the trees of the different species. Each age class included generations of trees in the stands. *Ai—Alnus incana L.; Aa—Abies alba* Mill.; *Cb—Carpinus betulus* L.; *Tilia—Tilia* spp. L.; *Pm—Pseudotsuga menziesii; Fs—Fagus sylvatica L.; Fe—Fraxinus excelsior* L.; *Qp—Quercus petraea* L.; *Ld—Larix decidua* Mill.; *Bp—Betula pendula* Roth; *Pa—Picea abies* Karst.; *App—Acer pseudoplatanus* L.; *Ps—Pinus sylvestris* L.; *Pn—Pinus nigra* Arn.; *Pst—Pinus strobus* L.; *Pt—Populus tremula* L.; *Rp—Robinia pseudoacacia* L.; *Qr—Quercus robur* L.; *Qru—Quercus rubra* L.; *Ulmus—Ulmus* spp. L.; *Ap—Acer platanoides* L.; *Pra—Prunus avium* L.; *Fo—Fraxinus ornus* L. A fairly large percentage (71%) of the stands were over 100 years old. Species of this age, and occurring in the largest proportions, were beech, fir, and spruce. Sessile oak, pine, larch, sycamore, and other species were additions to these, in smaller proportions.

The composition of the studied forest was 59% deciduous and 41% conifer. Due to the site conditions, the management unit allowed for the development of a large number of species. The natural spread of species was determined by the site conditions. However, the largest share was held by beech (48%), followed by fir (16%) and spruce (16%). Pine had an 8% share and larch 1% (Figure 2b).

The pines, larch, red oak, black locust, Douglas fir, and a percentage of the spruce were planted (accounting for 12% of the area of the management unit), with the other species coming from natural regeneration. The pines and larches were introduced 120 years ago, on rocky, calcareous land with superficial soil. They were rarely mixed with other species, and then only in stands close to the city of Braşov, being introduced for their aesthetic effect. Beech forms cheerful stands, thanks to the lighter foliage, the greater brightness, and the varied shapes of the trees. Together with fir, beech forms pleasant stands for visitors, due to the rich variation in shapes and colors. Variations in the proportions of the species in the mixture, and the presence of disseminated resinous species in the pure stands of beech, maple, sycamore, and cherry, reduced the monotony of the stands.

3.1.1. Target Composition

The basic criterion for the selection and promotion of species is the forest site itself, with favorable site conditions dictating the species being introduced. In this sense, information is required on the health and vitality of the trees and seedlings. In the last decade, there have been reports on the drying of pine trees. This drying has also manifested in fir, in forest sites located on the upper parts of sunny slopes, as well as in spruce, at altitudes below 1000 m. This has led to silvicultural interventions that have reduced the proportions of these trees in the last decade. Forest types identified in the management unit content and the economic criterion recommend maintaining a high proportion of fir, on favorable sites, especially since it is already quite well represented in the seedling composition. Thus, for the future, at the management unit level, there will be a tendency toward a composition of 49% *Fs*; 33% *Aa*; 12% *Pa*, *Ld*, *Pn*, and *Ps*; 3% *App*, *Ap*, and *Fe*; 2% *Qp*; and 1% *Ac*, *Cb*, *Pra*, *Fo*, and other deciduous (*Ulmus* spp., *Tilia* spp.) (Tudoran et al., 2021). With respect to the forest site criterion and the natural regeneration of the stands, the conditions for maintaining the stability of the stands have been created.

3.1.2. Tree Dimensions for Exploitability

In this forest, the production of wood mass must also be taken into account. According to the management plan, the cutting age has been set at 120 years in the silvicultural system age class forest for the beech, mixed fir–beech, and mixed beech–coniferous stands. The sessile oak stands also reach 140 years old. In forests with protective functions, such as these, these ages can be increased by up to 20 years. At the age of 120 years, on high-productivity sites, the beech reached a diameter of 45 cm in the stands and the fir a mean diameter of 52 cm (Figure 3).

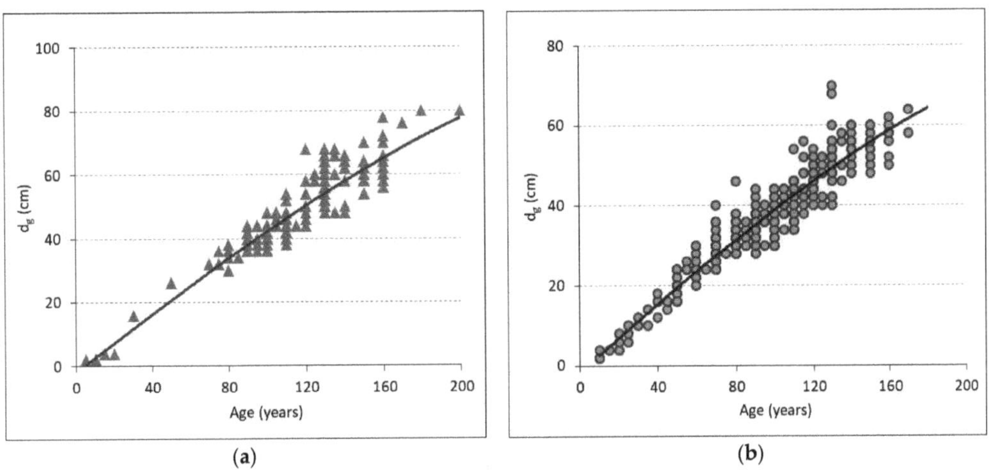

Figure 3. Diameter development, in terms of productivity, in (**a**) silver fir and (**b**) European beech stands.

Mean diameter when considering the basal area (i.e., dg) was estimated between 88 and 90% of the age variation, with the estimated parameters being significant ($p < 0.05$). The graphs show the mean diameters of fir and beech species, based on their age. The mean diameters provide information about the diameters of the trees that made up the highest percentage in the stands. These diameters are specific to stands located in high-productivity sites that have undergone interventions (the average density at the management unit level being 0.70). The diameter values were estimated with a mean absolute error of between 2.93 (beech) and 3.55 (fir) and a root mean square error of between 3.92 (beech) and 4.67 (fir), using a polynomial model of the form $y = ax^3 + bx^2 + cx + d$, with the parameters in Table 1.

Table 1. Parameter estimates.

Species	Coefficients	Parameter	Std. Error
Fir	a	-1.66×10^{-6}	4.13×10^{-6}
	b	7.73×10^{-5}	0.001192
	c	0.4478	0.100443
	d	−1.78391	2.600612
Beech	a	-2.65×10^{-6}	2.82×10^{-6}
	b	0.000199	0.000776
	c	0.409387	0.065054
	d	−1.33947	1.672826

In the model, x represents the species age. The model can be used for stands with age between 10 and 200 years (200 years for fir and 180 years for beech). The beech has a larger sample size compared to fir, which explains the reduced standard error values.

The young stands on which the model is based have a low proportion in the studied forest (Figure 2a). Thus, the statistical indicators lead to weaker arguments supporting the model. However, the model was considered as it fits the observed data and reflects the growth and development processes of the investigated stands. The d_g values predicted by the model have a normal variation related to age and site conditions.

3.1.3. Vertical Stand and Forest Structures

Vertical stand and forest structures are essential components pursued by forest management. The stands in the studied management unit have evolved toward various structures as a result of the silvicultural interventions carried out, but also because of their development. The most common was the multi-aged structure, which presents a wide range of diameter variation (Figure 4), in which two or three generations of trees can be distinguished.

An analysis at the entire-forest level showed that the stands had crowns in the overstorey on 72% of the forest area (Table 2). This is a favorable structure that fully satisfies the aesthetic requirements. Beech was the most well-represented species at all storey levels, whereas fir, as shown by the structure of the age classes (Figure 2b), was present only in the upper storey. Currently, fir is only at the seedling and sapling stages in the mature and exploitable stands. The overstorey is of most interest from both the aesthetic and economic points of view.

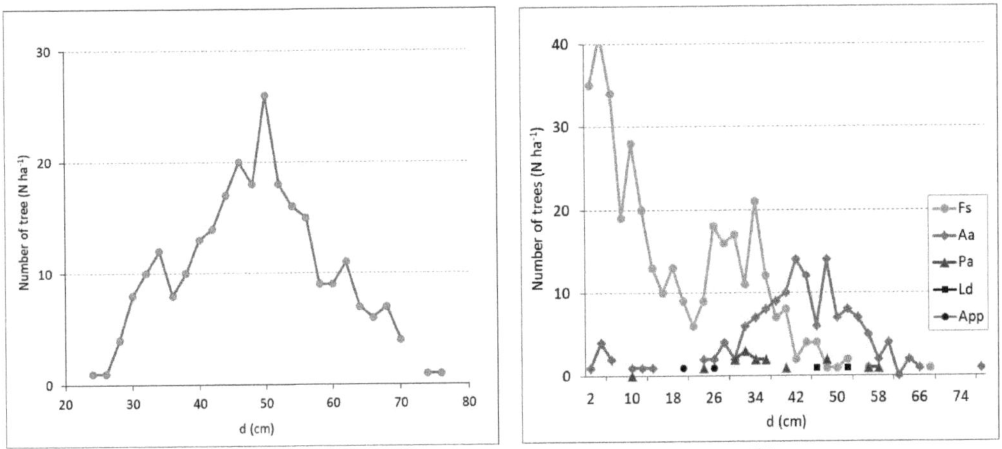

(a) (b)

Figure 4. Number of trees vs. diameter distribution in (**a**) a beech stand (aged 140 years) and (**b**) a mixed beech–fir stand (aged 120 years). *Fs—Fagus sylvatica* L.; *Aa—Abies alba* Mill.; *Pa—Picea abies* Karst.; *Ld—Larix decidua* Mill.; App–Acer pseudoplatanus L. The beech stand had a greater variation in diameter. This was a closed stand (with a stand crown density of 1.0). Although it was a pure stand, the sizes of the trees and straightness and quality of the trunks made an impression. This stand is accessible on the tourist route, and is greatly appreciated and much-frequented by tourists. A young generation of trees was already developing in the mixed beech–fir stand, the young firs most often being in the seedling stage and growing vigorously, already providing 40% coverage. This is an accessible stand, positioned on the tourist route, and offering an illustration of the transition between generations resulting from management measures designed through management plans and applied by the forest district.

Table 2. Characteristics of the forest, differentiated by canopy layer (overstorey, midstorey, and understorey).

Canopy Layer	Proportion of Species (% of Basal Area)	Basal Area (m^2 ha^{-1})	Growing Stock (m^3 ha^{-1})	Age (Years)	d_g (cm)	h_g (m)	h_p (m)
Understorey	45% *Fs*, 20% *Cb*, 13% *Pn*, 5% *Pa*, 4% *App*, 3% *Ld*, 6% *Aa*, 8% other species	0.74 (*Pn*)–15.62 (*Pa*)	7 (*Ld*)–72.7 (*Pa*)	5 (*Aa*)–95 (*Cb*)	2 (*Aa*)–24 (*Pn*)	2 (*Aa*)–12 (*Pa*)	0 (*Aa*)–5 (*Fs*)
Total	3	10.7	47.4	33	11.5	8.3	1.5
Midstorey	49% *Fs*, 11% *Cb*, 10% *Qp*, 7% *Pa*, 6% *Ps*, 3% *Fe*, 14% other species	14.94 (*Cb*)–39.48 (*Pa*)	144.8 (*Cb*)–374.1 (*Aa*)	30 (*Aa*)–160 (*Qp*)	14.0 (*Cb*)–60 (*Qp*)	13 (*Aa*)–24 (*Pa*)	0 (*Aa*)–16 (*Pa*)
Total	25	23.97	244.8	89	29.7	20.4	10
Overstorey	48% *Fs*, 21% *Aa*, 20% *Pa*, 4% *Ps*, 2% *Ld*, 2% *Pn*, 1% *Qp*, 2% other species	21.57 (*Qp*)–39.14 (*Pa*)	301.6 (*Qp*)–482.8 (*Aa*)	70 (*Aa*)–200 (*Pa*)	32.0 (*Fs*)–84.0 (*Pa*)	25 (*Aa*)–36 (*Pa*)	8 (*Ps*)–24 (*Pa*)
Total	72	30.5	348.5	118	49.0	29.0	17.2

Fs—Fagus sylvatica L.; *Cb—Carpinus betulus* L.; *Pn—Pinus nigra* Arn.; *Pa—Picea abies* Karst.; *App—Acer pseudoplatanus* L.; *Aa—Abies alba* Mill.; *Qp—Quercus petraea* L.; *Ps—Pinus sylvestris* L.; *Fe—Fraxinus excelsior* L.; Ld—Larix decidua Mill. Total—average per storey. The basal area of the stands was calculated through Bitterlich. The whole forest was storied. The trees with the highest mean heights were included in the upper storey ($h_g > 24$ m). They also had the largest share of the forest area, and 72% of the basal area, with respect to the volume of the forest. This storey was the best represented, and included trees with mean diameters greater than 37 cm.

3.2. Structural and Species Diversity

Species diversity is preferred in stands with recreational functions, and this was determined at both the management unit and stand levels. For the management unit, the

H-index values ranged between 1.63 (H2) and 4.19 (H1) (Table 3). The percentage of species was very unbalanced in the management unit, as indicated by the low E-index value (in the case of H2). In the stands, the highest H-index value (H1 = 1.85) came from an old stand composed of several generations (60–150 years). Species diversity was characterized by a maximum value of 1.54, which was determined from a 20-year-old stand, but one that was made up of five species. The diversity indices of the mean diameters and heights had values of 1.72 and were also found in species mixtures. At the level of the species and generations making up the stands, basal area had balanced proportions, as suggested by the high E-index values (95%).

Table 3. Shannon index per management unit.

Shannon Index	H1 (age-sp.)	H2 (sp.)	H3 (age)	H4 (d_g)	H5 (h_g)
	General (management-unit) level				
Value	4.19	1.63	2.76	3.17	2.76
Evenness (%)	76	51	78	85	76
Frequency of the variable	257	25	35	41	34
	Stand level				
Maximum value	1.85	1.54	1.50	1.72	1.72
Evenness (%)	95	95	93	96	96
Frequency	7	5	3	6	6

To calculate the Shannon index, the species and their tree generations (H1); only the species (H2); only the tree generations, regardless of species (H3); the mean diameter (H4); and mean height (H5) were introduced as variables. The proportion of the basal area of the variables was used. The management unit had 2923.6 ha and an average basal area of 28.2 m^2 ha^{-1}. The 25 species formed stands of different ages. In different stands, the species had different ages, and they were called generations of trees. If there is a 40-year-old fir in one stand and a 50-year-old fir in another, these generations would be actually age classes (i.e., denoted by "age"). In total, there were 257 of all the generations of all the species. This explains the H-index value of 4.19 at the entire-forest level.

At the crown level, the species diversity was greater in the understorey (Table 4). If the overstorey area was reduced, the biodiversity was also reduced, with the percentage of species in this storey being much different.

Table 4. Species diversity in the canopy layer at the management unit level.

Index	Proportion of Maximum Mean Height (h_g) of the Stands (%)					
	0–33	34–66	67–100	0–50	51–80	81–100
H2	1.85	2.0	2.15	2.13	2.1	1.64
E	77	80	86	81	78	66

The maximum h_g of the stands was 36.0 m. Increasing the understorey area changes the share of species in the lower storey through the participation of this storey in a larger proportion of middle-aged stands (h_g increases from 12 to 16 m). However, this analysis is much more edifying when carried out at the stand or group of stands levels, and is of interest for the location of recreation areas preferred by tourists.

The value of the H indices, in the case of all variables (species, generation, age, d_g and h_g), depends on their frequency (Figure 5a) and proportion. It is expected that in different species, the values of those variables differ. However, for example, in the case of H5, even if there are several species in the stand, if they have the same mean height, the value of H5 is equal to 0. On the contrary, at different heights, the value of H5 increases (Figure 5b).

For each variable (tree generations and species, only generation or species, d_g, and h_g), the proportion of the basal area was used. In general, the frequency of the variables favorably influenced the H-index values; however, the closest relationship was determined when the species frequency, in terms of generations, was used (R2 = 69%–86%).

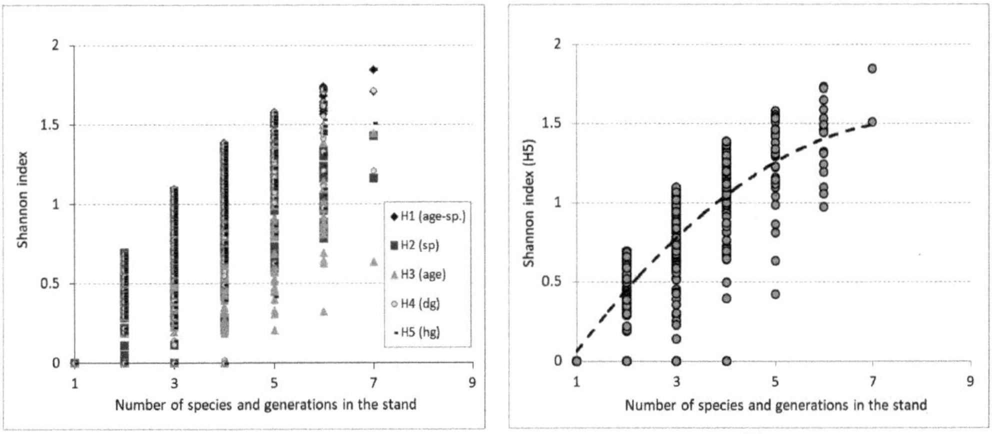

(a) (b)

Figure 5. Shannon index calculated at the management unit level for (**a**) the stand characteristics (426 stands) and (**b**) the mean height of the stands.

The characteristics of the stands (age, d_g, and h_g) had little influence on the H index (Figure 6a). However, an increase in diameter diversity (H4) may also cause an increase in height (H5) (Figure 6b). The same relationship was noted between H2 and H5.

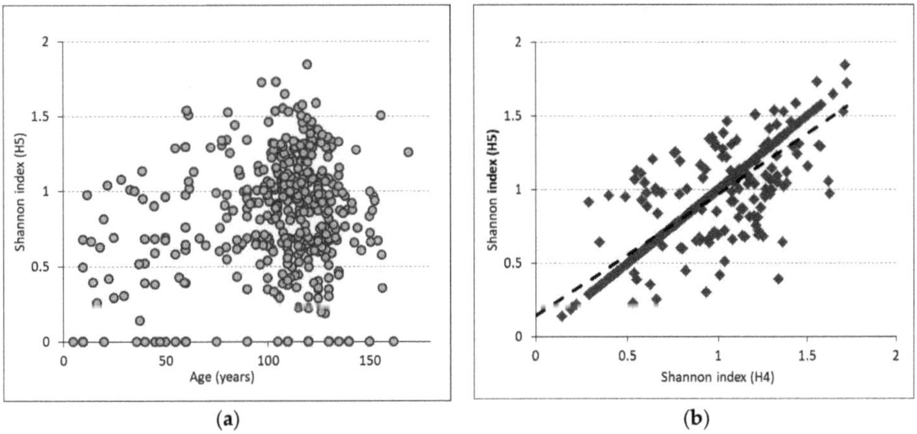

(a) (b)

Figure 6. (**a**) Variation in the Shannon index (H5) in relation to stand age, and (**b**) Shannon diversity calculated at the level of the mean diameter and height of the stands.

The age, mean diameter, and mean height of the stands did not have a significant influence on the H-index values by diameter (H4) or height (H5). However, 61% (273) of the stands had the same H-index values, indicating the diversity in diameter (H4) and height (H5). Thus, 70% of the H4 variation was reproduced by the relationship expressed in (b).

At the stand level, the indices had different values. A value of 0 for all indices was obtained in the case of pure and even-aged stands, with their values increasing as the structure became more complicated, horizontally and vertically, and the proportion of variables became more balanced. For example, in the sessile oak–beech stands (Figure 7a), although made up of five species, the percentage of sessile oak was 76%, which caused a reduction in H2, with beech stands (Figure 7b) showing H2 = 0. In the mixed beech–fir stands (Figure 7c), the mean height for each species and generation showed different values and, therefore, H5 had a higher value (1.06), and also a reduced value due to the much different proportions of species and generations.

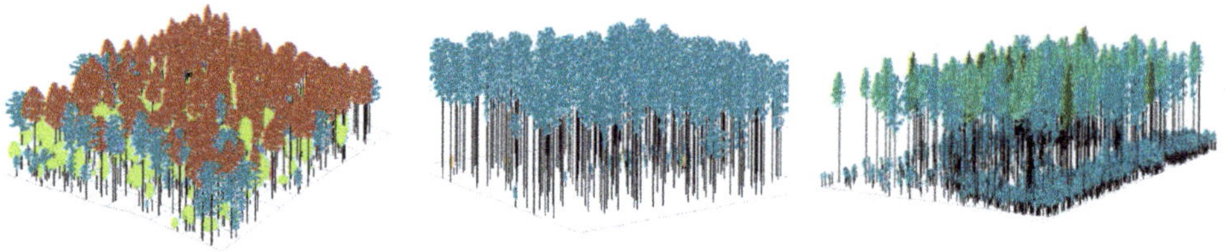

(a) (b) (c)

Figure 7. Structure of (**a**) mixed sessile oak–beech stands, (**b**) beech stands, and (**c**) mixed beech–fir stands. See Table 5 for the characteristics used.

Table 5. Characteristics used in Figure 7.

Stand	(a)	(b)	(c)
Basal area (m^2 ha^{-1})	34.6	52.0	36.0
Age limits (years)	90–120	110–150	90–120
	H index		
H1 (E)	0.77(48)	0.67 (97)	1.05 (76)
H2 (E)	0.77 (48)	0 (0)	0.87 (79)
H3 (E)	0.31 (45)	0.31 (45)	0.21 (31)
H4 (E)	0.77 (48)	0.67 (97)	1.06 (76)
H5 (E)	0.77 (48)	0.67 (97)	1.06 (76)

For the same stand, the H-index values differed in relation to the frequency of the variables and their proportion in the mixture. For the sessile oak–beech stands, the lowest H-index value was determined at the level of tree age, and the highest at the level of species, with sessile oak having a percentage of 74%. In the beech stands, being pure, the H-index value was 0, although the diversity H index was 0.67 in the structure of the stand because two generations of trees were present. In mixed beech–coniferous stands, the mixture of species was more balanced, with the value of the Shannon index reaching 1.06 (for diameter and height).

3.3. Management Decisions

In exploitable stands past the cutting age, the stand structure had closed vertically and the stand acquired more and more the character of an uneven-aged structure. These conditions were contributed to by the cuttings carried out to extract the dry coniferous trees, as well as gaps created by windfalls, after which an abundant youth of beech and fir became installed (Figure 4b). The management-planned volume established for harvesting from the stands was largely extracted by unplanned cuttings. Carrying out regeneration cuttings within the limits imposed in order to achieve a normal growing stock by age class is contradictory to the current state of the stands. Currently, at the management unit level, the stands that have reached cutting age occupy more than 50% of the forest area (most being beech and mixed beech–fir stands). These require interventions, although these stands have a special aesthetic value and should be maintained. Interventions have also been imposed by the presence of seedlings, which have developed large percentages of coverage in the stands. Thus, the regeneration cuttings of the group–shelterwood system, with a period of 40 years, carried out up to now, no longer conform to the characteristics of the current structures. Rational and sustainable management of the forest must satisfy both aesthetic demands and economic requirements. Thus, we consider that the current structures meet the conditions to be directed toward uneven-aged structures, which can satisfy both the requirements of visitors and of wood production. Of course, fellings would transform the character of the current structures toward the selection system. Additionally, not all stands lend themselves to the application of such cuttings. However, an alternation of structures, with different degrees of complexity (even- and uneven-aged), would be welcome and would satisfy the requirements of all categories of visitors. As one of the most intensively visited areas, transformation cuttings toward the selection system must be applied to those stands closest to the city of Brașov as a priority.

The H-index values provided information on the characteristics of the current structures of the stands. The most diversified structures were found in mixed beech–coniferous stands, which showed high index values. The H4 and H5 indices highlighted a high structural diversity of the stands at the tree diameter and height levels. Thematic maps were generated from these H-index values (Figure 8). Recreational activity areas, rest areas, and tourist routes in the forest can be located in relation to the H-index values and the visitors' preferences. The studied forest offers structural conditions that satisfy the various preferences of its visitors. The ridge trails provide the possibility of observing a varied range of structures, since the ridges are the boundaries that separate the stands and the stands have different structures. At the same time, these routes offer viewpoints out over the city and the surrounding mountains. The slope routes can be extended in different areas so that they cross stands with different biodiversity indices. Such trails are especially preferred by visitors who are passionate about the intimate environment of the forest. Routes with high biodiversity indices, affected as little as possible by silvicultural interventions, reveal the natural beauty of the forest and are the most preferred by naturalists and artists.

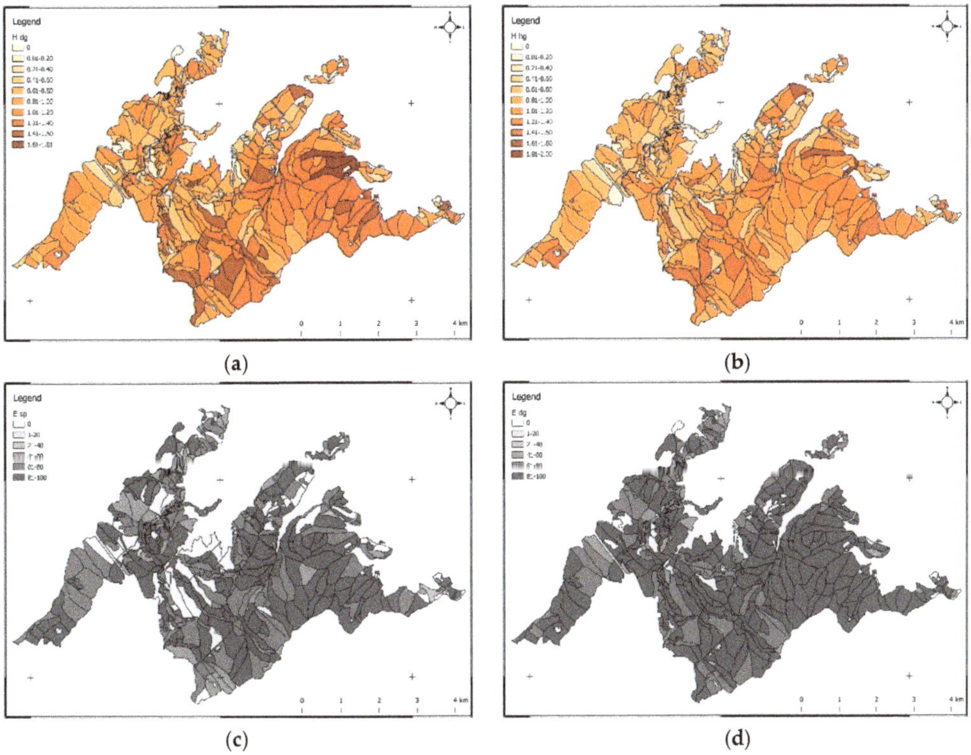

Figure 8. Stand distribution maps in relation to the (**a**) Hd_g values, (**b**) Hh_g values, (**c**) Esp. distribution, and (**d**) Ed_g values.

Silvicultural interventions must not produce visible changes in the structures of the stands, but should contribute to variations in the shapes and sizes of the trees and to the spatial groupings of trees of different sizes in order to ensure broad variety in the stands.

4. Discussion

4.1. Structural Diversity

The age class structures of the studied stands are favorable from an aesthetic point of view, but from the point of view of wood production, they are not characteristic of a normal growing stock. The stands are rich in species, resulting from the evolution of the stands and the management method applied. The largest share is held by beech, and its proportion will remain high in the future. In the future composition of the forest, however, the aim is to reduce the proportion of pines and spruce [34], these being more sensitive to climate warming. Beech remains the most stable species, more resistant to climate change than spruce [37]. The natural forest type has guided Romanian management plans for decades, but in recreational forests, the aesthetic effect generated by mixtures of species that can thrive in this forest site, that are favorable for wood production and preferred by visitors, must be capitalized on. Special effects, by contrast, can be created by clumps of one species in stands of another species, such as pine or larch in beech stands. Spruce, with its high and regular crowns, contrasts well with the rich crowns and shapes of beech trees. Thus, the aesthetic properties of stands can be increased by the introduction of other species into pure stands, even if they are disseminated [38,39]. The mixing of species also leads to an increase in the stability of stands, and it is well known that mixing species under changing climatic conditions can lead to a distribution of risk and to the stabilization of growth [37].

The aesthetic value of stands can be increased through their management at advanced ages. A combination of social and economic interests requires the management of stands up to such ages, but without exceeding a rational limit. This limit is guided by whether the trees are maintaining their vitality and capacity for regeneration, and the wood assortment harvested through regeneration cuttings is retaining its quality. Under the conditions in the investigated forest, the high proportion of exploitable stands that should undergo regeneration cuttings could affect the forest's ability to continuously provide ecosystem services. In a study carried out in northern Finland, in which the main trade-offs between tourism and forestry were analyzed, the percentage of old-growth stands was 11% of the total forest area and the average volume of old-growth stands was 76.9 m^3 ha^{-1}, with the age of these stands being 158 years [33]. In other studies, the optimal stand volume value for recreational activities is between 80 and 200 m^3 ha^{-1} [40].

The H index is a measure of diversity and takes into account species richness and evenness of species distribution [36], so its value increases with the number of species [41]. The determination of the H index at the level of stand characteristics, as component elements of the forest, led to different diversity index values. It is worth noting the high H-index values determined at the management unit level for the mean diameter of the stands (H4 = 3.17), with this calculation being based on the species and generations that made up the stands. A diversity of diameters also determines a diversity of heights. However, H5 was lower, at 2.76, than H4. This is because, in mixtures, the crowns of different species of different generations occur in the same plane and do not contribute to an increase in H5. The species diversity was highlighted by an H-index value of 1.63, resulting from a weight of very different species in the forest structure.

At the stand level, the highest H-index value indicating species diversity (H2) was determined to be 1.54, with the diameter (H4) and height (H5) diversities reaching 1.72. Both H4 and H5 were favorably influenced by an increase in the number of species (Figure 9).

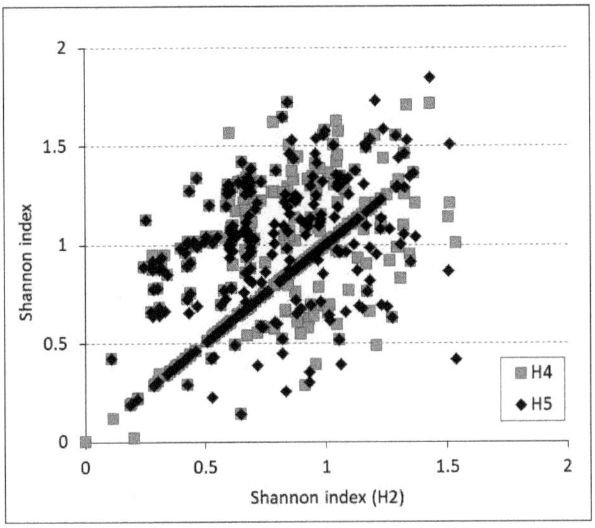

Figure 9. Relationship between diversity indices. In 49% (204) of the studied stands, it appeared that the increase in diversity in the stands in the vertical plane resulted from the increase in species diversity (H2).

At the stand level, the diversity H index showed the proportion of participation of species in the different height categories (0–50, 50–80, and 80%–100%) of the stands varied. Thus, in mixed Norway spruce–European beech stands, a participation of 90% of the trees in the first layer led to a diversity index of 1.00, with a balanced participation of the trees at those levels producing an index of 1.65 [42].

Following the application of different types of thinnings, the H-index values regarding species diversity increased from 0.53 (before thinning) to 0.64 (after thinning), and from 0.50 to 0.61 in the case of selective thinnings [43]. Forest site conditions and, of course, the characteristics of the stands also contributed to the change in diversity [41].

Diversity increases with stand volume. The natural factors that determine or affect diversity are altitude, slope, latitude, site quality, forest formation, stand volume, and quadratic mean diameter at breast height [41]. Quadratic mean diameter has a positive relationship with species diversity for large trees (d > 12.7 cm). For these trees, an average Shannon index value of 1.26 was determined, while for those with diameters of between 2.5 and 12.7 cm, this value was 0.89 [41].

4.2. Management Planning of Recreational Forests

Recreational forests fulfil multiple functions and, as such, their management must be designed to achieve the established objectives, since the forests are located in the most varied site conditions and are made up of different forest formations. Management regulations must be adapted to meet the recreational requirements as a matter of priority, without, however, disturbing the provision of other ecosystem services. Further expansion of tourism in traditionally forested areas could also compensate for the restrictions imposed on these forests to conserve recreation resources [33]. Therefore, along with the biodiversity of forests that fulfil production functions, landscape level should be of equal research interest [44]. The management of recreational forests also differs in relation to the specific function intensity of different recreational areas within the forest and the stand's conditions. When planning management measures, at stand and forest level, the following relationship should be considered: management goals–assigned functions–structure–management measures. This process is repeated every 10 years, when the management adjust its decisions to the new requirements and challenges.

A systematic management approach [45] involves creating recreation areas of interest to different categories of visitors. In areas where recreation has a higher priority, there are a number of other developments within the forest that alternate with the forest landscape, frequented by tourists. Overall, these areas differ from typical forest biocenoses. Management measures should particularly promote the aesthetic qualities of trees and species diversity. In the other recreational areas, alongside the aesthetic principle, importance is also attached to the other principles of forest management which lead to the preservation of the natural features of the forest in order to ensure multiple requirements. The measures promoted through the forests' management should aim to restore the natural stands and forest structure as a condition for their stability in order to consistently provide ecosystem services.

In management planning of recreational forests, easily observable biodiversity indicators such as structural variables (e.g., old stands, large trees, and canopy cover) are of particular importance [44]. In general, indicators that characterize stand structure are easier to assess by the managers, as management is applied at the stand level [44].

For the studied recreational forests, the provisions of the management planning can be summarized in the measures presented in Table 6.

Table 6. Management goals and silvicultural prescriptions for recreation forest.

Function Intensity of the Recreation Area	High-Intensity Area	Area of Moderate to Low Intensity
1. Management goals	Protection (recreation, soil, water, and biodiversity conservation)	Protection and production (timber, valuable timber—veneer, resonance)
2. Type of stand structure	Various structures with aesthetic effects (H index: at stand level: H2 > 1.5; H3, H4 and H5 > 0.3; E > 40 at level forest: H2 > 1.5; H3, H4 and H5 > 0.3; E > 40)	Multi-aged and uneven-aged structures (H index: at stand level: H2, H3, H4 and H5 > 0.7; E > 70 at forest level: H2, H3, H4 and H5 > 2.5; E > 70)
3. Silvicultural system	Cuttings to maintain the functional structure of the stands and the aesthetic effects of trees Tending operations	Selection system (cuttings of transformation to selection system) Group shelterwood system with regeneration period of 40–60 years (in mixed beech–coniferous stands and beech stands) Tending operations
4. Method of regeneration	Natural regeneration ± artificial regeneration by planting	Natural regeneration
5. Felling age of stands (cycle for forest) (years)	-	110–130: spruce stands 120–140: mixed beech–coniferous stands and beech stands 150–180 (200) (for valuable timber: veneer and resonance)
Rotation (years)/target diameter (cm)	10/-	10/60–100 (Pa and Aa); 60–90 (Fs)
6. Target composition	Forest type composition ± mixtures of species (with aesthetic qualities) to the site adapted	Forest type composition
7. Structure by age classes	-	Even distribution of stands of different ages (mosaic of stands of different ages) (in silvicultural system age class forest)
8. Future structure: by age classes by diameter classes	-	Normal distribution by age classes (in silvicultural system age class forest) Normal distribution by diameter class (in selection system)

The limits of felling ages refer to stands located in lower- to higher-productivity sites and are provided by technical norms on silvicultural systems of Romania (2000).

5. Conclusions

Effective management of recreational forests with multiple functions must take into account the requirements of all the factors that benefit from the ecosystem services of these forests. The recreational value of a forest can be expressed by the H index, applied to the structural characteristics of the stands and the forest. High H-index values are found in forests with complex horizontal and vertical structures. The management of stands toward specific compositions of natural forest types does not necessarily lead to an increase in species diversity. This is the case in beech and spruce stands. However, in such formations, the diversity of the stands can increase in the vertical plane. Given the current structural conditions of the stands, we recommend directing them toward uneven-aged structures, where the forest management undertakes transformation to selection system cuttings. This would harmoniously combine aesthetic and economic interests. During the application of the management plan, the cuttings could be dispersed throughout the forest and could have reduced intensities, thereby going unnoticed by visitors and creating the impression that the stands always remain the same. In complex, uneven-aged structures, the shapes of the different species and sizes of trees, in the presence of sunlight, would blend harmoniously and create an aesthetic appearance that would satisfy even the most demanding of aesthetic requirements. In terms of the wood production ratio, the advantages of the selection system are already known. Stand structures are in continuous transformation and so the values of their diversity indices vary during stand development. Structural diversity indices, however, offer the possibility of assessing and monitoring stand recreational values.

Author Contributions: Conceptualization, G.-M.T.; Data curation, G.-M.T., A.C., M.C. and A.-C.D.; Investigation, G.-M.T., A.C., M.C. and A.-C.D.; Methodology, G.-M.T.; Validation, G.-M.T., A.C., M.C. and A.-C.D.; Writing—original draft, G.-M.T.; Writing—review and editing, G.-M.T., M.C. and A.-C.D. All authors have read and agreed to the published version of the manuscript.

Funding: This research received no external funding.

Data Availability Statement: Not applicable.

Acknowledgments: The authors thank RPLP Kronstadt for giving us access to the forest database.

Conflicts of Interest: The authors declare no conflict of interest.

References

1. Fredman, P.; Romild, U.; Yuan, M.; Wolf-Watz, D. Latent demand and time contextual constraints to outdoor recreation in Sweden. *Forests* **2012**, *3*, 1. [CrossRef]
2. Bötsch, Y.; Tablado, Z.; Scherl, D.; Kéry, M.; Graf, R.F.; Jenni, L. Effect of recreational trails on forest birds: Human presence matters. *Front. Ecol. Evol.* **2018**, *6*, 175. [CrossRef]
3. De Meo, I.; Paletto, A.; Cantiani, M.G. The attractiveness of forests: Preferences and perceptions in a mountain community in Italy. *Ann. For. Res.* **2015**, *58*, 145–156. [CrossRef]
4. Blattert, C.; Lemm, R.; Thees, O.; Lexer, M.J.; Hanewinkel, M. Management of ecosystem services in mountain forests: Review of indicators and value functions for model based multi-criteria decision analysis. *Ecol. Indic.* **2017**, *79*, 391–409. [CrossRef]
5. Dudek, T. Recreational potential as an indicator of accessibility control in protected mountain forest areas. *J. Mt. Sci.* **2017**, *14*, 1419–1427. [CrossRef]
6. Sievänen, T. Recreation: Inventory, monitoring and management. In *Encyclopedia of Forest Sciences*; Burley, J., Evans, J., Youngquist, J.A., Eds.; Elsevier: Amsterdam, The Netherlands, 2004; pp. 958–964. [CrossRef]
7. Ciobotaru, A.-M.; Patel, N.; Pintilii, R.-D. Tree cover loss in the Mediterranean region—An increasingly serious environmental issue. *Forests* **2021**, *12*, 1341. [CrossRef]
8. Kim, D.; Avenzora, R.; Lee, J.H. Exploring the outdoor recreational behavior and new environmental paradigm among urban forest visitors in Korea, Taiwan and Indonesia. *Forests* **2021**, *12*, 1651. [CrossRef]
9. Gundersen, V.S.; Frivold, L.H. Public preferences for forest structures: A review of quantitative surveys from Finland, Norway and Sweden. *Urban For. Urban Green.* **2008**, *7*, 241–258. [CrossRef]
10. Edwards, D.M.; Jay, M.; Jensen, F.S.; Lucas, B.; Marzano, M.; Montagné, C.; Weiss, G. Public preferences across Europe for different forest stand types as sites for recreation. *Ecol. Soc.* **2012**, *17*, 22. [CrossRef]
11. Filyushkina, A.; Agimass, F.; Lundhede, T.; Strange, N.; Jacobsen, J.B. Preferences for variation in forest characteristics: Does diversity between stands matter? *Ecol. Econ.* **2017**, *140*, 22–29. [CrossRef]
12. Nigussie, S.; Liu, L.; Yeshitela, K. Indicator development for assessing recreational ecosystem service capacity of urban green spaces–A participatory approach. *Ecol. Indic.* **2021**, *121*, 107026. [CrossRef]
13. Sergiacomi, C.; Vuletić, D.; Paletto, A.; Barbierato, E.; Fagarazzi, C. Exploring National Park Visitors' Judgements from Social Media: The Case Study of Plitvice Lakes National Park. *Forests* **2022**, *13*, 717.

14. La Rosa, D.; Spyra, M.; Inostroza, M. Indicators of Cultural Ecosystem Services for urban planning: A review. *Ecol. Indic.* **2016**, *61*, 74–89. [CrossRef]
15. Pleasant, M.M.; Gray, S.A.; Fernandes, A.; Hunter, N.; Ford, D. Managing cultural ecosystem services. *Ecosyst. Serv.* **2014**, *8*, 141–147.
16. Jurmalis, E.; Libiete, Z. Developing a framework for characterizing recreational potential of forest areas using weighted criteria analysis. In *Research for Rural Development, Proceeding of the Annual 25th International Scientific Conference, Jelgava, Latvia, 15–17 May 2019*; Latvia University of Life Sciences and Technologies: Jelgava, Latvia; pp. 89–94. [CrossRef]
17. Oftedal, A.; Davenport, M.; Schneider, I.E.; Zerger, C.; Schreurs, B.; Vogel, M. An Ecosystem Approach to Recreation Location Quotients. *Forests* **2011**, *2*, 993–1012. [CrossRef]
18. Abildtrup, J.; Garcia, S.; Olsen, S.B.; Stenger, A. Spatial preference heterogeneity in forest recreation. *Ecol. Econ.* **2013**, *92*, 67–77. [CrossRef]
19. Termansen, M.; Zandersen, M.; McClean, C.J. Spatial substitution patterns in forest recreation. *Reg. Sci. Urban Econ.* **2008**, *38*, 81–97. [CrossRef]
20. Felton, A.; Felton, A.M.; Wam, H.K.; Witzell, J.; Wallgren, M.; Löf, M.; Sonesson, J.; Lindbladh, M.; Björkman, C.; Blennow, K.; et al. Forest biodiversity and ecosystem services from spruce-birch mixtures: The potential importance of tree spatial arrangement. *Environ. Chall.* **2022**, *6*, 100407. [CrossRef]
21. Pach, M.; Sansone, D.; Ponette, Q.; Barreiro, S.; Mason, B.; Bravo-Oviedo, A.; Löf, M.; Bravo, F.; Pretzsch, H.; Lesi´nski, J.; et al. Silviculture of mixed forests: A European Overview of Current Practices and Challenges. In *Managing Forest Ecosystems: The Challenge of Climate Change*; Springer International Publishing: Cham, Switzerland, 2018; Volume 31, pp. 185–253.
22. Battles, J.J.; Shlisky, A.J.; Barrett, R.H.; Heald, R.C.; Allen-Diaz, B.H. The effects of forest management on plant species diversity in a Sierran conifer forest. *For. Ecol. Manag.* **2001**, *146*, 211–222. [CrossRef]
23. Decocq, G.; Aubert, M.; Dupont, F.; Alard, D.; Saguez, R.; Wattez-Franger, A.; De-Foucalt, B.; Delelis-Dusollier, A.; Bardat, J. Plant diversity in a managed temperate deciduous forest: Understorey response to two silvicultural systems. *J. Appl. Ecol.* **2004**, *41*, 1065–1079.
24. Heyman, E. Analysing recreational values and management effects in an urban forest with the visitor-employed photography method. *Urban For. Urban Green.* **2012**, *11*, 267–277. [CrossRef]
25. Jiang, B.; Sun, C.; Mu, S.; Zhao, Z.; Chen, Y.; Lin, Y.; Qiu, L.; Gao, T. Differences in Airborne Particulate Matter Concentration in Urban Green Spaces with Different Spatial Structures in Xi'an. China. *Forests* **2022**, *13*, 14. [CrossRef]
26. Hoppa, A.; Sikorska, D.; Przybysz, A.; Melon, M.; Sikorski, P. The Role of Trees in Winter Air Purification on Children's Routes to School. *Forests* **2022**, *13*, 40. [CrossRef]
27. Mundher, R.; Abu Bakar, S.; Maulan, S.; Mohd Yusof, M.J.; Al-Sharaa, A.; Aziz, A.; Gao, H. Aesthetic Quality Assessment of Landscapes as a Model for Urban Forest Areas: A Systematic Literature Review. *Forests* **2022**, *13*, 991. [CrossRef]
28. Kinho, J.; Arini, D.I.D.; Abdulah, L.; Susanti, R.; Irawan, A.; Yulianti, M.; Subarudi, S.; Imanuddin, R.; Wardani, M.; Denny, D.; et al. Habitat Characteristics of *Magnolia* Based on Spatial Analysis: Landscape Protection to Conserve Endemic and Endangered *Magnolia sulawesiana* Brambach, Noot., and Culmsee. *Forests* **2022**, *13*, 802. [CrossRef]
29. Grebner, D.L.; Bettinger, P.; Siry, J.P. *Introduction to Forestry and Natural Resources*; Academic Press: Cambridge, MA, USA, 2014; pp. 173–197.
30. Aikoh, T.; Abe, R.; Kohsaka, R.; Iwata, M.; Shoji, Y. Factors Influencing Visitors to Suburban Open Space Areas near a Northern Japanese City. *Forests* **2012**, *3*, 155–165. [CrossRef]
31. Komossa, F.; van der Zanden, E.H.; Schulp, C.J.; Verburg, P.H. Mapping landscape potential for outdoor recreation using different archetypical recreation user groups in the European Union. *Ecol. Indic.* **2018**, *85*, 105–116. [CrossRef]
32. Pătrășcoiu, N.; Dima, V.; Teodoreanu, E.; Gherbanovski, E. *Cercetări Privind Influența Structurii Arboretelor Asupra Funcțiilor de Interes Social in Structuri Optime pentru Păduri cu Funcții de Protecție*; I.C.A.S.: Bucureşti, Romania, 1981; pp. 67–89.
33. Ahtikoski, A.; Tuulentie, S.; Hallikainen, V.; Nivala, V.; Vatanen, E.; Tyrväinen, L.; Salminen, H. Potential trade-offs between nature-based tourism and forestry, a case study in Northern Finland. *Forests* **2011**, *2*, 894–912. [CrossRef]
34. Tudoran, G.-M.; Cicșa, A.; Boroeanu, M.; Dobre, A.-C.; Pascu, I.-S. Forest Dynamics after Five Decades of Management in the Romanian Carpathians. *Forests* **2021**, *12*, 783. [CrossRef]
35. Tudoran, G.M.; Zotta, M. Adapting the planning and management of Norway spruce forests in mountain areas of Romania to environmental conditions including climate change. *Sci. Total Environ.* **2020**, *698*, 133761. [CrossRef]
36. Shannon, C.E.; Weaver, W. *The Mathematical Theory of Communication*; The University of Illinois Press: Urbana, IL, USA, 1949; pp. 1–117.
37. Pretzsch, H. Transitioning monocultures to complex forest stands in Central Europe: Principles and practice. In *Achieving Sustainable Management of Boreal and Temperate Forests*; Burleigh Dodds Science Publishing: Cambridge, UK, 2019; pp. 355–396.
38. Rucăreanu, N.; Leahu, I. *Amenajarea Pădurilor*; Editura Ceres: Bucureşti, Rmania, 1982; pp. 356–450.
39. Leahu, I. *Amenajarea Pădurilor*; Editura Didactică şi Pedagogică: Bucureşti, Romania, 2001; pp. 436–498.
40. Kellomäki, S. Recreational potential of a forest stand. *Silva Fenn.* **1978**, *12*, 5001. [CrossRef]
41. Polyakov, M.; Majumdar, I.; Teeter, L. Spatial and temporal analysis of the anthropogenic effects on local diversity of forest trees. *For. Ecol. Manag.* **2008**, *255*, 1379–1387. [CrossRef]
42. Pretzsch, H. *Forest Dynamics, Growth and Yield*; Springer: Berlin, Germany, 2009; pp. 225–328.
43. Marchi, M.; Paletto, A.; Cantiani, P.; Bianchetto, E.; De Meo, I. Comparing thinning system effects on ecosystem services provision in artificial black pine (*Pinus nigra* JF Arnold) forests. *Forests* **2018**, *9*, 188. [CrossRef]

44. Ćosović, M.; Bugalho, M.N.; Thom, D.; Borges, J.G. Stand Structural Characteristics Are the Most Practical Biodiversity Indicators for Forest Management Planning in Europe. *Forests* **2020**, *11*, 343. [CrossRef]
45. Hruza, P.; Pelikán, P.; Olišarová, L. Single-Track Bike Trails in the Moravian Karst as Part of Forest Recreation. *Forests* **2021**, *12*, 1601. [CrossRef]

MDPI AG
Grosspeteranlage 5
4052 Basel
Switzerland
Tel.: +41 61 683 77 34

Forests Editorial Office
E-mail: forests@mdpi.com
www.mdpi.com/journal/forests

www.ingramcontent.com/pod-product-compliance
Lightning Source LLC
Chambersburg PA
CBHW061808210926
51850CB00006B/648

* 9 7 8 3 7 2 5 8 6 9 8 4 8 *